ANALYTICAL MICROBIOLOGY

ANALYTICAL MICROBIOLOGY

Edited by

FREDERICK KAVANAGH

Eli Lilly and Company
Indianapolis, Indiana

1963

ACADEMIC PRESS • New York and London
A Subsidiary of Harcourt Brace Jovanovich, Publishers

ACADEMIC PRESS, INC.
111 Fifth Avenue, New York, New York 10003

United Kingdom Edition published by
ACADEMIC PRESS, INC. (LONDON) LTD.
Berkeley Square House, London W1X6BA

LIBRARY OF CONGRESS CATALOG CARD NUMBER: 63-12813

Third Printing, 1973

PRINTED IN THE UNITED STATES OF AMERICA

Contributors

Page numbers are shown in parentheses following names of contributors.

WILLIAM C. ALEGNANI, *Research Division, Parke, Davis & Company, Detroit, Michigan.* (271).

ORSON D. BIRD, *Parke-Davis Research Laboratories, Ann Arbor, Michigan.* (497).

K. E. COOPER, *Department of Bacteriology, University of Bristol, Bristol, England.* (1).

L. J. DENNIN, *Eli Lilly and Company, Indianapolis, Indiana.* (261, 283, 289, 303, 309, 313, 349, 369, and 489).

E. EIGEN, *Department of Microbiology, Colgate-Palmolive Company, New Brunswick, New Jersey.* (431).

MARGARET GALBRAITH, *Research Division, Parke, Davis & Company, Detroit, Michigan.* (271).

JOHN R. GERKE, *The Squibb Institute for Medical Research, New Brunswick, New Jersey.* (219 and 387).

ROLAND L. GIROLAMI, *Abbott Laboratories, North Chicago, Illinois.* (295 and 353).

HELENE NATHAN GUTTMAN, *The Haskins Laboratories, New York, New York; Department of Biological Sciences, Goucher College, Towson, Maryland; and Department of Medicine, Medical College of Virginia, Richmond, Virginia.* (527).

THOMAS A. HANEY, *The Squibb Institute for Medical Research, New Brunswick, New Jersey.* (219).

ROBERT HANS, *Research Division, Parke, Davis & Company, Detroit, Michigan.* (271).

FREDERICK KAVANAGH, *Eli Lilly and Company, Indianapolis, Indiana.* (125, 141, 249, 265, 283, 289, 313, 347, 361, 363, 369, 375, 381, 411, 415, 417, 519, and 675).

JOSEPH D. LEVIN, *The Squibb Institute for Medical Research, New Brunswick, New Jersey.* (365 and 387).

JOSEPH F. PAGANO, *The Squibb Institute for Medical Research, New Brunswick, New Jersey.** (219, 365, and 387).

G. D. SHOCKMAN, *Department of Microbiology, Temple University School of Medicine, Philadelphia, Pennsylvania.* (431 and 567).

* Present address: Sterling-Winthrop Research Institute, Rensselaer, New York.

J. S. SIMPSON, *The Midlands Counties Dairy Ltd., Birmingham, England.* (*87*).

HELEN R. SKEGGS, *Merck Sharp & Dohme Research Laboratories, West Point, Pennsylvania.* (*421, 521, and 551*).

Preface

During the last 20 years many millions of assays for antibiotic substances have been made by methods hastily devised early in the penicillin program. The methods were based upon rudimentary theory—there was no other—and they underwent occasional empirical modifications through the years. Theoretical bases of the methods have been developed during the last decade until now practice lags behind theory.

This volume was planned because the time has come to provide a firm theoretical foundation for methods of assay and to change current practices to fit the new foundations.

The first portion of this book is concerned with theory and the second portion with specific methods of assay. It has been written out of the experience of the authors, gives the methods they are using, and provides also an up-to-date survey of the literature. It has been designed to serve those who assay for amino acids, antibiotic substances, and vitamins by microbiological methods.

Principles developed in the chapter on diffusion assays and the chapter on photometric methods of assays may be used to increase understanding of those two fundamental types of assays. The chapter on the large plate assay provides a much needed account of principles and practices and should stimulate investigation of its advantages and disadvantages relative to the less accurate, less precise, and less efficient petri dish version so much used here. The chapter on automated assays introduces what promise to be the methods of the future and which are now receiving the bulk of effort devoted to assay development in our pharmaceutical industry. Instrumentation of automated methods is expensive by an order of magnitude greater than that for either of the older methods, and unless careful attention is paid to principles, both microbiological and chemical, excessively expensive. Automation makes greater demands upon the operator than the older methods and should not be considered unless appropriate personnel are available; it is not economically justified for all assay situations.

The second portion of the book contains specific methods of assay for common antibiotic substances, vitamins, and amino acids. Details have been limited to those that are essential and repetition has been kept to a minimum. Not all substances capable of assay by microbiological methods have been treated. The assumption was made that users of this work will be well enough educated in biochemistry, physical chemistry, analytical chemistry, and microbiology that basic concepts and operations in those fields would not require discussion.

Selection of most of the authors from the pharmaceutical companies of the United States and England was a consequence of the fact that the broadest experience seems to have been obtained in these industries and countries.

Academic microbiologists have shown little interest in the theory and practice of assaying by means of microbes. Theory is incomplete and practice can be improved as is indicated in several chapters. Contributions from microbiologists in our universities toward solutions of problems and training of students will be welcomed.

I hope that both those who do the assays and those who supervise the assayist, at one level or another, will learn from the book of the complexity of microbiological assaying and not expect an impossible accuracy. I also hope that this work will contribute to the quiet interment of the favorite excuse for inadequate design and poor execution of microbiological assays—microbiological variation.

I wish to acknowledge the help obtained during the last 9 years from the laboratory assistants in the assay laboratories of the Antibiotics Manufacturing and Development Division of Eli Lilly and Company. They were most patient, usually, in carrying out my suggestions and requests to deviate from long time practices. Together we were able to learn much about photometric assaying and to improve our operations. Particular thanks are due the management of the Division for encouragement and help during the editing of this book and for supporting my efforts toward improving instrumentation of assays.

F. KAVANAGH

December, 1962

Contents

The Theory of Antibiotic Inhibition Zones

K. E. COOPER

Microbiological Assay Using Large Plate Methods

J. S. SIMPSON

CHAPTER 3

Dilution Methods of Antibiotic Assays

FREDERICK KAVANAGH

CHAPTER 4

Elements of Photometric Assaying

FREDERICK KAVANAGH

CHAPTER 5

Automation of Microbiological Assays

THOMAS A. HANEY, JOHN R. GERKE, AND JOSEPH F. PAGANO

CHAPTER 6

Antibiotic Substances

Part I   Antibacterial Assays

6.1 Introduction

FREDERICK KAVANAGH

6.2 Bacitracin

L. J. DENNIN

6.3 Cephalosporin C

FREDERICK KAVANAGH

6.4 Chloramphenicol

ROBERT HANS, MARGARET GALBRAITH, AND WILLIAM C. ALEGNANI

6.5 Dihydrostreptomycin

FREDERICK KAVANAGH AND L. J. DENNIN

6.6 Erythromycin

FREDERICK KAVANAGH AND L. J. DENNIN

6.7 Fumagillin

ROLAND L. GIROLAMI

6.8 Hygromycin B

L. J. DENNIN

6.9 Neomycin

L. J. DENNIN

6.10 Penicillins

FREDERICK KAVANAGH AND L. J. DENNIN

6.11 Penicillin V

FREDERICK KAVANAGH

6.12 Polymyxin

L. J. DENNIN

6.13 Ristocetin

ROLAND L. GIROLAMI

Part II Antifungal Assays

JOHN R. GERKE, JOSEPH D. LEVIN, AND JOSEPH F. PAGANO

CHAPTER 7

Vitamins

7.1 Introduction

FREDERICK KAVANAGH

7.2 pH as Assay Response

FREDERICK KAVANAGH

7.3 Specificity

FREDERICK KAVANAGH

7.4 Biotin

HELEN R. SKEGGS

7.5 The Folic Acid Group

E. EIGEN AND G. D. SHOCKMAN

7.6 Agar Plate Assays for Pantothenic Acid, Inositol, and Pyridoxine

L. J. DENNIN

7.7 Pantothenic Acid and Related Compounds

ORSON D. BIRD

7.8 Riboflavin

FREDERICK KAVANAGH

7.9 Turbidimetric Assay for Thiamine

HELEN R. SKEGGS

7.10 Vitamin B_{12} and Congeners

HELENE NATHAN GUTTMAN

CHAPTER 8

Amino Acids

GERALD D. SHOCKMAN

CHAPTER 9

Glucose

FREDERICK KAVANAGH

CHAPTER 1

The Theory of Antibiotic Inhibition Zones

K. E. COOPER

Department of Bacteriology, University of Bristol, Bristol, England

I. Introduction. The Principles Involved

A. Historical

The qualitative use of zones of agar made inhibitory to the growth of one microorganism by diffusing substances produced by another organism was in use at least as early as 1885. Florey[1] quotes the work of Garré[2] on the inhibition of *Staphylococcus pyogenes* by *Pseudomonas fluorescens.* Fleming[3] used this method in 1922 for his work on lysozym, and it was his observation of the lysis of staphylococci in the neighborhood of a colony of *Penicillium notatum* that led to his work on penicillin[4] in 1929. It was natural therefore that Chain *et al.*[5] in 1940 should use inhibition zones to follow the activity of the penicillin they isolated and purified, and that it should be suggested that zone size should be a basis for quantitative comparison with standard penicillin for purposes of estimation. Abraham *et al.*[6] and Foster and Woodruff [7,8] investigated many of the factors influencing the results and compared the advantages and disadvantages of the method with the turbidimetric. Schmidt and Moyer[9] also examined the influence of experimental conditions on the assay, and from that time many papers have been published involving many modifications of techniques. Despite the great use made of these agar diffusion methods, it was not until the years after the war that any systematic work on the theoretical aspects involved was published in 1946 by Cooper and Woodman.[10]

Much of the work described in this chapter was done to elucidate the principles involved and not for the purposes of assay. The organisms, the antibiotics, the media, and the conditions used were chosen deliberately to obtain answers to particular questions involving principles regarded at the time as important. They are not necessarily the best for assay purposes, not even sometimes for the work in question, but they have enabled results under different conditions to be compared and quantitative results to be obtained which it is hoped will be of much wider application.

B. Principles

Consider the sequence of events which may occur when an antibiotic diffuses through agar in which bacterial cells capable of multiplying in

[1] H. W. Florey, *Brit. med. J.* 4427 (1945).

[2] C. Garré, *Korrespondenzbl. Schweiz. Aerzte* **17**, 385 (1887).

[3] A. Fleming, *Proc. Roy. Soc.* **B93**, 306 (1922).

[4] A. Fleming, *Brit. J. Exptl. Pathol.* **10**, 226 (1929).

[5] E. Chain, H. W. Florey, A. D. Gardner, N. G. Heatley, M. A. Jennings, J. Orr-Ewing, and A. G. Saunders, *Lancet* **ii**, 226 (1940).

[6] E. P. Abraham, E. Chain, C. M. Fletcher, H. W. Florey, A. D. Gardner, N. G. Heatley, and M. A. Jennings, *Lancet* **ii**, 177 (1941).

[7] J. W. Foster and H. B. Woodruff, *J. Bacteriol.* **46**, 187 (1943).

[8] J. W. Foster and H. B. Woodruff, *J. Bacteriol.* **47**, 43 (1944).

[9] W. H. Schmidt and A. J. Moyer, *J. Bacteriol.* **47**, 199 (1944).

[10] K. E. Cooper and D. Woodman, *J. Pathol. Bacteriol.* **58**, 75 (1946).

the medium are present. It will be evident that consideration must be given to the following.

1. The Laws of Diffusion

These describe the changes of concentration which take place in time at each point in the agar. Particular circumstances may render the laws of diffusion inapplicable, such as precipitation on the media, chromatographic absorption by constituents in the gel, solvent flow, and evaporation; these disturbances will be made evident by disagreement with the quantitative predictions of the simple theory.

2. The Laws of Growth

These describe the rate of production of cells and of cell constituents; they decide the rate of formation of constituents which may combine with antibiotics (cell receptors) and describe the relationship between cell division and metabolism, either or both of which may be inhibited by the antibiotics. Special cases may be concerned with adaptation and genetic changes.

3. The Laws of Absorption, Permeability, and Partition

These govern the way in which the antibiotic outside the cell reaches the affected cell receptors controlling metabolism. The effects of competitive absorption of constituents in the media need to be considered here.

4. The Laws of Specific Antibiotic Action

These are concerned with the chemistry of the effects of the antibiotic on metabolism. They may be concerned with the production of antagonists or the metabolic destruction of antibiotic or with the possibility of alternative metabolic pathways rendering less effective antibiotic action.

5. The Laws of Statistics

These enable an estimate to be made of the probability that differences in experimental results are due to chance variations. This allows estimates of probable error in determinations only if the many chance unknown factors influencing the final measurements of a sufficiently large series cancel each other, i.e., if the experiments are free from *bias*. Certainty in this cannot be achieved, but the more known factors are taken into account the less *unknown* factors may produce bias and be satisfactorily treated by statistical methods.

II. The Laws of Diffusion

A. Fick's Laws

The assumption that the laws of mechanics are applicable to molecules is justified by their explanation of many properties of gases, and of dilute solutions of solutes in solvents. Diffusion is the process by which the random movements of molecules transfer matter from one position to another. In gases and with solutes in dilute solution it leads to the uniform dispersion of the molecules throughout the containing vessel or solvent, to uniform pressures—due to molecular collisions on the walls of the vessel—to equal volumes containing equal masses and therefore equal numbers of molecules. The molecular agitation which is continually proceeding favors no particular direction. Uniform dispersion ensures that average velocities in three directions at right angles will therefore each be equal to zero. This will be true for any large number of molecules taken at random.

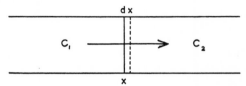

FIG. 1. Diffusion and concentration gradient. The concentration changes from C_1 to C_2 along the tube. The cross sectional area is a, and at x over the small distance dx the change in concentration is dc.

What is true for any randomly chosen groups of molecules at a particular time is also true for one group of molecules considered at different instants of time. Position, velocities, and kinetic energies may be considered with these properties resolved in any particular direction at any particular instant of time, and the laws of probability govern these chance events.

Consider a system which has not reached equilibrium (Fig. 1) in which two different concentrations are side by side (C_1 and C_2). In the cross-sectional area, a, between the two concentrations, over a small distance dx at right angles, a concentration gradient dc/dx will, over a small interval of time dt, control the rate of diffusion. If the amount of substance arriving at x is dm, it will be proportional to the area a, the time, and the concentration gradient:

$$dm = -Da\left(\frac{dc}{dx}\right) dt. \tag{1}$$

The constant of proportionality is defined as the diffusion coefficient D.

The number of molecules leaving the volume adx if the amount entering is dm will be

$$dm + \left(\frac{d(dm)}{dx}\right)_t dx.$$

The amount accumulating (in time dt) is therefore

$$-\left(\frac{d(dm)}{dx}\right)_t dx$$

and increase in concentration is

$$-\left(\frac{d(dm)}{dx}\right)_t dx/adx.$$

Therefore

$$\left(\frac{dc}{dt}\right)_x dt = -\left(\frac{d(dm)}{dx}\right)_t \frac{1}{a}.$$

From Eq. (1), differentiating in respect of dx,

$$\left(\frac{d(dm)}{dx}\right)_t = -Da\left(\frac{d^2c}{dx^2}\right)_t dt.$$

Therefore

$$\left(\frac{dc}{dt}\right)_x dt = Da\left(\frac{d^2c}{dx^2}\right)_t \frac{dt}{a};$$

eliminating a and dt gives

$$\left(\frac{dc}{dt}\right)_x = D\left(\frac{d^2c}{dx^2}\right)_t. \tag{2}$$

Equations (1) and (2) are two forms of Fick's law[11] and lead to the definition of the diffusion constant D as the amount of solute which would diffuse across unit area under a concentration gradient of unity in unit time if the rate were constant during that time.

B. Solutions for Linear Diffusion

1. From a Constant Concentration

The solutions of these differential equations are numerous and depend on the conditions imposed. Consider the special case in which a solution is *kept at a constant concentration* m_0 in contact with an infinite column of pure water or agar, which at time zero is free from the solute (Fig. 2). Diffusion will take place into this column. At time t the concentration at a distance x from the original junction is given by

[11] A. Fick, *Ann. Physik.* **170** (*Poggendorf's* **94**), 59 (1855).

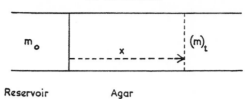

Reservoir   Agar

FIG. 2. Concentration of antibiotic diffusing from a reservoir into an agar gel in a tube of infinite length. At time 0 concentration for all values of x (distance from the agar-reservoir junction) is 0, and at time t it is $(m)_t$ at x. Concentration in the reservoir is kept constant at m_0 (by convection, stirring, or addition of antibiotic).

$$(m)_t = m_0 \exp\left(-x^2/4Dt\right) \qquad (3)$$

or

$$x^2 = 4Dt \ln\left(m_0/m\right) \qquad (4)$$

or

$$t = x^2/4D \ln\left(m_0/m\right). \qquad (5)$$

Cooper and Woodman[10] showed that this formula was applicable by following the progress of 1/500,000 dilution of crystal violet by its color intensity when m_0 was 1/500 in contact with nutrient agar. Plotting x^2 against time gave a straight line through the origin for as long as 400 hours (Fig. 3). The diffusion coefficient was 0.109 mm.^{2}/hour. A slight lag occurred during the first 2 hours accompanied by slight agar swelling, suggesting some movement of water and salts. Crystal violet is an ionized dye and its diffusion in pure water would create a potential gradient which would modify the rate of diffusion. However, as Lehner and

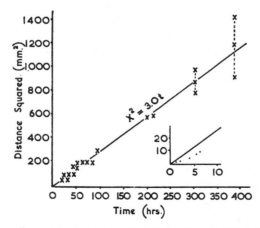

FIG. 3. Diffusion of crystal violet in cylindrical tubes at 15°C. The progress of a concentration of $m = 2 \times 10^{-6}$ w/v of crystal violet was matched colorimetrically, and the distance of this concentration from the surface of the agar measured at intervals. The concentration of the crystal violet in the reservoir was $m_0 = 2 \times 10^{-3}$ and the graph shows (distance)^2 against time. $D = 0.109$ mm.^{2}/hour in $2\frac{1}{2}\%$ agar.

Smith[12] pointed out, the presence of a few equivalents of sodium chloride, such as is present in the nutrient agar, prevents this and allows the use of a formula for neutral molecules.

Similar results in agreement with the formula were obtained for diffusion of crystal violet from a large cup punched out of the agar in the center of a petri dish. Gutters cut out of the agar would be expected to serve as reservoirs for linear diffusion methods, and, provided cups are above 8 mm. in diameter, good approximation to the expected results is obtained. Smaller cups require the formulas for radial diffusion described in the next section. Figure 4 shows x plotted against t for different concentrations of m_0 in the cup.

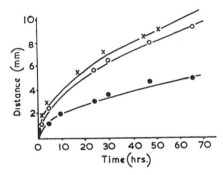

FIG. 4. Diffusion of crystal violet in agar plates from a cylindrical cup. The graph shows distance x of the edge of the visible color (concentration $m = 8 \times 10^{-5}$ w/v) plotted against time t for three concentrations m_0 in the cup.

$\times$——$\times$ $m_0 = 0.4\%$	Theoretical curve	$x^2 = 1.7t$
O——O $m_0 = 0.2\%$	Theoretical curve	$x^2 = 1.4t$
●——● $m_0 = 0.02\%$	Theoretical curve	$x^2 = 0.4t$

The maintenance of the constancy of m_0 in contact with the agar depends not only on a large reservoir of solution, but on constant *convection* to replace the solute at the surface lost by diffusion. If very narrow tubes are used, then the concentration at the agar junction falls immediately to $m_0/2$ and replacement of losses is maintained at this level by *diffusion* through the solution.

2. The Relation of Concentration of Diffusing Solute to Time at Varying Distances from Origin

If the concentration (log scale) at a point x in the agar is plotted against *time*, a hyperbolic curve results. (This concentration m is expressed as a fraction of that in the reservoir m_0 from which diffusion is taking place.) Each curve in Fig. 5 shows how the concentration increases

[12] S. Lehner and J. E. Smith, *J. Phys. Chem.* **40**, 1005 (1936).

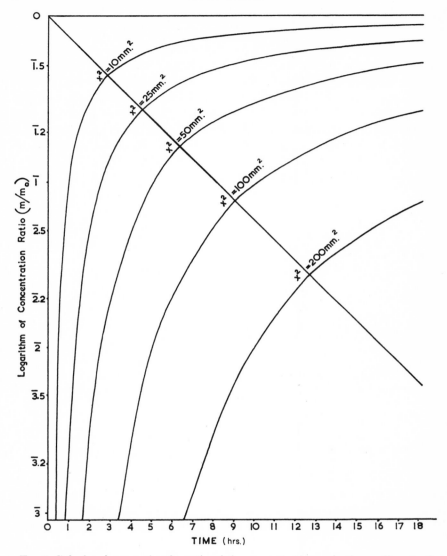

FIG. 5. Calculated curves for the ratio of the concentration m in the medium to the concentration $m_0 = 1$ in the reservoir at various distances x plotted against time for a substance of diffusion coefficient $D = 1.09$ mm.^{2}/hour. The concentration ratio is plotted on a logarithmic scale, and the curves are asymptotic to log $m_0 = 0$ and to $t = 0$ as the concentration in the agar is initially nil, and at infinite time equals the constant concentration maintained in the reservoir.

with time at a *particular distance* (x) from the edge of the reservoir. It increases rapidly at first, especially if concentration in the reservoir is high; then it increases at a gradually diminishing rate, approaching, but never reaching, that of the reservoir. When a particular concentration has been reached, provided that in the reservoir is maintained, it never falls. This is an important distinction from solutions of the diffu-

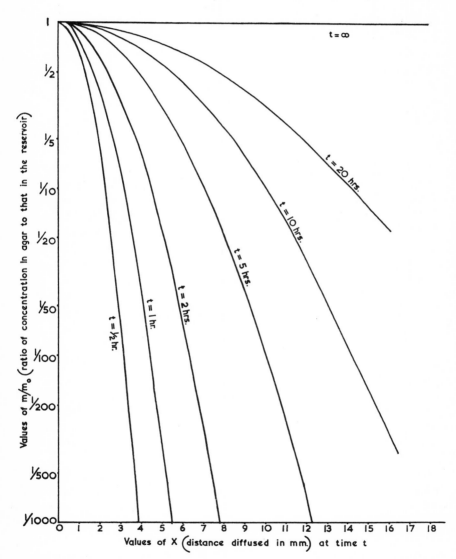

Fig. 6. The same figures as for Fig. 5 with the concentration ratio at various times plotted against distance.

sion laws concerning systems in which a constant amount occurs in the reservoir and the concentration falls after a time (Section B.3.e).

Concentration gradients are best shown by plotting concentration as ln (m/m_0) against distance. This gives a series of parabolas (Fig. 6) for the different times considered (plotting against x^2 gives straight lines). The gradients are steep so long as m is a small fraction of m_0, but when the concentration in the agar being considered is not very different from that in the reservoir the gradient becomes very shallow, i.e., increasing distance from the reservoir produces little difference. As time proceeds the gradient gets less and at infinite time concentration becomes uniform throughout.

The concentration gradient at the time of the formation of inhibition zones is an important factor determining the sharpness of the edge of the zone.

3. Inhibition Zones

a. *Conditions.* Though the laws of diffusion have been well known for years, their application to antibiotic inhibition zones was only suggested by Cooper and Woodman[10] in 1946 to give a quantitative explanation of the assay method that had been adopted for penicillin. Since the Oxford workers[6] first used the agar cup assay method as a standard for the estimation, many modifications had been suggested. The many details of the different techniques adopted influenced the results and a theoretical background was absent when a guide to the rationality of the methods was required.

The investigation of the theoretical background required the control of many variables which were not considered by those undertaking assays. Comparison with a standard under what was hoped were identical conditions was expected to eliminate the importance of these variables. This hope was not always justified as future work was to show. At that time also there was no proved pure standard available and the biological unit for penicillin was a measurement of zone size in comparison with a crude standard. Such units, though necessary in the circumstances, are very unreliable unless all the factors which may affect them are known and controlled.

Though the plate method was used for assays it presented difficulties in temperature control and numerous other irregularities. These were only overcome by the design of strictly comparative methods under extensive statistical controls. The agar tube method which, however, did not work satisfactorily with penicillin, offered greater possibilities of temperature control. Mitchison and Spicer[13] showed that streptomycin could be assayed satisfactorily by the use of tubes, and that the degree

[13] D. A. Mitchison and C. C. Spicer, *J. Gen. Microbiol.* **3,** 184 (1949).

of anaerobiosis produced did not affect the minimum concentration of streptomycin required to inhibit the growth of staphylococci. This method enabled Cooper and Gillespie[14] to investigate the effects of temperature on growth and inhibition, and Cooper and Linton[15] to compare the results with tubes and plates (Fig. 7). From this work it became evident that the tube method was best suited to elucidate the theoretical principles involved whatever its demerits as an assay method for some anti-

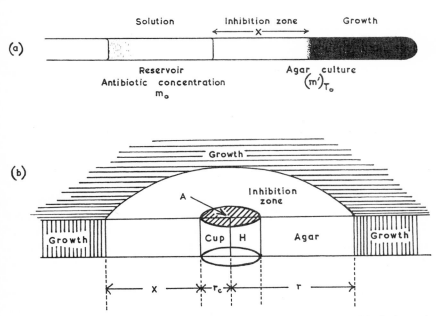

Fig. 7. (a) The formation of an inhibition zone by diffusion of antibiotic from the reservoir into the growing agar culture in a tube (linear diffusion). (b) The formation of an inhibition zone by diffusion of antibiotic from a cylindrical reservoir in an agar plate culture (radial diffusion).

biotics. A relatively large reservoir of antibiotic above the agar could be kept mixed (by convection) and a *constant concentration* maintained for the duration of the experiment. The use of thin-walled tubes gave immediate temperature control on immersion in a water bath. The conditions of linear diffusion from a constant concentration were thus achieved.

    b. The fact of a critical concentration. If an antibiotic solution of concentration m_0 is placed on the surface of an agar culture in a tube which has just been prepared and set, after mixing with a standard

[14] K. E. Cooper and W. A. Gillespie, *J. Gen. Microbiol.* **7**, 1 (1952).
[15] K. E. Cooper and A. H. Linton, *J. Gen. Microbiol.* **7**, 8 (1952).

inoculum, then incubation results in an inhibition zone extending a distance x from the surface Fig. 7(a). If it is assumed that the edge of a zone was determined by the arrival at this point of a concentration of antibiotic m' at a particular time T_0, then m' can be calculated, because from the laws of diffusion (substituting Eq. (4))

$$\ln m' = \ln m_0 - \frac{x^2}{4DT_0},$$
(6)

m_0 can be chosen over a wide range of values, and x measured. With constant conditions of inoculation, temperature, media, and timing arrangements of the experiment, $4DT_0$ can be kept constant. It is found that for values of $x > 3$ mm. the value of calculated m' is also constant. From the experiments on diffusion previously quoted slight deviations were expected for very small zones, due to disturbances of water and salt movements between antibiotic solution and agar culture.

Plotting values of x^2 against $\ln m_0$ gives a straight line intercepting the concentration axis at $\ln m'$ (Fig. 8). The slope of the line is determined by DT_0. (It will of course be modified if $\log_{10}$ is used instead of $\log_e$ by the factor 2.30.)

The value of m' is now called *critical concentration*. Concentrations below this produce no zones by diffusion methods, but it is important to realize that critical concentration is usually 2–4 times the *minimum inhibitory concentration* found by adding small inocula to liquid media. Such a dilution method will prevent growth at a lower concentration for a number of reasons, e.g., the small inoculum removes less antibiotic, and the antibiotic is in large excess of the organisms; it acts for a long time and fresh antibiotic is brought to the cells by convection. The growing culture of organisms in the agar tube at the zone edge, on the other hand, does not come into contact with the critical concentration diffusing towards it until after a time interval T_0 when it has multiplied to a population considerably greater than the inoculum. The critical concentration has to stop the next cell divisions or at least produce their inhibition in a very few generations to be effective.

Another factor, however, may exaggerate the difference between critical concentration and the minimum inhibiting concentration. If the tubes used are narrow, and if they are immersed so deeply in the water bath as to maintain an even temperature throughout the solution above the agar culture, then convection currents may be so minimized as to make *diffusion in the solution* the only method of renewing the antibiotic lost from the agar/liquid interface into the culture. In this case the concentration at the agar surface becomes $m_0/2$ instead of m_0, and as the sensitivity of the organism still requires the same critical concentration m', this will be achieved only by the use of apparently double strength

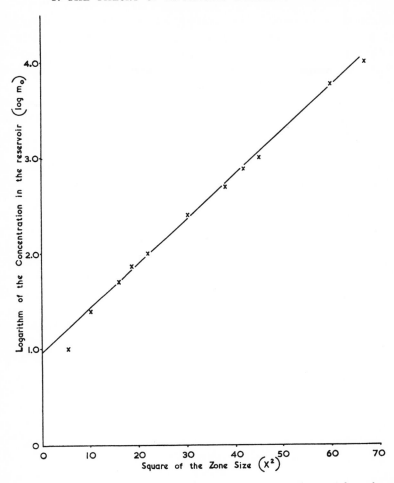

Fig. 8. Inhibition zone sizes and antibiotic concentration (tubes). Tube cultures of Staphylococcus prepared with inocula giving $N_0 = 5.46 \times 10^5$ organisms/ml. of agar were incubated at 36°C. Streptomycin solutions (of calcium chloride double salt) were added above the culture simultaneously with the commencement of incubation ($h = 0$). Sizes of the zones were read after 18 hours. Concentrations expressed as logarithm of micrograms per milliliter of streptomycin base. Critical concentration for strain in this medium was $m' =$ antilog 0.95 = 8.9 μg./ml.

solutions if m_0 is used in the graphs or for calculation purposes (see Höber[16]). It seems likely, however, that the longer the experiment proceeds the more likely are the concentration gradients within the liquid to be disturbed by convection, and if enough antibiotic is present the concentration m_0 will be renewed next to the agar.

[16] R. Höber, "Physical Chemistry of Cells and Tissues," p. 10. J. & A. Churchill, London, 1952.

Mitchison and Spicer[13] used $\frac{1}{2}m_0$ in their formulas because they consider the tubes are narrow enough to justify the assumption of renewal by diffusion only. In our experiments we try to maintain convection by only half-immersion of the solution, and in the early experiments on the diffusion of crystal violet large diameter tubes were used and the concentration m' in the medium followed by color matching. However, in Mitchison and Spicer tubes, it was shown that no experimental difference was detectable if the length of the solution was >1 mm. This suggests that exhaustion is not occurring.

    c. Alternative formulas. Mitchison and Spicer made an investigation into alternative diffusion formulas. They concluded that (converting their y into our symbol x, and their C and C_0 being given by $m' = C\sqrt{\pi}$ and $C_0 = m_0/2$) *when x is large*

$$\ln C_0 = \ln (2C\sqrt{\pi}) + \frac{x^2}{4DT}, \tag{7}$$

i.e., $\log m_0$ plotted against x^2 is a straight line, and *when x is small*

$$\ln C_0 = \ln (2C) + \frac{x}{\sqrt{(\pi DT)}}, \tag{8}$$

i.e., $\log m_0$ plotted against x is a straight line.

    Their graphs of experimental results for *Staphylococcus aureus* with from 4 to 400 μg. of streptomycin give good linearity with the x^2 plot. They suggest some deviation from linearity below 2 μg. streptomycin. From what has already been quoted above it would be expected that some experimental deviation would occur as a result of factors (water flow and salt movement) not considered by any of these formulas. Also, with concentrations of streptomycin approaching so closely to the critical, zone edges become much less definite. There seems no reason therefore for not regarding the formula of Cooper and Woodman as a satisfactory approximation. The precise interpretation of critical concentration depends on the experimental conditions, and it does not seem justified at present to equate it with minimum inhibitory concentration in liquid media. When the absorption of antibiotic by cells is considered, other reasons will be found to distinguish critical and minimum inhibitory concentration (see Section IV).

    Mitchison and Spicer assume their C (i.e., the precise concentration at point x and time t, arriving by diffusion) is the minimum inhibitory concentration (but use $C\sqrt{\pi}$ in place of the m'). Cooper *et al.*[17] suggested that at the zone edge a rapid absorption by the growing organisms is occurring, and that it would be more realistic to consider the precise time (t) as the moment when half the critical concentration (m') is

[17] K. E. Cooper, A. H. Linton, and S. N. Sehgal, *J. Gen. Microbiol.* **18**, 670 (1958).

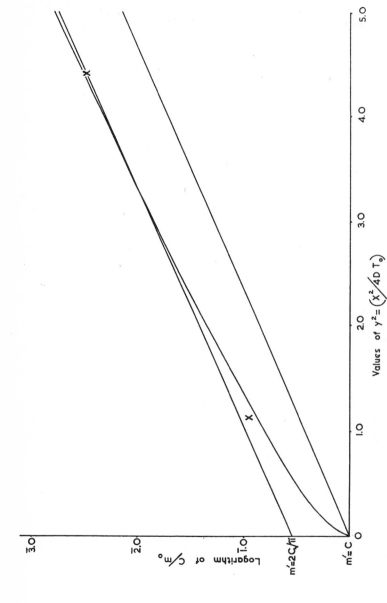

FIG. 9. Graph of log C/m_0 against $y^2 (= x^2/4DT_0)$ where $C/m_0 = \frac{1}{2} [1 - 2\pi^{-\frac{1}{2}} \int_0^y \exp(-y^2) dy]$ (curved line). For comparison graphs of log $m'/m_0 = y^2$ are plotted for $m' = 2C\sqrt{\pi}$ (upper line) and $m' = C$ (lower line). The former corresponds to the approximation of Cooper and Woodman[10] and the average of the experimental determinations of Cooper and Gillespie[14] are shown thus: $\times$.

absorbed and half is still in solution. Thus $m'/2$ will be a nearer approximation to the minimum inhibitory concentration C. This assumption enabled them to interpret the effects of the size of the cell population. This aspect will be discussed further when the laws of growth and of absorption are considered (Section III).

The degree of approximation used in the Cooper and Woodman[10] formula can be seen by examination of the graph (Fig. 9) comparing y^2 and $\log_{10} C/C_0$ where $y^2 = x^2/4DT$ and $m' = 2\sqrt{\pi C}$. The assumption of $C_0 = m_0$ (i.e., the reservoir concentration being kept constant by convection and addition of antibiotic) gives the theoretical diffusion formulas

$$C = C_0 \left[1 - \frac{2}{\sqrt{\pi}} \int_0^y \exp\left(-y^2\right) dy \right] = C_0[1 - \text{erf }(y)]. \tag{9}$$

Tables of the function erf (y) are available in Alexander and Johnson[18] and in Levy and Preidel.[19] The corresponding straight line formula is

$$(2.30) \log_{10} C_0/C = y^2 = x^2/4DT. \tag{10}$$

Deviation between the two only becomes important for small values of y. The differences in slope becomes less as y increases and is within the limits of experimental error (being dependent on DT). The average values for the experimental points obtained by Cooper and Gillespie[14] for streptomycin and staphylococci are inserted ($C_0 = 2$–64 μg., $m' = 0.75$ μg., $D = 0.84$, $T_0 = 5.25$).

d. Preincubation and critical time. When the addition of the antibiotic to the culture tubes is made simultaneously with the beginning of incubation which starts the growth of the organisms, the period of time which elapses before the edge of the zone is decided is T_0. If the organism is grown for a period of h hours before adding the antibiotic, the time of diffusion will be reduced to $T_0 - h$ if the zone is formed at the same period in the growth of the culture. Zones will thus be made smaller and a series of lines may be constructed by plotting $\ln m_0$ against x^2, each line representing a different value of h.

The results recorded in Fig. 10 show such lines. The fact that $\ln m'$ is independent of the value of h is further proof of the reality of the concept *critical concentration*. The time of diffusion $(T_0 - h)$ taken for the critical concentration to reach the edge of the zone is very different for each line; nevertheless, the same concentration (m') decides the edge of the zone. The formula

$$x^2 = 4D(T_0 - h) \ln (m_0/m') \tag{11}$$

[18] A. E. Alexander and P. Johnson, "Colloid Science," p. 236. Oxford Univ. Press, London and New York, 1950.

[19] H. Levy and E. E. Preidel, "Elementary Statistics," p. 155. Thomas Nelson & Sons, London, 1944.

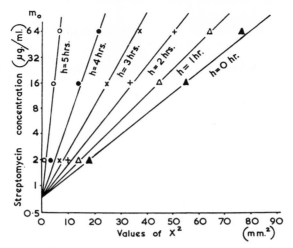

FIG. 10. The effect of preincubation on zones of inhibition of *Staphylococcus* by streptomycin at 27°. Concentration on a logarithmic scale is plotted against x^2 where x is the depth of inhibited zone. The hour of preincubation (h) represents the length of time before streptomycin was added to the incubating tubes for each set of determinations. Theoretical lines $m' = 0.75$ μg./ml., $D = 0.84$ mm.²/hour, and $T_0 = 5.36$ hour, $x^2 = D(T_0 - h) \ln m_0/m'$.

may also be written

$$h = T_0 - x^2/4D \ln (m_0/m'). \qquad (12)$$

This shows a straight line relationship between h and x^2 for constant concentration m_0. Figure 11 shows such a graph (of the same points as

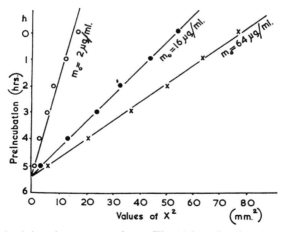

FIG. 11. Graph giving the same results as Fig. 10 but showing the constancy of T_0 for three concentrations of streptomycin and the method of its determination by extrapolation to $x = 0$, of lines obtained by plotting x^2 against h for each concentration.

in Fig. 10). When x^2 is reduced to 0, then $h = T_0$. It is evident that T_0 is independent of concentration, and that therefore we can correctly assume a *critical time* in the growth of the organisms when the edge of the zone is formed.

Critical time can be measured by making a series of preincubation experiments and measuring x^2 for one concentration of antibiotic. Additions of antibiotic to culture tubes during growth must be at the correct temperature and without removing them from the water bath. Similar results can be obtained with large cups in plates, but as Cooper and Linton[15] showed only under constant temperature conditions. The nature of critical time is best considered in relation to the laws of growth (Section III.A).

4. Diffusion from a Constant Amount

In the case of diffusion from a reservoir of a constant amount of antibiotic, the effects of the fall in concentration which occur from the loss into the agar will constantly lessen the concentration gradients upon which the diffusion depends. The curves corresponding to Fig. 5 will no longer be hyperbolas, nor those in Fig. 6 parabolas. The final concentration at infinite time will no longer be m_0 but $m_0 v/(V + v)$ where v is the volume of fluid in the reservoir and V the volume of the agar culture. If the latter is large the final uniform concentration will be very small ($m \rightarrow 0$ as V and $t \rightarrow \infty$), even if an initial high value of m_0 produces (especially near the reservoir) temporary high concentrations (m). The values of m/m_0 will, instead of approaching asymptotically the value 1, rise and then fall.

Instead of parabolas as in Fig. 6 we shall have a series of half-probability curves (Fig. 12) with the initial concentration at the apex falling with time, and the bases asymptotic to the final concentration. If V is finite, this is given by the volume relationship; if infinite ($V = \infty$), m becomes nil at infinite time. The graphs in Figs. 12 and 13 illustrate a hypothetical case calculated from $c/M = (2\sqrt{\pi Dt})^{-1} \exp(-x^2/4Dt)$ where c is the concentration at point x at time t and M is the total amount of antibiotic per unit area of surface of agar in the reservoir. The concentration (c) is assumed to be produced by the diffusion of $0.5 M$ into the agar and $0.5 M$ into the liquid. It is also assumed that M is concentrated initially at the surface in infinite concentration. This solution is for linear diffusion.[19a] When the time involved is short and the amount of antibiotic available from the reservoir (in tubes, or large cups or cylinders) is sufficient (M is large and c/M small) the slight falls in concentration at $x = 0$ that occur are unimportant. With small cups, beads, or disks these considerations are much more important, but as

[19a] W. Jost, "Diffusion in Solids, Liquids, Gases." Academic Press, New York, 1952.

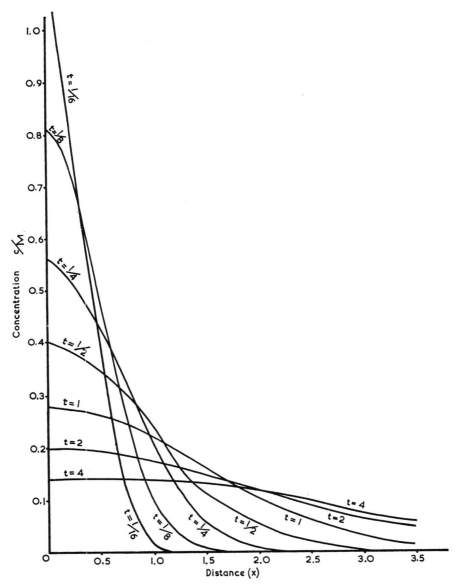

FIG. 12. The effect of a falling concentration of antibiotic in the reservoir on the concentration in the agar tube at successive intervals of time. Concentration (c) in the agar plotted against distance (x) from the formula $c/M = (2\sqrt{\pi Dt})^{-1}\exp(-x^2/4Dt)$. M is the quantity per unit area, of the antibiotic initially concentrated at the surface of the agar so that at time $t = 0$, $c = \infty$ at $x = 0$. Half this antibiotic diffuses into the agar and half into the liquid.

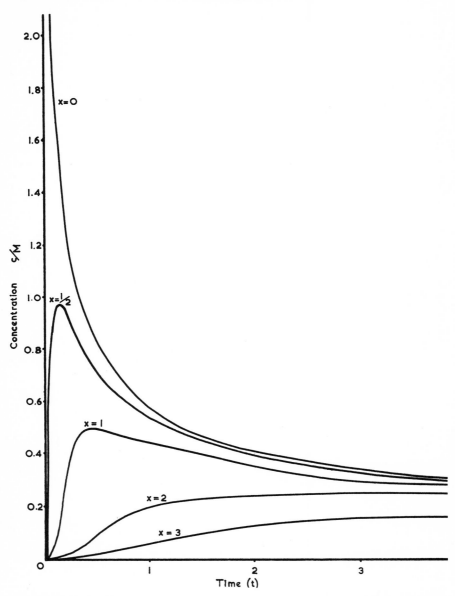

FIG. 13. Graph giving same calculations as Fig. 12 but concentrations plotted against time for different points in the agar.

these are cases most accurately depicted by the formulas for radial diffusion and not linear, we shall examine them in Section II.C.

In the cases concerned with linear diffusion depths of above a few millimeters of solution on the agar in tubes show no detectable deviations such as would occur with falling concentration (Mitchison and Spicer give >1 mm.). Cups in plates >8 mm. and disks >12 mm. seem adequate to allow the formula of Cooper and Woodman to be used for rapidly growing organisms. Applicability depends on the maintenance of a constant concentration and not on the amount of antibiotic. In the case of disks the volume of fluid held by the disk is of paramount importance, as this determines the concentration. The careful work of Loo et al.[20] fits the x^2 against $\log m_0$ formulas just as well as the formulas for radial diffusion, except for the smallest zones where $x < 3$ mm. and divergences would be expected for other reasons than falling concentration. Other examples are quoted by Kavanagh.[21]

C. Solutions for Radial Diffusion

1. A Comparison of Formulas

Finn,[22] writing on the Theory of Agar Diffusion Methods, quotes experiments illustrating the effect of cylinder diameter on the size of antibiotic inhibition zones when the cylinder size is $\leqslant 8.2$ mm. Table I

TABLE I

EFFECT OF VARIATIONS IN CUP DIAMETER
ON SIZE OF ZONE[a]

Cup i.d. (mm.)	Zone diameter (mm.)	x (mm.)
2.6	10.1	3.8
3.8	11.8	4.0
5.0	13.3	4.2
6.4	15.4	4.5
8.2	17.4	4.6

[a] Finn.[22]

gives his results for the effect of the same solution of penicillin on *Bacillus subtilis*. These cylinders are too small to expect the formula for linear diffusion to apply, and in confirmation of this the distance diffused (x) varies from 3.8 to 4.6 mm. as the cup size increases (2.6 to 8.2). Constant concentration was maintained by further additions of antibiotic as necessary (private communication, R. K. Finn).

[20] Y. H. Loo, P. S. Skell, H. H. Thornberry, J. Ehrlich, J. M. McGuire, G. M. Savage, and J. C. Sylvester, *J. Bacteriol.* **50**, 701 (1945).

[21] F. Kavanagh, *Advances in Appl. Microbiol.* **2**, 71 (1960).

[22] R. K. Finn, *Anal. Chem.* **31**, 975 (1959).

For such conditions, therefore, equations are required for radial diffusion. The differential equation of Fick for linear diffusion was

$$\frac{dc}{dt} = D \frac{d^2c}{dx^2}. \tag{2}$$

The corresponding one for radial diffusion is

$$\frac{dc}{dt} = D \left(\frac{d^2c}{dr^2} + \frac{1}{r}\frac{dc}{dr} \right). \tag{13}$$

Brimley,[23] considering the area of spots formed by diffusion in paper chromatography, gives a solution for the area:

$$a = \pi r^2 = 4\pi DT(\ln C_0 - \ln C' - \ln (2DT) + \ln \text{``}F\text{''}) \tag{14}$$

where C_0 is the initial concentration and C' the concentration at the visible edge; F is a complex integral containing a modified Bessel function, but according to Finn and Deindoerfer[24] this can be neglected when the cup diameter is small compared with the zone diameter. The formula thus becomes, where r is the radius of zone from the center of the cup,

$$r^2 = 4DT[\ln C_0 - \ln C' - \ln (2DT)]. \tag{15}$$

Vesterdal [25] considered the case of a solution of concentration C_0 on the center of a small area (A) and gave the solution of the diffusion problem as

$$C' = \frac{AC_0}{4\pi DT} \exp (-r^2/4DT) \tag{16}$$

which gives

$$r^2 = 4DT[\ln (AC_0) - \ln C' - \ln (4\pi DT)]. \tag{17}$$

Humphrey and Lightbown[26] used a solution by A. G. Liddiard for a quantity of antibiotic M, distributed from a small bead through a fine central column of agar in a plate of thickness H which was

$$C' = \frac{M}{H4\pi DT} \exp (-r^2/4DT) \tag{18}$$

which gives

$$r^2 = 4DT[\ln (M/H) - \ln C' - \ln (4\pi DT)]. \tag{19}$$

If the small cup of Vesterdal and the small cylinder of agar beneath the bead of Humphrey and Lightbown are considered identical, containing quantity M at a concentration C_0 in a volume AH, then $AC_0 = M/H$, and

[23] R. C. Brimley, *Nature* **163**, 215 (1949).

[24] R. K. Finn and F. Deindoerfer, 51st General Meeting Society American Bacteriologists, Chicago (May, 1951).

[25] J. Vesterdal, *Acta Pathol. Microbiol. Scand.* **24**, 273 (1947).

[26] J. H. Humphrey and J. W. Lightbown, *J. Gen. Microbiol.* **7**, 129 (1952).

these central volumes correspond to the initial concentration on the spots considered by Brimley. The three formulas differ only in the constant $-\ln 2$ (Brimley), $-\ln 4\pi$ (Vesterdal), $-\ln 4\pi$ (Liddiard). All take into account the falling concentration at the center (which is ignored by Cooper and Woodman) by the further addition of the term $-\ln Dt$, but none of them have been tested adequately by accurate determinations of the value of T to be used. Vesterdal made no attempt to determine critical time, though he noted that the sensitivities of organisms should not be compared by this method if they had different growth rates. Humphrey and Lightbown did attempt to measure critical time, but they averaged two methods of determination, one of which gave a value 1 hour shorter than the other. It seems unlikely that their method of removing the agar from which antibiotic was diffusing by cutting it out, would give a value for T sufficiently related to the experimental conditions used in their assays. Brimley was not concerned with antibiotic assays at all, but his theoretical considerations at least in their most complex form appear the most accurate.

All these formulas indicate that the relationship between the square of the zone radius and the logarithm of the concentration is a linear one. The smaller the cup or cylinder and the larger the zone, the less is the difference between r^2 and the x^2 of Cooper and Woodman. The real difference between these formulas for radial diffusion and that used for linear diffusion is the allowance made for falling concentration at the source. All these formulas differ only in the initial conditions assumed to develop their theoretical basis. Table II summarizes these.

2. The Effect of Falling Concentration

The effect of a small reservoir and falling concentration may be well seen in the analysis given by Vesterdal. The differences are best seen in the graphs for low concentration ratios (C_0 approaching C) and for rather low values of r.

The graphs of Vesterdal for radial diffusion are similar to Figs. 12 and 13 for linear diffusion. To show more clearly the effects of falling concentration, concentration has been plotted on an arithmetic instead of a logarithmic scale. Vesterdal's formulas differ from the formula used for linear diffusion in that

$$C = \frac{C_0 A}{4\pi DT} \exp(-r^2/4DT).$$

It has C_0A in place of M, the amount of substance, and $4\pi DT$ instead of its square root, but the general form of the curve is the same. Maximum concentration at a distance r is given by

$$C_{\max} = C_0 A / \pi r^2 e \tag{20}$$

TABLE II[a]

Author	Cooper and Woodman[10]	Mitchison and Spicer[13]	Vesterdal[25]	Humphrey and Lightbown[26]	Finn and Deindoerfer[24] and Brimley[23]
Concentration at diffusing edge[b]	$m' = m_0 \exp(-x^2/4DT')$	$C = C_0[1 - \pi^{-1/2} \int_0^y \exp(-y^2)\,dy]$ $y = x/\sqrt{4DT}$	$C = \frac{AC_0}{4\pi DT} \exp(-r^2/4DT)$	$C = \frac{M}{H4\pi DT} \exp(-r^2/4DT)$	$C = \frac{C_0}{2DT} \exp \frac{-r^2}{4DT}$ $\cdot \int_0^{r_c} [\exp(-r^2/4DT)] I_0^* \, dr$
Distance measured	$x^2/4DT = \ln m_0 - \ln m' + "F"$	$x^2/4DT = \ln C_0 - \ln C + "F"$	$r^2/4DT = \ln C_0 - \ln C + "F"$	$r^2/4DT = \ln M - \ln C + "F"$	$r^2/4DT = \ln C_0 - \ln C + "F"$
Difference "F"[b]	$"F" = 0$	$"F" = -\ln(x\sqrt{\pi}/\sqrt{DT})$	$"F" = -\ln(4\pi DT/A)$	$"F" = -\ln(4\pi DT'H)$	$"F" = -\ln(2DT)$
Amount of antibiotic	Large	Large	$C_0 \cdot A/H$	M	C_0 over A, $H \to 0$
Initial concentration	m_0	C_0 immediately becoming $C_0/2$ at $x = 0$	C_0	M over H, $A = 0$, $C_0 \to \infty$	C_0
Conditions in solution	Convection → uniform, constant concentration, at $x = 0$	Stationary, diffusing from constant concentration at $x = 0$	Diffusing from center falling concentration	Diffusing from center falling concentration	Diffusing from a falling concentration
Size of A	Large → ∞	∞	Small	Small	Small
Experimental systems	Tubes, gutters, and large cups	Tubes	Small cups	Bead on agar surface	Disk on surface

[a] $A = \pi r_c^2$. Area of interface $= 2\pi r_c H$. Volume of cup $= AH$. $C_0 = M/AH$. In the modified Bessel function, $I_0^* = I_0\left(\frac{r_c \cdot r}{2DT}\right) r$.

[b] Where the interpretation of symbols is slightly different (e.g., critical concentration, minimum inhibitory concentration, distance from center or edge of cup) the original symbols have been retained to avoid confusion, when they may be approximately equivalent (see Fig. 7(b)).

and the time this occurs by

$$T_m = r^2/4D. \tag{21}$$

The longer the distance, the later the maximum concentration and the lower its value.

If the conception of a critical time for the formation of *inhibition zones* is applied to these conditions, as was suggested by Cooper and Linton, preincubation would reduce the size of the zones. Graphs *analogous* to Figs. 10 and 11 can be prepared (see Figs. 14 and 15). The formula becomes

$$r^2 = 4D(T_0 - h) \left[\ln \frac{C_0 A}{C 4\pi D(T_0 - h)} \right]. \tag{22}$$

For a particular value of h (the preincubation time) this takes the form

$$r^2 = 4DT \left[\ln \frac{C_0}{C} + \ln K \right] \tag{23}$$

where K is constant though it includes $T_0 - h$. Thus for each value of h a straight line is obtained by plotting ln concentration against r^2. However, it intercepts the concentration axis at $\ln K$ where $K = (C/A) 4\pi D (T_0 - h)$. The bigger h is, the smaller is K, so that the lines do not give the value of C at the point analogous to m' (when $r^2 = 0$). Plotting h against r^2 when C_0 is constant does not give straight lines but slight curvature, but when $r^2 = 0$, $h = T_0$ independently of C_0.

Experimental investigations along these lines have been little done. They involve adding antibiotic to the cups during growth of the organism *without altering the temperature* as T_0 is very susceptible to temperature shock. Shock alone produces curvature even without a falling concentration of antibiotic in the cup (see Cooper and Linton[15]).

When shock was avoided, 8-mm. cups gave no curvature for plots of r^2 against h that was outside possible experimental error. Vesterdal made no such determinations of critical time when using 5-mm. cups. His experiments with penicillin assays by means of staphylococci gave double edged zones. These were attributed to cell populations containing sensitive and resistant cells which were shown experimentally to be able to produce such partial zones. The phenomena of secondary lysis produced by penicillin may have also played a part, but it seems reasonable to suggest that the existence of maximal concentrations such as are shown in Fig. 13 by curves $x = \frac{1}{2}$ and $x = 1$ would cause double zones. Organisms near the edge would be held above the inhibitory concentration C' for only part of the growth period. To obtain the sharp single-edged zones best suited to antibiotic assays it seems clear that *systems giving falling concentrations should be avoided.* Small cups, especially when used for low potency antibiotic are an obvious example. The small beads used by

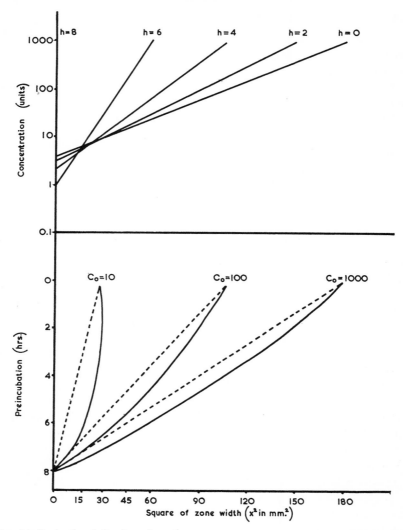

Fig. 14. Zone size (x^2) plotted against concentration (C_0) for plate cultures with cup reservoir of area A. Effect of falling concentration and preincubation. $x^2 = (2.30)4D(T_0 - h)(\log C_0/C - \log 4DA^{-1}(T_0 - h))$, $D = 1.09$ mm.^{2}/hour, $T_0 = 8.0$ hour, $C = 2.0$ gm., cup radius 2.0 mm.

Fig. 15. Graph giving same calculations as Fig. 14 but x^2 plotted against preincubation time (h) for each concentration.

Humphrey and Lightbown[26] are open to this same objection, but have an even greater one, namely, the impossibility of saying accurately when $T = 0$. The antibiotic is never actually distributed in an infinitely thin cylinder of height h nor is the quantity of antibiotic (M) present in an infinite concentration. When diffusion starts from the agar surface, as from beads or disks, it is at first a three-dimensional procedure, though it becomes two-dimensional as soon as the depth of the agar has achieved the concentration at the surface. This is apparently achieved *partly by water flow,* as Humphrey and Lightbown say the bead empties. This is another variable making the precise determination when $T = 0$ difficult.

A more complicated two-dimensional diffusion formula, such as that which Brimley used in paper chromatography, has been applied by Finn to the case of paper disks. It seems doubtful if the added refinements in the diffusion formula compensate for the greater difficulty in determining $T = 0$. The results quoted by Finn[22] and Loo *et al.*[20] fit the formula of Cooper and Woodman just as well, except for the smallest zone, when this kind of difficulty is at its maximum, but determinations of critical time with disks at *constant temperature* have not been made. Such work needs to be done. Similar remarks are applicable to the even more complex formulas suggested by Miyamura.[27]

D. Prediffusion

Diffusion formulas may also be tested by adding antibiotic to agar cultures for prolonged periods before the growth of the culture is started. This prediffusion of the antibiotic has to be done before inoculation, or at a temperature below the growth temperature of the organism. The value of T (the diffusion time) then becomes T_0 (the growth time) plus the hours of prediffusion. As the symbol h has been used for preincubation (when T becomes $T_0 - h$), if the time of prediffusion is regarded as $-h$, the same formulas will apply. However, the diffusion time from $-h$ to 0 will be at the low temperature (usually refrigeration) and requires the use of the diffusion coefficient (D_{t_1}) at the correct temperature, whilst after time 0, when incubation starts a corrected coefficient (D_{t_2}) must be used for the period 0 to T_0. As diffusion is proceeding during the change of temperature, allowance must be made for the prolonged time taken for plates in air to make this change. As a result prediffusion slopes and preincubation slopes will be different, and on projection to time $h = 0$ may produce discontinuity. Tubes in water baths are subject to much more rapid change and give more constant results (for the theory see Section E.2 and Figs. 18 and 19).

The larger size of prediffused zones offers greater ease of measurement, and may be of advantage with slowly diffusing antibiotics for assay purposes, despite the added difficulties of theoretical calculation. Calibration of standards is possible and comparative assays may be of increased accuracy. In these circumstances with weak antibiotics and small zones, plotting x instead of x^2 against log concentration may be justified if a

[27] S. Miyamura, *Antibiotics & Chemotherapy* **3**, 903 (1953).

better approximation to a straight line is obtained (see Mitchison and Spicer[13] and formula (8) in Section B.3.c for small values of x). But the longer prediffusion is used to obtain larger zones, the more likely are the results to fit the x^2 formula.

Prediffusion has also been used for exhibition zones produced for vitamin assays. (See Harrison et al.,[28] Assay of vitamin B_{12} by E. coli mutant.)

E. Factors Influencing the Diffusion Coefficient

1. Kinetic Theory

For "spherical molecules" of radius r_m in a solvent of viscosity η, the diffusion coefficient is given by the equation of Sutherland-Einstein

$$D = R\theta/6\pi\eta N r_m \tag{24}$$

where R is the gas constant, N is Avogadro's number, and θ the absolute temperature. It follows that D should vary with the absolute temperature and viscosity of the medium according to the equation

$$\frac{D_{t_1}}{D_{t_2}} = \frac{\theta_1}{\theta_2} \cdot \frac{\eta_2}{\eta_1}. \tag{25}$$

It also varies inversely as the radius of the molecule, if Stokes law for the frictional resistance offered by the small solvent molecules to the ideal large spherical diffusing molecule is obeyed ($f = 6\pi\eta r$). Einstein derived from Fick's law the value

$$D = R\theta/Nf \tag{26}$$

by equating the forces of diffusion and osmosis acting on a semipermeable membrane. These laws obtained by applying the kinetic theory of gases to dilute solutions will be expected to be approximations. For more concentrated solutions D will vary with concentration. These equations have been applied to the diffusion of protein in pure water, by assuming that the partial specific volume and hence the density (d) was constant, to calculate an ideal radius from the molecular weight, $M = \frac{4}{3}\pi r_m^3 N d$. This gives a relationship between molecular weight and diffusion constants, so that

$$\log D = -4.49 - 0.33 \log M \tag{27}$$

taking the viscosity of water at 20°C. as 0.01 e.g.s. units.[16] The results are very approximate, M calculated from D being about twice the value obtained by the ultracentrifuge.

Similar relationships were used by Perrin to explain the Brownian movement of particles of 1μ diameter under the microscope. This move-

[28] E. Harrison, K. A. Lees, and F. Wood, Analyst 76, 696 (1951).

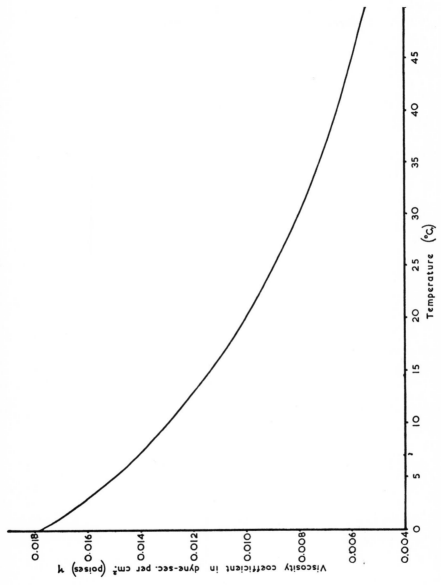

FIG. 16. The viscosity of water (η) plotted against temperature (°C.). Average of recorded values in the literature.

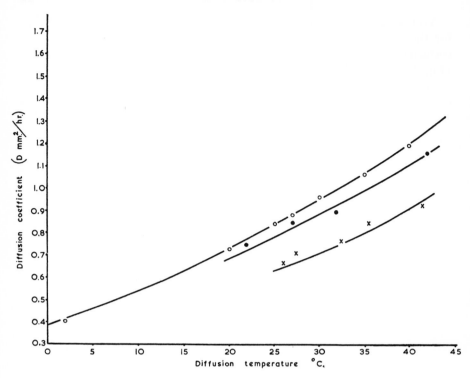

FIG. 17. The diffusion coefficient for streptomycin in three different media plotted against temperature and calculated from the size of agar inhibition zones.

●——●	Media I	1% agar broth half standard dilution,	salt ¼% pH 7.8
○——○	Media II	1% agar broth half standard dilution,	salt ½% pH 7.8
×——×	Media III	2% agar broth undiluted standard,	salt ½% pH 7.6

I, Cooper and Gillespie; II, Linton; III, Cooper, Linton, and Sehgal. Lines are the best values calculated from theoretical effects of temperature on diffusion in water.

ment is visible diffusion, and was used very successfully by Perrin[29] to calculate Avogadro's number, i.e., the number of molecules in the gram molecule. The value of the diffusion coefficients of penicillin and crystal violet led Cooper and Woodman[10] to conclude that the penicillin molecule had the smaller radius before its formula was known.

2. Temperature

The influence of a temperature rise on diffusion is thus twofold (Eq. (25)). The driving force of the diffusing molecules is proportional to their absolute temperature and the resistance of the solvent molecules expressed as the viscosity of the solvent is decreased.

[29] J. Perrin, "Atoms," translated by D. L. Mammick, p. 107. Constable, London, 1916.

Average figures from a number of sources are depicted on the graph for the viscosity of pure water (Fig. 16). The values for the diffusion coefficient of streptomycin, as used in formulas for zone sizes, are shown (Fig. 17) as an example of the effect on D.[17,30]

Figures 18 and 19 show the results of preincubation and prediffusion experiments with tubes. The ratio of the slopes of the prediffusion part to

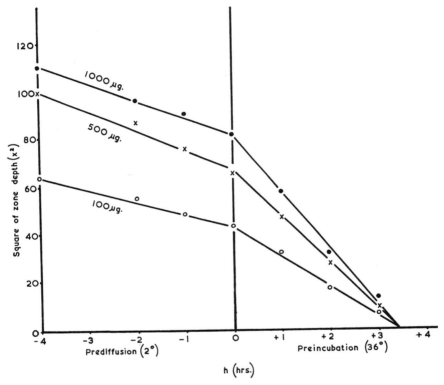

FIG. 18. The effect of prediffusion and preincubating on zone size for three concentrations of streptomycin/ml. Staphylococcal inocula $N_0 = 7.55 \times 10^5$/ml., $m' = 8$ μg./ml. Ratio of slopes of prediffusion curve at 2° to preincubation curve at 36° = 0.39. Theoretical ratio of $D_{2°}/D_{36°} = 0.37$. x^2 plotted against h.

the preincubation part for any one concentration is the same as the ratio of the diffusion coefficients at the two temperatures. Very good agreement with ratios calculated for pure water has been obtained for experiments with staphylococci and with *Klebsiella* for streptomycin. Critical time, T_0, depends on temperature but can be varied at will by using inocula of different sizes, as will be explained in the section on growth (see Section

[30] A. H. Linton, *J. Bacteriol.* **76**, 94 (1958).

　　　　　K. E. COOPER

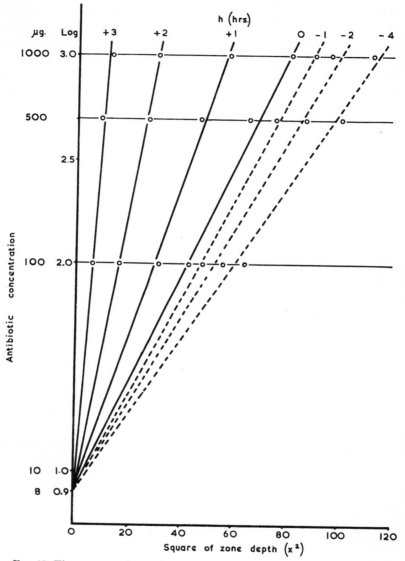

FIG. 19. The same results as Fig. 18. x^2 plotted against log concentration.

III.D.2). The product $D(T_0 - h)$ determines the slope of the log concentration against x^2 graph (see Eq. (11)), and from this D may be determined (see Section III.F).

3. Viscosity

With dilute solutions in weak agar gels it is normally sufficient to use the viscosity of water for calculations of D. However, the addition of viscous substances such as glucose or the use of strong agar may modify the viscosity until this becomes of major importance.

4. The Properties of the Agar Gel

Japanese agar from *Gelidium corneum,* or one from an East Indian seaweed *Eucheuma muricatum* have both been claimed as the original agar introduced to bacteriology by Frau Hesse. Since 1939 new seaweeds have been investigated and agars prepared from many sources. *Gelidium cartilagincium* in California, *Pterocladia* sp. in New Zealand, *Gracilaria confervoides* in Australia, and related species in India, Malaya, and Ceylon, and various Red seaweeds off the shores of Britain have been exploited (*Gigartina stellata, Chondrus crispus, Gracilaria* sp., and *Gelidium corneum*). The gelling properties of agars from these species differ considerably, and it is to be expected therefore that as mechanical strength, setting, and melting points differ, so will pore size, solubility, and the viscosity of the contained aqueous solution. A constant agar, highly purified, can be obtained only from firms making very large batches from constant sources of material. Some agars on the market differ in source and properties from batch to batch in a very unpredictable way. With the best agars, however, the precise method of preparation of the final medium has important effects, especially the amount of heating used in sterilization, and the age of the material, conditions of setting, strength and storage. Methods for testing agar are described by Marshall *et al.*[31] who made an investigation into British agars. The main constituent of agar is the calcium salt of a sulphuric acid ester of a complex polysaccharide. Hydrolysis yields galactose, and the main polysaccharide named agarose consists of long chains with both α and β linkages. The calcium may be replaced by hydrogen ions or by various metals.

Salts of Ca, Ba, Cu, Zn, and Ni affect the setting time of the gel; strong acids and alkalies, Fe, Al, Mn, and Ag cause coagulation of agar sols. Soluble fractions can be extracted at 3°, 22°, and 45° which are increased by aging, and osmotic pressure studies have shown that NaCl causes coiling of molecular chains in the gum agar, the interaction increasing with temperature. The intermicellar spaces in the agar gel contain such a sol and the viscosity will be affected by many agents, especially salts and sugars.

The agar gel possesses remarkable strength, even with a very high water content. Such structures are possible when chain molecules provide a molecular framework with junctions so that a reticulum with elastic

---

[31] S. M. Marshall, L. Newton, and A. P. Orr, "A Study of Certain British Seaweeds and Their Utilization in the Preparation of Agar," H. M. Stationery Office, London, 1949.

properties maintains shape. The contained water also forms a continuous system, so that the two interwoven molecular systems form one phase and not two. The agar network thus acts as a supporting frame for capillary columns of water branching through its substance, and in such dilute gels diffusion of solutes takes place as though in still aqueous medium (see Frey-Wyssling[32] for further accounts of gel structure).

Osmotic forces may produce swelling or shrinkage of the gel by imbibing or exuding water. Substances of considerable molecular size can pass into gels or be filtered through them, if the pore diameter of the gel is sufficiently large. Often however electric charge, active groups on the gel micelles, or affinity for solvent or other molecules, may complicate the result. Agar is fortunately chemically very inert and adsorption effects are usually weak or absent. It is therefore a very suitable medium for investigating simple diffusion.

Friedman[33,34] investigated the structure of agar gels by diffusion methods. Using urea, sucrose, and glycerine as diffusates in gels from 2 to 5% he obtained straight line relationships between increasing gel concentration and decreasing diffusion coefficient of the solute. The exact methods of preparing gels, especially the amount of preheating, were important factors modifying results. He used a formula which is obviously of very limited application to express the relationships found, and from which he calculated the average pore size of the gel. It was

$$D_{H_2O} = D_{gel}(1 + 2.4r/R)(1 + a)(1 + \beta) \qquad (28)$$

where r is the radius of diffusing molecule, R the average radius of pores of gel, a is a correcting factor for the influence of the gel on the viscosity of the contained liquid, and β is a correction for the mechanical blocking of the capillaries by solid agar. Extrapolating the values to zero concentration gave results that differed from pure water only by the factor a which could thus be calculated. The blocking factor β was given by $\beta = \sqrt[3]{(g/d)^2}$ where g equals grams of gel per cubic centimeter and d the gel density (i.e., inverse partial specific volume). Radius r was given by

$$D = \frac{R\theta}{N} \frac{1}{6\pi\eta r} \qquad (24)$$

and the remaining factor $(1 + 2.4r/R)$ the best approximation derived from the fall of bodies in capillary tubes. This obviously does not apply when $r > R$, for it is well known that diffusibility suddenly ceases when the gel concentration reaches this point. Urea gave pore size for agar 2%

[32] A. Frey-Wyssling, "Sub-microscopic Morphology of Protoplasm and Its Derivatives," p. 46. Elsevier, Amsterdam, 1948.

[33] L. Friedman, *J. Am. Chem. Soc.* **52**, 1311 (1930).

[34] L. Friedman and E. O. Kraemer, *J. Am. Chem. Soc.* **52**, 1295 (1930).

as 2.9 mμ, and 5% as 0.74 mμ. However, glycerine gave nearly double these values, a result Friedman ascribes to its effect on the solvation of the gel. Glycerine, lactose, and sucrose all increased the diffusion rate of urea, though alcohol reduced it. Neurath et al.[35] showed that the diffusion of serum albumen was retarded by sucrose, urea, and guanidine HCl by some 10% higher amount than would be expected by the measured relative viscosity.

5. The Effect of Sugars

The effects of sugars are complex and first received notice as a result of assays of antibiotics in lozenges containing them. Incorporation of carbohydrate in the assay medium was suggested to compensate for the high results obtained. However, too high a concentration in the medium gave excessively low results. Furthermore, not only was sensitivity of the organism to antibiotic often altered, but the slopes of the assay graphs ceased to be parallel to the standard, and precision was adversely affected.

Results with sugars[36] included the effects of glucose in the medium and the test solution over a range of concentrations from nil to 25%. The organism was *Bacillus subtilis* and the method the cylinder plate method of assaying penicillin (see Fig. 20).

If the same percentage concentration of glucose is *in both* agar medium and in the penicillin solution being assayed, then the size of the zones from 2 to 25% glucose is decreased rather more than expected from the effect of the increased viscosity on the diffusion coefficient:

$$D_G/D_W = \eta_W/\eta_G. \tag{29}$$

This agrees with Neurath's experience previously alluded to. Results from 0 to 2% varied from -10 to $+20\%$; usually somewhat above normal. Friedman[33] as previously stated attributed the increased diffusion of urea in the presence of carbohydrates to effects on the solvation of the gel, altering pore size and frictional resistance. In the case of antibiotic zones the slight increase in size with low concentrations of glucose may be due not only to effects on the diffusion coefficient but to effects on growth. Changes in pH may occur in addition to effects such as increased oxygen consumption. Bond[37] working with *Staphylococcus aureus* observed increased growth and definition at the zone edge, and indeed Leisegang rings with glucose. It must be remembered that with organisms that utilize the sugar destruction occurs outside the zone. With inhibited growth within the zone a concentration gradient will be created at the

[35] H. Neurath, G. R. Cooper, and J. O. Ericson, *J. Biol. Chem.* **142**, 249 (1942).

[36] E. G. Jefferys, Personal communication (1952).

[37] C. R. Bond, *Analyst* **77**, 118 (1952).

edge, and more sugar becomes available to the growing organisms by diffusion from the zone.

With glucose *in the medium* but none in the penicillin solution, very similar results are obtained from 0 to 6.2% glucose. Above this to 25% further decrease in zone size did not occur as might have been expected from viscosity considerations. However, with high concentrations of glucose some would be lost by back diffusion to the penicillin solution.

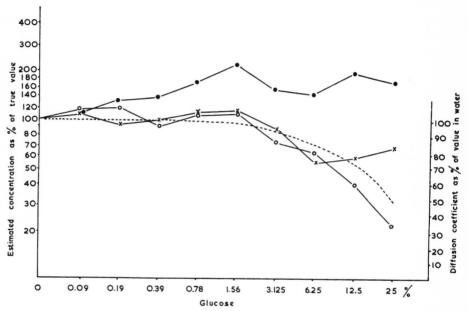

Fig. 20. Percentage deviation of average of assays of penicillin (10, 5, and 2.5 units) from true value without glucose (= 100) when glucose is (a) incorporated in both penicillin and media ○——○ (b) with glucose in media only ×——× and (c) with glucose in penicillin only ●——●. The percentage deviation is plotted on a logarithmic scale against doubling concentrations of glucose. The expected per cent value of D on an arithmetic scale is shown to indicate the expected variation in curve (a) if the effect was entirely due to the changed viscosity. ———, Results of Jeffreys.[36]

Also, the great osmotic difference would be expected to cause the gel to imbibe the water containing the penicillin. Both these factors would lessen the effect of the sugar.

With glucose *in the penicillin* solution and none in the medium a very different state of affairs exists. Here sugar will diffuse into the medium and following the same laws as the diffusing antibiotic, a low concentration will accompany the latter. The concentration accompanying the critical concentration in the agar will be determined by the ratio of the diffusion coefficients of the antibiotic and sugar and the logarithms of the initial concentrations. From 0 to 2% glucose gives a steady increase in the size of zones, and if estimates of the antibiotic are made by comparing with a standard containing no carbohydrate potencies over 200% can be obtained. From 2 to 25% results fluctuate between 150 and 200%, the depressing effect of increased viscosity being only slight as the concentrations at the time the zone edges are formed are much lower than initially present in the cup.

As no figures are available in which T_0 had been determined under these conditions, detailed calculation cannot be given. However, if one accepts the increased diffusion rate with low concentrations as due to an effect on the agar, there seems nothing in these results that might not be expected.

The papers of Bond [37] and Sykes et al.[38] should be consulted for further details and for the effects of other sugars. It would also be interesting to contrast the use of organisms utilizing a sugar with one which fails to do so. Sehgal [39] incorporated glucose (from 0.25 to 3%) in media used for the action of staphylococci and *streptomycin*. He found no effect on the sensitivity or growth rate of the organism, no change in the size of antibiotic zones below 1.5% and the diminished zones from 1.5 to 3.0% were entirely explained by the effect of the changed viscosity on the diffusion coefficient.

6. Salts

Apart from the effects of salts on the gel structure and intermicellar sol, many other properties are influenced. Salts capable of more rapid diffusion than the antibiotic are necessary to prevent the development of an electrical potential gradient by the diffusing antibiotic ions. The diffusion formula for neutral molecules can be applied only to ionic substances in the presence of excess salt. In its absence diffusion is reduced by the potential gradient and formulas involving the mobility of the ions must be used. This does not normally arise in bacteriological media or buffers.

It is easy to see that combination of salt ions with either the gel or the organism may affect the absorption of the antibiotic, and modify its effect on the growth of the organisms. These aspects will be considered in later sections. Considering at present only the diffusion of the antibiotic, apart from direct combination and precipitation of the latter, it must be remembered that soluble complexes of salts and antibiotic are often formed. The calcium chloride double salt of streptomycin contains two molecules

[38] G. Sykes, D. G. Lewis, and B. Goshawk, *Proc. Soc. Appl. Bacteriol.* **14**, 40 (1951).

[39] S. N. Sehgal, Ph.D thesis. Department of Bacteriology, University of Bristol, England, 1957.

of the base, and even if it is this large molecule and not the ion that diffuses, it does so more slowly than would be predicted by its molecular weight. This suggests either hydration of the antibiotic or the gel or both. Altering the salt concentration may therefore modify the diffusion coefficient.

F. Measurement of Diffusion Coefficients

A method of measuring diffusion coefficients in agar media was given by Humphrey and Lightbown.[26] Other methods are described in textbooks of physical chemistry.[16,19a,40] However, for purposes concerned with assay or bacterial sensitivities in agar media the most relevant determinations will be those calculated from the slopes of the lines plotting log concentration against the appropriate function of zone size. (Normally x^2 or r^2 when $h = 0$; see Fig. 8). This gives DT_0, and if critical time T_0 is determined by the methods given (see Section II.B.3.d), D can be calculated for the particular antibiotic at the temperature of the experiments in the particular media used.

III. The Laws of Growth and Multiplication

A. Critical Time

As a result of the study of the effects of preincubation and prediffusion it has been seen that there exists a *critical time* in the growth of the culture when the position of antibiotic inhibition zone is decided. This time T_0 can be measured experimentally by varying the period of preincubation (h) and applying the equation (see Eq. (11)):

$$T_0 - h = x^2/4D \ln (m_0/m').\tag{30}$$

For a particular concentration of antibiotic in the reservoir (see Fig. 11), as m_0 and m' are constant, this equation can be simplified to

$$T_0 = Kx^2 + h.\tag{31}$$

Plotting x^2 against h gives a straight line graph intercepting the axis when $x = 0$ at $h = T_0$.

Investigations into the significance of critical time have been very few. This is surprising because it is obvious that the growth of the organism is just as important as the diffusion of the antibiotic. The abrupt change from growth inhibition within the zone to continued growth outside the zone, occurring as it normally does during the early logarithmic period of growth, merits investigation.

[40] A. E. Alexander and P. Johnson, "Colloid Science," p. 239. Oxford Univ. Press, London and New York, 1950.

Critical time is independent of the concentration of antibiotic in the reservoir. It is true that at this time it is always the same critical concentration m' that has reached the zone edge, but the concentration gradient of the diffusing antibiotic varies greatly. We shall find that the growth rate of the organism can also vary widely with conditions, and a knowledge of the factors controlling growth is thus indispensable to an understanding of the formation of inhibition zones.

B. Viable Count and Time

As the object of nutrient media is to produce bacterial multiplication and the use of antibiotics is to cause this to cease, it seems logical to

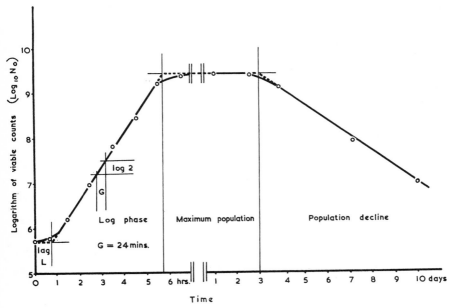

FIG. 21. The phases of the growth curve. The logarithm of viable counts in broth plotted against time. $G = 24$ minutes, $L = 48$ minutes. Time in the left-hand graph in hours and in the right-hand graph in days.

embark on an investigation into their effects on the organism by studying *viable* counts. Consider the course of events in a population of many organisms (averaging the results of individual cells). Bacteria, under satisfactory conditions, divide by fission in a constant time (G) so that the population increases by geometrical progression, or if plotted on logarithmic paper the increase with time will be constant for each interval (on Fig. 21 = log phase):

Generations:	0	1	2	3	4	... n
Cells (1):	1	$\rightarrow$ 2	$\rightarrow$ 4	$\rightarrow$ 8	$\rightarrow$ 16	... 2^n
Cells (N_0):	N_0	$2N_0$	$4N_0$	$8N_0$	$16N_0$	... $N_0 2^n$
Time T:	L	$L + G$	$L + 2G$	$L + 3G$	$L + 4G$	... $L + nG$

from which we see

$$N = N_0 2^n \tag{32}$$

and

$$T = L + nG \tag{33}$$

where N = number of viable cells at time T

N_0 = number of viable cells at time 0 and by extrapolation at time L

n = number of generations since time 0 each taking time G.

Instead of starting regular cell division from time 0, the constant L has been introduced because at the moment of inoculation into a new medium conditions are rarely satisfactory enough to allow growth with a constant generation time to proceed. A *lag phase* results during which the steady state is gradually achieved; for this requires a balanced composition of active enzymes in the cells, a steady ingestion of nutrients and excretion of waste products. The result is a temporary cessation or slowing down of cell division, with or without some cell destruction from the moment of inoculation.

When the steady state is achieved,

$$N = N_0 2^n \tag{32}$$

or

$$\log_2 (N/N_0) = n. \tag{34}$$

Plotting the logarithm of the cell population against time gives therefore a straight line, and this phase of growth is called the logarithmic phase. It is mathematically convenient to consider this as starting from a population N_0 (the size of the inoculum), but because of the existence of the lag phase the point $N = N_0$ and $T = L$ may or may not lie on the experimental growth curve. If it does not, it can be found by extrapolation as shown in the graph (see Fig. 21). Therefore

$$T = L + nG = L + G \log_2 (N/N_0) \qquad \text{(combining (33) and (34)).}$$

When $T = T_0$, let $n = n'$ and $N = N'$, then the critical time when the zone is decided is

$$T_0 = L + n'G = L + G \log_2 (N'/N_0) \tag{35 \& 36}$$

or, if the logarithms to the base 10 are preferred,

$$T_0 = L + 3.32G \log_{10} (N'/N_0), \tag{37}$$

but

$$T_0 - h = x^2/4D \ln (m_0/m') = x^2/4D2.30 \log_{10} (m_0/m'). \tag{11}$$

Therefore

$$x^2 = [4D2.30 \log (m_0/m')][L - h + 3.32G \log (N'/N_0)]. \qquad (38)$$

This formula was suggested by Cooper[41] as it explained the variations in size of streptomycin inhibition zones with temperature[14] and was later shown to explain the variations due to alterations in the size of the inoculum.[17,30]

It will be seen that *critical time* is the time required for the population to increase to a particular value N' which we shall call the *critical population*. This population is achieved after a definite number of generations (n') of the organism depending on the number of organisms in the inoculum. Critical time will thus vary with any factors which may alter the lag period (L), the generation time (G), the size of inoculum (N_0), or the critical population (N').

C. Temperature and Growth Rate

Generation time has been used as a measure of growth rate, because of interest in the number of viable organisms, and G is easily defined as the time required to double the number of cells during the *logarithmic period of growth*. It is important to notice that the *instantaneous growth rate* as defined by the chemist interested in the rate of syntheses of molecules is *not* $1/G$. If consideration is given to the dry weight of bacterial protoplasm (w) instead of cell numbers and the growth rate (K) is expressed as a multiple of the weight of cell material already present

$$\frac{dw}{dt} = Kw \qquad (39)$$

$$\int_{w_0}^{w} \frac{dw}{w} = K \int_{L^*}^{t} dt$$

where L^* is the end of the metabolic lag period (it will be seen later that this is not identical with L the lag period for cell division);

$$\log_e (W/W_0) = K(t - L^*)$$
$$W = W_0 e^{K(t-L^*)} \qquad (40)$$

where W_0 is the weight of cells at times 0 and L^*, and W is the weight at time t.

Consider the time taken to double the weight, i.e., when $W = 2W_0$, $t = t_1$, Eq. (40) becomes $2W_0 = W_0 e^{K(t_1-L^*)}$ or $\ln (2W_0/W_0) = \ln 2 = K(t_1 - L^*)$. The time $(t_1 - L^*) = \ln2/K$ equals the generation time. Therefore

[41] K. E. Cooper, *Nature* **176**, 510 (1955).

$$G = 0.693/K \tag{41}$$

(see Hinshelwood [42] and Brody[43]).

The influence of temperature on the rate of a chemical reaction can be expressed by the Arrhenius equation. This is an approximation sufficiently accurate for temperatures applicable to biological systems. In its integrated form the constant K determining rate of reaction is given by

$$K = Ae^{-E/R\theta} \quad \text{or} \quad \ln K = \frac{-E}{R\theta} + \ln A \tag{42}$$

where E is related to the energy difference between the normal and the activated molecule, A is a constant of integration, θ the *absolute* temperature, and R the gas constant. Plotting the logarithm of the reaction rate against the reciprocal of the absolute temperature gives therefore a straight line.

The effective rate of growth of the cell protoplasm can be regarded as the rate of synthesis less the rate of degeneration[42]:

$$A_1e^{-E_1/R\theta} - A_2e^{E_2/R\theta}. \tag{42a}$$

The second term is negligible at low temperatures, but becomes overwhelmingly important at high temperatures and at an intervening temperature a maximum will be obtained. Expressing the growth rate as a percentage of this maximum value and plotting this on a logarithmic scale against the reciprocal of the absolute temperature gives a graph made up of two straight lines, one ascending and one descending with a combined overlapping portion as a curve (see Fig. 22).

Variation of the constants involved when considering organisms of different species is responsible for the varying temperature ranges and optimum temperatures of growth. Usually in highly nutrient media the two straight lines are adequate to express these temperature relationships. This means that each line results from the control of the system by one limiting enzyme, a state of affairs well known in a series of consecutive reactions, where the overall rate is determined by the slowest link in the chain. With varying conditions, growth may be limited by different enzymes over different temperature ranges, in which case breaks will occur in the linear relationship. For a fuller discussion see Johnson *et al.*[44]

In determining growth rates, viable counts must be made over relatively short periods, not more than a few generations if chemical changes

[42] C. N. Hinshelwood, "The Chemical Kinetics of the Bacterial Cell." Oxford Univ. Press, London and New York, 1946.

[43] S. Brody, "Bioenergetics and Growth," p. 502. Reinhold, New York, 1945.

[44] F. H. Johnson, H. Eyring, and M. J. Polissar, "The Kinetic Basis of Molecular Biology," p. 232. Wiley, New York, 1954.

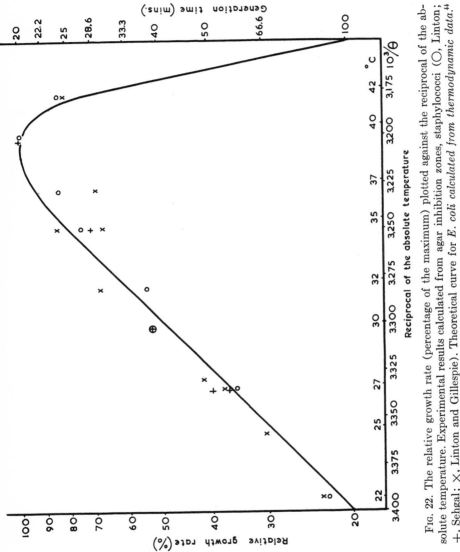

FIG. 22. The relative growth rate (percentage of the maximum) plotted against the reciprocal of the absolute temperature. Experimental results calculated from agar inhibition zones, staphylococci (O, Linton; +, Sehgal; ×, Linton and Gillespie). Theoretical curve for *E. coli* calculated from *thermodynamic data*.[44]

in the environment are to be avoided. If these occur growth rate may be altered. Fortunately in the case of antibiotic-inhibition-zones with the normal sizes of inocula, the critical times involved are only a few hours. Moreover, with increase of temperature they are reduced, and the generation times involved may be expected to give the straight line relationship up to near the maximum multiplication rate. It is interesting that in the case of staphylococci the points plotted on the graph show very good agreement when the generation times are calculated from the size of antibiotic zones, on the assumption that the critical population remains constant. Indeed, determinations from viable counts in broth show the influence of heat on the enzymes at a somewhat lower temperature, presumably because the organisms were held for relatively longer periods at the experimental temperature in order to determine the slope of the line against time.

All the points on the graph (Fig. 22) were determined from experiments in which whether in broth or agar the organisms had been held at 45°C. for a few minutes and then cooled from 45°C. to room temperature before placing in a water bath at the temperature of the experiment. This was in order to keep the viable count determinations in broth as nearly similar to the antibiotic experiments in agar as possible. It is not of course possible to determine generation times in agar under these conditions sufficiently accurately by direct examination. However, the experiments on the influence of size of inocula (Section III.D.2) at 40°C. do not suggest any change in the value of critical population, so it seems justified to use its constancy to calculate G from the zones at different temperatures using a constant inoculum.

Staphylococcus is very similar in the lower range of temperatures to *Escherichia coli* in its growth rate. The curve for *E. coli* derived theoretically from thermodynamic constants by Johnson and Lewin[45] to fit their experimental data is shown for comparison (Fig. 22).

It is important to realize that the time an organism is held at 45°C. (during cooling of the agar) has an effect on the generation time of the organism when it is subsequently put at the growth temperature.[46] Prolonged holding by increasing enzyme destruction increases the subsequent generation time. Even 10 minutes has been shown to be sufficient to produce visible changes in the granulation of the cytoplasm (though these are still reversible) by examination under the electron microscope.[47] Conditions for inoculation and setting of the agar must therefore be carefully standardized in assay work.

The values of T, T_0, and L in the formula used obviously depend on

[45] F. H. Johnson and I. Lewin, *J. Cellular Comp. Physiol.* **28**, 23 (1946).

[46] F. H. Johnson and I. Lewin, *J. Cellular Comp. Physiol.* **28**, 47 (1946).

[47] C.-G. Hedén and R. W. G. Wyckoff, *J. Bacteriol.* **58**, 153 (1949).

the precise moment chosen to be represented by $T = 0$. In assay work this is normally the moment when the antibiotic is added and the already inoculated culture is placed at the correct temperature for growth. This is not the moment of inoculation as is usual in methods in liquid media, because the inoculation is made into liquefied agar at 45° which must be cooled and set. If the time (0) is standardized at 10 minutes or $\frac{1}{4}$ hour after inoculation, the organisms will have recovered from this initial shock, so that full metabolism (at least in satisfactory media) will have started, but no cell division will be taking place.[48] Under these conditions in the experiments of Cooper, Linton *et al.* it was found that the lag period in cell multiplication had a constant ratio to the generation time of the organisms even at different temperatures so long as the medium was of constant composition. The ratio was altered if the medium or the organism were changed or if the age or conditions of culture of the inoculum were modified. With carefully controlled conditions, however, since

$$T_0 = L + n'G \tag{35}$$

and L/G was constant,

$$T_0/G = L/G + n' = n'' \tag{43}$$

where n'' is a new constant. It was found (Eqs. (33)–(36)) that just as $n' = \log_2 (N'/N_0)$ so

$$n'' = \log_2 (N''/N_0) \tag{44}$$

and this could be used in the formula and L eliminated. When $h = 0$, then

$$T_0 = L + G \log_2 (N'/N_0) = G \log_2 (N''/N_0) = x^2/4D \ln (m_0/m'). \tag{45}$$

It will be seen that when the inoculum

$$N_0 = N', \qquad \text{then} \qquad T_0 = L = x^2/4D \ln (m_0/m').$$

Therefore a zone occurs (diffusion having taken place during the lag period) but when $N_0 = N''$, then $T_0 = 0 = x$. No zone is formed whatever the value of m_0. N' is the *critical population* which forms the zone edges only at the end of the lag period or after cell division. N'' is a completely *inhibitory population* which even at time 0 abolishes zone formation.

Table III shows the range of values obtained for L and G in broth and for T_0, n', n'', and m' in two different media for streptomycin and chloramphenicol with staphylococci.[14,17,39,49] Within the limits of experimental error L/G was constant for all temperatures. The values of n' and n'' are also approximately constant in each medium, though the

[48] O. H. Sherbaum, *Ann. Rev. Microbiol.* **14**, 291 (1960).

[49] A. H. Linton, M.Sc. thesis. Department of Bacteriology, University of Bristol, England, 1949.

TABLE III

EFFECT OF TEMPERATURE ON GROWTH CONSTANTS (*Staphylococcus* MAYO)

	Observed interpolated broth results (minutes)			Agar medium[a] I (hours)			Agar medium II (hours)			Values of m'			
										Strepto-mycin		Chloram-phenicol	
				T_0	n'	n''	T_0	n'	n''				
Temp.	G	L	L/G	obs.	calc.[b]		obs.	calc.[b]		I[c]	II[c]	I[c]	II[c]
22	89	180	2.0	9.0†	4.0	6.0				1.25			
25	64	120	1.9	7.0*	4.5	6.4	9.0*	6.4	8.2			7.9	7.5
26	52	102	2.0				8.0†*	7.2	9.2		4.0		7.7
27	50	90	1.8	5.5†	4.8	6.6				0.75			
27.5	48	84	1.75				6.5†*	6.4	8.0		4.0		8.4
30	38	66	1.7	4.25*	5.0	6.7	4.75*	5.8	7.5			13	8.9
32	33	60	1.8	3.5†	4.5	6.4	4.0*	5.7	7.4	0.90			10.0
32.5	33	57	1.8				4.25†	6.1	7.9		4.0		
35	29	50	1.7				3.5*	5.3	7.0				11.3
35.5	29	48	1.7	2.8*	4.25	5.95	3.5†	5.7	7.4		4.0	16	
37	28	45	1.6	2.5†	3.7	5.3				0.75			
38.5	27	42	1.6				3.5*	6.2	7.8				13.0
40	27	39	1.5	2.4*	4.0	5.45						20	
41.5	26	37	1.5				3.5†*	6.7	8.1		3.2		16.0
42	25	36	1.4	2.2†	3.9	5.35				1.0			
Avg.	—	—	1.7	—	4.3	6.0	—	6.3	8.0	0.93	3.8	—	—

[a] Beef Heart Infusion (neat).

Minced lean beef heart (freed from gross fat)　　　500 gm.
Distilled water　　　　　　　　　　　　　　　　850 ml.

Extract in steamer for 2 hours, filter through gauze, and squeeze out. Add 0.5% sodium chloride, the peptone and adjust the pH to about 0.3–0.5 above the final pH required, according to the agar to be subsequently used. (Bacto agar is neutral, Oxoid slightly acid, and Davis more acid.) Make up to 1000 ml. Boil to precipitate phosphates, filter through paper pulp. Add agar and if necessary adjust final pH. Sterilize by autoclaving.

[b] $n' = T_0 - L/G$; $n'' = T_0/G$. $n'' - n' = 1.7 = L/G$. *Chloramphenicol. †Streptomycin.

[c] *Medium I:* 1% Bacto agar, $\frac{1}{4}$% NaCl, $\frac{1}{2}$ strength meat extract, 1% Peptone, pH 7.8. *Medium II:* 2% Oxoid agar, $\frac{1}{2}$% NaCl, neat meat extract, 1% Peptone, pH 7.6. Inoculum: 18 hour culture of *Staphylococcus* Mayo 1/1050 dilution ($N_0 = 0.5 - 1.0 \times 10^6$; $\log_2 N_0 = 19.0$ to 20.0).

inoculum size was only standardized by the dilution of an 18-hour culture and the results were obtained in different laboratories over a period of some years. T_0 values were the same with these two antibiotics though dependent on the medium (and the organism). Values of m' for streptomycin showed no significant change with temperature though those for chloramphenicol increase with temperature increase. The sensitivity of the organism (m') was very dependent on the nature of the medium, and this will be discussed in Section III.D.4.

D. The Effect of the Inoculum

1. Cell Mass, Cell Number, and Age of Inoculum

If the conditions of inoculation result in the full metabolic rate being achieved by $T = 0$, then antibiotic absorbing centers, or reacting substances, are being synthesized during the lag period of the viable count growth curve. It is the mass of cell substance which determines the amount of antibiotic used. The number of cells at the beginning of the lag period, which is required to deal with the critical concentration m' is therefore much greater (N'') than the number of cells at the end (N'). That variations in cell size occur during the lag period and early log period of growth is well known.[50,51] The constancy of L/G in the above experiments depends on the constant rate of protein synthesis from time 0, and the eventual achievement of a rate of cell division equal to the doubling rate for protein. It is probably closely related to the temperature shocks inherent in the method of preparing the agar culture.[48]

If the conditions of inoculation are so changed that a metabolic lag occurs (this corresponds to the time lag of Monod [52]) in which no antibiotic reacting centers are being synthesized, then *preincubation* during such a lag will not alter the size of the zones. Certain experiments of Linton[49] using young inocula (6-hour culture) had to be preincubated for a further 6 hours before any diminution in zone size occurred with further preincubation. The true value of T_0 could only be obtained by measuring the time (h) from the moment when preincubation reduced the size of x (see Fig. 23). With old inocula (72-hour culture) the usual h against x^2 graph was obtained. The values of x^2 were however larger than with the normal age for inoculation (16–24 hours). This would be expected if the viable count and thus N_0 was less. No extensive investigation has been made of the effect of the age of the inoculum in which cell mass and viable counts have been determined. For a general discussion of the factors affecting growth see Brody,[43] Hinshelwood,[42] Dubos,[50] Henrici,[51] Monod,[52] Sherbaum,[48] and Hawker et al.[53]

With standard conditions of inoculation under which L/G was constant extensive investigations were made by Sehgal [39] and by Linton.[30] Strepto-

[50] R. J. Dubos, "The Bacterial Cell," p. 139. Harvard Univ. Press, Cambridge, Massachusetts, 1945.
[51] A. T. Henrici, *J. Infectious Diseases* **38**, 54 (1926).
[52] J. Monod, *Ann. Rev. Microbiol.* **3**, 371 (1949).
[53] L. E. Hawker, A. H. Linton, B. F. Folkes, and M. J. Carlile, "An Introduction to the Biology of Micro-organisms," p 226. Arnold, London, 1960.

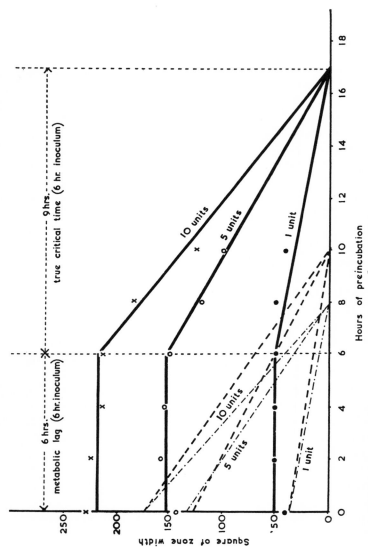

FIG. 23. The effect of age of the inoculum on zone size and critical time. With young inocula (6 hours) a *metabolic* lag may be introduced so that preincubation fails to reduce zone size until lag is ended. True critical time should be measured from the end of this period. Results for the action of penicillin on surface cultures of *Staphylococcus* (Oxford H strain) inoculated with broth cultures 6 hours, ———; 24 hours, —·—·—; and 72 hours, ——— old. Zone size (x^2) plotted for 10, 5, and 1 unit of penicillin against time of preincubation (h).

mycin and chloramphenicol have been used over a full range of temperature for growth of *Staphylococcus* and *Klebsiella,* and with wide ranges
of inoculum size and of concentration of the antibiotics. These will be
discussed in the next section.

2. The Effect of Inoculum Size on Critical Time and Zone Size

Table IV gives results at one temperature (35°C.) of varying the size
of the inoculum. The validity of Eqs. (43)–(45) is tested by using the
value $\log_2 N'' = 27.0$ to calculate the theoretical generation time from

$$n'' = \log_2 N'' - \log_2 N_0 \qquad (44)$$

and

$$G = T_0/n''. \qquad (43)$$

T_0 was determined for each inoculum size in the way previously described
and it will be seen how accurately the generation time remains constant
and how closely this agrees with determinations in broth by the viable
count method. Similar results were also obtained when the antibiotic
was changed from streptomycin ($m' = 1.6$ μg./ml.) to chloramphenicol
($m' = 12.9$ μg./ml.), the value of G being the same to within a minute.

With very heavy inocula critical time becomes reduced to under the
hour, and the zone edge is decided before cell division has occurred. Results become difficult to read and are not very accurate. It was possible
to establish with certainty that small zones could be formed with sizes
of inocula between N' and N'' during the lag period. With very light
inocula the zone edge becomes diffuse as the colonies become isolated
and 3×10^3/ml. is about the lower limit for reading zones. The smaller
the size of the inoculum, the larger the zone, because more cell divisions
(n'') were necessary to reach the critical population and the longer critical time (T_0) allowed the critical concentration of antibiotic to diffuse
a greater distance.

If the values of T_0 are plotted against $\log_2 N_0$, then a straight line
should be obtained in accordance with Eq. (45) intercepting the axis
when $T_0 = 0$ at $\log_2 N''$ (see Fig. 24).

Agreement with theory was obtained when streptomycin was used
on staphylococci[17] or on *Klebsiella*.[30]

Equation (45) also predicts that the square of the zone size x, should
for any one concentration of antibiotic m_0 give a straight line when
plotted against $\log_2 N_0$, and that when $N_0 = N''$, then $x = 0$ independently of what concentration is used. Agreement within the limits of experimental error was also obtained in these cases (see Fig. 24).

Agreement was also obtained with a resistant staphylococci ($m' =
7$ μg./ml.) as well as with a more sensitive one ($m' = 2.5$ μg./ml). De-
viations from the straight line for x^2 against $\log N_0$ occurred with chlor-

TABLE IV

The Effect of Inoculum Size on Critical Time in a Standard Medium on Staphylococcus

N_0/ml.	$3.27 \cdot 10^3$	$6.55 \cdot 10^3$	$1.64 \cdot 10^4$	$3.27 \cdot 10^4$	$8.09 \cdot 10^4$	$2.02 \cdot 10^5$	$4.04 \cdot 10^5$	$6.55 \cdot 10^5$	$8.09 \cdot 10^5$	$1.28 \cdot 10^6$	$1.40 \cdot 10^6$	$3.50 \cdot 10^6$	$7.00 \cdot 10^6$	$1.40 \cdot 10^7$	$1.60 \cdot 10^7$	$2.14 \cdot 10^7$	$3.21 \cdot 10^7$	$6.41 \cdot 10^7$
$\log_2 N_0$	11.7	12.7	14.0	15.0	16.3	17.6	18.6	19.3	19.6	20.3	20.4	21.7	22.7	23.7	23.9	24.3	24.9	25.9
n''	15.3	14.3	13.0	12.0	10.7	9.4	8.4	7.7	7.4	6.7	6.6	5.3	4.3	3.3	3.1	2.7	2.1	1.1
T_0 (hours)	6.45	6.3	5.57	5.2	4.9	4.2	3.6	3.2	3.3	3.2	2.95	2.6	2.3	1.7	1.4	1.3	1.04	0.75
G (minutes) (calc.)	25.3	26.4	26.5	26.0	27.5	27.1	25.7	25.0	26.8	28.5	26.8	29.4	32.1	31.1	27.2	29.1	30.0	(41.4)

$m' = 1.6$ μg.; $\log_2 N'' = 27.0$; $n'' = \log_2 N'' - \log_2 N_0$; $G = T_0/n''$. *Streptomycin: temperature 35.0°C., medium III.* Average generation time in agar (excluding $T_0 < 1$ hour) = *27.7 minutes.* Average value of determinations in broth by viable count 29 minutes. Similar determination with chloramphenicol ($m' = 12.9$ μg./ml.) gave $G = 28.2$ minutes.

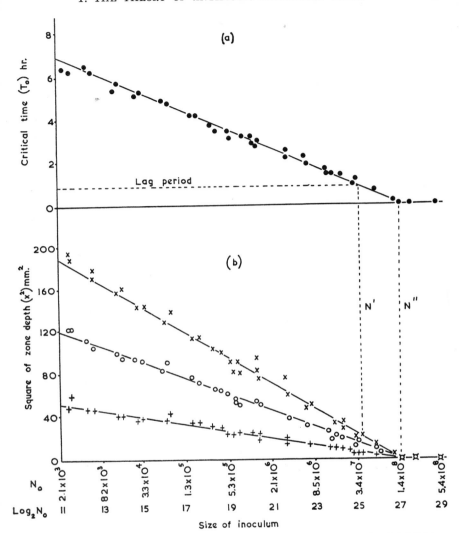

FIG. 24. The effect of inoculum size on critical time and the depth of inhibition zones produced by the action of streptomycin on *Staphylococcus* (Mayo 6473) in tubes at 35°. (a) Critical time (T_0) plotted against inoculum size ($\log_2 N_0$). The points are extrapolated values of h obtained by plotting zone size (x^2) against hours of preincubation (h) since when $x^2 = 0$, then $h = T_0$. Three lines for different concentrations of antibiotic ($m_0 = 10, 100, 1000$ μg./ml.) were extrapolated and the average value used for each size of inoculum (●). (b) Zone size (x^2) determined in triplicate experiments for m_0 values of 10 μg./ml. (+), 100 μg./ml. (○), and 1000 μg./ml. (×) when $h = 0$ plotted against inoculum size. N' is the *critical* population which all inocula less then N' achieve at time T_0 when the zone edge is formed. N'' is the *inhibitory* population or the size of inoculum ($N_0 = N''$) which reduces T_0 and x^2 to nil. Theoretical lines are with the values (35°). $D = 1.06$ mm.²/hour, $m' = 1.8$ μg./ml., $L = 0.81$ hour, $G = 0.45$ hour, $\log_2 N'' = 27.0$.

amphenicol,[54] and the reason for these will be discussed later (Section V.A.3).

3. Temperature, Critical Population (N'), and Inhibitory Population (N'')

It has been seen that when the inoculum size equals the critical population $N_0 = N'$, then the critical time equals the lag period before the onset of cell division $T_0 = L$. This is only true if at time 0 full metabolism has been established. Then L/G is constant.

If these conditions are observed, then graphs similar to Fig. 24 can be obtained at different temperatures. The changed values of L and G will alter T_0 as was seen in Table III. The slope of the critical time graph in Fig. 24 is determined by G. The difference between the critical and inhibitory populations is determined by L/G. In the experimental work done with the two strains of staphylococci and with *Klebsiella*, both the critical population and the inhibitory population were *constant* at temperatures from 20° to 40°C. The time of onset of cell division was thus proportional to the generation times at each temperature, and zone sizes with streptomycin were also as expected from the growth rates. This was true with sizes of inocula so small that 14 generations were required to reach the critical population, and with sizes of inocula so large that no cell division took place before the position of the zone edge was decided. Despite the variation in generation time from 20 to 84 minutes and of critical time from 0 to 15 hours, the critical and inhibitory population of the organisms remained constant. They are obviously fundamental measurements of the bacterial populations which determine with how much antibiotic they can deal. The values of N' and N'' were unaltered for the resistant mutant staphylococci (which had the same growth rate as the prototype) though the critical concentration had been trebled. The values N' and N'' for *Klebsiella* were less than half those for the staphylococci suggesting that the larger organism could deal with the antibiotic by the use of half the number of cells, i.e., each cell can deal with twice as much antibiotic. These experiments were all in a standard medium. With further work it became evident that certain changes in the medium could affect these cell characteristics and the effects of media composition must be considered if any valid comparison between one organism and another is to be made.

4. Inhibitory Population and the Medium

No systematic investigation has been made of the effect of each media constituent on the constants in the formula. Values in a number of dif-

[54] K. E. Cooper, A. H. Linton, and S. N. Sehgal, *Intern. Congr. Microbiol., 7th Congr., Stockholm,* Abstr. 19c, p. 331 (1958).

TABLE V

THE EFFECT OF MEDIA COMPOSITION ON INHIBITION ZONES. STREPTOMYCIN ON STAPHYLOCOCCUS AT 35°C[a]

Media	I (dilute)	II (routine)	III (standard)	IV (special)	V (deficient)	VI (salt)
Agar	1% Bacto	2% routine[b]	1% Bacto	1% Bacto	1% Bacto	1% Bacto
Meat extract	$\frac{1}{2}$ strength	full strength	$\frac{1}{2}$ strength	Nil	Nil	Nil
Peptone	1% routine[c]	1% routine[c]	1% Bacto	1% Bacto	0.5% Bacto	1% Bacto
Salt (added)	0.25%	0.5%	0.5%	0.5%	0.5%	3.5%
pH	7.8	7.6	7.8	7.8	7.8	7.8
L in media less agar[d]	50 minutes	50 minutes	50 minutes	—	—	—
G in media less agar[d]	29 minutes	29 minutes	29 minutes	—	—	—
G in agar[e]	29 minutes	27 minutes	28 minutes	36 minutes	56 minutes	45 minutes
n'' when $\log_2 N_0 = 20.0$[f]	6.0	7.5	6.5	5.8	5.0	5.8
$\text{Log}_2 N''$[g]	26.0	28.0	27.0	25.8	25.0	25.8
m'[h]	0.9 μg./ml.	4.0 μg./ml.	2.0 μg./ml.	2.0 μg./ml.	2.0 μg./ml.	12.6 μg./ml.

[a] $G \log_2 (N''/N_0) = T_0 = x^2/4D \ln (m_0/m')$. The average values for all experiments available are given, excluding zones $x < 3$ mm. Each value of x obtained in triplicate at least.

[b] Certain batches of some agars were antagonistic at 2% but not 1% strength.

[c] Peptone batches also varied in routine media, especially in their salt content.

[d] Average results. Range ± 6 minutes by viable counts.

[e] Calculated from $T_0/n'' = G$. T_0 is experimentally determined from at least 9 points, 3 concentrations (m_0) for 3 times of preincubation (h) at each inoculum size (N_0).

[f] $n'' = \log_2 (N''/N_0)$ value at standard inoculum for each concentration m_0 and T_0 for a range of inocula sizes N_0 from $\log_2 N_0 = 13$ to $\log_2 N''$.

[g] Extrapolated.

[h] Sensitive *Staphylococcus*. Streptomycin salt effect.

ferent media have been obtained, and those capable of valid comparison are summarized in Table V. Further work has also been done on the effect of salt concentration which markedly affects the absorption of streptomycin. This will be considered in Section IV.B.3, but it will be seen that increasing salt concentration increases the critical concentration m'.

With enriched media containing meat extract, the generation time G was not affected by the other changes examined. When meat extract was omitted (special media IV) G was prolonged. Further lessening of nutrients by halving the peptone (deficient media V) markedly increased G. A large increase in salt (compare salt media VI with IV) also increased G.

The lowest value for the inhibitory population $(\log_2 N'' = 25.0)$ was obtained with the media most deficient in nutrients (V) and the highest value $(\log_2 N'' = 28.0)$ with the richest medium. Unfortunately, no estimates were made of the size of the organisms or the mass of the cells.

Work has been done with continuous or intermittent growth cultures on the mass per cell achieved in balanced growth in different media (see papers by Schaechter et al.[55,56]). In their experiments in many differ-

[55] M. Schaechter, O. Maaløe, and N. O. Kjeldgaard, *J. Gen. Microbiol.* **19**, 592 (1958).

[56] M. Schaechter, O. Maaløe, and N. O. Kjeldgaard, *J. Gen. Microbiol.* **19**, 607 (1958).

ent media the rate of protein synthesis per unit ribonucleic acid was nearly the same at all growth rates. Variations in the mass per cell depended on the number of nuclei per cell as well as on changed ratios of mass per nucleus.

The results obtained for the factors influencing the inhibitory population of staphylococci for streptomycin may not be identical for other organisms and other antibiotics. Much more information is required regarding the combining power of cell material with antibiotic. Some of this can be considered in the section on absorption of antibiotics, as this is the first stage of the reaction. It should be evident however that critical and inhibitory population will also be concerned with the precise mechanism of action of antibiotic. The same results cannot be expected with antibiotics which act on (a) cell wall synthesis, (b) cytoplasmic synthesis, or (c) nucleic acid synthesis. In a further section the mechanism of antibiotic action must be considered.

Accurate evaluation of N'' in different conditions may thus provide valuable information relevant to the elucidation of the action of the antibiotic.

E. Other Factors Concerned with Growth

Little work has been done with organisms having long generation times. *Mycobacterium tuberculosis* gives critical times measured in days instead of hours and such results as have been obtained suggest that a similar number of generations occur before the zone edge is formed, as with more rapidly growing organisms. No accurate determinations of critical populations with inocula of known sizes have been made, but there is no reason to think that the same laws do not apply.

In experiments with Lowenstein Jensen's media, Hedges[57] showed that a reversible absorption of streptomycin on some media constituent occurs. This slows down the apparent rate of diffusion, and changes the slope of the normal assay curve. The progress of the antibiotic through the medium is thus probably governed more by the laws of absorption chromatography than diffusion. Extrapolation to find the critical concentration gives results of the expected values for the sensitivity of the organism.

Such experiments as have been done with tubercle bacilli have necessarily been with surface cultures, because of the strictly aerobic growth of this organism. With staphylococci under such conditions zones are much larger than with colonies growing only *within* the agar.[15] This is to be expected, because antibiotic will be able to reach the surface colonies from the agar beneath where no organisms contribute to the

[57] A. J. Hedges, Private communication, 1960.

inhibitory population. Investigations into the effect of inoculum size with surface cultures were made by Ericsson et al.,[58] but are not accompanied by determinations of viable counts. It is evident that an inhibitory population exists, and that smaller inocula give larger zones. In fact, the logarithmic dilution curves of Ericsson approximate to the expected results. Accurate calculations cannot be made on the published results as he was using a dried disk method in which though the quantity of antibiotic present was known, the effective concentration achieved was unknown. Investigations by Mayr-Harting[59] have shown that the distances between surface colonies on media greatly influence the number of viable organisms, but no accurate determinations for surface populations have been made during the early hours of growth under the conditions of these assays.

The conditions under which the test organism has been trained, stored, cultivated, and prepared for inoculation may affect the lag period and the growth rate. Investigations by Hinshelwood [42] and Dagley et al.[60] give many factors such as age, washing, and media constituents which should be standardized if constant results are to be obtained. The more nutritiously exacting the organisms, the more difficult it is to achieve reproducible results. The more successful the attempt to achieve minimum generation times and lag periods, the more rapid the accumulation of metabolic products. The use of added pH buffers is advisable.

Probably the factor most difficult to control, and often least considered, is the availability for growth of the constituents of the atmosphere. Redox potential and carbon dioxide may both have profound effects on growth and on antibiotic activity. Aeration of inocula and of cultures affects both, and often changes the size of zones. The presence of CO_2 in air markedly affects the size of inhibition zones produced with gonococcus cultures. The tube method of producing inhibition zones is unsatisfactory with penicillin and staphylococci giving very diffuse edges, though clear cut zones are obtained with surface cultures on plates.

The testing of anaerobic organisms introduces another difficulty. It is difficult to achieve anaerobiosis (in an anaerobic jar for instance) and at the same time have rapid temperature control of cultures so that time 0 when growth starts may be accurately defined. The determination of generation times and critical times for anaerobes at specific temperatures thus offers experimental difficulties and is probably responsible for our ignorance in this field.

[58] H. Ericsson, C. Högman, and K. Wickman, Scand. J. Clin. & Lab. Invest. 6 (Suppl. 11) 26 (1954).
[59] A. Mayr-Harting, J. Hyg. 45, 19 (1947).
[60] S. Dagley, E. A. Dawes, and G. A. Morrison, J. Gen. Microbiol. 4, 437 (1950).

IV. The Laws of Adsorption and Partition

A. General Principles

When a bacterial cell is immersed in an antibiotic solution the antibiotic must, if it is to interfere with metabolism and growth, act on certain vital cell receptors. Specific action is normally accounted for by specific combination with particular cell constituents, such as enzymes, coenzymes, essential nutrients or intermediates of metabolism. Combination at least at certain concentrations may also take place with other cell constituents, such as proteins, lipoids, or particular cell products. This so called nonspecific combination may not contribute to antibiotic action, and must be distinguished from the primary specific reaction which leads to cell inhibition and even death.

The cell is a complex structure, and we may in different instances be concerned with different features. The permeability of the cell wall, and the cell membrane (protoplast); diffusion through the cytoplasm, combination with cytoplasmic granules or with nuclear structures; may be involved. The combinations may be in the first instance reversible, or they may become irreversible. The degree of "fixation" may vary from simple solution in a cell constituent to firmer forms of union involving van der Waals forces, salt linkages or to the formation of chemical bonds. A simple solution will lead to differing concentration within and outside the cell, and the partition will be governed by solubility laws. Frey-Wyssling[32] classifies the linkages existing in cytoplasm into (1) homopolar cohesive bonds, (2) heteropolar cohesive bonds, (3) heteropolar valency bonds, and (4) homopolar valency bonds. We have thus the possibility of many types of union from purely physical attractions to chemical combination, and the distinction is by no means clear cut. It is convenient to speak of removal of antibiotic from the external solution by the cell as adsorption though this may be on cell surfaces or on internal cell structures.

The relationships between concentration in the external medium and the amount removed by cells, proteins, carbohydrates, or lipoids have been studied extensively for chemicals such as the hydrogen ion, salts, dyes, tannin agents, and drugs. The scientific literature of textiles, leather, color chemistry, plastics, and pharmacology has much relevant information, but there are less studies available for antibiotics. The biochemistry of enzymes is probably the most relevant and the reader should consult standard textbooks in these fields for the very numerous references to the physical chemistry of relevant systems. Baldwin[61] gives an account

[61] E. Baldwin, "Dynamic Aspects of Biochemistry." Cambridge Univ. Press, London and New York, 1952.

of Michaels' theory of enzyme substrate combinations and inactivations. Cohn and Edsall [62] give a fuller account of the properties of proteins. Höber[16] and Alexander and Johnson[18] have already been referred to. The action of drugs on cells has been specially considered by Clark,[63] and he points out that many of the alternative formulas suggested for adsorption phenomena (Eqs. (47)-(49)) cannot be experimentally distinguished over the range of concentrations available for biological experimentation. More recently, however, the use of radioactive tracers has enabled more extensive investigations to be made.[64] The use of electrophoresis by determining the electric charge on a cell with adsorbed ions of antibiotic has also enabled adsorption to be measured.[65,66] Hinshelwood [42] has considered the theoretical aspects of the application of the Langmuir adsorption isotherm to disinfection. Johnson *et al.*[67] have shown how the necessity for adsorption of a different number of molecules per cell to produce a lethal effect would affect cells with different numbers of receptors available, and modify the shape of the disinfection curves against time.

The alternative adsorption formulae mentioned by Clark[63] were:

$$\sigma = KC + K_2 \qquad \text{Henry} \qquad \text{(linear)} \qquad (46)$$

$$K_3\sigma = \log (K_4C + 1) \qquad \text{Weber-Fechner} \quad \text{(logarithmic)} \qquad (47)$$

$$\sigma = K_5C^n \qquad \text{or} \qquad \log \sigma = \log K_5 + n \log C$$
$$\text{Freundlich} \qquad \text{(double logarithmic)} \qquad (48)$$

$$\sigma = K_AC/(1 + K_AC) \qquad \text{Langmuir} \qquad \text{(saturation)}. \qquad (49)$$

Both Eagle *et al.*[64] for penicillin and McQuillen[66] for streptomycin give adsorption curves which fit the theoretical Langmuir isotherms. Humphrey and Lightbown[26] used a formula of the Freundlich type, but in the range 5–70% saturation the differences between the formulas are negligible. The inhibitory concentrations were much lower than saturation concentrations and would lie within this range. Work by Borzani and Vairo[68] with dyes also support such formulas, though most of their results were obtained with dead cells.

The Langmuir adsorption curve has a theoretical foundation based on

[62] E. J. Cohn and J. T. Edsall, "Proteins, Aminoacids and Peptides." Reinhold, New York, 1943.

[63] A. J. Clark, "The Mode of Action of Drugs on Cells." Arnold, London, 1933.

[64] H. Eagle, M. Levy, and R. Fleischman, *J. Bacteriol.* **69**, 167 (1955).

[65] K. McQuillen, *Biochim. et Biophys. Acta* **7**, 54 (1951).

[66] K. McQuillen, *Biochim. et Biophys. Acta* **6**, 534 (1951).

[67] F. H. Johnson, H. Eyring, and M. J. Polissar, "The Kinetic Basis of Molecular Biology," p. 458. Wiley, New York, 1954.

[68] W. Borzani and M. L. R. Vairo, *J. Bacteriol.* **80**, 574, 572 (1960); **76**, 251 (1958); *Stain Technol.* **35**, 77 (1960).

the increasing saturation of cell receptors by the antibiotic. If the prosthetic group or protein receptor is designated by P, and the antibiotic or combining substrate by S, then

$$[P] + [S] \underset{K''}{\overset{K'}{\rightleftharpoons}} [PS] \tag{50}$$

represents the reversible equilibrium between the concentrations concerned. If the velocity constants of the combining and dissociating reactions are K' and K'', then at equilibrium if concentration $S = C$, $PS = \sigma$ the fraction of receptors occupied and $P = (1 - \sigma)$ the fraction of receptors left free, then

$$K'(1 - \sigma)C = K''\sigma \tag{51}$$

or

$$\frac{\sigma}{1 - \sigma} = \frac{K'C}{K''} = K_A C.$$

Therefore

$$\sigma = \frac{K_A C}{1 + K_A C}. \tag{52}$$

When C is small, the amount absorbed $(\sigma) \propto C$ since $\lim C \to 0$ is $\sigma = K_A C$. When C is large, the amount absorbed is independent of C since $\lim C \to \infty$ is $\sigma = 1$.

This type of quantitative relationship fits very well the initial reversible absorption of penicillin and streptomycin by staphylococci. Subsequent events however may be much more complex. The quantitative explanation of time action and time concentration curves offers many complications in heterogeneous systems. Not the least of these is that with dividing unicellular organisms the individual organisms have differences in age, size, generation time, and resistance to environment. Bearing this in mind, it is still possible at least in some cases to consider the average effect and explain in greater detail the formation of the edge of antibiotic inhibition zones.

B. The Relationship of Critical Population and Critical Concentration

1. Minimum Inhibitory Concentration (C)

Minimum inhibitory concentration may be defined as the concentration of antibiotic in the external medium which just inhibits cell division of a normal cell. Kavanagh[69] has pointed out the need for the use of *small inocula* because of the presence of some organisms of greater resistance in large inocula. Another reason for small inocula is so that the antibiotic is in large excess of the amount adsorbed, when the concen-

[69] F. Kavanagh, *Bull. Torrey Botan. Club* **74**, 309 (1947).

tration is near the minimum that is effective. The possibility of the occurrence of mutations of greater resistance will increase with time, so ideally the time of examination should be short.

Therefore minimum inhibitory concentration will be used here for the concentration which just achieves bacteriostasis in the early hours of growth with a standard inoculum. With streptomycin and the staphylococci used in experiments depicted in Fig. 24, $C = 0.9$ μg. at 35° when $N_0 = 10^6$ organisms/ml. Figure 25 shows the value obtained at 27°

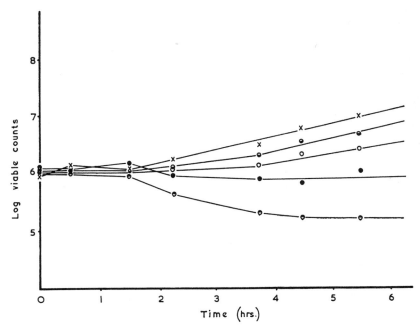

Fig. 25. The growth of *Staphylococcus* in broth at 27°C in the presence of streptomycin to determine the minimum inhibitory concentration ($C = 1.0$ μg./ml.). Concentration of streptomycin = nil $-$ $\times$, 0.50 μg./ml. $-$ $\ominus$, 0.75 μg./ml. $-$ O, 1.0 μg./ml. $-$ ●, 2.0 μg./ml. $-$ ◓ (Sehgal[39]).

($C = 1.0$ μg.) for another strain of *Staphylococcus*, which gave $C = 0.75$ μg. at 30° and 35° and $C = 0.5$ μg. at 40°.

2. Critical Concentration (m')

It has been assumed, for reasons of simplicity, that the streptomycin arriving at an organism by diffusion acts instantaneously, and if the concentration is less than m', is without effect; but if greater, it inhibits further cell divisions. Such an assumption is obviously a mathematical approximation. Adsorption of antibiotic by cells takes time, and the further steps in the inhibition process will take further time. The absorption of streptomycin and penicillin is a rapid process taking no more than a few minutes and it may be assumed that in the diffusion process the previous building up of the antibiotic concentration by the earlier arrival of concentrations less than m' compensates to some extent for the time necessary for subsequent events. It seems that these inhibitory mechanisms therefore operate within a short time of concentration m' being achieved, and that cell division is then stopped after one or very few divisions. The shortest possible generation time of normal organisms under optimal conditions is some 20 minutes. This is enough time for the establishment of an adsorption equilibrium to be accomplished.

Let a be the amount of antibiotic removed from the medium by one coccus.

Then the amount of antibiotic removed by the inhibitory population N'' is $(N''a)$ per milliliter.

Let it be assumed that at the moment the zone edge is determined the amount of antibiotic which remains in the medium in equilibrium with the amount absorbed is the minimum inhibitory concentration (C).

Then the critical concentration determined by diffusion experiments is given by

$$m' = N''a + C \qquad (53)$$

or

$$a = \frac{m' - C}{N''}. \qquad (54)$$

This gives the amount adsorbed per coccus. Adsorption formulas usually are expressed as the amount adsorbed by 1 mg. dry wt. of the organism, and if the number of organisms per milligram is N_w, then the amount of adsorbed antibiotic $[PS] = \sigma = N_w a/\text{mg.}$:

$$\sigma = N_w a = N_w \frac{m' - C}{N''} \qquad (55)$$

or

$$N'' = N_w(m' - C)/\sigma. \qquad (56)$$

If σ is expressed by one of the equations (47), (48), or (49), according to the experimentally determined results of adsorption experiments, then N'' can be calculated.

Humphrey and Lightbown[26] give for a sensitive strain of staphylococci in streptomycin the result $\sigma = 20.4 \, C^{0.24}$, and if the value of $N_w = 3 \times 10^9$ organisms/mg. dry wt. quoted in the tables for Brown's opacity tubes is used, the value for the minimum inhibitory concentration of the strain of *Staphylococcus* Mayo was $C = 0.9$ μg. when the critical concentration $m' = 1.8$ μg./ml.;

$$N'' = 3.10^9(1.8 - 0.9)/20.4(0.9)^{0.24} = 1.36 \times 10^8$$

or $\log_2 N'' = 27.00$.

This value is identical with the results of experiments on the effect of inoculum size.[17] The fact that this result is of the right order of magnitude suggests that the assumptions upon which its derivation was founded should be investigated experimentally in other cases.

3. The Effect of Salts on the Critical Concentration and Absorption

The amphoteric nature of proteins is well known not only in their combining powers with hydrogen and hydroxyl ions, but in their capacity to form salts with both anions and cations. The importance of heavy metal ions and of phosphates and citrates, for example, in modifying the activities of many enzymes are well known. Most dyes, stains, many antiseptics, and a number of antibiotics have a salt structure. The power of the larger ion to form a salt linkage with acidic or basic protein side chains is often responsible for their fixation. In these cases, however, a reversible equilibrium usually exists at least until other reactions supervene, and a competitive equilibrium thus exists with the other ions of similar charge present in the aqueous phase.

Thus a way of lessening the effectiveness of the antibiotic receptors is by competitive ions in the medium. Waksman[70] summarizes the antagonistic effects of a large number of salts on streptomycin. The results of different workers are sometimes somewhat different, probably largely as a result of the different, often unknown salts, present in their media, in addition to the one being investigated. Peptones, agar, meat extracts, and pH adjustments all contribute to these salt effects. The basal salt content of the final medium should be known before the effect of added salts can be accurately investigated. The matter is not even as simple as this, because many "nonspecific" proteins, prosthetic groups, etc., may function as receptors for salt ions that are not receptors for more complex and specific antibiotic. These will occur often in unknown quantity in the media, but they may also be produced by the syntheses of the bacterial cell. The age of the cell and of the population have both been said to affect resistance to some antibiotics.

Control of as many known factors as possible does enable the action of particular salts to be investigated. The remarkable thing is how consistently the critical and minimum inhibitory concentrations can be reproduced *in a particular medium*. Differences are discernible in the effect of each ion in different media. The competition is not only with the antibiotic but with other ions as well. It follows that determination of sensitivity of an organism against an antibiotic like streptomycin means little,

[70] S. A. Waksman, "Streptomycin," p. 204. Ballière, Tindall & Cox, London, 1949.

unless the ionic environment is accurately specified, and this applies not only to critical concentrations by diffusion methods but to minimum inhibitory concentrations by incorporation methods.

Results obtained by Linton[71] and Sehgal [39] for the effect of magnesium in different media on the critical concentrations of streptomycin for a number of organisms are given in Fig. 26. Most of the results obtained show a simple direct relationship between the amount of ion (in this case Mg^{++}) added and the critical concentration. This suggests a straightforward competitive relationship for cell receptors between Mg^{++} and streptomycin. Analogous results can be obtained with other ions differing in degree, and involving both anionic and cationic effects, so it is not surprising that in some media the relationship is more complex. With some media the straight line relationship was not obtained and the effect of the high Mg concentration was considerably modified. Even in such a case Sehgal showed that though critical concentration was considerably modified the values of critical time for the complete range of inocula sizes was unchanged. Thus the growth rate was unaffected. The straight line relationship between x^2 and $\log_2 N_0$ was maintained for each Mg^{++} concentration (range 40–4040 p.p.m.) though the zones were reduced in size due to the increase in m'. With high concentrations of NaCl (3.0–5.5%) effects on the generation time (G) increased T_0, and the value of the diffusion coefficient (D) was also modified.

Quantitative experiments on the absorption of streptomycin by the staphylococci under the above varying conditions have not been made, nor has the relationship between critical concentration and minimum inhibitory concentration in the presence of Mg^{++} been sufficiently investigated. The work quoted by Berkman et al.[72] and Waksman[70] shows that analogous changes occur in the minimum inhibitory concentration of streptomycin in the presence of salts as have been found in the critical concentration. The relative constancy of the inhibitory population and the variability of the critical concentration suggests that in accordance with Eq. (56) the difference in m' and C compensate for altered absorption and that this equilibrium requires large alterations in external concentration in the presence of salts for its maintenance. Experimental evidence for this is at present only qualitative but does emphasize how important media composition is in determining the sensitivity of an organism to an antibiotic.

The importance of pH in influencing ionization of proteins and the

[71] A. H. Linton, Ph.D. thesis. Department of Bacteriology, University of Bristol, England, 1954.

[72] S. Berkman, R. J. Henry, R. D. Housewright, and J. Henry, *Proc. Soc. Exptl. Biol. Med.* **68**, 65 (1948).

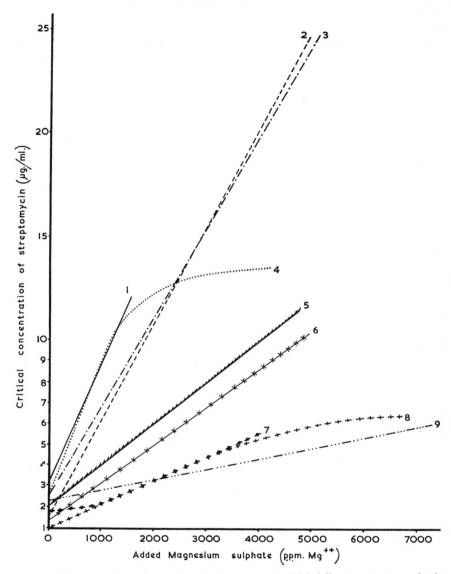

Fig. 26. The sensitivities of *Staphylococcus* and *Klebsiella* to streptomycin in different media to which magnesium sulphate has been added. The values of the critical concentration (m') determined by extrapolating zone sizes (x^2) for different concentrations ($\log m_0$) to $x = 0$ are plotted against the concentration of added magnesium ions (Mg^{++} in parts per million). Based on experiments of Linton[49,71] and Sehgal.[39] *Staphylococcus,* graphs 2 and 4. *Klebsiella,* all others. Nutrient agar media, 1, 2, and 3. Peptone agar, all others. 0% salt graphs 3, 7, and 9. 0.5% salt, all others.

availability of receptors for other ionic groups is well recognized and its influence on the activity of the saltlike antibiotics well known. Its importance in considering the adsorption of salts and their competition with antibiotics is obvious. Some of the specificity of antibiotics for different types of cells can be explained by studying dissociation constants. For a fuller discussion of ionization effects see Albert[73] and Work and Work.[74]

4. Resistant Strains and Adsorption

Berkman et al.[72] showed that the adsorption of streptomycin by staphylococci reaches equilibrium rapidly and that streptomycin could be completely eluted by sodium chloride. The amount of streptomycin adsorbed and recovered was identical with resistant and susceptible strains though the concentrations necessary in the external medium to inhibit growth might be as different as 1 and 1000 μg. streptomycin/ml. This agrees with the analogous results of Eagle et al.[64] with radioactive penicillin. It suggests that the resistant strains have numerically less receptors available for the antibiotic or that the dissociation from these receptors has been modified so that an increased external concentration is necessary to cause the adsorption of the required amount of antibiotic. The change from sensitive to resistant strains may thus be due to loss of receptors or to modification of their affinity for the antibiotic (by a change in dissociation constants, blocking or permeability changes, for example).

Cooper et al.[17] compared a resistant staphylococcus ($m' = 7.1$) with a sensitive one ($m' = 2.0$) at different temperatures over a wide range of inocula sizes. The resistant mutant used was chosen because it had the same growth rate as the prototype over the temperature range investigated. The values of critical time, critical population, and inhibitory population were unchanged. This suggests in agreement with Eagle et al.[64] that the higher critical concentration required to inhibit the more resistant strain is necessary to achieve the adsorption of the same *amount* of antibiotic per cell as for the sensitive strain.

Studies of the absorption of polymyxin E by Bliss et al.[75] by Salton[76] by Few and Schulman[77,78] showed that from a particular concentration of the antibiotic, whether low or high, much more was removed by sensitive organisms than by resistant ones. Only about $\frac{1}{12}$th saturation with polymyxin was necessary to kill sensitive organisms according to Schul-

[73] A. Albert, "Selective Toxicity." Methuen, London and Wiley, New York, 1960.

[74] T. S. Work and E. Work, "The Basis of Chemotherapy." Oliver and Boyd, London, 1948.

[75] E. A. Bliss, C. A. Chandler, and E. B. Schoenbach, Ann. N. Y. Acad. Sci. 51, 944 (1949).

[76] M. R. J. Salton, J. Gen. Microbiol. 5, 391 (1951).

[77] A. V. Few and J. H. Schulman, Nature 171, 644 (1953).

[78] A. V. Few and J. H. Schulman, J. Gen. Microbiol. 9, 454 (1953).

man *et al.*[79] so these statements do not necessarily contradict the assumption that inhibition of sensitive and resistant strains may be accomplished by approximately the same amount of absorbed antibiotic as found by Eagle *et al.*[64] for penicillin.

It must be remembered that many different mechanisms are known which produce resistance to antibiotics, and resistance is not necessarily concerned with adsorption changes. The production of enzymes destroying the antibiotic outside the cell or inside the cell, the provision of alternative synthetic pathways for substances blocked by the antibiotic, the production of antagonists to the inhibitory agent are a few examples of other mechanisms. Some of these will be considered in Section V, if they are known to affect the size of inhibition zones.

5. Dead Cells and Adsorption

If the cells are killed by the antibiotic the question of what happens to the antibiotic is of obvious importance. Penicillin, for example, that has combined with cells is destroyed and is no longer available for further action. Streptomycin adsorbed on living cells can be eluted by salt unchanged, and is then available for antibiotic action. Can dead cells influence the size of inhibition zones?

Sehgal [39] investigated this in the case of streptomycin and staphylococci. He attempted to kill staphylococci without destroying the streptomycin adsorbing receptors, by using four methods of destruction. These were:

(a) heating a suspension of washed cells in glass distilled water at 60° for $\frac{1}{2}$ hour,

(b) heating a suspension of washed cells in glass distilled water at 55° for 1 hour,

(c) treating washed cells with normal hydrochloric acid and then neutralizing,

(d) treating washed cells with normal lactic acid and then neutralizing.

With an inoculum size of 1.39×10^7 organisms/ml., the addition of 8.0×10^7 dead organisms/ml. killed by any of these methods produced no change in zone size, critical time or critical concentration.

The addition of the broth from a filtered culture also made no difference to the size of zones. The filtrate was a 20-hour broth culture passed through a Gradacol membrane (average pore size 0.24 μ) after rejecting the first 10 ml. This was added to standard agar medium III. Thus the products of the growth of the inoculum, whether dead organisms or soluble material, could not be shown to affect the results.

[79] J. H. Schulman, B. A. Pethica, A. V. Few, and M. R. J. Salton, *Progr. in Biophys. and Biophys. Chem.* **5,** 41 (1955).

Welsch[80] makes the statement "the addition of heat killed bacteria in large or very large numbers does not modify the results of streptomycin titration, although heat killed bacteria absorb the antibiotic as well as live bacteria." The last part of this statement seems difficult to reconcile with the first part. He also states that autolysate from several million cells does not interfere with streptomycin action.

In this laboratory Mayr-Harting[81] has shown that the addition of dead cells of *Escherichia coli* (Section C.6) to the live inoculum reduces the size of inhibition zones produced by a colicine if the cells were killed at 60°C. for 40 minutes, but has little or no effect if the cells were killed at 100°C. for 1 hour.

Humphrey and Lightbown[26] considered the amount of adsorption of penicillin by *Bacillus subtilis* was too small to account for the zone edge formation, and detected low penicillinase activity which they thought was most likely to be responsible for zone edge formation. No quantitative work has been done on the relationship between zone size with different sizes of inocula and penicillinase production.

In the case of aureomycin they failed to detect either "aureomycinase" activity or adsorption though no details are given of the methods attempted. Much more investigation is required to see whether the principles shown to apply in the case of streptomycin can be applied to other cases. The mechanism of destruction of the organism and of removal of the antibiotic must both be considered quantitatively and their effects on critical time, critical concentration, and critical population determined before the principles so far described are discarded in any particular case.

V. Mechanisms of Antibiotic Action

A. Antibiotic Destruction or Deviation

So far we have followed the antibiotic diffusing from the reservoir through the agar medium until the critical concentration enabled a particular amount to be incorporated or adsorbed on each cell in the growing population. This adsorbed quantity was regarded as having the *specific effect* of so interfering with normal metabolism as to inhibit further cell divisions within the area of the zone. Quantitative agreement with theory cannot be expected unless the whole of this amount exerts its specific effect. If nonspecific adsorption occurs, antibiotic will be diverted by receptors not essential to the life of the cell; or if metabolic reactions destroy some of the antibiotic in the cell then more will be needed, ex-

[80] M. Welsch, *Intern. Symposium Chem. Microbiol. 1st Symposium* W.H.O. p. 173 (1952).

[81] A. Mayr-Harting, Unpublished.

ternal concentrations will have to be increased, and the size of inhibition zones will be reduced. The importance of such reactions will probably be related to the amount of cell substance or metabolism, and therefore vary with the size of the inocula and the time required for zone formation.

1. Penicillinase

Penicillin resistant strains of staphylococci isolated from clinical cases (but not usually laboratory resistant mutants) owe their resistance to the production of penicillinase. It has been shown that the individual cells are as sensitive to penicillin as nonpenicillinase producers, and small inocula can thus be readily killed. Large inocula are able to produce enough penicillinase to protect the cells which are then able to produce still more. Zone formation is thus abolished by increasing size of inocula much more readily than would be expected from the sensitivity of the cells. The influence of size of inoculum in liquid media has been studied by many workers, but little quantitative work is recorded on penicillinase and agar zones.[82-85] Many other penicillinase producing organisms show this inoculum effect. Iland [86] reported a marked inoculum effect with a virulent human strain of *Mycobacterium tuberculosis* but not with the avirulent strain of Woodruff and Foster.[87] Hedges[88] developed a quantitative exhibition zone method for the measurement of diffusible penicillinase by incorporating inhibitory amounts of penicillin in poured plates containing staphylococci. By the use of prediffusion methods penicillinase assay was developed by measuring exhibition zones of growth produced by the destruction of the inhibitory penicillin by diffusing penicillinase. The concept of a critical time for the formation of the zone was established.

2. Chloramphenicol

It has already been noted in Section III.C (see Table III) that results obtained with chloramphenicol and staphylococcus produced identical values of T_0, n', and n'' with a standard inoculum as with streptomycin. This was true for temperatures from 22° to 42°, but was only tested for one size of inoculum. The m' value did change with temperature increasing as this rose in contrast to the relative constancy of the critical concentration for streptomycin.

[82] W. M. M. Kirby, *J. Clin. Invest.* **24,** 165 (1945).
[83] S. E. Luria, *Proc. Soc. Exptl. Biol.* **61,** 46 (1946).
[84] M. Barber, *J. Pathol. Bacteriol.* **59,** 373 (1947).
[85] K. R. Eriksen and D. Hansen, *Acta Pathol. Microbiol. Scand.* **35,** 169 (1954).
[86] C. N. Iland, *J. Pathol. Bacteriol.* **58,** 495 (1946).
[87] H. B. Woodruff and J. W. Foster, *J. Bacteriol.* **49,** 7 (1945).
[88] A. J. Hedges, Ph.D. thesis. Department of Bacteriology, University of Bristol, England, 1959; *J. Appl. Bacteriol.* **23** (2), 269 (1960).

Sehgal made an extended investigation of the effect of inoculum size on chloramphenicol zones produced at 35°. The graph obtained by plotting T_0 against $\log_2 N_0$ gave a straight line identical with streptomycin shown in Fig. 24. The graph for x^2 for three concentrations of chloramphenicol should theoretically consist of three straight lines similar to those shown in the lower half of Fig. 24, calculated from Eq. (45), substituting the diffusion coefficient of chloramphenicol (2.06 mm²/hour at 35°) for that of streptomycin. Excellent agreement was obtained for light inocula ($\log_2 N_0 = 13$ to 20) but for heavier inocula the experimental points fell somewhat below the theoretical line. Both the values of N'' and the numerous determinations of m' were constant, despite the fact that the zones were smaller than expected. The only "constant" in the formulas not experimentally determined was the diffusion coefficient (D) as determinations of T_0 were made and had the expected values. In the case of streptomycin this was thought to be independent of the size of inocula at 35°.

Rearranging Eq. (45), it is possible to determine D from zone measurements for each inoculum size:

$$D = \frac{x^2}{T_0} \frac{1}{4 \ln (m_0/m')}. \tag{57}$$

The results are shown in Fig. 27 in which it is evident that heavy inocula slow down the diffusion of chloramphenicol. If allowance is not made for this, and estimates are made of the amount of chloramphenicol in the reservoir from the sizes of the zones using the normal value of D, these estimates will be low with heavy inocula, though correct with light. It is evident that reaction is removing or preventing some of the chloramphenicol from exerting its inhibitory effect on cell division when heavy inocula are used.

There are a number of possible explanations for this effect. Simultaneously with the specific adsorption of chloramphenicol producing its inhibition, there may occur nonspecific adsorption not contributing to the inhibition, or there may be actual destruction of chloramphenicol by metabolic reactions of the cells, or neutralization of the antibiotic by products from the cells.

A number of workers have shown that many organisms can destroy chloramphenicol under suitable conditions.[89-93] It is evident that these organisms contain receptor enzymes for chloramphenicol concerned in

[89] G. N. Smith and C. S. Worrel, *J. Bacteriol.* **65,** 313 (1953).

[90] G. N. Smith and C. S. Worrel, *Arch. Biochem.* **28,** 232 (1950).

[91] G. N. Smith and C. S. Worrel, *Arch. Biochem.* **24,** 216 (1949).

[92] F. Egami, M. Ebata, and R. Sato, *Nature* **167,** 118 (1951).

[93] G. N. Smith, *Bacteriol. Revs.* **17,** 19 (1953).

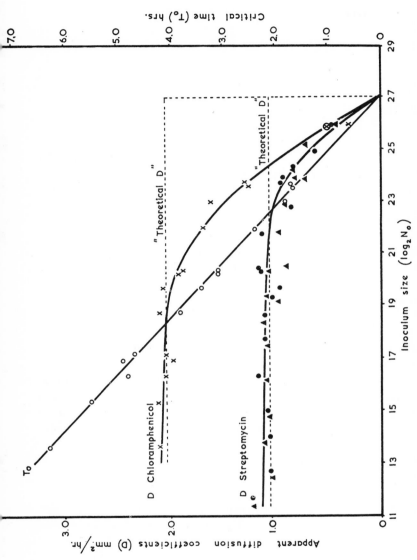

Fig. 27. The *apparent* diffusion coefficient for chloramphenicol ✕——✕ and for streptomycin ▲——▲ in agar tubes with different sizes of inocula ($\log_2 N_0$). The values of critical time (T_0) ○——○ are also plotted against inoculum size for the *Staphylococcus* used. The "theoretical values" for D if no adsorption on organisms occurred until the zone edge was reached would be at 35°, for chloramphenicol, $D = 2.05$ mm.²/hour; streptomycin, $D = 1.06$ mm.²/hour, ————. Calculations from the results of Sehgal[39] and Linton.[71]

reactions leading to destruction of the antibiotic as well as the specific receptors concerned with its inhibition of cell division. With increasing numbers of cells more antibiotic will be removed from the medium and this will have the effect of slowing down diffusion. It has already been shown that in the case of the specific receptors there exists an inhibitory population capable of (at least temporarily) completely stopping further diffusion and thus abolishing zone formation altogether. In the same way the critical population achieved by cell division forms the zone edge by stopping further diffusion. As the simplest assumption possible it has been considered that the effect of inoculum on the cells produced before the critical time was reached could be ignored as it produced such insignificant adsorption as not to affect the result, and that after the critical time adsorption was so complete as to lower the free antibiotic below its minimum inhibitory concentration. Such a state of affairs is an obvious approximation. In the case of chloramphenicol it is evident that additional adsorption occurs and slows down diffusion in experiments with heavy inocula.

However, a re-examination of the results published by Cooper *et al.*[17] shows that with very heavy inocula a similar effect is evident even in the case of streptomycin especially at low temperatures. A study of the *apparent* diffusion coefficient thus reveals an aspect of the mechanism of zone formation which has wider implications.

3. The Apparent Diffusion Coefficient and Inoculum Size

The published graphs for streptomycin on staphylococci show a tendency for the experimental points on the graph of x^2 against $\log_2 N_0$ to fall below the theoretical straight line when $\log_2 N_0 \geqslant 23.0$. This is increasingly evident at lower temperatures $27°$ and $30°$. It was at first thought to be experimental error due to the increased difficulty of reading the zone edge with very heavy inocula. The deviations are consistent, and when D is calculated for different concentrations (m_0) at each inoculum size (N_0) the results agree with each other. The average values of D for three concentrations at $35°C$ are plotted on Fig. 27.

It is important to realize that such calculations of D do not indicate a true variation in real diffusion coefficient, but that their *apparent value* is the result of the slowing of diffusion by a lowering of the concentration of the antibiotic as it proceeds through the agar, as a result of its removal by the inoculum and the growing cells. The resultant figure is also an *average value* during the time interval 0 to T_0. The antibiotic will start diffusing from the reservoir at the fastest rate consistent with the size of the inoculum. As the cell population increases diffusion will get less rapid, until when the inhibitory population N'' is reached it ceases altogether. Within the zone the position is further complicated by the later diffusion of concentrations greater than the critical, and the earlier diffusion of concentrations less than the critical. These will have inhibitory effects which by lessening the number of cells will make the effect on D less important. It is thus only as the zone edge and the critical time are approached that the effect on diffusion becomes marked. This is why a sharp zone edge is possible, and why the crude all or none assumptions of antibiotic action can give an approximate quantitative picture of events.

B. Specific Antibiotic Action

This chapter is concerned with the general theory of inhibition zones and detailed discussion of the effects of specific differences in antibiotic action lies beyond its scope. Though the general principles enumerated have been applied to zones produced by substances as widely different as crystal violet and penicillin, it must be expected that in certain cases or circumstances differences in the mechanism of antibiotic action may lead to differences in the utilization and fate of the antibiotic. This would be expected to influence the initial quantitative relationships between antibiotic and cells leading to the formation of the zone.

Many references will be found attached to the articles of Newton,[94] Gale,[95] Lacey,[96] and Knox[97] in the symposium on the stragegy of chemotherapy which are concerned with these specific mechanisms. It would seem that differences of this kind will be most likely to affect the balance between critical concentration and critical population by having effects upon the adsorption of the antibiotic. Investigations of the factors concerned in determining m', N', and T_0 would therefore be particularly worthwhile if correlated with knowledge of the mechanism of action.

VI. Some Biological Considerations Affecting the Use of Statistical Methods

A. Biological Assay

Most textbooks of pharmacology include chapters on biological assay, that is, on the use of tests of potency for the assessment of the quantity

---

[94] B. A. Newton, in "The Strategy of Chemotherapy" (S. T. Cowan and E. Rowatt, eds.), p. 62. Cambridge Univ. Press, London and New York, 1958.

[95] E. F. Gale, in "The Strategy of Chemotherapy" (S. T. Cowan and E. Rowatt, eds.), p. 212. Cambridge Univ. Press, London and New York, 1958.

[96] B. W. Lacey, in "The Strategy of Chemotherapy" (S. T. Cowan and E. Rowatt, eds.), p. 247. Cambridge Univ. Press, London and New York, 1958.

[97] R. Knox, in "The Strategy of Chemotherapy" (S. T. Cowan and E. Rowatt, eds.), p. 288. Cambridge Univ. Press, London and New York, 1958.

of an active principle in a preparation. There are usually three stages in the characterization, isolation, and purification of an active principle:

(1) The recognition that some substance or mixture of substances is responsible in a preparation for a pharmacological reaction, and the attempt to assess the *potency* of different preparations by their capacity to produce a measurable reaction.

(2) The production of a stable, active but chemically undefined, preparation which can serve as a *standard* with which to compare the potency of other preparations.

(3) The preparation of a chemically defined *pure standard*, a known weight of which is responsible for the pharmacological action required, with which the unknown preparation may be compared by use of a potency test.

The use of antibiotic inhibition zones for purposes of assay may fall into any of these three categories. As Miles[98] states, "most of the troubles in assay arise from the conception of potency as the amount of a substance that has a certain effect on a biological system." We have seen that many environmental factors and substances *in addition to the amount of antibiotic* may influence the effect, i.e., the size of the inhibition zone.

It is true that as knowledge of the assay of an unknown antibiotic increases these factors come more and more under experimental control. Comparison with a suitable standard can be made by a careful design of experiment so that as many of the factors as possible likely to affect zone size may be made similar for the standard and the unknown. Only after this has been done as completely as possible is it justifiable to consider the remaining unknown influences as accidental and small enough to justify the application of statistical theory to the results. Statistical theory is concerned with the effect of chance fluctuations and assumes a canceling out of numerous small accidental errors. It is concerned with the probability that in a particular series of determinations the error is likely to lie within certain limits of the mean.

The usual way to ensure the degree of accuracy of a measurement is to repeat the measurement a number of times. For many practical purposes physical measurements need only to be duplicated (checked). Biological measurements are more variable, even when a biological standard has been agreed. When the standard of length was a cubit (defined as the length of the king's forearm) the standard was liable to change. When the unit for the toxicity of digitalis was the amount required to kill a cat, or even 50% of a population of cats, uniformity was impossible to achieve at different places or different times. It is important to notice

[98] A. A. Miles, *Intern. Symposium Chem. Microbiol. 1st Symposium* W.H.O., Geneva, p. 131 (1952); Bull. W.H.O. 6, p. 131 (1952).

that the inaccuracies involved in such a unit cannot be abolished by simply increasing the number of cats used at a particular place and time. Even if the population of cats to be used is biologically specified in a sufficiently accurate manner (genetics, nutrition, weight, age, sex, etc.) and the unknown factors canceled out as much as possible by random sampling within such a population, a potency test alone (category 1) gives an accuracy that as Miles[98] states, is a myth. The use of two such random samples of a specialized population for comparison of an unknown with a *suitable standard* will give more reproducible results. That is to say, repeat measurements, if standard and unknown are strictly comparable, will be scattered round a mean in a way that can be specified by statistical theory. Measurements of the variance or standard deviation by repeat experiments will enable the probable error to be ascertained. Even so, if the unknown contains other active principles, stimulants or antagonists different from the standard, strict comparison is impossible.

What is true for cats and digitalis in the previous paragraph is also applicable to bacterial populations and antibiotics. Nevertheless, approximate assays are necessary during the development of an antibiotic, and discrepancies in results may be tested and used to indicate the purity or otherwise of preparations. When complete purity is achieved biological assay often becomes unnecessary though sometimes it is more sensitive and convenient than chemical assay.

B. Antibiotic Assay

1. Principles

For a full account of statistical method in biological assay, textbooks such as that of Burn *et al.*[99] and Finney[100] should be consulted. It is beyond the scope of this chapter to deal with statistical theory. Technical details of the various recommended methods will be dealt with in other chapters. There are many published accounts of assay methods and their statistical assessment.[9,21,101-107] The requirements for a valid biological

[99] J. H. Burn, D. J. Finney, and L. G. Goodwin, "Biological Standardization." Oxford Univ. Press, London and New York, 1950.

[100] D. J. Finney, "Statistical Method in Biological Assay." Griffin, London, 1952.

[101] D. C. Grove and W. A. Randall, "Assay Methods of Antibiotics." Med. Encyclopaedia, New York, 1955.

[102] Pharmaceutical Society of Great Britain, "Antibiotics, A Survey of Their Properties and Uses." Pharm. Press, London, 1952.

[103] K. A. Brownlee, C. S. Delves, M. Dorman, C. A. Green, E. Greenfell, J. D. A. Johnson, and N. Smith, *J. Gen. Microbiol.* **2**, 40 (1948).

[104] K. A. Brownlee, P. K. Loraine, and J. Stephens, *J. Gen. Microbiol.* **3**, 347 (1949).

[105] E. Greenfell, B. J. Legg, and T. White, *J. Gen. Microbiol.* **1**, 171 (1947).

[106] L. F. Knudsen and W. A. Randall, *J. Bacteriol.* **50**, 187 (1945).

[107] E. J. de Beer and M. B. Sherwood, *J. Bacteriol.* **50**, 459 (1945).

assay have been summarized by Miles[98] and by Kavanagh.[21] It may be of use here to point out some of the indications (that can be elicited by a suitable design of experiments) suggesting that the conditions for a valid assay are not being fully satisfied.

Whichever of the diffusion formulas (see Table II) is applicable, and however the time (T_0) in these formulas is expressed in relationship to the growth of the organism, we are concerned with the estimation of the amount or concentration of the antibiotic $(M, m_0, \text{ or } C_0)$ by measurements of the inhibition zones $(x \text{ or } r)$, i.e., the effect of a "dose" in producing a "response." This relationship is determined under particular experimental conditions devised by the assayist for a *standard antibiotic*. The unknown antibiotic should be treated in exactly the same manner, and *on the assumption* that the relation between dose and response is the same for both the ratio of the dilution producing the same effect as the standard is determined. It is found that the deviations obtained on making repeat assays are scattered in the most symmetrical way about the mean if the dose is expressed as the logarithm. It is thus usual to plot the function of the zone size $(x^2, r^2, x, \text{ or } r)$ against the logarithm of the dose (dilution, concentration, or amount).

The mathematical *conditions for a valid assay* are:

(1) Log dose plotted against response (a measure of zone size) should give a line of the same form (preferably straight) for both standard and unknown.

(2) The ratio of the response for high and low dilutions of the standard should be the same as for the unknown (the ratio of high to low dilutions chosen being the same). In the case of a straight line relationship between log dose and response, the two lines should be parallel.

If these conditions are to be fulfilled the following consequences must be considered:

(a) The first condition is usually satisfied by plotting log concentration or dilution against x^2 or d^2, but in some cases as has been seen, x or d may be a better approximation to the straight line. Which should be used ought to be ascertained for the range of dilutions to be employed for the assays.

(b) The second condition depends on the constancy of the slope of the log dose/response line, that is, on DT. It has been seen that T (the time of diffusion of the antibiotic up to the moment the zone edge is formed) is equal to $(T_0 - h) = (G \log_2 (N''/N_0)) - h$. It is thus evident that $D, G, N_0, N'', \text{ and } h$ must be the same for standard and unknown if the second condition is to be rigorously satisfied. Expressing this statistically the differences between the values of these variables for standard and unknown must be small and accidental. The design of the experiment must attempt to randomize the inevitable deviations.

(c) The other part of Eq. (45) relating zone size and log dose is $\ln (m_0/m')$ from which it follows that m_0 and m' for the standard be *strictly equivalent* to m_0 and m' for the unknown.

It is necessary to examine how far it is possible to control these factors influencing the accuracy of antibiotic assay.

2. Factors Influencing Accuracy

a. *Heterogeneity of standards and/or samples.* The term "strictly equivalent" has been used for the relationship between standard and unknown. If both were identical chemically pure active principles responsible for the inhibition there would be no need for the assay. If the unknown sample contained only the same active principle as the pure standard, but was mixed with *completely inert material* then the assay properly controlled should be valid. Dilutions of standard and unknown would have their activity differing by a constant ratio. Such a comparison should be *independent of the biological system used.* Any organism that was inhibited by the antibiotic should give the same ratio of activity of unknown to standard. Because organisms (and even different strains of one organism) differ in m', G, or N'', the absolute zone sizes would differ but not the assay results.

What is meant by *completely inert material?* When Schmidt et al.[108] changed from *Staphylococcus aureus* to *Bacillus subtilis* as the organism for the assay of penicillin they found a variation of potency ratio. Investigation led directly to the discovery that penicillin contained at least two substances Penicillin G and K. The sensitivity of the two organisms differed for each substance, and the effect of dilution thus varied with the ratio of G to K. Many other antibiotics have since proved to be mixtures of closely related but differing substances, the streptomycins, the bacitracins, the polymyxins, and the gramicidins for example. Early standards were thus often heterogeneous. Unless unknown samples assayed against them had exactly the same ratio of active substances, results with different organisms did not agree. Discrepancies of this kind are not always evident by diffusion methods if the diffusion rate or activity of the second constituent is very different. Inhibition zones may only detect the most rapidly diffusing or the more active constituent.

Inertness does not depend solely on the absence of another antibiotic. Any material which affects viscosity, if present in sufficient amounts, can alter D. Other substances may also affect the growth rate, and any such modification of the action of the antibiotic in the unknown which was absent or different in the standard will affect the assay. It has been seen how glucose and other sugars led to inaccuracies in the estimation of penicillin in lozenges (Section II.E.5).

[108] W. H. Schmidt, G. E. Ward, and R. D. Coghill, *J. Bacteriol.* **49**, 411 (1945).

Because of the importance of pH, the presence of buffer substances, acids or basic substances, likely to diffuse into the medium and modify its pH will affect the result.

It has also been seen that various salt ions, especially with multiple ionic charges, may compete with and modify the adsorption and action of some antibiotics, such as streptomycin.

In general any antagonistic substance, whether competitive or non-competitive, should be absent from the assay sample or its amount compensated for in the standard. Needless to say stimulant or symbiotic effects are equally important.

Not only substances affecting the direct action of the antibiotic may be of importance, the size of the zone is equally dependent on the growth rate of the organism. Growth promoting or growth inhibiting substances by altering G will also cause lack of parallelism of the lines for the unknown and standard. Effects of this kind are most likely to disturb assays if organisms are grown on poor media where growth rate may be limited by some constituent present only at a low level. Enriched media lessen the chances of such effects but increase the chances of the presence of substances antagonistic to the antibiotic. Information may therefore be obtained by comparing assay values obtained in a number of different media.

b. Temperature. It has been seen that temperature has a direct effect on the diffusion coefficient, and an indirect effect on this by its effect on the viscosity of water. Of even greater importance is its effect on the growth rate of the organisms which directly alters the time required for a particular size of inoculum to reach the critical population. In all these ways temperature affects zone size and unless the changes of temperature necessarily occurring when growth is initiated are deliberately controlled to ensure that zones formed by standard and unknown are similarly treated, error in assay is inevitable.

Cooper and Linton[15] investigated a number of factors concerned with temperature of petri dishes during the earlier hours of incubation and showed how these influenced the size of zones. Radiation, convection, and conduction all played a part, and the specific heat and mass of the materials, glass, and agar have to be taken into account. The main part of the heating process followed Newton's law of cooling. The arrangement and number of plates in the incubator were very important. Packed plates in the middle of a pile might require 5 hours to reach within a degree of the incubator temperature. Plates placed directly on a metal shelf had the most rapid heating. The thickness (and therefore amount) of the agar had a marked effect.

Brownlee *et al.*[104] recommended the use of large rectangular plates for the assay of both streptomycin and penicillin. In these, the position of the dilutions of standard and unknown could be arranged to minimize the effects of any difference in position of the standard and unknown dilutions. They also in their arrangement compensated for differences in the time of filling the cup with antibiotic, thus eliminating variations in assay due to noncoincidence of the time at which diffusion and incubation start ($h = 0$). The arrangements were based on the Latin square designs.[109,110] They reduced the range of the error (95% confidence limits) to 96 to 104% for a single assay of streptomycin.

c. Other influences. At first sight a far better way of dealing with temperature fluctuations such as occur with plates would be to change the method of assay to one using tubes which by immersion in a water bath can be made to reach rapidly the desired temperature for growth. This method advocated by Mitchison and Spicer for the assay of streptomycin is the one we have used to investigate the formation of inhibition zones. It is, however, of limited application and cannot be applied to penicillin because the edge of the inhibition zone becomes too diffuse to read. This may be due to the lowering of oxygen tension in the depths of the tube, though Mitchison and Spicer[13] showed this was unimportant in their medium with streptomycin. Linton[71] has shown that by incorporating a concentration of streptomycin in agar which just fails to inhibit growth in the depths of the tube, it is possible to produce an inhibition zone near the surface by the diffusion of O_2 into some media, but not others. No such effect was obtained in any media tested with chloramphenicol.

The tube method also abolishes evaporation from the surface of the agar, but it does not abolish water flow into or out of the agar, unless osmotic equilibrium is maintained between antibiotic solution and media. The best solvent in which to dissolve standard and unknown should take this into account at the same time as pH effects are considered.

The disk method is one in which water flow is of particular importance. The *dry disk* depends on capillary attraction to imbibe water and dissolve the antibiotic, and with small disks at any rate it is the amount rather than the concentration of antibiotic that determines zone size. The use of *wet disks* saturated by antibiotic solution of known concentration is another method depending on the volume of fluid retained by the disk as it is used to transfer antibiotic to the agar.

The cup method as pointed out by Brownlee *et al.*[104] offers some advantages over the cylinder method as the distance diffused is more certain. The bead method seems more related to the disk than the cylinder, as it contains a small quantity of antibiotic. The solution usually flows

[109] R. A. Fisher, "The Design of Experiments." Edinboro. Oliver and Boyd, 1942.
[110] R. A. Fisher and F. Yates, "Statistical Tables for Biological, Agricultural and Medical Research." Edinboro. Oliver and Boyd, 1943.

out of beads and cylinders into the agar and the precise moment which ought to be considered as $T = 0$ seems less certain.

All these methods can be standardized in such a way as to treat standard and unknown in analogous ways. Accurate assay should be possible if careful techniques are devised for their use.

VII. Other Uses of Diffusion Methods

A. Surface Cultures

If agar plates are inoculated only on their surface and antibiotic allowed to diffuse through the agar medium beneath, inhibition zones are produced around the antibiotic reservoir. Conditions are different in a number of respects from those produced by poured plate cultures. The kind of surface growth produced by different species of organisms varies enormously from the thin spreading growths of *Clostridium tetanus* or of *Proteus* to the thick heaped up growths of staphylococci or human tubercle bacilli. Some of these growths are promoted, others inhibited, by oxygen. The standardization of the experimental conditions such as oxygen, carbon dioxide, humidity, or pH is more difficult to achieve. The influence of inoculum size on the rate of multiplication is more difficult to control, and its effect in producing at first isolated colonies and later confluent growth more difficult to measure.

The diffusion of the antibiotic through the uninoculated agar which at first appears as simple linear or radial diffusion becomes complicated in the third dimension as antibiotic is removed upwards by the culture growing on the surface. It should perhaps be emphasized that the use of uninoculated agar below or above an inoculated layer always introduces this third-dimensional complication. If the uninoculated layers are thin, the results will approach those to be expected if the inoculum were distributed throughout all the layers. Surface culture introduces another complication by the antibiotic having to be absorbed into the surface colonies through the lower layers of growth to the organisms heaped up above away from contact with the agar. The flow of nutrient and waste products through solid colonies has been little studied and accurate determination of the numbers of organisms in such growths are not available. The effect of dead cells among the living or the extracellular structures such as capsules or flagella on the diffusion of substances through the colonies has been scarcely investigated. Some colonies absorb dyes from the media, others do not, but any knowledge of such facts is quite empirical and theoretical prediction of results impossible.

Despite this ignorance it can be shown in many cases that the formation of such inhibition zones is determined in an analogous manner to those already discussed. That is to say, they show the existence of a critical concentration which is effective in producing an inhibition zone if it reaches the growing cultures within a certain critical time. The critical concentration and the critical time can be experimentally determined exactly as described for poured plate cultures under standard conditions. Equation (6) may then be applied for linear diffusion from a constant concentration. For the other conditions discussed in Section II, one of the formulas summarized in Table II may be chosen to suit.

The relationship between critical time and the laws of growth described for poured cultures in Section III has not been worked out for surface cultures though it seems probable that a critical population exists even though it may depend on the spatial relationship of the organisms to the diffusing antibiotic in a third dimension, or the thickness of the confluent growth and of the agar plate. The work of Mayr-Harting[59] on the numbers of organisms in surface colonies shows how this is dependent on the distance between colonies. No accurate determinations have been made during the early hours of growth, but it is evident that the effects of inoculum size may be more complicated than with deep cultures.

B. Exhibition Zones

1. Penicillinase

In Section V.A.1 reference has been made to exhibition zones produced by diffusing penicillinase into plates containing an inhibitory concentration of penicillin and viable organisms. With careful adjustment of the antibiotic concentration to the sensitivity of the organism it can be shown that a critical time exists before which the organism can recover and multiply if the penicillin is destroyed. After the critical time death occurs and incubation of the plate fails to produce growth. The conditions are thus the reverse of those leading to the formation of inhibition zones. Growth occurs near the reservoir where the critical concentration of penicillinase arrives before the critical time, and is inhibited in areas further away.

As such a system is concerned with a diffusing *enzyme* the relationship between enzyme concentration and interference with the destructive action of the penicillin on the growing cells may be more complex than it is with diffusing antibiotic. Much further work is required to establish the constancy or inconstancy of critical concentration and critical time and the effect of conditions on these.

2. Vitamins

The production of basal media including all the known food requirements of an organism in a form which enables a particular essential requirement to be omitted at will has enabled the production of growth by adding the essential requirement to such a deficient medium to be used as a means of estimating the amount added. The use of such a deficient medium for agar culture and the addition of the essential nutrient by diffusion sometimes produces sharply defined exhibition zones.[111–113] With other nutrients incubation results in growth which gradually diminishes from the reservoir to the periphery, the amount of which gradually increases with further incubation. In the former case there exists a critical time of incubation during which the arrival of the nutritional requirements can initiate cell multiplication but after which death occurs. In the latter case it appears that cells can remain in a resting state for a long period and still recover if nutriment is supplied; no short critical time exists. Knowledge of the metabolism of the resting state of bacterial cells is too incomplete to allow a prediction as to the category into which a particular foodstuff will fall. Experimental determination of whether critical time can be measured by the effect of preincubation and prediffusion alone will decide whether assay is possible by the exhibition zone method. A further complication must be taken into account. It is that the growth response of microorganisms to vitamins may differ at different temperatures.[114]

Research work on the streptomycin dependent mutant of *Escherichia coli* forms an interesting link between exhibition zones and the inhibition zones produced with sensitive strains. Little quantitative work has been done, though a critical time does exist with this system.

3. Antagonistic, Additive, and Synergistic Effects

Diffusion methods have been used by many workers to determine the combined action of antibiotics on an organism. Dye[115] describes results obtained by qualitative or at most semiquantitative methods which demonstrate effectively additive, antagonistic, and synergistic effects. Paper strips impregnated with agar containing antibiotic were set at right angles on the surface of the agar plate. The zone of inhibition will form an edge parallel to the antibiotic containing strip. Only if the action of the antibiotic is completely independent will the two edges meet at right angles at the corner (no effect). If the same antibiotic is placed in each strip an additive effect will result in a rounded corner. Synergism, such as is produced by two inhibitors competing for a single enzyme, will interfere with growth over a further area beyond the curved edge of the additive

[111] E. Harrison, K. A. Lees, and F. Wood, *Analyst* **76,** 696 (1951).
[112] W. F. J. Cuthbertson, H. F. Pegler, and J. T. Lloyd, *Analyst* **76,** 133 (1951).
[113] A. L. Bacharach and W. F. J. Cuthbertson, *Analyst* **73,** 334 (1948).
[114] G. J. Kasai, *J. Infectious Diseases* **92**(1), 58 (1953).
[115] W. E. Dye, *Antibiotics Ann.* **1955/56,** 374 (1956).

effect. Antagonism (diffusion of acid from one strip and base from the other for example) will extend growth toward the corner where the strips meet.

Quantitative work needs to be done on the effect of incorporating sublethal concentrations of an antibiotic in the medium on the zone sizes due to diffusing antibiotics. Determinations of the effect of such competing systems on critical concentration, critical time, and critical population have not been made. Even the effect of diffusing oxygen on the action of incorporated streptomycin can be demonstrated by agar tube experiments.[71] The observations suggest that many interesting results could be obtained and the action of many agents on cell receptors might thus be elucidated. Such work might shed light on the effects of impurities in media and in antibiotic preparations used for assay purposes.

VIII. Sensitivity Tests

The determination of the sensitivity of an organism to an antibiotic by means of the size of inhibition zones on agar plates is a very popular method in clinical laboratories. As usually carried out the test is of doubtful significance, and disagreement in the reports on strains submitted to different laboratories only too often occurs. Often control strains of high and low resistance to each antibiotic are not used. Reliance is placed on the absolute size of the zone produced by the unknown strain. Even if there is careful control of inoculum size, and of time and temperature conditions such results depend, as has been seen, on many other factors such as pH, salts, and peptones. There seems to be little realization that the size of zones depends not only on these factors, but, in addition to the *sensitivity* of the organism to the antibiotic, on the *growth rate* of the organism. Any difference in growth rate (whether as a result of environmental conditions or of genetic constitution) between the unknown strain and a known control will modify zone size, quite apart from differences in sensitivity. Freshly isolated strains from clinical material often have rigid nutritional requirements, differing from those of laboratory adapted strains, and media deficiencies may lead to slower growth and larger zones for such newly isolated strains. Resistant mutants often have different growth rates and if slower will also appear more sensitive than they really are if zone size is interpreted purely in terms of sensitivity. If plates are piled in an incubator a plate in actual contact with the incubator shelf will have much smaller zones than one in the middle of a pile, and may then be erroneously reported as comparatively resistant. The production of antagonists or antibiotic destroying enzymes by an organism may produce a false picture if this is not taken into account in laboratory tests. A penicillinase producing *Staphylococcus* will, if used

as a heavy inoculum, show complete inability to produce an inhibition
zone, due to destruction of penicillin by heavy growth and enzyme synthesis. The individual cells may be fully sensitive as shown by tests
with small inocula, and in early cases of infection such strains may in
clinical practice respond to penicillin though reported by inadequate
laboratory tests as "resistant."

It should be evident from the foregoing that the same care is necessary
in making sensitivity tests by inhibition zone methods as is used in antibiotic assays, that adequate control strains should be included in any
experimental design under the same conditions as the unknown, and that
in interpreting the results of tests the effects of different growth rates
be borne in mind.

In what has been said above the determination of sensitivity against
one antibiotic has been considered. Things are even more misleading
when the attempt is made to judge sensitivity on the basis of zone size
produced by different antibiotics. It is almost a platitude to point out
now that the size of the zone depends on both sensitivity and diffusion
rate. The corollary that at least two concentrations of each antibiotic
must be used to define the slope of the log concentration against the
zone size graph in order to determine sensitivity is certainly not acted
on sufficiently. A hypothetical example may best illustrate this point.

Consider two antibiotics: A, with diffusion coefficient $D = 1.09$
mm.^{2}/hour; B, with diffusion coefficient $D = 2.18$ mm.^{2}/hour (A might
well be streptomycin and B chloramphenicol).

It is desired to find the sensitivity of a strain of an organism. (Suppose the sensitivity to be $m' = 1$ for A and $m' = 10$ for B.) Then for A:

$$x^2 = 4(2.30)1.09 \log m_0 = 10 \log m_0;$$

for B:

$$x^2 = 4(2.30)2.18 \log (m_0/10) = 20 \log (m_0/10).$$

The graphs for x^2 against $\log m_0$ are shown in Fig. 28. If three different firms issue disks containing different amounts of antibiotic equivalent
to (1) 10 units, (2) 100 units, and (3) 1000 units, then the disks from:

(1) will show no zone with B, and $x^2 = 10$ mm.^{2} with A

(2) will show $x^2 = 20$ mm.^{2} for both A and B

(3) will show $x^2 = 40$ mm.^{2} for B and only 30 for A.

If reports are made on the basis of one set of disks only, then:

(1) will show the organism is resistant to B but sensitive to A

(2) will show that both are equal, and

(3) will show that the organism is more sensitive to B than to A;
clearly these are discrepant reports.

Only the full graph shows the true relative position of the antibiotic
to this organism. Diffusion rate is important in both laboratory and

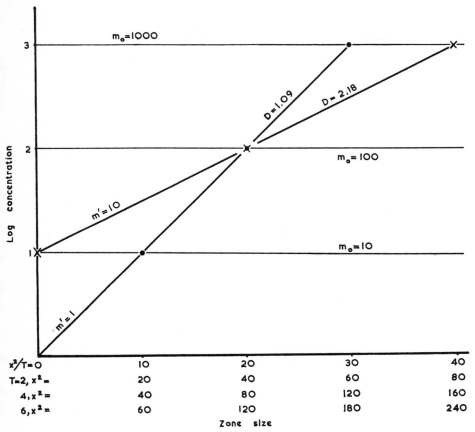

FIG. 28. Zone size for two antibiotics of different diffusion rate ($D = 1.09$ and 2.18). The zone sizes at $m_0 = 1000$ μg./ml. are in reverse order of magnitude to that obtained for $m_0 = 1$ μg./ml. and equal for $m_0 = 10$ μg./ml. for an organism which is ten times as sensitive to antibiotic A ($m' = 1$) as B ($m' = 10$). Sensitivity can thus only be obtained by extrapolation, not by absolute zone size. Log m_0 is plotted against x^2/T as the absolute sizes depend on T which is very dependent on temperature.

clinical assessment of the possible value of an antibiotic, but should not be confused with sensitivity.[116]

IX. Summary of Antibiotic Zone Formation

In the previous sections a large number of factors have been discussed which can influence the size of antibiotic inhibition zones. These factors

[116] A. H. Linton, *J. Med. Lab. Technol.* **18**, 1 (1961).

have been analyzed as far as possible with the existing data into effects upon the factors in the formula (Eq. (45)) relating critical time to the growth of the organism on the one hand and to the diffusion of the antibiotic on the other:

$$T_0 = L + G \log_2 (N'/N_0)$$

$$T = x^2/4D \ln (m_0/m').$$

When experimental conditions are such that growth and diffusion start simultaneously $T = T_0$; if there is an interval between them $T = T_0 - h$ where h is the time of preincubation or $-h$ is the time of prediffusion.

When $h = 0$, then the relationships expressed above may be represented graphically by combining the two expressions for time and placing the graph for growth above the graph for diffusion (see Fig. 29).

In the top part of the graph the population of the growing culture (outside the zone of inhibition) is expressed as multiples of the inoculum size on a logarithmic scale ($\log_2 (N/N_0)$). In the bottom part of the graph the zone size (x^2) is shown for concentrations of antibiotic in the reservoir (m_0) expressed as multiples of the critical concentration (m').

Consider the course of events. As incubation proceeds after a lag time (L) the population increases at a rate determined by the doubling time (G). When it reaches the critical population N', critical time has been reached and the position of the zone edge is determined. Expressing N' as a multiple of N_0 a vertical line can be drawn representing critical time for the particular growth conditions concerned. Lines are shown for two temperatures (27° and 35°C.).

The progress of the diffusion of the critical concentration m' is shown in the lower graph the slope being determined by the concentration in the reservoir (m_0) and the diffusion coefficient D. Lines are shown for three concentrations (10, 100, and 1000 times the critical) and for two rates of diffusion as determined by the temperatures.

After the critical time (for the appropriate temperature), antibiotic is adsorbed so that the critical concentration is lowered below the minimum inhibitory concentration and the organisms will proceed with their growth. Before the critical time the diffusing antibiotic is sufficient to inhibit growth. Thus the zone edge is formed where the diffusion graph cuts the line of critical time. This line, as has been shown, is determined by the achievement of an absolute population (critical population N'). How long is taken to achieve this depends on the inoculum size and the growth curve. Critical population is thus seen to be the key to the formation of inhibition zones. In Section IV.B, it has been suggested that it is the amount of antibiotic absorbed by this critical population that determines the zone edge and that this amount adsorbed on the cell receptors is in equilibrium with the minimum inhibitory concentration

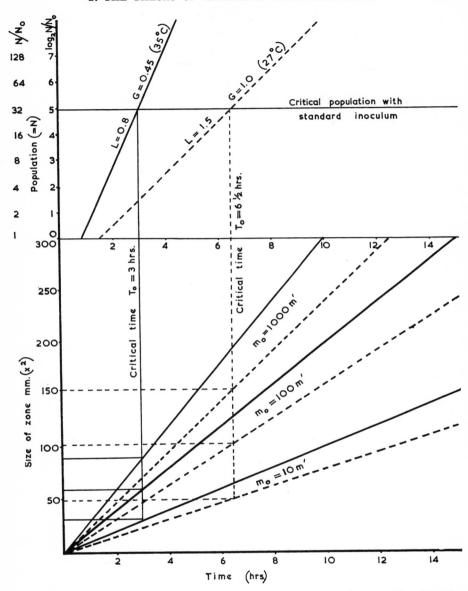

FIG. 29. The combination of the relation of the critical time $T_0 = L + G$ $\log_2 (N'/N_0)$ of the organism on the one hand (upper graph) with its relation to diffusion $T_0 = x^2/4D \ln (m_0/m')$ on the other. Zone sizes (x^2) are shown for three concentrations (m_0) of antibiotic at two different temperatures; —— 35°, $D = 1.09$ mm.²/hour, and ———, 27°, $D = 0.9$. A critical population $N' = 38.9 \times 10^6$ is shown, that is 32 times the inoculum size, but the graph can be used for any other size of inoculum. The values of L and G at the two temperatures are those for *Staphylococcus* (for viable counts outside the zone edge) and D, m_0, and m' are for streptomycin.

outside the cells. In a diffusing system the critical concentration has to provide both adsorbed and free antibiotic.

ACKNOWLEDGMENT

We wish to record our thanks to H. J. Washer, Associate of the Institute of Medical Laboratory Technology, for the preparation of all the figures in this chapter.

CHAPTER 2

Microbiological Assay Using Large Plate Methods

J. S. SIMPSON

The Midlands Counties Dairy Ltd., Birmingham, England

I. Introduction

The past 5 years has seen, in this country, a tidying up process among microbiological assays generally. The methods developed many years ago have been found to be satisfactory but their application as precise analytical tools has taken, and will yet take, a good deal of time. Confidence in microbiological assay has increased in recent years and the microbiologist must assume the responsibility of developing that confidence among his colleagues to such a point that microbiological assay can be readily accepted not only as an essential part of analytical and production control but also as an invaluable technique for investigational work. These three aspects present different problems for microbiological assay and in this paper the requirements and considerations governing these are discussed but they represent the author's personal opinion.

The microbiological assay laboratory has, with the advent of modern antibiotics and chemotherapy and with a fuller knowledge of the nutritional requirements of bacteria, become of first-class importance in the research, development, production and analytical control of antibiotics, vitamins, amino-acids, and other substances.

There are three main methods of microbiological assay by which the potency of samples and standard solution may be compared: (1) dilution methods, (2) turbidimetric, titrimetric, and gravimetric methods, and (3) diffusion methods.

This paper is concerned mainly with diffusion methods and which have been the subject of much experimental work in laboratories in the U.K. during the past few years. Even so, it is very necessary to be aware of the advantages and disadvantages of methods other than by diffusion techniques. There is little doubt that provided certain observations are valid diffusion methods are much to be preferred.

These observations are listed: (1) estimated potency of the sample, (2) volume and nature of sample, (3) number of samples received for assay, (4) approximate required standard error of assay, and (5) quality of training of assay laboratory staff.

All these factors should be carefully considered before the choice of method of assay is made.

The disadvantages of dilution, turbidimetric, titrimetric, and gravimetric methods are now well recognized. They are, however, summarized: (1) successful assay depends largely on accurate forecast of potency, (2) a low throughput of samples only is possible, (3) comparatively low degrees of precision unless increased replication and close spacing of dilution levels are employed, (4) extensive incubator and bench space is required, (5) difficulties in preparing an assay design to avoid or reduce sources of error, (6) critical statistical analysis may be most complicated, (7) results in some cases may not be available for 1 to 5 days.

Long experience of these methods of assay has emphasized the disadvantages; in many cases, however, no other method is available. Diffusion methods are now preferred wherever possible and during the past few years in the U.K. the large plate methods have been firmly established as the method of choice. The use of petri dish methods is still preferred in some countries, notably in the U.S.A., but there seems little doubt that once a knowledge of the large plate methods has been acquired these will rapidly replace the petri dish methods. The author was for seven years head of the Micro-biological Section of a large company in the U.K. and has had experience of the various methods of assay, particularly diffusion methods both petri dish and large plate up to 1958.

Large plate methods of assay of antibiotics, vitamins, amino-acids, and of other substances have been in use in many laboratories of the U.K. for some 10 years. It requires 3–4 years to develop sufficient knowledge of the method and to train staff adequately, whereas the last few years has seen the various refinements being added and, in particular, a sufficient knowledge of statistical analysis, assay design, and a greater degree of personal interest in the methods by the various skilled and unskilled workers involved. It would seem reasonable now to assume that "high precision" assays may yet be within our reach although there is still much to learn yet of low and medium precision assays.

A failing of most microbiologists in the past has been of talking rather too glibly of high or even medium precision assays. This defensive attitude of mind has been largely due to the colleagues of a microbiologist assuming or expecting rather more from a microbiological assay than it can give. However, this failing on both sides is now rapidly disappearing in this country and while we must be continually on our guard to ensure that complacency does not enter into our routine work, it is perfectly reasonable now to feel that we have a method of assay which will stand up to much more searching criticism than any other method yet employed.

The history of the development of the large plate method of assay in the U.K. laboratories is probably well known. The disadvantages of the petri dish methods of assay had been realized for some time and following on the work of Brownlee *et al.*[1,2] in 1948 and 1949 in which large glass plates were first employed but without the use of balanced statistical designs, Lees and Tootill,[3-5] in their laboratories, laid the basis of the assay as used today. Their work, based mainly on the assay of benzyl penicillin in fermentation liquors, represented the most important advance in this field for many years and while many of their designs have now been replaced by alternative ones, nevertheless, the basic techniques and observations still apply. In particular, their work on the various factors affecting the size and shape of inhibition zones of bacterial growth whether on petri dishes or large plates, has been of value. It is very important in developing large plate methods, whatever the design employed, that these factors are clearly understood and appreciated and that the assay design selected eliminates or reduces the effect of as many of these as possible. The paper by Adamson and Simpson[6] to the International Symposium of Microchemistry lists the many factors involved.

In the light of the factors discussed by Lees and Tootill and of those additional factors now known to be important and mentioned previously in this paper, a large plate assay design can be developed to meet the majority of these needs. In some few cases all needs can be satisfied and this small group represents the potential "high precision" assay group. The uses and applications of large plates in antibiotic and vitamin assays have been discussed in four papers by Simpson.[7-10]

II. Requirements for Accurate Assay

A. Estimated Potency of the Sample

It is fundamental in any analytical procedure that it is preferable for a sample and standard to be weighed, extracted, and diluted by the same operator. Thus, variables due to cleaning and type of glassware, laboratory conditions, use of pipets, burets, and chemical glassware diluents, storage of the test solutions, and of balances are all eliminated. If ex-

[1] K. A. Brownlee, C. S. Delves, M. Dorman, C. A. Green, E. Greenfell, J. D. A. Johnson, and N. R. Smith, *J. Gen. Microbiol.* **2,** 40 (1948).

[2] K. A. Brownlee, C. S. Delves, M. Dorman, C. A. Green, E. Greenfell, J. D. A. Johnson, and N. R. Smith, *J. Gen. Microbiol.* **3,** 347 (1949).

[3] K. A. Lees and J. P. R. Tootill, *Analyst* **80,** 95 (1955).

[4] K. A. Lees and J. P. R. Tootill, *Analyst* **80,** 110 (1955).

[5] K. A. Lees and J. P. R. Tootill, *Analyst* **80,** 531 (1955).

[6] D. C. M. Adamson and J. S. Simpson, *Proc. Intern. Symposium Microchem., Birmingham, 1958,* p. 184 (1960).

[7] J. S. Simpson and K. A. Lees, *Analyst* **81,** 562 (1956).

[8] J. S. Simpson, *J. Med. Lab. Technol.* **13,** 267 (1955).

[9] J. S. Simpson, *J. Med. Lab. Technol.* **13,** 474 (1956).

[10] J. S. Simpson, *Analyst* **82,** 210 (1957).

traction procedures are not carried out by assay staff, then solutions prepared elsewhere should be submitted at the maximum potency possible. Assistants develop a more personal interest in the assay when they realize that the complete assay is in their hands. An accurate forecast of potency of the sample is essential for successful assay, the further apart are the actual and estimated potencies, the greater will be the error. As a working rule if the ratio of estimated potency to actual potency exceeds the "dose ratio" of the assay then a new assay should be carried out. There is a real danger in using diffusion methods of assay for low potency samples as the methods are less 'sensitive' than tube or flask methods and as a result invalidity of assay together with poorer precision will follow. For this reason the large plate assay of amino acids is generally not possible.

B. Volume and Nature of Sample

Successful assay is always best performed on samples of large volume. The pipetting of 10-ml. aliquots is well known to be less subject to error than that of smaller aliquots and it is and has always been the practice in these laboratories to use large volumes for the preparation of dose level solutions, both of sample and standard. All pipets should be rinsed in the test solution by drawing up sufficient to rinse the interior of the pipet and expelling it to waste before pipetting the chosen aliquot. All flasks should be rinsed in the diluent to be employed. The nature of the sample is often of importance, particularly with clinical or veterinary investigational work. Assays of body fluids such as serum and pleural fluids often require to be carried out on small volumes of material. Similarly, the preparation of standard solutions to be used for the assays require that the active factor be prepared in material similar in nature to the test materials, otherwise variations in zone sizes will occur between sample and standard due to the failure to maintain as a constant the volume of test solutions applied to a plate.

C. Number of Samples Received for Assay

This factor often governs all other considerations and unless control is exercised two main sources of error will occur. A laboratory employing almost exclusively one type of design and size of plate will merely alter the daily work load per assistant to meet the requirement. Gross daily variations should be avoided at all costs and nothing should be allowed to interfere with the normal throughput of work with its known standard error of assay. Personal interest tends to be lost in the face of heavy pressure of work. Work study considerations generally tend to break down the assay procedure whereby each assistant helps in the entire process and again personal interest is lost. A laboratory aware of the problem will have various assay designs and a system whereby acceptable standard errors are indicated by the supplying units. More time can then be given to those assays requiring rather lower standard errors than those requiring rather higher. There is quite obviously no point in assaying, for example, prospective standards under the same conditions as animal feeding stuffs.

I have found that a steady daily flow is most difficult to achieve in a busy pharmaceutical house and yet every effort must be made to achieve this if successful assay at all levels is sought after. It requires a diplomatic approach to assure two analysts that their samples do not necessarily require the same standard error of assay.

D. Approximate Standard Error of Assay Required

This has been largely discussed under the preceding paragraph. Suffice it to say that a work load based on considerations of assay laboratory staff and facilities is more likely to be successful and realistic than one based on production or control conditions.

E. Quality of Training of Assay Laboratory Staff

It has already been stated that the successful assay, whatever the precision, is based very firmly on the ability of staff to develop personal interest, appreciation, and understanding of the techniques. In many laboratories it has always been the practice for all samples to be assayed independently by each of two assistants against independently prepared standards. This procedure has the obvious advantage that consistent unsatisfactory work by any operator is readily detectable. A further advantage is the reporting of the mean of two results and this combined with a test of consistency which will be described later does ensure that a result can be reported confidently and within certain limits of error.

Training of staff is of first rate importance and should again be of personal interest to the microbiologist. New staff should be trained on low precision assays and graduate subsequently to medium and then to high precision assays. The reverse procedure should be applied rigidly to staff producing consistent unsatisfactory high precision assays. It has always been my aim to develop a sense of responsibility in the unqualified staff necessary for routine assays: when I succeed I find that the personal interest taken in the assay technique is in no small extent conducive to a low standard error.

F. Additional Factors Requiring Supervision

The following additional factors require constant supervision: (1) choice of test organism, (2) preparation of bacterial suspensions, (3) preparation of assay, maintenance, and inoculation culture media, (4) preparation and the pouring of assay plates, (5) application of the test solutions to the plate, (6) incubation and reading of plates, (7) computation and reporting of results.

All these factors can exert an influence on an assay and failure to eliminate variation in them will inevitably lead to less satisfactory assay procedures. It is popularly believed that the use of balanced designs eliminates all variables other than that due to the potency of the sample. This is not so and it is fundamental in any successful assay that technical efficiency, personal interest, and balanced design are all interrelated and dependent on each other.

A balanced design is one in which all controllable variables have been accommodated and in respect of large plate methods of assay the following effects must be controlled.

1. Condition of Test Organism

This is less likely to be a variable of one large plate than of many petri dishes, tubes or small plates. Variables due to inadequate washing and sterilizing are not likely to occur.

2. Assay Medium

Clearly, the pouring of say 250 ml. of medium from one bottle into one plate is less likely to introduce variables than multiple pourings. Certainly the seeded bacterial population must be more consistent with removal of such variables as temperature of molten medium, pH of medium and of bottles, moisture content, etc.

3. Depth of Seeded Agar

Use of previously leveled large plates is now accepted as almost entirely removing the variable of depth of agar—for so long recognized as a variable with petri dishes and with liquid medium in test tubes.

4. Time of Application of Test Doses

Length of incubation cycle. Temperature of incubation. Again there is less likely to be a variable evident when a large plate is in use as against stacks of petri dishes or racks of tubes.

It is well to be aware of variables remaining even when large plates are employed. Errors due to drift and to edge and time effects remain in a number of assays particularly of the incomplete or quasi-Latin type. In balanced designs of the true Latin type these latter variables should be eliminated.

Clearly, the use of such designs should not imply automatically that elimination of certain errors has been dealt with. Use of the design is only one part of the process, and unless technical skill, understanding, statistical knowledge, and correct interpretation of results are also applied, no design, no matter how balanced, will produce the required result. It is most necessary to apply all these skills and understandings at one and the same time to maintain the standard of each at all times.

The problem of balanced designs requires special considerations and has occupied a good deal of time in our efforts to establish sound assay procedures. In this respect the work of Lees and Tootill has led to a greater knowledge of the statistical needs and requirements of a microbiological assay. Many of the designs employed some years ago have now been abandoned and replaced by more practical ones. In this respect the assistance of Daly[11] expressed in many personal communications has been of particular significance and the development of high precision assays in two separate laboratories is proceeding along lines indicated by him. The degree of technical efficiency required to undertake some of these assays has not yet been achieved and will obviously take some time. Valuable experience in the assay techniques and particularly in the assay design and analysis is being gained by those involved. I do not in any way accept the principle that staff trained only in assays of low or medium precision should undertake the high precision assays. This simply leads in practice to a greater number of assays being rejected after statistical analysis. I prefer the approach of training selected staff for this work and attempt to develop a sense of responsibility and appreciation of the differences between the various assays. It is not considered advisable on economic grounds to have staff assigned to one type of assay only and I prefer to have all staff trained to undertake medium precision assays and to select those few assistants deemed capable of the higher precision work, as and when it is required. The microbiologist should develop sufficient knowledge of statistics to enable him to appreciate which factors cause invalidity of assay and also to have a working knowledge of the errors involved in the use of different assay designs. Preparation of assay design, work sheets, and preliminary analysis is generally outside the scope of most microbiologists and the advice of a competent and interested statistician should be employed. Once the initial stages of development are complete the statistician's advice is less frequently required.

III. Assay Designs

A. Latin Square

Numerous assay designs have been employed in many laboratories and all have their place under certain conditions. It is not considered neces-

[11] C. Daly, Personal communications (1959–1961).

sary to discuss all of these but only those that are used fairly frequently or could be employed under certain conditions by the author.

1. 4×4 (2 + 2) Latin Square Design

This is used for the assay of one sample and one standard at two levels each and allows four zones for each level in a true Latin square design where each treatment appears once in every row and in every column. The plate and design size is $7\frac{1}{4} \times 7\frac{1}{4}$ inches and the 95% limits for the average of two such assays is regarded as ±9%.

2. 6×6 (3 + 3) Latin Square Design

This is used for the assay of one sample and one standard at three levels each and allows six zones for each level in a true Latin square design. The plate and design size is 9×9 inches and the 95% limits for the average of two such assays is regarded as ±5%. This assay is of particular value in checking the dose/response slopes of new assays, products, or formulations since it includes checks for linearity and parallelism.

3. 6×6 (2 + 2) Latin Square Design

This is used for the assay of one sample and two standards or two samples and one standard at two levels each and allows six zones for each level in a true Latin square design. The plate and design size is 9×9 inches and the 95% limits for the average of two such assays is regarded as ±7%. This assay is carried out on the same plates, designs, etc., as for the (3 + 3) assay above. The two dose level assay, although carried on a true Latin square design and including a complete check on parallelism of two straight lines, does not allow for a full check of parallelism with reference to two lines having different curvature.

4. 8×8 (2 + 2) Latin Square Design

This is used for the assay of two samples and two standards or three samples and one standard at two levels each and allows eight zones for each level in a true Latin square design. The plate and design size is 12×12 inches and the 95% limits for the average of two such assays is regarded as ±3%. This assay is carried out on the same plates as that previously described. It is particularly suitable for the assay of 2 or 3 samples having the same nature, e.g., storage samples, and where strict comparisons of potency are required. The improved precision makes this design a useful step between the medium and high precision assays.

There are four types of 8×8 Latin square designs used for higher precision assays. Table I shows the nature of these and the tests of validity which are incorporated in each.

TABLE I[a]

	Dose levels	Test of validity	
1. S_1, S_2, versus T_1, T_2	2	Weighings:	parallelism
2. S_1, S_2 versus T, U	2	Weighings of standard:	parallelism
3. S versus T, U, V	2	Parallelism	
4. S versus T	4	Linearity:	parallelism

[a] The notation is probably familiar; S_1, S_2 refers to two "weighings" (or dilutions, etc.) of the standard.

These are much simpler than the quasi-Latin square designs to compute and complete tests of validity are easily carried out using the statistical technique of analysis of variance.

5. 8 × 8 (4 + 4) Latin Square Design

This is used for the assay of one sample and one standard at four levels each and employs eight zones for each level in a true Latin square design. The plate and design size is 12 × 12 inches and the 95% limits for the average of two such assays is regarded as ±2%. This assay is carried out on the same plates and designs as for the (2 + 2) assay described above. This assay has been selected as that most likely to be successful as a "high precision" assay. The size of plate, type of design, and certain technical data are exactly as for the routine medium precision assays. The (4 + 4) assay allows for excellent checks for linearity and parallelism. Statistical analysis is not too involved and can be carried out by assay staff with a minimum of training. It is with this assay that much experimental work and staff training is at present being carried out and while much remains to be done before it can be regarded as a satisfactory "high precision" assay, it would appear that it offers most hope of achieving that ideal. This assay has provided an excellent example of my belief that extra training is essential and that the assay design itself yields the additional precision. Initially, the majority of assays of this type were invalid for various reasons (mainly nonparallelism), and it was soon obvious that additional strict supervision of plate-pouring techniques, weighing, dilution, plating out, and plate reading procedures were required. These, too, take time to develop in staff trained previously only in medium precision techniques. It is, however, anticipated that the additional supervision, skill, and added personal interest and appreciation will produce a majority of valid high precision assays. I would emphasize that I was under no illusion regarding the long term required before such assays can be confidently employed. A typical design is given in Fig.

1. The work sheets for a penicillin sample assayed according to design number 1 are given in Figs. 2 and 3. The dilution used in computing D (Fig. 3) is that expected to be 1 unit/ml. and is 180,000 in this example.

LATIN SQUARE DESIGN 1								QUASI-LATIN SQUARE DESIGN 16							
3	4	8	1	7	6	5	2	5	1	3	7	14	10	12	16
2	7	3	8	5	1	4	6	8	4	2	6	9	13	15	11
8	1	5	2	4	3	6	7	14	7	5	16	12	1	3	10
4	2	1	6	3	5	7	8	2	11	9	4	15	6	8	13
7	8	2	5	6	4	3	1	3	10	12	1	5	16	14	7
6	5	7	3	8	2	1	4	15	6	8	13	2	11	9	4
1	3	6	4	2	7	8	5	12	16	14	10	3	7	5	1
5	6	4	7	1	8	2	3	9	13	15	11	8	4	2	6

FIG. 1. Typical Latin square and quasi-Latin square designs.

6. 9 × 9 (3 + 3) Latin Square Design

This is used for the assay of two samples and one standard or one sample and two standards at three levels each and allow nine zones for each level in a true Latin square design. The plate size and design is as for the 8 × 8 types. This assay has not been much used and has been now replaced with the 6 × 6 assay previously described. The assay can also be used to accommodate eight samples and one standard using a quasi-Latin square design, but again little evidence to support use of this assay is available.

7. 12 × 12 (3 + 3) Latin Square Design

This is used for the assay of two samples and two standards at three levels each and allows 12 zones for each level in a true Latin square design. The plate size and design is 18 × 18 inches and the 95% limits for the average of two such assays is regarded as ±2%. This assay is the "high precision" assay used by some laboratories and is that originally recommended by Lees and Tootill. I do not prefer this assay to the 8 × 8 (4 + 4) design as plate size and exposure are excessive, an involved statistical analysis is required, and, most important of all, the strain and tedium of plating out the 144 solutions introduces human errors which render statistical analysis most difficult. However, the assay can be satisfactorily employed but is probably best dealt with by staff fully trained and employed on this assay alone and with the analysis being carried out by a statistician.

J. S. SIMPSON

HIGH PRECISION ASSAY

8 x 8 LATIN SQUARE (4 + 4)

Sample No. 441	Type Penicillin	Read by R.E.D.	Computed by J.R.	Date 16.5.61
Plate No. 3	Diluted by YB	Definition Good.	Checked by R.E.D.	Rack No. 34
Design No. 1	Plated by Y.B.	Artefacts —	Weight of sample	Factor of standard
			99.6	1.01136

	Low level ∝ 3		∝ 2		∝		High level
Tube Number	S4	4	S3	7	S2	2	S1 5
Standard level	0.5		1.0		2.0		4.0
Tube Number	T4	1	T3	6	T2	8	T1 3
Sample dilution	360,000		180,000		90,000		45,000

	1	2	3	4	5	6	7	8
	17.6	23.5	25.3	18.2	25.7	20.4	21.2	22.8
	17.2	23.2	25.0	17.6	25.6	20.5	21.0	22.1
	18.0	23.1	25.3	18.1	24.7	19.5	20.2	22.5
	17.9	22.5	24.2	18.6	24.7	19.0	20.4	22.3
	16.8	22.3	24.4	17.3	24.7	19.0	19.5	21.9
	16.4	22.2	24.2	17.9	25.0	19.6	19.7	21.7
	16.4	22.0	24.3	17.5	24.4	19.3	19.7	21.2
	17.0	22.8	23.7	17.0	24.6	18.9	19.8	21.5
Treatment total	137.3	181.6	196.4	142.2	199.4	156.2	161.5	176.0

									Rows	
	21.2	20.4	18.2	17.6	25.7	25.3	22.8	23.5	174.7	3,879.27
	20.5	21.0	25.0	17.6	22.1	23.2	25.6	17.2	172.2	3,773.86
	25.3	23.1	22.5	24.7	19.5	20.2	18.0	18.1	171.4	3,729.94
	17.9	24.2	24.7	22.5	20.4	18.6	19.0	22.3	169.6	3,642.80
	21.9	16.8	22.3	24.4	17.3	19.0	19.5	24.7	165.9	3,505.13
	17.9	25.0	19.6	21.7	22.2	16.4	24.2	19.7	166.7	3,535.99
	24.4	17.5	19.7	19.3	16.4	21.2	22.0	24.3	164.8	3,455.08
	22.8	21.5	17.0	19.8	23.7	24.6	17.0	18.9	165.3	3,476.19
Columns	171.9	169.5	169.0	167.6	167.3	168.5	168.1	168.7	1350.6	28,998.26

Fig. 2. Laboratory work sheet of Latin square design No. 1 for assay of one sample of penicillin. The zone diameters in the upper table were abstracted from the lower table. The test solution numbers are given at the top of the table. In lower table of zone diameters, the zone diameters correspond to the test solution numbers given in design 1 (Fig. 1).

Preparations	Dose levels	Treatment totals	i→ (1)	(2)	(3)	(4)	(5)	(6)	(7)
S	1	199·4	—1	3	1	1	3	—1	1
	2	181·6	—1	1	—1	—3	1	1	—3
	3	161·5	—1	—1	—1	3	—1	1	3
	4	142·2	—1	—3	1	—1	—3	—1	—1
T	1	196·4	1	3	1	1	—3	1	—1
	2	176·0	1	1	—1	—3	—1	—1	3
	3	156·2	1	—1	—1	3	1	—1	—3
	4	137·3	1	—3	1	—1	3	1	1
$Li \equiv$ Sum of products			—18·8	388·8	0	—3·4	5·4	3·0	—2·8
$di \equiv$ Divisor			64	320	64	320	320	64	320
$Li^2/di \equiv$ Sum of squares			5·5225	472·3920	0	0·0361	0·0911	0·1406	0·0245

L_1 Difference between preparations

L_2 Slope

L_3 Quadratic curvature

L_4 Cubic curvature

L_5 Difference in slope (i.e. lack of parallelism)

L_6 Difference in quadratic curvature

L_7 Difference in cubic curvature

$$C = \text{Correction factor} = \frac{(\text{total of all zones})^2}{64} = \frac{1350 \cdot 6}{64} = 28{,}501 \cdot 8806$$

$$S_R = \frac{\text{Sum of squares of rows}}{8} - C = \frac{228{,}109 \cdot 88}{8} - 28{,}501 \cdot 8806 = 11 \cdot 8544$$

$$S_C = \frac{\text{Sum of squares of columns}}{8} - C = \frac{228{,}029 \cdot 46}{8} - 28{,}501 \cdot 8806 = 1 \cdot 8019$$

$$S_T = \frac{\text{Sum of squares of treatments}}{8} - C = \frac{231{,}840 \cdot 7}{8} - 28{,}501 \cdot 8806 = 478 \cdot 2069$$

$$S = \text{Sum of squares of all zones} - C = 28{,}998 \cdot 26 - 28{,}501 \cdot 8806 = 496 \cdot 3794$$

$$s_2 = \text{Residual mean square} = \frac{S - (S_R + S_C + S_T)}{42} = \frac{4 \cdot 5162}{42} = 0 \cdot 1075$$

$$M = 1 \cdot 5052 \times \frac{L_1}{L_2} = -0.07287 = \bar{1}.92713$$

$$D = \text{Antilog } M \times \text{Dilution} = 0.8454 \times 180{,}000 = 152{,}172$$

$$P = \frac{D \times \text{Factor of Standard}}{\text{Weight of sample}} = 1527.83$$

$$SE(M) = \frac{12 \cdot 04s}{L_2} = 0.0102$$

$$SE(P) = 2 \cdot 3026 \times SE(M) \times P = 36.1756$$

The 95% limits of P are given by $\pm 2 \times SE(P) \pm 72.3512$

FIG. 3. Continuation of the work sheet of Fig. 2 showing the method of computation and information obtainable by this method of assay.

B. Quasi-Latin Square

8 × 8 (2 + 2) Quasi-Latin Square Design

This is used for the assay of six samples and two standards at two levels each and allows four zones for each level in an incomplete Latin square design where each treatment appears in four of the eight rows and four of the eight columns. The plate and design size is 12 × 12 inches and the 95% limits for the average of two such assays is regarded as ±7%. This widely used assay forms the routine procedure for medium precision assay.

The daily number of samples received for assay dictate that true Latin squares cannot be employed and this assay has been accepted as a reasonable compromise having regard to the number of samples, staff, facilities available, and acceptable standard error. Trained operators can each deal with four plates daily and with a staff of 10 assistants the assay in duplicate of over 100 samples daily is possible. It is difficult to be dogmatic about throughput as so much depends on the amount of work required and such points as the nature of the material for assay may exert a marked influence on throughput. The figure of 100 samples in duplicate for 10 assistants daily is in the light of the work of one laboratory. However, it is important not to overemphasize high throughput as it is now readily accepted that the natural consequence of tedium and routine practice with no diversions often brings a false sense of efficiency. The human body generally reacts against routine. Tedium and precision make poor companions. The necessity for high throughput of samples on a microbiological assay has now been reduced by the development of reliable chemicophysical methods for many of the antibiotics.

C. Incomplete Block

1. 11 × 5 (3 + 1) Incomplete Block Design

This is used for the assay of eight samples at one level each and one standard at three levels. The plate size and design is 13 × 7 inches and the 95% limits for the average of two such assays is regarded as ±9%. This assay is one of low precision but is frequently used. It is particularly suitable for samples of low potency or low volume or both and where one level only can be employed. The use of a three-level standard enables a check for linearity to be carried out, but the method of potency determination is mainly graphical.

2. 13 × 4 (3 + 1) Incomplete Block Design

This is used for the assay of 10 samples at one level each and one standard at three levels. The plate size and design is 15 × 6 inches and the 95% limits for the average of two such assays is regarded as ±11%. This assay is not preferred to the 11 × 5 design and has been used only when a large number of samples justifies the rather poor precision. Both assays have been used for the assay of antibiotics in clinical material and particularly, in blood serum samples from animals. Low potency animal feeding stuffs may also be suitable for these assays, but whenever possible the 8 × 8 (2 + 2) quasi-Latin square assay is employed and is preferred.

IV. Microbiological Assay of Penicillin

To illustrate the techniques employed in microbiological assays by large plate methods, those developed for benzyl penicillin and riboflavin and using the 8 × 8 quasi-Latin square design (moderate precision) are detailed.

A. Apparatus Required

(a) Water bath operating at 50°C (±2°C), fitted with a mechanical stirrer and of sufficient depth to ensure that the water level is above that of the medium in the bottles.

(b) Autoclave operating at 15 lb./sq. inch.

(c) Hard steel punches with a diameter of 8.5 mm. for punching agar cups.

(d) Sterile dissecting needles.

(e) Dropping pipets fitted with rubber teats—specially designed having a platinum tube 0.045-inch external and 0.032-inch internal diameter and about ¾ inch long. The platinum tube must not protrude into the glass barrel. The pipets are stored in 70% I.M.S. when not in use.

(f) A guide frame (or semiautomatic punching machine) for punching out the assay cups is desirable although, for a small number of plates, careful hand punching is adequate. Precautions must be taken to ensure that all cups are punched vertically. Suction methods to remove the agar disks are not preferred to hand removal with sterile dissecting needles.

(g) Leveling screws and spirit levels. These are used to level plates accurately before pouring of the assay medium.

(h) Large plates. Plate glass ¼ inch thick, free from distortion, and purchased with rounded edges and corners, should be used for the base of the plates. With four pieces of plate glass ½ inch wide and ¼ inch thick, make a frame for the surface of the plate. Alternatively, aluminium frames 12 inches square by ⅜ and ½ inch may be used. Fix these to the base of the plate with a suitable adhesive taking care to ensure a perfect seal between the frame joints and the frame and the base since absorption of antiseptics, etc., may otherwise interfere subsequently.

(i) Plate cover. An aluminium cover can be used with a slot designed to expose one row of holes only. This keeps most of the plate covered while the agar disks are being removed and while plating is carried out, and helps to prevent the formation of artefacts. A convenient size is made from sheet aluminium 13 inches wide by 22 inches long. Edges of ⅜ inch are turned over at each side to fit over the plate and a slit 1 inch wide is cut across the middle to ¾ inch from each edge.

In laboratories where contamination of plates by antibiotic dust is no problem, plain sheet metal or glass plate lids may be used.

(j) Assay designs. These are employed to ensure that all sample and standard solutions are applied to the plates in a randomized manner, and to eliminate errors due to drift, to edge and time effects, variation in depth of agar, and in the volume of seeding inoculum. Thus, estimates of potency are normally obtained which are unbiased by these various effects, leaving only a small residual error inherent in microbiological assay due to imperfect replication of the zone diameters. The number of designs should be sufficient to ensure that each is used once daily only by all available operators. Different designs must be used for any replicate determinations on a given sample or standard.

(k) The following notes give some information on the types and sources of supply of some of the apparatus and materials used.

Opacity tubes supplied by Burroughs Wellcome Limited.

Test organism from the National Collection of Industrial Bacteria.

Plating-out pipets are prepared locally as required.

Plate glass and strips may be obtained from Messrs. Pilkington, St. Helens, Lancashire, England.

Aluminium frames may be obtained from Messrs. Bulpitt & Sons Limited, Swansea Works, Birmingham, England.

Plate lids should be of sheet aluminium gage 13 or 14.

Adhesives. The following are suitable:

(i) Titebond Adhesive supplied by Messrs. Surridges Limited, Beckenham, Kent.

(ii) Araldite Adhesive 103 and Hardener 951 supplied by Aero Research Limited, Duxford, Cambridge.

Calipers. Needle point calipers reading to 0.1 mm are necessary. Chesterman No. 1075 calibrated in millimeters and modified by grinding down to give needle points have been found to be suitable. One pair should be allocated to each operator.

Leveling Screws.

Assay Designs and Work Sheets—prepared as required. All designs should be prepared only with qualified statistical assistance.

Reading boxes on projection apparatus.

B. Preparation of Assay Plates

(a) Check the temperature and operation of all water baths, etc.

(b) Melt the required number of bottles of assay medium by steam heat and transfer to a water bath maintained at 50°C (±2°).

(c) Place the plates on a level bench, or, if no such bench is available, level each plate by placing it on three leveling screws and adjusting these with the aid of spirit levels placed at right angles on the surface of the plate. Leave the leveling screws undisturbed until after the plate has been poured and the medium has set.

(d) Swab each plate with alcohol (70% I.M.S.) or with alcohol (70% I.M.S.) containing 4% hydrochloric acid, then with ether, dry in an incubator, and flame the surface of the plate and the under surface of the lid.

(e) Allow the plates to stand for a few minutes and repeat the last process. Replace the lid.

(f) Inoculate each bottle of cooled assay medium with well-shaken spore suspension of *Bacillus subtilis* (NC 1 B 8535) which may be prepared as in Chapter 6.8, section VC. The inoculum will vary with each batch of suspension but generally 0.3 ml. of a $\frac{1}{10}$ dilution for 225 ml. of medium is adequate. Mix thoroughly, taking care not to form air bubbles.

(g) Taking precautions to keep the plate sterile, pour the medium into each plate using a circular motion to ensure the even distribution of medium over the plate. Remove any air bubbles by quick application of a Bunsen flame but avoid excessive flaming as this gives the medium a streaky appearance. Care should be taken to ensure that medium is not allowed to fall on the frame of the plate as capillary attraction thus caused will produce irregular depth of assay medium. Rubber bungs may be used to raise the lids until the medium has solidified.

(h) Allow the poured medium in each plate to set with the plate still remaining on the leveling screws. Replace the lid.

(i) Transfer each plate, when set, to a refrigerator at +4°C. and leave inverted until required for punching out the cups. To avoid excessive condensation, the plates should not be placed in close proximity to the freezing chamber.

(j) Place the plate over the design to be used and, using a guide frame, cut out the required number of cups with a hard steel punch so that each cup appears directly over a number of the design.

(k) Using a sterilized dissecting needle and taking care not to damage the cups, remove the disks of agar and place them in a closed container until sterilized when they may be discarded.

Replace the flamed aluminium lids and place the prepared plates in an inverted position in a refrigerator until required.

C. Preparation of Standard Solution

The internal working standard is a crystalline preparation of sodium benzyl penicillin of low moisture content which has been standardized against the British National Standard or the International Standard or both. Each sealed ampoule should clearly indicate the batch number, potency, and date of preparation. All flasks, bungs, distilled water, and buffer solution used in the preparation of sample or standard solutions for assay should be sterile.

Transfer approximately 150 mg. of the working standard to a clean, dry weighing pot. Dry this, with a second dry weighing pot, at 100°C for 3 hours. Allow to cool in a small desiccator. Weigh the empty weighing pot. Transfer the dried material to the tared weighing pot and re-weigh. Add a few milliliters of buffer solution to the weighing pot. Transfer this with the aid of a funnel to a 250-ml. Grade A volumetric flask. Rinse the weighing pot several times with phosphate buffer solution, add the washings to the flask, and then add 0.5 ml. of chloroform. Dilute to volume with buffer solution. Stopper and mix well. Calculate the strength of the solution using the dry weight of the standard. Label the flask with details of potency, date of preparation, standard number, etc. Store at +4°C. This solution should not be used after 2 days. Remove approximately 50 ml. from this solution daily and allow it to stand at room temperature for 2 hours before use. Dilute 20.0 ml. of this solution in a 200-ml. volumetric flask with phosphate buffer and from this prepare daily working standards containing approximately 1.0 and 2.0 units/ml. by diluting 10.0 ml. to 500 ml. and 1000 ml., respectively, with phosphate buffer. (A solution containing approximately 0.5 unit/ml. may be required for low potency samples, checks for parallelism or high precision assays.) From the actual strength of the solution calculate a "factor" by which the strength is removed from 1.0 and 2.0 units/ml. In certain circumstances dose levels of 1.0 and 4.0 units/ml. may be required.

D. Preparation of Sample Solutions for Assay

1. Solids

Accurately weigh about 100 mg. of the sample in a clean, dry weighing pot. Transfer the sample to a volumetric flask of suitable size with buffer solution and dilute to volume. Insert a bung and mix the contents thoroughly. Calculate the approximate potency of the solution in units per milliliter from the stated potency of the sample, and prepare further dilutions containing about 1.0 and 2.0 units/ml. in graduated flasks, using phosphate buffer as diluent.

Note: (i) Three weighings by different operators are normally employed for each sample and the potency of the sample is given as the mean of all valid observations.

(ii) With stable solids and when sufficient time is available, it is an advantage to carry out duplicate assays on successive days.

(iii) If the solid material is difficult to dissolve in phosphate buffer, a small amount (10 ml.) of a suitable solvent should be employed to ensure solution of the material before dilution in phosphate buffer.

2. Water-Miscible Solutions

Pipet a suitable volume, preferably 20 ml. and not less than 5 ml., into a graduated flask, dilute to volume with phosphate buffer, and mix well. From this solution, prepare dilutions containing 1.0 and 2.0 units/ml. in phosphate buffer. Insert bungs into each flask and mix thoroughly.

Note: Duplicate assays on one day by separate workers are usually sufficient. The result is expressed as the mean of the two valid observations.

3. Lozenges and Tablets

Allow at least five lozenges or tablets, if available, to dissolve in approximately 500 ml. of phosphate buffer solution and then dilute to 1 liter with the buffer solution. From this prepare dilutions in phosphate buffer to contain about 1.0 and 2.0 units/ml. Insert bungs into each flask and mix thoroughly.

Note: Duplicate assays from the same primary solution on one day by different workers are usually sufficient. The result is expressed as the mean of the two valid observations.

After dilution, transfer the solutions to dry test tubes and place these in position in the assay racks. For example, 16 solutions are required for an 8×8 quasi-Latin square and it is convenient to place the 8 strong solutions in the upper tier numbered 1, 3, 5, . . . , 15, and the 8 weak solutions in the lower tier numbered 2, 4, 6, . . . , 16, with the position of the standard at random.

E. Plating-Out of Standard Controls and Sample Solutions Using an 8×8 Quasi-Latin Square

Plating-out should be commenced as soon as possible after dilution and on no account should a rack of sample solutions be allowed to stand without giving the standard solutions the same treatment.

Select a design and stand the assay plate on it so that the randomized solution numbers appear through the cups punched in the plate. Mark the top of the plate clearly and enter the plate number, design number, and details of the assay on a work sheet.

Remove the dropping pipet from its container of alcohol (70%

I.M.S.), rinse it thoroughly in phosphate buffer, and dry the outside of the pipet barrel.

Rinse the pipet 3 times in the test solution and introduce 4 drops into the appropriate assay cup using the plate cover if desired. Discard the solution remaining in the pipet and rinse 3 times with buffer. Avoid immersing the glass barrel of the pipet in either the buffer or sample solution, otherwise variations of drop size will occur.

Similarly, plate out all the solutions in strict row and column order starting at row 1, working from left to right and taking care, if no plate cover is being used, to avoid splashing the plate while rinsing the pipet. Once started, plating-out should not be interrupted as it is important that the solutions are placed in the cups at regular time intervals. Replace the aluminium lid as soon as plating-out is complete.

Carefully transfer the plates to an incubator at 30°C, avoiding any overflow of the penicillin solution from the cups, as this will result in irregular response zones. Then incubate overnight.

F. Measurement of Zone Diameters

Remove the plates from the incubator and mount them in the reading boxes with the top of the plates in their correct positions. Beginning with

FIG. 4. Measuring the diameter of zones using an illuminated reading box and Chesterman vernier scale calipers.

the top left-hand zone, measure the diameters of the zones on the surface of the medium to the nearest 0.1 mm. with needle-point calipers (Fig. 4). Dictate these figures to a second operator for entry on the appropriate work sheet.

Alternatively, for routine work a projection method may be used which will give results to 0.25 mm. and more recently to 0.1 mm.

If possible, in order to avoid personal bias, the person reading or recording the plate should not have prepared the plate concerned on the previous day and should not be in a position to see the design or assay sheet relating to the plate being read.

G. Estimation of Sample Potencies by 8 × 8 (2 + 2) Quasi-Latin Design

The computation sheet for this example is given in Fig. 5.

(1) Sum all high doses for each sample.

(2) Sum all low doses for each sample.

(3) Calculate the "treatment total" by summing the total of high and low doses for each sample.

(4) Subtract the treatment total for the standard solution from the treatment total for the sample. This will give "$\pm D$."

(5) Sum the totals of all the sums of the high doses.

(6) Sum the totals of all the sums of the low doses.

(7) Subtract the total low doses from the total high doses $= B$.

(8) Determine $4/B$.

(9) Multiply $4/B$ by the log of the dose ratio $= C$.

(10) Multiply $D \times C = M$.

(11) Antilog $M = P$.

(12) Multiply P by the dilution employed to yield 1.0 unit/ml. and then multiply by the factor of the working standard to obtain sample potency. The use of a large carefully prepared nomogram may be employed where a large number of routine assays are involved.

H. Validity of the Assay

The above example of computation of potencies is that followed when validity is not in doubt. Tests for validity are illustrated in the following example taken from another laboratory. Sixteen solutions made up of eight preparations each at the two dose levels of 4 units/ml. and 1 unit/ml. are required. These are given numbers 1 to 16, where 1 and 2 refer to the high and low doses of preparation A, 3 and 4 refer to the high and low doses of preparation B, etc. One of these preparations is a standard or control of known potency. Which of 16 solutions goes into which of 64 cups is determined beforehand from a given design. A number of designs of which the following is typical are at present available:

MICROBIOLOGICAL ASSAY LABORATORY, GREENFORD

8 x 8 QUASI LATIN SQUARE ASSAY

Rack No: 2

Sample No.	3	8	10	11	12	13	N.P. 30	C.S.
							STANDARDS	
Dilutions Primary	1 100,000	3 250,000	5 600	7 600	9 600	11 600	13 2·0 units/ml	15
Secondary	2 200,000	4 500,000	6 1200	8 1200	10 1200	12 1200	14 1·0 unit/ml	16
Weight							Factor 0.91566	

Date	17. 5. 1961	Read by	Y.B.
Type	Penicillin on B. subtilis	Computed by	R.E.D.
Plate No.	4	Design used	X̄
Diluted by	J.R.	Checked	J.R.
Plated by	J.R.	Factor of Standard	0.89168

Zone Definition	Good.
Artefacts	1

Sample No.	3		8		10		11		12		13		N.P.		C.S.		Slope Total	
Slope	11·8		8·6		9·8		10·7		10·5		10·1		9·2		9·8		80·5	= B
	1	2	3	4	5	6	7	8	9	10	11	12	13	14	15	16	Totals of high and low dilutions	Difference = B
	23·5	20·6	23·6	21·1	23·8	21·3	23·7	21·3	23·5	20·8	23·5	20·3	23·5	21·5	23·5	20·6		
	23·8	20·4	23·5	21·4	23·7	21·3	23·7	21·3	23·6	20·7	23·6	21·1	23·4	21·1	22·8	21·1		
	23·4	20·6	23·2	21·2	23·8	21·6	23·8	20·3	23·5	20·7	23·4	21·4	23·3	20·8	23·2	20·8		
	23·2	20·5	23·6	21·6	23·7	21·0	23·5	21·1	23·5	21·4	23·3	20·9	23·0	20·6	23·5	20·7		
Sum high dose →	93·9		93·9		95·0		94·7		94·1		93·8		93·2		93·0		751·6	80·5
Sum low dose →		82·1		85·3		85·2		84·0		83·6		83·7		840		83·2	671·1	
Treatment total →	176·0		179·2		180·2		178·7		177·7		177·5		177·2		176·2		$\frac{4}{B}$ × Log. Dose Ratio = C	
Sample treatment total − Standard treatment total = D →	−0·2		+3·0		+4·0		+2·5		+1·5		+1·3		+1·0		S			
D x C →	$\bar{1}$·997		0·0449		0·0598		0·0374		0·0224		0·0194		0·015				0·01495	
Antilog →	0·9931		1·109		1·148		1·09		1·053		1·046		1·035					
Antilog x Dilution x (Standard level = E) →	198,620		554,500		1,378		1,308		1,264		1,255		1·035					
E x Factor of Standard = Sample Potency →	177,105		494,437		1,229		1,166		1,127		1,119		1·008					

FIG. 5. Laboratory work sheet for an 8 × 8 quasi-Latin square design accommodating six samples and two standards at two levels each.

11	8	13	3	10	5	16	2
4	12	6	15	7	9	1	14
16	3	10	2	11	8	13	5
7	15	1	14	6	12	4	9
13	5	11	8	16	2	10	3
6	9	4	12	1	14	7	15
10	2	16	5	13	3	11	8
1	14	7	9	4	15	6	12

Eventually "zones of inhibition" are formed round these cups and their diameters are measured.

The following analysis is then carried out on these zone diameters to determine validity of the assay and residual mean square. Let

$$SS_T = x_1^2 + x_2^2 + \ldots + x_{64}^2 - \tfrac{1}{64}(T)^2$$
$$SS_R = \tfrac{1}{8}(r_1^2 + r_2^2 + \ldots + r_8^2) - \tfrac{1}{64}(T)^2$$
$$SS_C = \tfrac{1}{8}(c_1^2 + c_2^2 + \ldots + c_8^2) - \tfrac{1}{64}(T)^2$$
$$SS_P = \tfrac{1}{8}(p_1^2 + p_2^2 + \ldots + p_8^2) - \tfrac{1}{64}(T)^2$$
$$SS_D = \tfrac{1}{4}(d_1^2 + d_2^2 + \ldots + d_{16}^2) - \tfrac{1}{64}(T)^2$$
$$SS_{R/P} = \tfrac{1}{64}(r_O - r_E)^2$$
$$SS_{C/P} = \tfrac{1}{64}(c_O - c_E)^2$$
$$SS_S = \tfrac{1}{64}(d_H - d_L)^2$$

Where the x's are the 64 zone diameters: the r's, c's, and p's are the totals for rows, columns, and preparations, respectively: T is the grand total; the d's are the totals for the 16 solutions; r_O and r_E are totals for odd and even rows, c_O and c_E are totals for odd and even columns; d_H and d_L are totals for the high and low dose levels; p_s refers to the standards. Then calculate:

Deviation from parallelism mean square:

$$(S_p^2) = \tfrac{1}{5}[SS_D - (SS_P + SS_S + SS_{R/P} + SS_{C/P})]$$

Residual mean square:

$$(s^2) = \tfrac{1}{36}[SS_T + SS_{R/P} + SS_{C/P} - (SS_R + SS_C + SS_D)]$$

$$\text{Let } F = S_p^2/s^2.$$

If $F < 2.5$, the assay is valid.

Estimation of potencies:

The seven (six in the example given) samples are compared with the standard and expressed as percentages. These percentage potencies are $R_1, R_2, \ldots, R_7$, and are calculated from the formulas

$$R_1 = 100 \text{ antilog } M_1$$
$$R_2 = 100 \text{ antilog } M_2,$$

etc., where

$$M_1 = 2.4084 \frac{p_1 - p_s}{d_H - d_L} \quad \text{or} \quad 2.4084 \frac{p_1 - \bar{p}_s}{d_H - d_L}$$

if two standards are used,

$$M_2 = 2.4084 \frac{p_2 - p_s}{d_H - d_L} \quad \text{or} \quad 2.4084 \frac{p_2 - \overline{p}_s}{d_H - d_L}, \text{etc.}$$

The standard errors of the potencies are given by:

(a) $277.28 \sqrt{s^2/SS_s}$ % . . . where one standard is used

(b) $240.14 \sqrt{s^2/SS_s}$ % . . . where two standards are used as in the numerical example.

The 95% limits can be taken as approximately twice the standard errors.

$$C = \frac{T^2}{64} = \frac{(1308.5)^2}{64} = 26{,}752.6914$$

$$SS_T = (22.5)^2 + (18.5)^2 + \ldots + (19)^2 - C = 209.0586$$
$$SS_R = \tfrac{1}{8}[(163.00)^2 + (163.00)^2 + \ldots + (164.25)^2] - C = 0.3086$$
$$SS_C = \tfrac{1}{8}[(163.75)^2 + (163.50)^2 + \ldots + (163.50)^2] - C = 0.2148$$
$$SS_P = \tfrac{1}{8}[(160.25)^2 + (164.75)^2 + \ldots + (162)^2] - C = 3.8086$$
$$SS_D = \tfrac{1}{4}[(87.25)^2 + (73)^2 + \ldots + (74)^2] - C = 206.9961$$
$$SS_{R/P} = \tfrac{1}{64}(655.00 - 653.50)^2 = 0.0352$$
$$SS_{C/P} = \tfrac{1}{64}(655.25 - 653.25)^2 = 0.0625$$
$$SS_s = \tfrac{1}{64}(711.25 - 597.25)^2 = 203.0625$$

$$M_1 = 2.4084 \left(\frac{-1.75}{114.00}\right) = -0.036971 \quad \therefore \quad R_1 = 91.83\%$$

$$M_2 = 2.4084 \left(\frac{2.75}{114.00}\right) = 0.058097 \quad \therefore \quad R_2 = 114.3\%$$

$$M_3 = 2.4084 \left(\frac{4.00}{114.00}\right) = 0.084505 \quad \therefore \quad R_3 = 121.4\%$$

$$M_4 = 2.4084 \left(\frac{0.75}{114.00}\right) = 0.015845 \quad \therefore \quad R_4 = 103.7\%$$

$$M_5 = 2.4084 \left(\frac{3.25}{114.00}\right) = 0.068661 \quad \therefore \quad R_5 = 117.1\%$$

$$M_6 = 2.4084 \left(\frac{3.50}{114.00}\right) = 0.073942 \quad \therefore \quad R_6 = 118.5\%.$$

We calculate $S_p^2 = 0.00546$; $s^2 = 0.04547$.

The required standard errors are given by

$$\text{S.E.} = 240.14 \sqrt{\frac{0.04547}{203.0625}} \%$$

$$= 3.6\%.$$

The required limits are thus given by

$$R_1 \ldots \pm 7.2\% \text{ of } 91.8\% = \pm 6.6\%$$
$$R_2 \ldots \pm 7.2\% \text{ of } 114.3\% = \pm 8.2\%$$
$$R_3 \ldots \pm 7.2\% \text{ of } 121.4\% = \pm 8.7\%$$
$$R_4 \ldots \pm 7.2\% \text{ of } 103.7\% = \pm 7.5\%$$
$$R_5 \ldots \pm 7.2\% \text{ of } 117.1\% = \pm 8.4\%$$
$$R_6 \ldots \pm 7.2\% \text{ of } 118.5\% = \pm 8.5\%.$$

I. Reporting of Routine Assay Results from Several Designs

It has been shown that the 8×8 quasi-Latin square is capable of yielding a standard error of 5% or slightly less.

On the basis of this error, the maximum permissible range of a number of replicate determinations is determined solely by the number of replicates and the acceptance of a definite probability figure, usually 0.95.

Accordingly, the following tabulation gives the maximum permissible range as a per cent of the mean potency for the number of determinations made.

Number of determinations	2	3	4	5	6	7	8
Range as % of mean value	13.8	16.6	18.2	19.3	20.2	20.9	21.5

Thus, if the observed range for the group of determinations is less than that indicated by the tabulation, the results will in general be mutually consistent, and the mean may be returned with considerable confidence as an unbiased final figure where 95% limits are given by $\pm (10/\sqrt{n})$ % where n is the number of determinations made.

If the observed range is greater than that indicated by the tabulation, another determination should be carried out in duplicate; if the observed range for the total number of observations is less than that indicated by the tabulation, the mean of the results should be returned.

The various approximate errors governing the results are returned as shown in Table II.

V. Microbiological Assay of Riboflavin

A. Design and Application to Large Plate Method

A large plate assay method for assaying seven samples simultaneously and yielding a standard error of $\pm 8\%$ has been developed. Linearity of the response lines is assumed and checks on the accuracy of the two dilutions employed are omitted.

The potency of the sample is determined by comparison with that of an Internal Working Standard. The zones of exhibition produced with the specified test organism are in direct relation to the logarithms of the concentrations of riboflavin in the test solutions.

It is now generally accepted that commercially available dehydrated cultural media are satisfactory for microbiological assay of vitamins,

TABLE II

No. of assays	P (%)
8 × 8 Quasi-Latin square (2 + 2 assay)	
1	0.95 ± 10
2	0.95 ± 7
3	0.95 ± 6
4	0.95 ± 5
5	0.95 ± 4.5
8 × 8 Latin square (2 + 2 assay)	
1	0.95 ± 3
2	0.95 ± 2
3	0.95 ± 1.5
4	0.95 ± 1.0
6 × 6 Latin square (2 + 2 assay)	
1	0.95 ± 7
2	0.95 ± 6
3	0.95 ± 5
4	0.95 ± 4
6 × 6 Latin square (3 + 3 assay)	
1	0.95 ± 5
2	0.95 ± 3.5
3	0.95 ± 3
4	0.95 ± 2.5
12 × 12 Latin square (3 + 3 assay)	
1	0.95 ± 2
2	0.95 ± 1.4
3	0.95 ± 1.2
4	0.95 ± 1

For the assay of solutions containing less than 1.0 μg./ml. of riboflavin, tube assay methods must be employed.

B. Special Apparatus Required

Items as listed in the method previously described for benzyl penicillin together with a centrifuge capable of attaining 3000–5000 r.p.m. are required.

C. Test Organism—*Lactobacillus casei* 773 (N.C.I.B. 8010)

Freeze dried cultures are available. These cultures are reconstituted every 3 months in glucose broth and, if morphologically pure, are inoculated into glucose-yeast-agar stabs. After incubation at 37°C for 18 hours all cultures are kept in the refrigerator. From these "master" cultures further subcultures are prepared as required.

D. Media

1. Glucose-Yeast-Agar Stab Medium

Difco yeast extract	3.0 gm.
Dextrose A.R.	3.0 gm.
Agar agar (Davis)	3.75 gm.
Glass distilled water to	300 ml.

Dissolve the ingredients by autoclaving at 15 lb./sq. inch for 20 minutes, adjust to pH 6.8 (±0.1), and steam for 10 minutes. Filter through paper pulp in a Buchner funnel, and distribute in 10-ml. amounts into suitable containers. Sterilize at 15 lb./sq. inch for 20 minutes and allow to set as stabs. Store at $+4°C$.

2. Double Strength Assay Medium

Photolyzed peptone solution	200 ml.
l-Cystine solution	50 ml.
DL-Tryptophan solution	50 ml.
Dextrose	40 gm.
Xylose	2 gm.
Yeast supplement	40 ml.
Sodium chloride	10 gm.
Sodium acetate (hydrated)	20 gm.
Ammonium sulphate	6 gm.
Adenine, guanine, and uracil solution	20 ml.
Xanthine solution	20 ml.
p-Amino benzoic acid solution	40 ml.
Pyridoxine solution	2 ml.
Ca-d-pantothenate solution	2 ml.
Nicotinic acid solution	2 ml.
Inorganic salt solution A	10 ml.
Inorganic salt solution B	10 ml.
Glass distilled water to	1000 ml.

Filter the medium through Whatman No. 41 filter paper, adjust to pH 6.8 (±0.1), and distribute 115-ml. amounts into 12-oz. screw capped bottles. Sterilize at 4 lb./sq. inch for 30 minutes or preferably, time permitting, by free steam for 20 minutes on each of three successive days.

3. Double Strength Agar Medium

Agar agar (Davis)	25 gm.
Glass distilled water to	1000 ml.

Dissolve the agar by autoclaving, filter through paper pulp, and distribute 115-ml. amounts into 12-oz. screw capped bottles. Sterilize at 15 lb./sq. inch for 20 minutes.

4. Single Strength Inoculum Medium with Riboflavin

Double strength assay medium	500 ml.
Riboflavin solution	20 ml.
Glass distilled water to	1000 ml.

Filter the medium through Whatman No. 41 filter paper, adjust to pH 6.8 (±0.1), and distribute in 100-ml. amounts into 20-oz. screw capped bottles. Sterilize at 4 lb./sq. inch for 30 minutes or preferably, time permitting, by free steam for 20 minutes on each of three successive days.

5. Single Strength Inoculum Medium (without Riboflavin)

Prepare as for the single strength inoculum medium above, omitting the riboflavin. Expose the medium to light, to ensure that it is completely free from riboflavin.

6. Preparation of Stock Solutions

a. Photolyzed peptone solution. Dissolve 40 gm. Difco Bacto-Peptone in glass-distilled water to 250 ml. Dissolve 20 gm. sodium hydroxide in glass-distilled water to 250 ml., mix the two solutions, and allow to stand for 24 hours in the light at room temperature. For at least half of this time expose the solution to the light of a 100-watt lamp at a distance of about 18 inches. Neutralize with glacial acetic acid (about 28 ml.) and add 11.6 gm. of sodium acetate (hydrated). Dilute to 800 ml. with glass-distilled water, filter, and sterilize by free steaming. Store at 4°C. This solution normally keeps for 14 days, but if a precipitate develops it should be discarded.

b. Yeast supplement. Dissolve 100 gm. Difco yeast extract in glass-distilled water to 500 ml. Dissolve 150 gm. basic lead acetate in glass-distilled water to 500 ml., mix the two solutions, and adjust the pH to approximately 10.0 with concentrated ammonium hydroxide solution. Filter and acidify the filtrate to litmus paper with glacial acetic acid. Bubble hydrogen sulphide through until all the lead is precipitated. Filter again and dilute to 1000 ml. with glass-distilled water. Preserve under toluene and keep at +4°C. It is stable for about 3 months.

c. 1-Cystine solution. Add 2.0 gm. of 1-cystine to 50 ml. of glass-distilled water and boil. Add 4 ml. of concentrated hydrochloric acid gradually to the boiling solution. Cool and dilute to 500 ml. with glass-distilled water. Preserve under toluene at +4°C. This solution keeps indefinitely.

d. DL-Tryptophan solution. Add 2.0 gm. DL-tryptophan to a few milliliters of glass-distilled water and boil. Add concentrated hydrochloric acid drop by drop until solution is complete. Cool, dilute to 500 ml. with glass-distilled water, and store at +4°C. This solution keeps indefinitely.

e. Adenine, guanine, and uracil solution. Add 0.1 gm. of each to glass-distilled water, boil, and add concentrated hydrochloric acid drop by drop until the solution is complete. Dilute to 100 ml. This solution keeps for about 14 days at 4°C.

f. Xanthine solution. Dissolve 0.1 gm. of xanthine in a little concentrated ammonia solution and dilute to 100 ml. with glass-distilled water. This solution keeps for about 14 days at 4°C.

g. Calcium-d-pantothenate solution. Dissolve 0.1 gm. of Ca-d-pantothenate in 100 ml. of glass-distilled water. Store at 4°C. This solution keeps for about 14 days at 4°C. Dilute 1 to 100 with glass-distilled water before use.

h. Pyridoxine solution. Dissolve 0.244 gm. of pyridoxine hydrochloride in 100 ml. of glass-distilled water. This solution keeps for about 14 days at 4°C. Dilute 1 to 20 with glass-distilled water before use.

i. p-Amino benzoic acid solution. Dissolve 0.1 gm. of p-amino benzoic acid in 2 ml. of glacial acetic acid and dilute to 100 ml. with glass-distilled water. This solution keeps for about 14 days at 4°C. Dilute 1 to 10 with glass-distilled water before use.

j. Nicotinic acid solution. Dissolve 0.1 gm. of nicotinic acid in 100 ml. of glass-distilled water. This solution keeps for about 14 days at 4°C. Dilute 1 to 10 with glass-distilled water before use.

k. Inorganic salt solution A. Dissolve 25 gm. each of potassium dihydrogen phosphate (KH_2PO_4) and dipotassium hydrogen phosphate (K_2HPO_4) in 250 ml. of glass-distilled water. This solution keeps indefinitely at 4°C.

l. Inorganic salt solution B. Dissolve 10 gm. of magnesium sulphate ($MgSO_4 \cdot 7H_2O$), 0.5 gm. of manganese sulphate ($MnSO_4 \cdot 4H_2O$), and 0.1 gm. of ferric chloride (anhydrous) in 250 ml. of glass-distilled water. Add 5 drops of concentrated hydrochloric acid to prevent precipitation. This solution keeps indefinitely under toluene at 4°C.

m. Riboflavin solution. Dissolve 50 mg. of riboflavin in a little glass-distilled water containing 1 ml. of glacial acetic acid. Dilute to 1000 ml. with glass-distilled water. Store at 4°C. This solution keeps for 14 days.

E. Preparation of Seeding Inoculum

Add 2–3 ml. of sterile normal saline to a fresh stab culture of *Lactobacillus casei*, and suspend the growth by using a sterile wire. Inoculate the suspension into the riboflavin single strength inoculum medium and incubate with the bottle lying flat and with the cap loosened, for 16 to 18 hours at 37°C. Transfer the culture to sterile centrifuge tubes and wash 3 times with sterile saline, centrifuging at about 3000 r.p.m. after each wash. Inoculate the final deposit into one bottle of the riboflavin free inoculum medium and incubate for 16 to 18 hours at 37°C. Wash the culture as before and resuspend the deposit in sufficient sterile saline so that the opacity matches Wellcome opacity tube No. 10. Add 10 ml. of this seeding inoculum to each bottle of assay medium which has been previously melted and cooled to 50°C.

F. Preparation of Plates

Level the plates, previously cleaned by any established potash/acid process, and sterilize as described previously. Melt one bottle of the double strength agar medium by autoclaving at 15 lb./sq. inch for 20 minutes. Place this, together with one bottle of the double strength assay medium, in a water bath at 50°C and allow the temperature of each to come to 50°C. Mix the two bottles of the medium, add 10 ml. of the seeding inoculum, and mix again. Pour the medium into one of the prepared plates and allow to set. Transfer to a refrigerator until required for punching out the cups, then proceed as described previously.

G. Preparation of Standard Solutions

Weigh about 100 mg. of standard riboflavin in a tared, dry weighing bottle. Dry at 100°C for 3 hours, allow to cool in a desiccator and re-weigh. Transfer the weighed material to a sterile 1000-ml. flask, add 5.0 ml. of glacial acetic acid, and dilute to volume with glass-distilled water. Keep at +4°C. A fresh primary standard solution should be made every 14 days. Prepare daily working standards as required by diluting the primary solution to approximately 2.0 and 1.0 μg./ml. Calculate the factor by which these solutions differ from these exact strengths. Protect from light at all times.

H. Preparation of Sample Solutions

Pipet a suitable amount, preferably at least 5 ml., into a graduated flask, dilute to volume with glass-distilled water, and mix well. From this solution, prepare dilutions in glass-distilled water to contain approximately 2.0 and 1.0 μg./ml. Insert dry, sterile rubber bungs and mix thoroughly.

After dilution, transfer the sample and standard dilutions to dry, sterile test tubes and place these in an assay rack. Store this in a dark cupboard until required for plating out.

I. Assay Conditions

1. Incubation: 37°C for 16 to 18 hours.
2. Plating out of standard controls and sample dilutions, see Section IV.E.
3. Measurement of zone diameters, see Section IV.F.
4. Estimation of sample potencies, see Section IV.G.

J. Reporting of Assay Results

On the basis of a 5% standard error for the 8×8 quasi-Latin square, the maximum permissible range of a number of replicate determinations is determined solely by the number of replicates and the acceptance of a definite probability figure, usually 0.95. Accordingly, the following tabulation gives the maximum permissible range as a per cent of the mean potency.

Number of determinations	2	3	4
Range as per cent of mean value	22	27	29

If the observed range is greater than that indicated by the tabulation, another duplicate determination is carried out. If the mean then satisfies the above test, this result is reported.

VI. Interpretation of Results

In the future much more attention will need to be paid to the interpretation of results and in particular to the detection of invalidity in any determination.

Successful assays require an objective method to determine whether an estimate conforms to certain standards of validity and the error to be expected from the results obtained. In this respect the work of Wood,[12] Finney,[13] and Lees and Tootill has been of particular value and, more recently, that of Daly in various communications. Even so, statistical methods are of limited value since validity of assay is no proof of accuracy. It is well known that many factors can interfere with an assay, and yet do not invalidate the result statistically. This is particularly true for antibiotic assays in the presence of a second antibiotic or other antibacterial substance, or in secondary production material and keeping quality samples and where linearity or parallelism of the response of the antibiotic under assay may not necessarily be affected.

Deviations from linearity or parallelism generally indicate invalid assays and care should always be taken to check the nature of all dose/response slopes at intervals. All too frequently an assay of the $(2 + 2)$ or $(11 + 3)$ type is carried out without adequate information to justify its use. Ideally, all new assays and those on unknown materials, new formulations, and the like should be carried out on a $(3 + 3)$ basis and only when parallelism and linearity of the curves for sample and standard have been confirmed should the $(2 + 2)$ assay be used. Even so, certain samples give initial assays that appear valid, but during

[12] E. C. Wood, *Analyst* **78,** 451 (1953).
[13] D. J. Finney, "Statistical Methods in Bio-assay." Griffin, London, 1952.

storage some changes occur which give rise to invalidity. A microbiologist would be most rash, having used $(2 + 2)$ assays only, to deny invalidity of his assays in the face of substantial evidence to the contrary.

Sound experimental work, correct assay design, and statistical analysis are the basis of any successful assay and for them there is no substitute. This aspect is all too frequently ignored or incorrectly applied to the assay design in its early stages. Time spent in assay development and control is often not included and is relegated to a subordinate place in a busy microbiological unit. The emphasis tends to be on throughput of samples. This point has been emphasized in the design of collaborative assays with other laboratories and this subject is now arousing some interest in the United Kingdom.

The advent of large plate methods of assay has emphasized the shortcomings of other methods of assay and a more sound analytical approach is gradually being developed. Examination at all stages is more critical and valid in large than in small plate methods and for this reason they have constituted the main line of approach and I prefer to use them whenever possible.

VII. Sources of Errors

During the past 5 years critical analysis of microbiological assays in many laboratories has detected many possible sources of error and every effort has been made to incorporate improvements in the large plate methods to remove or reduce them. Even so, some sources of error still remain and it is with these that further work remains to be done. These factors are worthy of discussion here.

A. Techniques of Weighing and Diluting

Errors occurring here are most difficult to discover, prove, and correct. The human inability to carry out a correct technical procedure without change is well known and there appears to be no satisfactory substitute to sound training, constant supervision of equipment and techniques, and a good personal relationship with staff. The importance of the latter is frequently underestimated, although its effect on personal interest in the work in hand has been amply demonstrated. We believe this is more important in the case of microbiological assay than in most other analytical procedures.

B. Techniques of Plating-Out

With large plate methods it is essential to maintain as a constant the volume of test solutions, both sample and standard, that are applied to give a test zone. Experience has shown this to be most difficult to achieve even by the most skilled operators and the problem still remains as one requiring further work. The use of as large volumes as possible reduces the error considerably; ideally this should be 0.5 ml. or greater. The use of paper disks, cylinders, or fish-spine beads is not recommended except in certain circumstances and we have preferred the cup plate procedure throughout. With large numbers of plates the tedium of "plating out" contributes a very real human error to the assay. No satisfactory automatic measuring devices for small volumes have yet been found, and this remains a real problem with this type of assay.

C. Reading of Zone Diameters

It is undoubtedly true that errors due to reading procedures are often greater than appreciated, and this is especially so with staff unskilled in this particular technique. Our preference has always been for the use of Chesterman 1075 vernier scale calipers reading directly on to the surface of the medium. All plates are read only by supervisory scientific staff and no person may read a plate who was in any way involved directly with its preparation. For high precision assays two sets of readings are taken by different operators and each separate reading is used for the computation. The mean of the two sets of readings is finally taken if they are mutually consistent and if not, a third reading is taken. We have found that skilled staff using vernier scale calipers are capable of reading assay plates quickly and accurately and that with such experienced staff the error is consistently less than 2%. Much thought has been given to other methods of reading and it does now appear as if the problem will at last be resolved.

The many projection machines seen in this country have not inspired confidence in this method of reading. A. B. Kabi (Stockholm) manufactures a machine which appeared to be very much more satisfactory than any others previously seen. It represented the result of close collaboration between engineers and scientific staff and was excellently designed, compact, speedy to use, and accurate in reading. Its reproduction here was not satisfactory.

The use of photoelectric methods has always been thought to be the solution to the problem and American workers in this field have now developed such an instrument. If satisfactory, it will obviously replace existing methods and reading errors should be further reduced to a point where errors are negligible.

D. Designs

It is well known that the position of a zone on a plate will influence the size of the resulting zone of inhibition or exhibition of growth and as a result the use of balanced Latin, quasi-Latin, lattice square, and incomplete block designs has been accepted as the only possible approach to an unbiased determination. Errors due to drift, edge effect, time effect, medium thickness, pH, and Eh variables are all eliminated or reduced depending on the choice of design. We have found it very important to have all designs prepared with qualified statistical assistance and to take precaution to ensure that sufficient designs are available so that none is employed more than once in a day. In this way operators do not "select" the designs to be used and do not use the designs in any way other than that prescribed.

E. Computation of Sample Potencies

The use of nomograms for potency computations is not favored and we much prefer to use calculating machines at all stages and to make approximations *only* of the final result. Nomograms correctly prepared and sufficiently large to ensure elimination of errors due the draughtsmanship can be used, but nothing is gained, either in time or accuracy, over the use of a reliable calculating machine such as the Marchant Model DRX. The use of slide rules constitutes another source of error as does the preparation of graphs. Many analysts are apparently quite unaware of the accumulated effect of making approximations at each stage of computation. For this reason we permit only the final reported result to be approximated to that potency satisfied by the standard error and probability of result, which are reported together with the potency estimate.

F. Internal Standards

The use of internal standards has become a routine procedure for those laboratories carrying out large numbers of assays, although the use of these is recommended with certain reservations. The choice of material to be set up as the standard, its chemical and microbiological analysis, the ultimate potency, methods of drying, filling out, sealing, storage, and above all the stability of the standard under the imposed storage conditions require most careful considerations and supervision. The use of National or International standards should obviously be restricted to collaborative work and to control of internal standards. Recently there has been a marked improvement in the interchange of internal standards between organizations and this has been of real help in accurately assessing the stability of standards.

In this organization all internal standards required for routine microbiological assay are prepared completely in one unit and distributed as requested to all units, many overseas. The chemical analysis is in the hands of competent analysts and the microbiological potency is determined only after carrying out large numbers of assays of the Latin square design, usually of the 12×12 type, each of which is analyzed by competent statisticians. The mean of all valid assays only is taken. Various National and International standards are used only as reference standards. Similar procedures apply to obtain stability data and to determine whether observations about a standard from a unit are correct or not.

The importance of good quality culture maintenance and assay media, maintenance and preparation of bacterial cultures and suspensions, maintenance of adequate and reliable standard materials, sterile techniques, and the correct use of equipment and glassware are all well known to be as important in successful large plate assay as in other microbiological methods of assay. It is not proposed to discuss these further. Laboratories in the U.K. have a marked preference for the use of spore suspensions as assay test organisms; this is in contrast to American workers who appear to prefer daily culture techniques often with nonsporing organisms.

G. Bacterial Standardization

Mention has been made previously of use of Wellcome opacity tubes in the standardization of bacterial suspensions. It is realized that the bacterial population per large plate per day should be sensibly constant as the diameter of zones of inhibition is governed largely by such population. In the U.K. the use of Wellcome opacity tubes for such standardization and for determining the approximate cell population has become a routine procedure. A series of opalescent sealed tubes have been carefully standardized against bacterial counts using conventional methods and using a wide variety of microorganisms. The unknown suspension is pipetted into an empty, dry tube of similar pattern and the resulting turbidity is matched against the sealed standards, the bacterial population being given by reference tables under the appropriate bacteria and standard tube number. Similar techniques are adopted in other countries. The suppliers in the U.K. are listed under the equipment required for an assay.

VIII. Validity of Assay

On the assumption that the factors previously discussed are fully understood and that a suitable assay design to eliminate or reduce these sources of error has been employed, the microbiologist is still faced with the problem of determining the validity of the assay. As validity is not proof of accuracy, it follows that the lack of validity only serves to make the problem even worse.

It is well known that a linear relationship between the mean response and logarithmic dose is the most convenient for plate assay methods. However, under the stricter conditions of a high precision assay a line which appears to be straight on cursory examination may prove to be curved to a greater or lesser degree. A minimum of three dose levels is required to prove this.

We should always be careful to define validity in these circumstances. We really mean that the extent of the observed lack of parallelism or the extent of curvilinearity are such that they could be accounted for by the intrinsic variability of the zone diameters.

Microbiological assays are called "dilution assays." This means that the test preparation behaves as though it were simply a dilution (or concentration) of the standard preparation in a diluent that is completely inert in respect of the response used. In testing for validity, the important feature being tested is parallelism—because the above statement implies that the test preparation will have the same dose-response relationship as the standard. Since it is usually found that the response bears a linear relation to the log of the dose level over a certain range of dose levels, it means that the test preparation and standard preparation should give parallel dose-response lines. If they do not, we say the assay is invalid. This invalidity may be due to a number of things—errors in dilution or weighing, a wrong initial estimate of the test preparation potency so that the dose-response relation is outside the linear range, the presence of an impurity in the test preparation, etc. A detailed account of types of invalidity may be found in Chapter 4 of D. J. Finney's "Statistical Methods in Biological Assay."

Invalidity is tested for by examining whether the extent of the differences in parallelism or departures from linearity, etc., are within the limits we expect from a knowledge of the residual error, i.e. the intrinsic variability of the responses (zone diameters, say) after taking into account row effects, column effects, dose level effects, etc. This is done by means of the statistical technique known as Analysis of Variance. This residual error determines the confidence limits to place on the estimates of potency, i.e., the more variable the zone diameter, the less precise are the estimates. This leads to the paradox that an assay could be invalid if carefully carried out, but valid if carelessly carried out. It would probably be better practice if we were all to cease using the term "validity of assay" and state that we were unable to demonstrate invalidity.

It is also very important, in all assays and more particularly in those involving higher precision, that parallelism of sample and standard response lines should be evident. Again, at least three dose levels are necessary to prove linearity and parallelism of the two dose-response slopes.

It is of fundamental importance to determine, before carrying out the assay, that a linear portion of the range of dose levels is employed. This requirement is generally ignored or at most assumed, but must, for high precision assay, be carried out in full.

In addition to linearity and parallelism it is necessary to consider also the slope of the dose-response line, the definition of the resultant zones of inhibition or of exhibition of growth, and the spacing of the dilution levels. All require consideration together and the microbiologist must make the best compromise possible from these. He should bear in mind that the precision of the assay is also dependent on the ratio of the slope to the standard error of the zone.

To satisfy the strict statistical requirements of a high precision assay it is necessary for the microbiologist to assure himself that so far as he can ascertain, the design is valid. Invalidity in design must give invalidity in analysis.

The consistency of successive daily results should not be taken as proof of validity although it is evident that valid assays do show such consistency. Such repetition, particularly when involving fresh weighings of sample and standard, is of particular value in obtaining higher precision where the percentage standard error for the assay of independent samples is then represented by

$$\pm \sqrt{\frac{\text{Standard error for single determination}}{\text{No. of true individual determinations}}}.$$

In our experience the development of high precision assays merely requires that the errors listed in the preceding points should be eliminated as far as is possible. These errors, together with those previously described for medium precision assays, are known and may be acceptable for low and medium precision assays. Even so, it is important to realize that in all microbiological assays variations in zone size do occur and indeed are inevitable. A successful assay recognizes this and attempts to reduce its effect by randomization and replication, the only methods of approach. Any other method generally introduces bias into the determinations. We are frequently dismayed at the degree of reproducibility and consistency of results claimed by some microbiological laboratories. It is only too easy to discover that bias has entered into the determination, but it is all too difficult to demonstrate this to such microbiologists. The use of large plate methods has done more than any other method to eliminate bias from microbiological assays and to develop a real knowledge of the problems and challenge that successful assay presents.

It is as yet too early to finally detail methods for high precision large plate assay methods since substantial improvements in equipment are quite obviously needed before even considering such an approach. It is however clear that certain staff have the flair and technical skill to undertake such work and that the statistical advice of the past few years has not been wasted. An important feature which promises well for such assays is that due to easing of pressure of work on assay laboratories more time is now available to develop the required skills and knowledge.

The newer antibiotics do not generally appear to require methods of microbiological assay as an alternative except for stability purposes.

The application of recently described methods to large plate methods of assay does not appear to present problems. An argument is often put forward that new methods involve few samples initially and that these do not warrant large plate methods. This is quite unrealistic and there is undoubtedly every reason to use large plate methods in preference and particularly is this so in the development stage of an assay. The number of samples is no point in favor of other methods of assay. As stated previously the dose level following extraction is the only factor likely to require serious consideration.

The methods of Pike and Sulkin[14] for meat adulteration and the work of Matthews[15] and of Ouchterlony[16] with agar diffusion methods for demonstrating antigen-antibody precipitation reactions have all been satisfactorily applied to large plates.

The rapid assay of penicillin in milk samples as described by Berridge[17] and the assay of trace metals described by Nicholas[18] have also been adopted although both methods have the disadvantage of relative insensitivity for agar diffusion techniques generally.

Work on hydrolytic enzymes described by Dingle et al.[19] and assay of trypsin, diastase, and penicillinase, and lipase have not been critically examined but there seems no reason why large plate methods should not be employed.

Sacks[20] has described a most useful method by agar diffusion on large plates to determine the pH limits for growth of microorganisms and to determine the influence of pH on the antibacterial activity of substances. This method has been most satisfactory and is undoubtedly of real value.

It is believed that the long quoted statement that "microbiological assay and precision have nothing in common" is being replaced by a somewhat grudging agreement that the opposite is true. The challenge rests with the microbiologist and his staff and there seems every possibility that the days when "high precision" microbiological assays can be carried out as required are not too far removed. The large plate method of assay should supply the means of achieving this.

[14] R. M. Pike and S. E. Sulkin, *J. Lab. Clin. Med.* **49**, 657 (1957).

[15] P. R. J. Matthews, *J. Med. Lab. Technol.* **15**, 95 (1958).

[16] O. Ouchterlony, *Acta Pathol. Microbiol. Scand.* **32**, 231 (1953).

[17] N. J. Berridge, *J. Dairy Research* **23**, 336 (1956).

[18] D. J. D. Nicholas, *Proc. Intern. Symposium Microchem., Birmingham, 1958*, p. 205 (1960).

[19] J. Dingle, W. W. Reid, and G. L. Solomans, *J. Sci. Food Agr.* **4**, 144 (1953).

[20] L. E. Sacks, *Nature* **178**, 269 (1956).

CHAPTER 3

Dilution Methods of Antibiotic Assays

FREDERICK KAVANAGH

Eli Lilly and Company, Indianapolis, Indiana

Simple dilution methods, agar dilution and serial dilution, are described in this chapter. The serial dilutions methods are given in great detail with typical answers. A serial dilution method not strictly antibacterial is given in Section III on Antiluminescent Assay. This chapter contains information on media, propagation of bacteria, selection of assay bacteria, errors of pipetting, and errors of microbiological origin.

I. Streak-Plate Method

In the streak-plate method,[1,2] the antibacterial substance is diluted in nutrient agar, the agar poured into petri dishes, and the surface of

---

[1] J. W. Foster and H. B. Woodruff, *J. Bacteriol.* **46**, 187 (1943).
[2] S. A. Waksman and H. C. Reilly, *Ind. Eng. Chem. Anal. Ed.* **17**, 556 (1945).

the agar streaked with bacteria. After a period of incubation, the highest dilution of the antibacterial substance that inhibits the growth of bacteria on the surface of the agar is chosen as the end point and the activity is computed from this dilution. Several species of bacteria can be streaked on the same plate. The dilution steps are usually rather large and the activities so obtained have a large systematic uncertainty. The method is useful in various surveys, however, because the approximate activity against as many as 10 organisms can be measured on one plate with relatively small expenditures of material and labor.

Any organism which will grow on an agar plate can be used in this method. The sensitivity of the test depends upon the organism, pH, composition of the medium, and temperature of incubation. In antibiotic surveys, tests using synthetic and natural media may not even detect the same class of compounds.

II. Serial Dilution Assay

A. Introduction

The serial dilution assay is simple in principle and an easily performed method of considerable utility. Several dilutions of the antibacterial substance in small tubes are inoculated with a test organism, incubated, and the lowest concentration of the substance which causes apparently complete inhibition of growth of the organism is taken to be the minimum inhibitory concentration. The activity of the compound is computed from the minimum inhibitory concentration and may be reported as micrograms of substance per milliliter. For many purposes, the answers so obtained are as useful as those given by the more laborious plate or turbidimetric assays. One good operator can do all of the work required to put on 60 samples in a working time of about 4 hours including reading and recording the answers. No elaborate equipment is needed. The details for doing three variants of the serial dilution method with different accuracies were given by Kavanagh.[3] The two methods described here are the simple twofold serial dilution method and the arithmetic. The serial dilution method gives a series of dilutions which form a geometric series in which the answer lies between a concentration equal to and one-half as much as the one reported. Supposedly it could be any value in between. Thus the serial dilution method has such a large inherent uncertainty that it is not suited to precise determination of quantity. It could never be used for control work where an error of 10% would be considered large. Nonetheless, it is a useful method for guiding chemical operations of the all-or-none sort. If the antibiotic is one that is purified by methods that

[3] F. Kavanagh, *Bull. Torrey Botan. Club* **74**, 303 (1947).

effect large changes in purity, then the serial dilution method is adequate.

The activity of an antibiotic substance measured by the serial dilution method is reproducible over a long period of time if attention is paid to certain details which include sensitivity of the test organisms, constancy of composition of medium, inoculum size, temperature of incubation, and pH of the medium. All of these factors can be controlled.

B. Test Organism

Any organism (bacteria, fungi, algae, protozoa) which will grow in a liquid medium can be used in a serial dilution type of assay. Such common gram-positive bacteria as *Bacillus subtilis* and *Staphylococcus aureus*, such gram-negative bacteria as *Escherichia coli, Klebsiella pneumoniae, Pseudomonas aeruginosa,* and *Salmonella enteritidis,* and such acid fast bacteria as *Mycobacterium smegma, M. phlei,* and *M. avium* have been used in serial dilution assays. For general laboratory use, the organism should be nonpathogenic for man. With suitable precautions, the method is used to measure the sensitivity of strains of pathogenic organism to antibacterial substances.

Representative of fungi which have been employed in testing activity of antibacterial substances by serial dilution methods are:

Aspergillus niger, Chaetomium globosum, Memnoniella echinata, Penicillium notatum, Phycomyces Blakesleeanus, Sacharomyces cerevisiae, and *Trichophyton mentagrophytes.*

Assays with fungi are done in the 1-ml. test to be described later. Inoculum usually is a spore suspension, consequently the method measures activity in terms of either spore germination or mycelium growth depending upon which response is the more sensitive.

The strain of bacteria used depends upon the object of the test, which may be: the determination of the potency of culture liquids and the fractions obtained from them in the process of purification; the determination of the characteristics of the active substance by measuring its activity against standard strains of bacteria; the determination of the activity against bacteria resistant to other antibacterial substances. Most tests are made to determine the potency of various preparation.

A satisfactory bacterial culture is one which produces a population most of whose members are inhibited by a concentration of an antibacterial substance less than twice as great as the concentration which causes an appreciable decrease in number. Whether or not a bacterial species is a satisfactory test organism depends upon the strain and the substance used.

Suitability of a strain is tested by first determining the minimum inhibitory concentration of the antibiotic for a concentration of bacteria of about 1000 cells per milliliter. Then increasing concentrations of antibiotic substance and bacteria are added to pour plates (test medium solidified with 1.5% agar) and the number of colonies counted after an incubation period of 48 hours. Concentration of antibiotic should be increased in small steps (2 ×) and concentration of bacteria in large steps (10 ×). When the Heatley strain of *Staphylococcus aureus* was tested by this plating method one cell in 40 million was resistant to 2.5 times the minimum inhibitory concentration of penicillin and a very great many were resistant to twice the inhibitory concentration of streptomycin. These results show why the Heatley strain of *S. aureus* is excellent for measuring penicillin but useless for streptomycin. The strain contains much too great a proportion of its population resistant to appreciable concentrations of streptomycin. The resistant individuals may grow out in tubes containing a higher concentration of antibiotic than the end-point tube. A strain of *Staphylococcus* free from resistant individuals or *Klebsiella pneumoniae* (ATCC 9997) can be used for streptomycin assays. This principle should be followed in selecting the assay organism for any antibiotic by any turbidimetric method. In the streak and cup-plate methods, the resistant organisms form isolated colonies in the zone of inhibition and can be ignored in evaluating the test. In the dilution tube method, these organisms cause turbidity where only clear tubes should be and make uncertain the interpretations of the test.

The minimum inhibitory concentration of an antibacterial substance may increase greatly with an increase in the concentration of bacteria used in the test because of the concomitant increase in the number of bacteria resistant to it. When large inocula are used, the bacteria that actually function in the test are the few that are resistant and not the greater part of the population that are sensitive to the active substance. This is one reason that small inocula (1000/ml.) are used in the methods described here.

C. Media

The media used in testing antibacterial substances fall into two groups, those used for propagation of the bacterial cultures and those used in the dilution tests. All of the 30 species and strains of bacteria, except the Photobacterium, maintained in one laboratory grew satisfactorily on the Bacto-A.C. medium used in the form of broth and agar slants. The medium contains per liter: 5 gm. dextrose, 3 gm. malt extract, 3 gm. Bacto-Beef Extract, 3 gm. Bacto-yeast extract, and 20 gm. Proteose Peptone No. 3. The pH is 6.3 after sterilization. The medium for the agar slants is made by adding 15 gm. agar to one liter of the A.C. broth. The *Mycobacteria* are maintained in a modified Kirchner medium in which they form a uniform suspension without a surface pellicle.

Weekly transfers of the bacteria (other than *Mycobacterium*) from agar slants to freshly sloped A.C. agar slants are made. The tubes are incubated at the appropriate temperature for 24 or 48 hours, depending upon the rate of growth, and are then stored at 11°C until needed. A complete set of cultures transferred at monthly intervals is kept in the refrigerator at 4°C. The *Mycobacteria* are grown in the modified Kirchner growth medium and fresh transfers are made every 3 days. The tubes are kept at 36°C until transfers are made to fresh broth. *Bacillus subtilis* and *B. mycoides* are grown at 30°C, and all other bacteria are grown at 36° to 37°C.

The fungi are grown on an agar medium made by adding 15 gm./liter of agar to the assay medium or on 2% malt agar.

In general, the assay media are not as rich as the A.C. medium and the growth of the bacteria is slower than it is in the A.C. medium. The assay media should not contain appreciable amounts of substances which inhibit the action of the antibacterial substances.

Assays with *Staphylococcus aureus*, *Escherichia coli*, *Klebsiella pneumoniae*, and *Pseudomonas aeruginosa* are performed in beef extract (B.E.) medium which contains 5 gm. of Bacto-Peptone and 3 gm. Bacto-Beef Extract per liter of solution. The pH after sterilization is 6.8. The B.E.D. medium is made by adding 5 gm. of dextrose to each liter of B.E. medium before sterilizing; it is used for assays with *Bacillus subtilis* and *B. mycoides*. The pH is 6.6 after sterilization. The mycobacteria are maintained in the "growth medium" and tested in the "test medium" the composition of which follows: asparagine, 5 gm.; glycerol, 20 gm.; lecithin (egg), 0.1 gm.; Tween 80, 0.1 gm.; Na_2HPO_4, 3 gm.; KH_2PO_4, 4 gm.; $MgSO_4 \cdot 7H_2O$, 0.6 gm.; sodium citrate, 2.5 gm.; ferric ammonium citrate, 0.05 gm.; and distilled water to make 1 liter. The growth medium is the same as the test medium except the lecithin and Tween 80 are each 0.5 gm. The pH is 6.6 before sterilizing.

The spores of the fungi listed above all germinate and grow in a simple assay medium composed of: glucose, 50 gm.; Neopeptone (Difco), 2 gm.; thiamine, 5 mg.; KH_2PO_4, 1.5 gm.; $MgSO_4 \cdot 7H_2O$, 0.5 gm.; and distilled water, 1000 ml.

All media are sterilized in the autoclave at 121°C. for 15 minutes.

D. Test Inoculum

The inoculum for most of the bacteria is prepared by transferring a large number of bacteria (a loopful) from an A.C. agar slant to a tube of A.C. broth and incubating it for 6 hours at 36°C. The tubes are shaken occasionally to aerate and to promote growth. *Bacillus subtilis* and *B. mycoides*, however, are grown for 24 hours at 30°C. The 1- or 2-day-old cultures of the mycobacteria in the Kirchner's broth are used as the inoculum.

The 24-hour cultures of *B. subtilis* and *B. mycoides* are diluted 1000 times in B.E.D. broth. The 24- or 48-hour cultures of mycobacteria are diluted 50 times in the modified Kirchner's test-medium. The 6-hour A.C. broth cultures of the other bacteria are diluted one million times in the B.E. medium to give a concentration of bacteria of from 500 to 2000/ml.

E. Incubation Temperatures

The tests with *B. mycoides* and *B. subtilis* are incubated at 30°C, though the sensitivity of *B. mycoides* and *B. subtilis* to the antibacterial substances is increased considerably by incubation at temperatures lower than 30°. The tests with all other bacteria are made at 36° to 37°C. Tests with fungi are incubated at 25° or other suitable temperatures until good growth occurs in the tubes with the highest dilutions of test substance.

F. Glassware

Serial dilution tests are done in 12 or 13 × 75 mm. Kahn tubes with serological pipets. The tubes are packed inverted in small baskets, sterilized, and stored in them. The tubes remain sterile for a week or two if kept in a clean cabinet. The tubes are used without cotton plugs. The occasional contamination usually comes from the fingers and not from the air. The frequency of such contamination is no greater for unplugged than for plugged tubes. Cotton plugs are used in the tubes if the test organism is pathogenic.

The pipet can be a standard 1-ml. serological pipet graduated in steps of 0.1 or of 0.01 ml. or a special pipet graduated to contain at 0.5 and 0.66, and 1.00 ml. The special pipet is used to make dilution in steps of 16 (0.66 ml. + 10.0 ml. of diluent), a convenient dilution interval in serial dilution assays. When a 1-ml. serological pipet is used as a wash-out pipet, considerable error in the dilution accumulate by the 6th tube of a series and may be quite large by the 10th tube. The pipet contains about 0.53 ml. at the 0.50 mark. This error probably could be avoided by drawing the liquid up to the 0.53 ml. mark (0.47 ml. indicated volume, 0.50 actual volume) instead of to the 0.50 graduation. An occasional gross error results from drawing the liquid to a mistaken graduation. This error is more frequent with pipets graduated in steps of 0.01 ml. than those with 0.1-ml. graduations. This error does not occur with the special pipet mentioned above because there are only three widely spaced graduation marks on the pipets. The 0.50-ml. volume of medium needs to be measured with an accuracy of 0.02 ml. This can be done with an automatic pipetting machine, a 5-ml. measuring pipet, or a "Cornwall Pipetting Outfit" (Becton, Dickinson and Company). Whatever the equipment used, constant care must be exercised to obtain the desired accuracy.

G. Preparation of Sample

Since the growth of the bacteria may be inhibited by increased acidity of the nutrient solution and the activity of the antibacterial substance may be greatly influenced by the acidity of the medium, the pH of the samples should be adjusted to about pH 6 before assaying. The solutions to be assayed are diluted with B.E. broth or water until the activity is less than 32 dilution units. Frequently it is necessary to assay solutions that contain large amounts of organic solvents. Saturated aqueous solutions of chloroform or ethyl acetate did not inhibit *Staphylococcus aureus*, even in the first tube. Concentrations of ethyl alcohol less than 5% did not inhibit *S. aureus*, *Bacillus mycoides*, *Escherichia coli*, and *Klebsiella pneumoniae*. However, there is always the possibility of synergism between the organic solvent and a subinhibitory concentration of the antibacterial substance. The solutions of the antibacterial substances must be prepared under aseptic conditions or sterilized before assaying. Acidic substances and neutral substances usually can be filtered through Seitz or sintered-glass filters without loss by adsorption. Thermostable basic substances may be sterilized by heating rapidly to boiling. The "Swinny Filter" is a very useful form of the Seitz filter because samples of between 1 and 5 ml. can be filtered with a loss of less than 0.5 ml.

H. Principle of the Dilution-Tube Test

The principle of the test is very simple. Several dilutions of the antibacterial substance in tubes are inoculated with the test bacterium, incubated, and the lowest concentration of the substance which causes apparently complete inhibition of the growth of the bacteria is taken as the inhibitory concentration. From this concentration is calculated the activity of the active substance.

The tests are usually read after 16 to 18 hours, after 24 hours, and after 42 hours of incubation. The concentration of bacteria and the composition of the test media are chosen so that good turbidity will develop in 16 hours except with the mycobacteria which are incubated for 24 hours before the first reading is made. Whether or not a large change occurs on incubation beyond 16 hours depends upon the active substance and the strain of bacteria. The time of incubation that gives reproducible assays depends upon both the antibacterial substance and the strain of bacteria and can be determined by repeated assay of the same solutions. Some substances may show at the end of 42 hours of incubation only one-eighth as great an activity as that at the end of 16 hours of incubation.

1. Nonpathogenic Organisms

The test used with relatively nonpathogenic organisms, and therefore used for most of the assays, requires the minimum of manipulation. Racks that hold two or three rows of 10 or 12 tubes each are filled with sterile, unplugged 12–13 × 75 mm. Kahn tubes, and 0.50 ml. inoculated broth are added to each tube by a Cornwall Pipetting Outfit or by a 5-ml. pipet. A 0.50-ml. volume of the sterile antibacterial substance is added to the first tube by means of a sterile 1-ml. serological pipet. The contents of the tube are mixed thoroughly and 0.50 ml. are transferred to the next tube. The contents of this tube are mixed, 0.50 ml. are transferred to the next tube, and so on to the end of the row. The solution of the antibacterial substance undergoes a dilution of 2 times in each step as follows:

Tube No.:	1	2	3	4	5	6	7	8	9	10
Dilution:	2	4	8	16	32	64	128	256	512	1024

The racks of tests are put in the appropriate incubator and left for about 16 hours, after which they are removed, shaken to aerate the solutions and to suspend bacteria that may have settled to the bottom of the tube. The number of clear tubes (tubes without the slightest trace of turbidity) is counted and the racks are returned to the incubator. The number of clear tubes is counted again after 24 hours and after 42 hours of incubation.

Shaking the tubes at the end of 16 hours of incubation aerates the solutions. This aeration may cause such a burst of divisions of the bacteria that one or two tubes that were clear at the end of 16 hours will be quite turbid after 20 hours of incubation.

If, after incubation, the first five tubes are clear and the sixth turbid, the antibacterial solution is said to be active at a dilution of 32, or to have an activity of 32 dilution units per milliliter when tested against the organism used in the assay. For most purposes, the activity of the solutions can be expressed in terms of the highest dilution that gives a clear tube. Tubes with incomplete inhibition are never used in computing activities unless the active substance is one which does not inhibit the growth of all of the bacteria. When it is available, a standard solution made from the pure substance is included in each set of assays, and the results are reported in terms of the standard. Usually the pure substance is neither available nor necessary.

2. Pathogenic Bacteria and Fungi

Tests with pathogenic bacteria and fungi are done in a 1-ml. geometric-series dilution test as follows: To the tubes are added 0.50 ml. of broth, the serial dilution is performed as above, and 0.50 ml. of inoculated broth is added to each tube. The racks of tubes are incubated, and the clear tubes are counted. The dilution for a tube is twice that given in the dilution scheme. The same number of bacteria or spores are added to each tube in this procedure in contrast to the previous one in which the number of organisms per tube is essentially constant only after the third tube in the series.

TABLE I

MINIMUM INHIBITORY CONCENTRATION OF ANTIBACTERIAL SUBSTANCES[ab]

Substance	B.m.	B.s.	S.a.	E.c.	K.p.	P.a.	M.s.
Aspergillic acid	2	4	4	62	13	1000	16
Cassic acid (Rhein)	4	8	8	1000	500	>250	30
Chloramphenicol	2	2	2	1	1	16	8
Chlorotetracycline	0.004	0.008	0.016	0.25	0.03	8	0.03
Citrinin	32	16	16	>1000	—	—	250
Dihydrostreptomycin	0.25	0.5	0.03	0.25	0.13	4	1
Enniatin A	32	16	8	>500	>500	>500	4
Gliotoxin	0.25	0.25	0.15	25	6	500	4
Helvolic acid	4	16	1	>1000	4	—	>32
Hydrogen peroxide	32	4	8	10	5	8	4
Kojic acid	2500	620	1250	2500	620	5000	310
2-Methyl-1,4-naphthoquinone	12	3	1.7	220	28	>400	36
Mycophenolic acid	500	250	250	500	>1000	>1000	250
Neomycin	0.25	0.03	0.016	0.5	0.25	1	1
Nordihydroguarietic acid	32	0.32	16	64	32	—	64
Oxytetracycline	0.25	0.13	0.13	0.5	0.13	8	0.25
Patulin	16	4	8	8	8	125	1
Penicillic acid	32	8	16	64	64	1000	32
Penicillin G	30	0.03	0.016	14	110	500	450
Penicillin X	30	0.06	0.03	14	240	500	470
Streptomycin	0.13	0.25	0.03	0.13	0.13	4	1
Streptothricin	4	0.25	0.016	0.016	0.016	<4	1
Tolu-p-quinone	4	1	1	25	13	125	2
5-Methoxy tolu-p-quinone	16	2	0.5	64	32	250	128
Viomycin sulfate	0.25	0.06	0.06	0.13	0.13	8	—

[a] Units: mcg./ml.

[b] The abbreviations represent: *Bacillus mycoides* ATCC 9634, *Bacillus subtilis* ATCC 6633, *Escherichia coli* ATCC 9637, *Klebsiella pneumoniae* 9997, *Mycobacterium smegma* ATCC 10143, *Pseudomonas aeruginosa* ATCC 10145, and *Staphylococcus aureus* ATCC 9144 (Heatley strain).

A few examples[4] of the antibacterial activity of antibiotics of historical interest measured by a serial dilution method against seven bacteria are given in Table I. These results illustrate the great variations in sensitivities of the bacteria and in the specific activities of the antibacterial substances. All of the bacteria are laboratory strains that were used for

[4] F. Kavanagh, *J. Bacteriol.* **54**, 761 (1947); and unpublished data.

many years in the search for new antibiotic substances. The therapeutically important antibiotics (chloramphenicol, chlorotetracycline, neomycin, oxytetracycline, penicillin G, and streptomycin) are among the most active compounds in the table. The reason *Pseudomonas aeruginosa* infections are so refractory to treatment is obvious.

3. Arithmetic-Series Tests

The arithmetic-series type of dilution test is a slight modification of the 2-ml. test described by McKee et al.[5] and Donovick et al.[6] In this test, volumes of a suitably diluted antibacterial solution ranging from 0.02 to 0.10 ml. in increments of 0.01, are added by a Kahn, or, better, by a 0.2-ml. measuring pipet to empty 12 × 75 mm. Kahn tubes. Then 0.5 or 1 ml. of inoculated broth is added from a 5-ml. pipet or a Cornwall Pipetting Outfit. After incubation, the highest dilution which gives a clear tube is taken as the activity of the solution. The dilution is obtained from the following scheme for the 1-ml. test:

Tube no.:	1	2	3	4	5	6	7	8	9
ml. sample added:	0.1	0.09	0.08	0.07	0.06	0.05	0.04	0.03	0.02
Dilution:	11	12.1	13.5	15.3	17.7	21	26	34.3	51
Relative conc.:	0.091	0.082	0.074	0.065	0.057	0.048	0.038	0.029	0.02

The systematic uncertainty varies from 10% of the dilution for the first tube, to 19% for tube 5 and to 39% for the last tube. Hence the solutions are diluted before assay so that the first cloudy tube will fall between tubes 2 and 6. While neither the dilution increment nor the percentage of uncertainty is constant for this test, tests can be devised in which either is constant.[3]

4. Results Obtained by the Two Test Procedures

Two types of assay procedures with two different degrees of uncertainty have been described. The arithmetic-series test presumably is the more accurate, and the values obtained by it will be considered to be correct. Results obtained by assaying the same solutions by the two test procedures are shown in Table II. All of the antibacterial compounds had a purity of 90% or better.

What is the usefulness of the tests? The geometric-series dilution test is used with solutions which may show activity between rather wide limits, and when a precision greater than 50% is not needed. The results of the geometric-series dilution test are used to estimate the dilution which will bring the activity within the range of the more precise arithmetic test. The type of test should be chosen after considering the precision needed.

[5] C. M. McKee, G. Rake, and A. E. O. Menzel, *J. Immunol.* **48**, 259 (1944).

[6] R. Donovick, D. Hamre, F. Kavanagh, and G. Rake, *J. Bacteriol.* **50**, 623 (1945).

TABLE II

MINIMUM INHIBITORY CONCENTRATIONS IN MICROGRAMS PER MILLILITER
OBTAINED BY TWO DILUTION METHODS

Substance	Bacterium	Type of test	
		Geometric	Arithmetic
Penicillin G	*S. aureus* H	0.031	0.021
Penicillin G	*E. coli*	28	16.9
Penicillin X	*E. coli*	14.6	12.5
Streptomycin	*K. pneumoniae*	0.125	0.069

I. Errors

Although millions of serial dilution tests on antibacterial substances have been performed, few discussions of the errors of mechanical and biological source have been published.[3] It is easy to show that there are large pipetting errors in a geometric-series dilution test and that sometimes resistant organisms can cause significant errors. The latter errors will be discussed first.

In about one-half of the assays of streptothricin solutions with *Staphylococcus aureus* H. by the geometric-series dilution method, one or more tubes showing a good growth of bacteria will be followed in the series by at least one tube showing no evident growth of the bacteria although it contained a lower concentration of streptothricin than the ones showing growth. Thus there are two end points in one test, one of which may be at a concentration of streptothricin from 4 to 16 times as great as the other. The bacteria in the out-of-place tubes will be found to be quite resistant to streptothricin. It is evident that the occurrence of out-of-line tubes at the high dilution end of the series of clear tubes could make the activity of the solution appear to be one-half or less of its real activity. The only satisfactory way to eliminate this error is to use a strain of bacteria which does not contain (or form) resistant forms.

Because it is not obvious, a serious error in the dilution tests is caused by errors in measuring the volumes of liquids. The errors, like the test, form a geometrical series, and a small error raised to the tenth power, as it is in a ten-tube geometric-series dilution test, becomes a large error. This can be demonstrated very simply by doing a ten-tube test with one pipet and repeating the test with ten pipets. When this was done the end point was the ninth tube in the first test and the eighth tube in the second test. Too frequently, if the dilution for the arithmetic-series test is computed from the activity obtained from a geometric-series dilution test in which the end point was the sixth tube or more, the range of the arithmetic-series test will be missed. Usually the dilution is too great.

Both bits of evidence indicate that the geometric-series test shows the activity to be greater than it is. Where are the sources of error and how large are they?

Since antibacterial tests are a poor way to discover errors in diluting, another method was sought which would have high precision even after the test-solution had been diluted 1000 times. By doing a dilution test with 5 N HCl and titrating the acid in each tube with a 0.1 N NaOH, the required precision was obtained. Water, 0.50 ml., was put in each tube with a measuring pipet and a geometric-series dilution test was performed, using a 1-ml. serological pipet. Racks of tubes and a tube of acid were also given to two experienced technicians who used their own pipets and their usual technique since they did not know that the test was not the usual antibacterial test. The results are given in Table III. By

TABLE III

ACTUAL DILUTION IN GEOMETRIC-SERIES DILUTION TEST
OBTAINED FROM TITRATION OF ACID

Tube no.:		1	2	3	4	5	6	7	8	9	10
Theoretical dilution:		2	4	8	16	32	64	128	256	512	1024
Actual dilution:	W.J.	1.98	3.62	7.15	13.5	25	54				
	W.J.	1.92	3.74	6.95	13.2	25	45				
	C.A.	1.90	3.7	6.7	13.5	24	24				
	C.A.	1.91	3.7	7.0	13	26	42				
	F.K.	1.93	3.52	6.9	12.5	21					
	F.K.	2.14	4.02	7.53	14.1	25.7	49.2	95.7	184	337	660
Computed for pipet of:	F.K.	1.94	3.76	7.30	14.2	27.5	53.4	103	200	387	750

weighing the water removed, it was found that the dry EXAX Blue Line serological pipet used by F.K. removed 0.54 ml. when the liquid was drawn up to the 0.50 mark. The pipet was found by the acid titration method to remove and deliver 0.53 ml. when washed out in a test. When 0.53 ml. is added to 0.50 ml., the dilution of the solution added is 1.94. Hence the dilution in each succeeding tube is 1.94 raised to a power equal to the number of the tube (see Table III).

The dilutions actually obtained in the first five tubes were nearly those computed for the pipet of F.K. and considerable deviation from the theoretical 2^n relationship began at the sixth tube. The error in pipetting probably could be eliminated by removing 0.47 ml. instead of 0.50 ml. of sample each time. But there are two other errors that can not be so easily avoided. One is the error in measuring the volume of broth put into the tube by the automatic syringe. In the above samples, the water volumes were measured with an error of ≤ 0.01 ml. The automatic syringe used for measuring the volume of the broth can cause two errors in volume, one the systematic error resulting from incorrect setting of the length of its piston stroke, and the second, a random one, caused by leakage of the poppet valves. In a series of 10 successive measurements of a 0.5 ml. volume, the average deviation from the mean was 0.02 ml., the largest deviation was 0.06 ml. The negative errors were fewer and larger than the positive errors. The systematic error probably can be reduced to about 0.02 ml. by weighing 2 ml. of solution delivered by the syringe. A systematic error of 0.03 ml. superimposed upon the random error causes an error of dilution equivalent to 1 tube in a 9- or 10-tube geometric-series dilution test. A 5-ml. measuring pipet was used to measure the 0.5-ml. quantities of broth with a mean volume delivered of 0.496 and an average deviation of 0.006 ml., an error in volume so small as to be insignificant.

Another error is a hidden one that makes the dilutions less than theoretical and is caused by adherence of high potency material to the wall of the pipet above the 0.50 ml. mark; this is finally washed down into a high dilution tube where it makes a large decrease in dilution. The small amount of liquid that adheres to the outside of the tip of the pipet also decreases the dilution. Thus the mechanical errors make the substances assayed seem more active than they are by making the dilution in a tube considerably less than it really is. Positive error in measuring the volume of broth added by the automatic pipet increases the dilution and tends to offset the errors of pipetting. The reason dilutions computed from end points obtained in the range of tube 6 to tube 10 are in serious error is now obvious. Since we want to use the geometric-series dilution test because of its convenience, we either dilute the solutions assayed so that the end point falls within the first five or six tubes or we use the results only as a guide in diluting for other tests. A pipet calibrated to contain 0.50 and 1.00 ml. would eliminate the pipet errors. Such a pipet, calibrated from the tip, can be had on special order.

Samples in the arithmetic test are measured with a 0.1- or 0.2-ml. measuring pipet graduated in units of 0.01 ml. or a Kahn pipet graduated in 0.001 ml. For measurements of the volumes delivered by the Kahn, the 1-ml. pipet, and the 0.1-ml. pipet, the meniscus was set by mechanical means so that the error of setting was not greater than the width of a calibration line. Volumes delivered by a Kahn and a 0.1-ml. measuring pipet at steps of 0.05 ml. were measured by weighing the water delivered:

Range (ml.):	0–0.05	0.05–0.10	0.10–0.15	0.15–0.20
Kahn pipet:	0.0526	0.0486	0.052	0.0506
measuring pipet:	0.051	0.050		

The accuracy of the Kahn pipet is not very great. If the 0.10–0.20 volume were used in the 1-ml. test, the dilution would be 10.7 as compared with the theoretical 11.0. If the 0.05 volume were measured into the sixth tube, the dilution would be 20.0 as compared with the theo-rectical 21.0. Neither volume is in error enough to make a one-tube differ-ence. The error of 0.05 ml. in measuring the volume of the broth in the tube probably would not make a 1-tube difference in the test.

If the activity of the sample is reported in terms of a standard of nearly the same activity, the errors of dilution presumably will be the same for both the sample and the standard. Then, when the concentration of the antibacterial substance in the sample is computed from the ratio of the activities of the sample and the standard and the concentration of the standard, the error resulting from mechanical imperfections will be rela-tively small.

III. Antiluminescent Assay

A. Introduction

Many antibiotic substances and chemicals, especially quinones, quench the luminescence of species of *Photobacterium*. Antiluminescent activity is of interest because it may be independent of antibiotic activity. The test is quick and simple to do but does require a 15°–20°C dark room for incubating and reading the test. The history of the assay, details of application, discussion of mechanism, and application to a number of antibacterial substances will be found in the article by Kavanagh.[7]

B. Test Organism

Photobacterium fischeri. Some strains give a considerably greater lu-minescence than others. Stock cultures were maintained in an artificial sea-water broth in which the bacteria were viable, though not lumines-cent, for at least 6 months at 15°C, and on the modified Egorova-Yarmo-link[8] agar (E-Y agar) on which they were luminescent for about 1 month and viable for more than 6 months at 15°C. The modified E-Y agar contains: NaCl, 30 gm.; Bacto-peptone, 10 gm.; asparagine, 5 gm.; K_2HPO_4, 1 gm.; $MgSO_4 \cdot 7H_2O$, 0.5 gm.; agar, 20 gm.; water, 1000 ml.; pH after sterilization 6.9.

C. Test Medium

Test medium is artificial sea-water broth of the following composition: NaCl, 26.7 gm.; KCl, 0.7 gm.; $CaCl_2$, 1.2 gm.; $MgCl_2 \cdot 6H_2O$, 5.1 gm.; $MgSO_4 \cdot 7H_2O$, 6.8 gm.; Bacto-peptone, 2 gm.; water, distilled, 1000 ml.; pH 6.2–6.6.

[7] F. Kavanagh, *Bull. Torrey Botan. Club* **74**, 414 (1947).
[8] A. A. Egorova, and L. Yarmolink, *Mikrobiologiya* **14**, 265 (1945).

D. Test Inoculum

A flask containing 100 ml. of medium is inoculated with bacteria taken from an agar slant or with several milliliters of solution from a tube of sea-water broth which had grown for 48 hours or longer at 15°–20°C. The inoculated flask is incubated at 15°–20°C for 48 hours with occasional shaking to aerate the solution and to make the suspension of bacteria brightly luminescent. Growth and luminescence of the bacteria may be better in the 15°–20°C range than at lower and higher temperatures. Incubation of culture and test should be at a temperature near the optimum. Vigorous sidewise shaking of the racks of tubes just before reading a test provides the oxygen needed to develop full luminescence.

E. Concentration of Bacteria

The concentration of the bacteria can influence decidedly the minimum antiluminescent concentration of the active substance. However, this is important only when absolute values are needed.

F. Serial Dilution Test

The geometrical-series[9] dilution test is used with *P. fischeri*. In each tube (12 × 75 mm.) is placed 0.5 ml. of the artificial sea-water broth; the serial dilutions of the antibacterial substances is made, and, when all of the dilutions of a set of tests are made, 0.5 ml. of a 48-hour culture of the test organism is added as rapidly as possible to each tube. The racks of tubes are shaken to mix and to oxygenate the solutions and are placed immediately at 15° to 20°C. After the tests had been at 15° to 20°C for the appropriate time, the racks are shaken to aerate the solutions, and the number of nonluminescent tubes are counted. The reading of the test is done with the unaided, dark-adapted eye in an absolutely dark room. The dilution of the antibacterial substance is calculated from the number of nonluminescent tubes, the first tube representing a dilution of 4.

The activities are expressed as the minimum antiluminescent concentration in micrograms per milliliter. The antiluminescent activities of several antibiotics and antibacterial chemicals available in 1947 as well as the antibacterial activities[10] against *Staphylococcus aureus* and *Escherichia coli* are given in Table IV. Activities of more compounds and at other times of incubation and a discussion of the results will be found in the article by Kavanagh.[7]

Antiluminescent activity seems to be unrelated to a particular type of antibacterial activity. Bactericidal concentrations of active substances

[9] F. Kavanagh, *Bull. Torrey Botan. Club* **74**, 303 (1947).
[10] F. Kavanagh, *J. Bacteriol.* **54**, 761 (1947).

TABLE IV

MINIMUM ANTILUMINESCENT AND ANTIBACTERIAL CONCENTRATIONS
OF ANTIBACTERIAL SUBSTANCES[a]

Antibacterial substance	Antiluminescent			Antibacterial	
	Photobacterium fischeri			S.	E.
	10 min.	3 hr.	24 hr.	aureus	coli
Aspergillic acid	—	2	2	4	62
Biformin	9	0.34	0.17	0.3	1.7
Cassic acid (Rhein)	64	32	16	8	1000
Citrinin	256	32	16	16	>1000
Dihydrostreptomycin	>400	>400	100	0.03	0.25
Gliotoxin	32	8	2	0.12	25
Hydrogen peroxide	12	6	6	8	10
Kojic acid	>2500	>2500	>2500	1250	2500
2-Methyl-1,4-naphthoquinone	3.3	3.3	3.3	1.7	200
1,4-Naphthoquinone	0.4	0.4	0.4	8	25
Patulin	256	2	0.5	8	8
Penicillic acid	128	16	4	16	64
Penicillin G	>500	>500	>500	0.016	16
Pleurotin	32	8	16	1	>500
Spinulosin	64	125	125	64	250
Streptomycin	>400	>400	100	0.03	0.13
Sulfanilamide	750	375	3000	3000	—
Tolu-p-quinone	1	1	2	1	25

[a] Units: μg./ml.

may be considerably higher than the antiluminescent concentration. With
one substance, tolu-p-quinone, there was exact proportionality between
minimum inhibitory concentration of the quinone and the bacteria. Thus
it seems that it is the number of molecules per bacterial cell and not the
concentration of tolu-p-quinone that determines inhibitions of luminescence but not the antibacterial action against *Staphylococcus aureus* or
Photobacterium fischeri.

CHAPTER 4

Elements of Photometric Assaying

FREDERICK KAVANAGH

Eli Lilly and Company, Indianapolis, Indiana

I. Introduction

Photometric methods of antibiotic assay have been used since the earliest days of the penicillin program. Assays were needed at each step in the production, isolation, purification, and investigation of pharmacological activity of penicillin. Many methods were devised, yet none was satisfactory for all purposes. A method might excel in one particular and fall short in others. A quick and sensitive method was needed for estimating penicillin in blood, a quick method for following production of penicillin in fermenters, and an accurate but not necessarily rapid method was required for standardization of production lots. A good discussion of the many methods, their advantages and disadvantages, is given by Heatley.[1] He reviewed the literature and practices up to 1947 and described numerous methods including turbidimetric and diffusion methods.

Methods in which the effect of antibiotics upon growth of a test organism in liquid medium is measured photometrically have several advantages. Among these are rapidity, ease of operations, objective measurement of response to the drug, absence of diffusion effects, and accuracy. The disadvantages are minor ones, including complexity of equipment and requirements that the sample not be grossly contaminated and not contribute measurable color to the assay medium.

The photometric method of assay is very simple in principle: the test substance is added to a suspension of the test organism in a nutrient medium, the mixture incubated, and the response of the test organism measured. The test organism may be any organism that will give a uniform suspension. Bacteria, protozoa, fungi, yeasts, and algae have been used. The turbidity of the suspension is measured after appropriate incubation. Other responses such as total number, dry weight, number of viable individuals, total nitrogen, pH, titratable acidity, carbon dioxide output, and oxygen consumption have been employed. Although these latter responses are not those of a turbidimetric method, the details of the test other than measuring the response may be the same. Discussion of the

[1] N. G. Heatley, Chapter 3, *in* H. W. Florey, E. Chain, N. G. Heatley, M. A. Jennings, A. G. Sanders, E. P. Abraham, and M. E. Florey, "Antibiotics," Vol. I, 2 vols., 1774 pp., Oxford Univ. Press, London and New York, 1949.

theoretical aspects of turbidimetric assays will be for bacteria responding to antibiotics and vitamins.

The first turbidimetric antibiotic assay seems to be that of Foster[2] who described a 16-hour test for penicillin activity. The method apparently was abandoned later in favor of the plate method. The publications of McMahan,[3] Joslyn,[4] and Lee et al.[5] in 1944 described practical rapid assay methods which were the forerunners of current methods. Articles published since 1944 indicate that little advance has been made since the publication of McMahan.

The microbiological elements of photometric assaying discussed in this chapter are included under the general headings of inoculum, media, incubation, and response of the bacteria to the test substance. Understanding of turbidimetric methods also requires consideration of the influences of physical factors upon turbidity and calibration of the photometers used to measure turbidity. The merits of commonly used expressions of the dosage-response curves are discussed as well as the related topics of design and validity of the assays.

II. Physical Factors Influencing Turbidity

A. Concentration of Cells

The bacteria commonly employed in assaying have cell dimension ranging from 0.5 to 3μ. The angular distribution of light scattered by particles of this size (approximately equal to the wavelength of the light) is described by the exact but complex formulae of Mie. The angular distribution of light is anisotropic, with much greater scattering in the forward direction (direction of propagation of the irradiating light) than in either the backward of the 90° directions. The photometer used to measure the optical density of a suspension responds to scattered light as well as to collimated light making the measured optical density less than the true value.

Measurement of the optical density, D, of a series of dilutions of a given suspension affords a method of discovering the extent of which the "particle concentration" form of Beer's law is obeyed for a particular instrument and suspension. When the law applies, s, in the expression $D = sC$, will be constant and a plot of D against the corresponding number of particles per milliliter, C, will be a straight line. D will vary

[2] J. W. Foster, *J. Biol. Chem.* **144**, 285 (1942).
[3] J. R. McMahan, *J. Biol. Chem.* **153**, 249 (1944).
[4] D. A. Joslyn, *Science* **99**, 21 (1944).
[5] S. W. Lee, E. J. Foley, and J. Epstein, *J. Biol. Chem.* **152**, 485 (1944).

with and be proportional to the thickness of the layer of scattering substance. The thickness is included in the term s.

Dreosti[6] designed a photometer for measuring the optical density of scattering media. It followed the theoretical relationship over an optical density range of 5. His approach influenced subsequent designers of nephelometers and photometer for measuring particle concentration. Mestre[7] constructed a photometer after the principles of Dreosti and found that a suspension of *Escherichia coli* followed the form of Beer's law for suspensions over a concentration range of 64, or a density range of 0.477. Departure from linearity, decreasing s, became apparent above $D = 0.5$ (600 million *E. coli* cells/ml.) and increased with increasing density of the suspension.

Longsworth[8] built a photometer which admitted more scattered light and departed more from linearity than the instrument of Mestre. He expressed the relationship between optical density D and the concentration C by the equation

$$D = \alpha C - \beta^2 C^2.$$

He applied the equation to measurements of suspensions of India ink, *Saccharomyces cerevisiae*, and *Lactobacillus acidophilus*. This parabolic equation is valid at least up to $2 \cdot 10^9$ cells/ml. of *L. acidophilus*, $0.5 \cdot 10^9$ of *Staphylococcus aureus* or $1.3 \cdot 10^9$ of *Salmonella gallinarum*. The numerical values of α and β depended upon the concentration scale, which was entirely arbitrary, since the concentration of each stock solution was taken as unity. The ratio, $\rho = \beta/\alpha$, was independent of the scale.

If optical density and concentration were proportional, β, and hence ρ, would be zero. The deviation of ρ from zero is a measure of lack of proportionality between D and C and is a property of the design of the instrument. If the design were such that no forward scattered light reached the photocell, ρ would be zero and D and C would be proportional. The Dreosti design requires the use of corrected lenses and long optical path. Most designers prefer uncorrected lenses and short optical paths even though the subsequent ρ is not even approximately zero. When ρ is not zero, an extra operation is introduced, that of converting the measured optical density into corrected optical density. The extra labor is worthwhile when use of the corrected optical density linearizes the assay curve.

Longsworth and several other investigators corrected the optical density for deviation from Beer's law. The corrected optical densities were used in subsequent calculations involving cell members, cell mass, etc. Alper and Sterne,[9] in their excellent and neglected study of the photoelec-

[6] G. M. Dreosti, *Phil. Mag.* [7] **11**, 801 (1931).

[7] H. Mestre, *J. Bacteriol.* **30**, 335 (1935).

[8] L. G. Longsworth, *J. Bacteriol.* **32**, 307 (1936).

[9] T. Alper and M. Sterne, *J. Hyg.* **33**, 497 (1933).

tric measurement of opacity of bacterial cultures, prepared a calibration curve relating instrument response to cell concentration determined by counting in a hemocytometer.

I prefer a calibration line with ends fixed by the extremes of the scale of the photometer and to convert scale values into units of relative concentration by means of a calibration curve. The calibration curve corrects the photometer readings for deviations from Beer's law. The method of preparing the calibration curve is given in Section III of this chapter.

Another solution to the problem of achieving linearity in a small instrument was to design a forward scattering turbidimeter so that its output was a linear function of concentration of cells (at least in the range encountered in assay work). The instrument, as constructed, also computed and indicated the cell concentration of any tube as percentage of the concentration of a selected tube of a standard curve. The ability to linearize and to compute simplifies the construction of both inhibition and growth curves.

B. Size of Cells

The assumption has been made to this point in the discussion of the measurement of turbidity that the size and contents of the cells in the population of a given species did not appreciably affect the measurement of concentration.

Numerous publications indicate that young cultures have larger and more varied sizes of cells than old cultures. Clark and Ruehl [10] measured the sizes of 37 species of bacteria at close time intervals and found that the majority reached their maximum size at about 6 hours under the growth conditions used. The size became more uniform and smaller as the cultures aged so that the cultures presented typical textbook pictures after 24 hours. Cells of *Staphylococcus aureus* at 4 to 6 hours were twice the diameter of the old cells. Cells of *Escherichia coli* from a 2-hour culture were more than 3 times as long as those from a 24-hour culture. The length of many of the rod forms changed much more than the diameter.

The relationship between cell size and photometric number had been reported in two papers not mentioned by either Mestre or Longsworth. Novel photoelectric photometers were used by both groups of investigators. The curves given by Pulvertaft and Lemon [11] suggest that the opacity of *E. coli* cells was a maximum at 2 hours and relatively constant for cells grown longer than 5 hours in beef broth. Alper and Sterne measured the opacity of *Salmonella gallinarum* growing in beef broth at pH 7.2–7.8 and made direct counts with a haemocytometer. They realized

[10] P. Clark and W. H. Ruehl, *J. Bacteriol.* **4**, 615 (1919).
[11] R. J. V. Pulvertaft and C. G. Lemon, *J. Hyg.* **33**, 245 (1933).

that the response of their photometer was nonlinear with respect to cell concentration and prepared a calibration curve using a culture 24 hours old. "Thus the opacity of the culture at any age was defined by the number of 24-hour-old organisms necessary to give the same scale reading." They found that the specific opacity (opacity per cell) was a function of the age of the culture, with young cells much more opaque than old (24

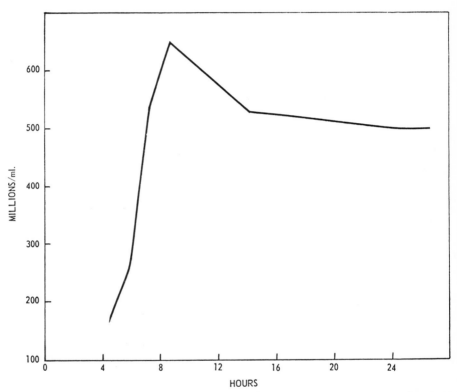

Fig. 1. Concentration of cells of *Salmonella gallinarum* from cultures of different ages needed to give opacity equal to 5×10^8/ml. of cells from a 24-hour-old culture (Alper and Sterne[9]).

hours) cells as is shown in Fig. 1. Wilson[12] measured the sizes and opacity of young and old cultures of *Bacillus aertrycke* (mutton) and concluded that opacity measures total volume of protoplasm. He found the specific opacity to be proportional to the volume of the cells computed from measurements of diameter and length of the cells. If this relation holds for the cultures of Alper and Sterne, their young cells (4.7 hours) were 7 times the volume of the old cells (24 hours).

[12] G. S. Wilson, *J. Hyg.* **25**, 150 (1926).

Hershey[13] determined the viable count, the nitrogen content of 10^9 organisms, and the photometric count of suspensions of young (3.5 hours) and old (24 hours) cells of *Escherichia coli*. The turbidity was expressed in terms of a calibration made with a 24-hour-old culture. The nitrogen contents were the same, the photometric counts were about the same, and the plate count of the young culture was one-third of the older culture. These results indicated that young cells were larger than old cells and that the photometric method measured cell mass, not cell number. The observations of Longsworth on *Lactobacillus acidophilus* growing at a pH constantly controlled at 6.0 support those of earlier investigators. Cells 8 hours old were larger than cells 18–72 hours old. The 18- to 72-hour-old cells had constant photometric size.

Pritchard[14] wrote that the dry weight of 20-hour-old cells of *Lactobacillus leichmannii* in a suspension and the scale of the Spekker photoelectric absorptiometer were nearly proportional (at least up to O.D. 1.12). His data show, contrary to his statement, that the Spekker follows the same equation but with somewhat less deviation from linearity than the other instruments (Section III.A). Schaechter *et al.*[15] state that optical density was proportional to the dry weight of *Salmonella typhimurium* irrespective of the cell size (optical density of 0.10 corresponds to 17 to 18 μg. dry weight/ml.). They worked in the region of small optical densities where deviation from a linear relationship between optical density and cell mass is small.

The results of these investigators indicate that the photometric method, when suitably corrected for instrument response, measures cell mass, not cell number. The method gives an accurate measure of cell number only when the cells are all of the same size as in the older cultures (and when grown in the presence of certain antibiotics?).

The drawings of Duguid[16] of the morphological changes of *Escherichia coli* with aging of the culture in the absence and presence of penicillin show the changes in cell size with age and penicillin concentration. Whereas in the absence of penicillin the cells decreased in size with aging, in the presence of penicillin the cells elongated and became swollen and distorted. With high concentrations of penicillin, some of the cells ruptured to leave ghosts (cell walls) and others assumed swollen forms quite unlike the normal short rod form. The enlarged cells obtained in the presence of penicillin would have greater opacity than normal cells of the same age. Consequently, cell concentration measured in terms of normal cells would be larger than the true concentration by several fold.

[13] A. D. Hershey, *J. Bacteriol.* **37,** 285 (1939).
[14] H. Pritchard, *Analyst* **76,** 155 (1951).
[15] M. Schaechter, O. Maaløe and N. O. Kjeldgaard, *J. Gen. Microbiol.* **16,** 592 (1958).
[16] J. P. Duguid, *Edinburgh Med. J.* **53,** 401 (1946).

These observations on rod-shaped bacteria indicate that cell mass and not cell number determines the optical density of the suspension. If the calibration curve is made with old (small) cells, then the photometric cell concentration of young cultures will be larger than the actual concentration. Any other factor, e.g., penicillin, that causes enlargement of bacteria will also cause an increase in the apparent concentration measured photometrically. This inherent error of measuring cell concentration may be responsible for the curvature of the penicillin dosage-response curves (Fig. 14) at high penicillin concentrations. Any factor which promoted cell division to give smaller than average size for the particular age of the culture would cause the photometric concentration to be smaller than the true cell concentration. The influence of cell size on apparent concentration measured photometrically must always be kept in mind in interpreting curves constructed on the basis of such measurements.

The inherent bias caused by different cell sizes distorts some of the dosage-response curves but does not cause practical difficulties in assaying antibiotics. The growth curves obtained in assay of vitamins and amino acids will not be subject to distortion caused by cell sizes if the population density is limited by exhaustion of the factor assayed and not by a short time of growth.

Alper and Sterne observed that the size of the inoculum as well as the age of the culture influenced the size of the cells of a young culture of a given age. With a small inoculum ($2 \cdot 10^6$/ml.), the cells of a 3-hour culture were 7 times as opaque as the cells from a 24-hour-old culture (the standard). A seeding of $130 \cdot 10^6$/ml. gave 3-hour cells which were as opaque as the standard. Both seedings grew to a terminal population of $500 \cdot 10^6$/ml. in 24 hours. This observation is important when inocula are "standardized" by photometric measurement. Age and seeding must be considered in estimating the population density from the turbidity of the suspension.

C. Opacity of Cells

Bacterial cells may be optically empty, opaque, or gradations between the two extremes. Mestre, who experimented with algae and bacteria, believed that the scattering of light by living cells is in the main due to reflection and refraction at interfaces where there are different indices of refraction. "The appearance of suspensions of microorganisms under the microscope suggests that they should be regarded as suspensions of microlenses, of shapes, sizes, and transparencies characteristic of the organism and its physiological state. Because of the low ratio (probably less than 1.04) of indices of refraction of the organisms and of their suspending medium, these microlenses would have a low surface reflectance and be of relatively extremely long focal length. The distribution of the flux incident on a single cell might consequently be expected to be characterized by an extraordinary anisotropy with almost all of the light scattered in a forward direction and only slightly deviated from its original direction." [7] About 10 times as much light is scattered in the forward direction by *Staphylococcus aureus* as at right angle to the incident light beam. Longsworth also observed that *Lactobacillus acidophilus* was optically empty in contrast to yeast which was well filled with material. His measurements with *L. acidophilus*, yeast, and India ink indicated that much more light was scattered in the forward direction by *L. acidophilus* than by the suspension of optically opaque particles. Measurements of the optical density of opaque particles deviated less from Beer's law than the measurements on the optically empty cells. Any change in opacity would be reflected in the measurement of turbidity. Ideally, a suspension should be measured in terms of a calibration curve prepared with organisms of the same opacity. The opacity affects the slope, β^2, of the equation, $S/C = \alpha - \beta^2 C$; apparently, the size of the cells does not affect the slope for the sizes encountered in assaying. The influences of opacity and shape and size of cells on measurement of turbidity need investigation by modern methods.

Mager *et al.*[17] observed an increase of as much as 100% in turbidity of living gram-negative bacteria with increase in salt or sugar concentration. It seems to be a pure osmotic effect. They did not know the cause and suggested that changes in the state of swelling of the cytoplasm caused the change in turbidity. An increase in opacity caused by the high salt concentration would decrease the deviation from Beer's law with a consequent increase in turbidity. These observations on gram-negative bacteria emphasize the necessity of careful control of environment if turbidity measurements are to have meaning. Gram-positive bacteria can not be plasmolyzed even by high salt concentrations and did not change turbidity with change in salt concentration.

Longsworth wrote that increase in the apparent thickness of the cell wall tends to increase the optical density of the suspension. Thickening of the cell wall occurs during aging.

III. Calibration of Photometers

A. Photometers and Their Use

The photometers most often mentioned, when any are, in the literature of the last 20 years are the Lumetron Model 402E and the Klett-Summerson. The Lumetron Model 402E is popular with the drug companies

[17] J. Mager, M. Kuczynski, G. Schatzberg, and Y. Avi-Dor, *J. Gen. Microbiol.* **14,** 69 (1956).

and is used in the Food and Drug Administration. Beckman spectrophotometers and the Bausch & Lomb Spectronic 20 are used now for special applications.

The Klett-Summerson instrument and the spectrophotometers have optical density scales. The reading of the scales are nearly proportional to the concentration of the bacteria (Fig. 2). The Lumetron photometer

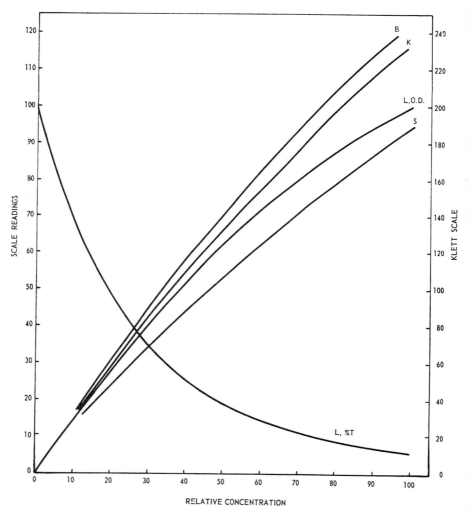

Fɪɢ. 2. Responses of four photometers to *Staphylococcus aureus* suspensions of accurately known relative concentrations. Beckman Model B (B); scale in 100 × O.D. units, 590 mμ. Bausch & Lomb Spectronic 20 (S); scale in 100 × O.D. units 600 mμ. Klett-Summerson photoelectric colorimeter, 66 filter (K). Lumetron 402E (L, O.D.) with scale proportional to optical density (100 = 1.00 O.D.). Lumetron 402E (L, %T.) with percentage transmittancy scale.

has a 100-division linear scale and reads percentage transmittancy. It is used in two modes of operation. In one, it is used as a balanced two-cell instrument, and the reading is obtained from the scale of the slide wire at electrical balance. The balancing is a slow operation because of the rather long time constant (4 seconds) of the galvanometer. The operator tends to turn the potentiometer faster than the galvanometer can follow and may guess at the balance point. There is no control over the number of potentiometer divisions representing the effective ends of the assay curve. As can be seen from Fig. 2, the scale is not a favorable one; too much of it is wasted on a region of cell concentrations (<10) of little interest. Potentiometric balancing seemed too slow when a great many tubes were to be read. A faster method of reading was devised and used generally by the antibiotics producers. The Lumetron photometer was used in an electrically unbalanced condition, and adjusted so that the highest standard read 90 on the galvanometer and the lowest standard read about 10 (100-division scale, 80 mm. long). The standard curve was measured and plotted on graph paper and the unknowns read from the graph. Reading the galvanometer scale was faster than making a careful potentiometric balance. All too often the operators shook the tubes, poured the samples into the cuvettes, and, rather than wait for the galvanometer to come to rest (it did move slowly), guessed at its rest point, emptied the cuvette, poured in the next sample, etc. Samples could be read very rapidly. A consequence of the emphasis on speed was errors larger than they need be. The galvanometer scale was short, the line of the optical pointer was broad, the rest point uncertain, and the sample was full of air bubbles. Anyone concerned with accuracy who knew no other turbidimetric assay would consider all turbidimetric methods to be inherently inaccurate. Somewhere the purpose of an assay laboratory was lost; it is to obtain accurate answers, not to process the maximum number of samples many of which would be reassayed the next day because of unacceptably high errors. The rapid reading method described above probably was as responsible as any single factor for the widely held belief that turbidimetric assays are less accurate than plate assays. The belief in inherent inaccuracy retarded the development of accurate turbidimetric assays for many years.

The Beckman Model B spectrophotometer is a single photocell instrument and is unsuited for rapid measurement of a large series of turbidities. The Spectronic 20 has an effective optical density scale much too short for high precision assays. None of the instruments is as mechanically stable as is desirable; the zero points shift too much during a long series of measurements.

The first large step toward improving the precision of the turbidimetric assays at Eli Lilly and Company was the conversion of the Lumetron

Model 402E photometers into self-balancing potentiometric instruments. Only the optical system of the photometers was used. The entire resistance of the lamp rheostat was wired permanently in series with the lamp to increase the life of the lamp. The speed of the self-balancing potentiometer was made slow (20 seconds for full scale) to give the air bubbles time to rise out of the light path. The converted photometers made with Minneapolis-Honeywell parts had optical density scales. Other instruments converted later used the Leeds and Northrup Speedomax H as the self-balancing potentiometer and had 100-division linear scales. These instruments have operated many millions of times in the last 6 years with only a minimum of servicing. The angular position of the potentiometer shaft is a measure of the turbidity and has been digitized by a Datex shaft position encoder and translator for transmission to a tape or card punch for automatic recording. The recorded information can be further processed by a computer.

The converted photometer has the disadvantage shared with all other commercial photometers, that the scale reading must be converted into relative concentration of cells by means of a calibration curve. Every conversion introduces a small error. The linear turbidimeter mentioned earlier avoids conversion and also computes the response (ratio of tube with antibiotic to zero tube) automatically when needed as in assay of antibiotic substances.

The curves in Fig. 2 show that different instruments responded slightly differently to the different concentrations of *Staphylococcus aureus*. The ratio, ρ, was computed for three of the instruments (see Section II.A for meaning of ρ). It is a property of the geometry of the instrument and should be a constant. It was 0.047 for Lumetron Model 402E instruments calibrated with suspensions of *S. aureus*, *Lactobacillus* sp., *Klebsiella pneumoniae*, or a unicellular green alga. It was 0.027 for the Klett-Summerson photometer, 0.041 for a Beckman Model B spectrophotometer, and about 0.02 for a Spekker absorptiometer (computed from data given by Pritchard, see Section II.B). The values of ρ indicate that the Spekker absorptiometer responded slightly less to forward scattered light than any of the other instruments. The differences in ρ values are of no significance in assaying. The ρ is much too large for its effect on the calibration curve to be ignored.

The Bausch & Lomb Spectronic 20 has been coupled to a digital voltmeter to provide a signal to a card punch proportional to the percentage transmittancy of the suspension. The cards are fed to a computer provided with a predetermined standard curve in its program. The successful application of this automatic computation of an assay requires that the standard curve be reproducible within narrow limits from day to day and that the relation between concentration of antibiotic and per cent transmittancy be linear. If these conditions do not obtain, the error of an assay may be larger than supervisors of some assay laboratories would accept.

B. Air Bubbles and Shaking Errors

The air bubbles formed and suspended in a broth suspension of bacteria by the usual vigorous shaking needed to suspend the bacteria completely interfere seriously with accurate measurement of turbidity. The measuring instrument responds to air bubbles as well as to bacteria. The bubbles rise in the suspension and the larger ones will be out of the optical path within 30 seconds, a length of time few operators who pride themselves on their speed will wait before measuring the turbidity. Thus the usual practice of measuring immediately after shaking contributes a substantial and variable error to the measurements. The retention and, perhaps, formation of air bubbles seems to be much greater for heavy suspensions of bacteria than for light ones.

Shaking the tubes causes three types of errors. One error results if the tubes are not shaken vigorously to suspend bacteria collected on the bottom of the tube; the turbidity is low. A second and large error comes from air bubbles in a vigorously shaken tube when it is measured immediately after shaking. The third and smallest error is caused by sedimentation of the bacteria in a tube vigorously shaken and then allowed to stand to allow air bubbles to rise. This error is small if the tubes stand only 15 minutes.

An example of the different errors is given in Table I. A *Staphylococcus*

TABLE I

EFFECT OF SHAKING ON APPARENT RELATIVE CONCENTRATION (R.C.)

Treatment	R.C.
Not shaken	39.7, 37.8
Shaken, read at once	58.7, 58.7
Shaken, allowed to stand for 15 minutes	47.0, 47.5

aureus suspension that had been incubated for 4.5 hours in an assay, killed, and cooled was measured without shaking, immediately after shaking, and 15 minutes after shaking. Two tubes of each treatment were measured. The unshaken tubes were about 20% low. The tubes measured immediately after shaking were 20% high and rapidly decreased toward and finally reached the values of 47.5 and 46.5, respectively. Waiting for the readings to stabilize is rather irksome and is avoided by shaking and then waiting 15 minutes before measuring.

A sample of uninoculated broth was shaken, poured in the cuvette of a linear-forward-scattering nephelometer, and measured periodically beginning at 5 seconds. The scale reading for the zero tube of an assay would be about 100. The same experiment was done with a suspension of cells in broth (see tabulation).

Time	(sec.): 5	15	20	30	45	(min.): 1	1.5	2	22
Broth	26	6.3	4.0	2.4	1.8	1.5	1.0	0.6	0
Cells	—	56.5	53.7	52.5	51.6	51.2	50.8	50.6	49.6

Other experiments show that bubbles clear from broth suspensions and water suspensions at the same rate.

If the turbidity were measured by an automatic instrument programmed to sample the reading for only a short time, errors caused by air bubbles might not be noticed. The magnitude of the error depends upon the speed of response of the photometer and the waiting time. A fast instrument response leads to larger and more variable errors than a slow response.

C. Flow Birefringence

Turbidity of suspensions of rod forms of bacteria is more difficult to measure accurately than turbidity of suspensions of cocci because of flow birefringence resulting from movement of the rods. The movement is caused by motion imparted to the cells during filling of the cuvette and from thermal agitation caused by absorption of heat from the light beam. The combination of these two sources of motion causes the photometer to reach its balance point slowly and erratically as the movement slowly dies out. The following tabulation was taken during the calibration of a Lumetron Model 402E with *Lactobacillus leichmannii* suspension of relative concentration 60. The time course of the optical density of

Time	(sec.): 5	10	30	45	(min.): 1	1.5	2	2.5	3	4	5
Scale reading:		74.5	74.5	71	67.5	66	65.5	64.7	66.4	65.3	65.0	65.0

the solution is *typical* of such suspension. What is the correct reading? The 1.5-minute values were used in constructing the calibration curve. This is an intolerably long time to wait; one rack of tubes (40) would require 1 hour for measurement. This particular difficulty in measuring turbidity reduces the accuracy of the usual *L. leichmannii* assay for vitamin B_{12} considerably below that attainable.

The same suspension was measured in a forward scattering turbidimeter (see tabulation). The scale reading obtained at the end of 30 seconds

Time (sec.):	5	15	30	45	(min.):	1	1.5	2	2.5	3
Scale reading:		33.4	33.8	33.0	32.9		32.8	32.7	32.9	33.0	33.1

is nearly close enough to the 3-minute reading to be satisfactory for assaying. Even in this instrument which was designed to minimize circulation of suspension in the cuvette, a longer time than we like was needed to obtain a reading with a small flow error.

The impracticability of accurate, quick measurement of turbidity of long rod forms of bacteria in the usual photometer is one reason for preferring cocci as test organisms, or at worst, short rod forms. The accuracy and convenience of measuring the turbidity should be considered when a new assay is being developed.

D. Calibration of Photometers

Toennies and Gallant[18] realized that for some problems in bacterial physiology absolute magnitudes of bacterial populations are not required and the necessary information will be conveyed by relative figures which are directly proportional to bacterial quantity. They knew that optical density of a bacterial suspension was only approximately proportional to the concentration. They introduced the use of adjusted optical density (AOD) computed by the equation of Longsworth from the observed optical density. The AOD is a linear function of concentration of bacteria and is the optical density which the instrument would have measured had it not been responsive to forward scattered light. Their way computing AOD is awkward and the AOD has no advantage over the concept of relative concentration. For use of AOD in amino acid assays see Chapter 8.

The photometric measurement of cell multiplication rates, of growth, or the application of the log-probability relationship (Section VIII.A) requires that the photometer (or nephelometer, or spectrophotometer) be calibrated at least in relative terms. The important and difficult calibration is the establishment of the relation between the instrument response and the concentration of cells. As is shown later in this section, the relation is nonlinear for all instruments except a specially designed nephelometer and even it must be adjusted for linearity. The relative calibration is all that is required for assay purpose. Once it is obtained, the approximate absolute calibration is easily made.

The general *shape* of the calibration curve was the same for *Staphylococcus aureus*, *Escherichia coli*, *Klebsiella pneumoniae*, *Lactobacillus* sp., and *Scendesmus basiliensis* when measured in an instrument that gave

[18] G. Toennies and D. L. Gallant, *Growth* **13**, 7 (1949).

the parabolic relation [Eq. (1)] between scale reading (optical density) and concentration of bacteria. Calibrations performed with *Staphylococcus aureus* suspensions could be used with other bacteria without appreciable error. The shape of the curve was the same for living and heat killed *S. aureus*.

The calibration curve is constructed by measuring the instrument response to several carefully prepared suspensions of killed *S. aureus*. To do this, prepare a very heavy suspension of *S. aureus* by heavily inoculating several liters of broth and incubating in the 37°C. water bath for 4 hours with frequent shaking of the flasks to aerate. Kill the bacteria by steaming the flasks for about 30 minutes. Centrifuge the suspensions, wash the centrifugate once with water. Suspend the bacteria in about 100 ml. of water, and shake vigorously. Now let stand several hours and carefully decant about 80 ml. of the suspension. Add a small amount of merthiolate or several milliliters of formalin to preserve the suspension.

The assumption will be made in the following discussion that the highest concentration will measure 1.00 on the optical density scale or 10% on the percentage transmittancy scale. This concentration is high on the scale of many instruments and need not be exceeded in practical assaying. Dilute carefully with water a sample of the heavy suspension prepared above, measure, and estimate the dilution needed to read 1.00 on the optical density scale. Repeat if necessary, and by this method of successive approximations prepare about 1 liter of the O.D. 1.00 suspension. Always keep in mind the necessity of having a uniform suspension free from air bubbles. One way of ensuring this is to shake the bottle vigorously and let it stand for 15 minutes and before measuring. Measure with volumetric pipets samples of the O.D. 1.00 suspension and dilute with water in a volumetric flask. If the instrument follows the parabolic equation, five concentrations will suffice to establish the curve. If it does not, then prepare concentrations at 10% intervals. Assign 100 to the concentration that gives a reading of O.D. 1.00, and prepare concentrations of 20, 40, 60, 80, from the 100 suspension. Set the instrument to O.D. of 0.00 with water and measure the standard suspensions. Pouring the sample into the cuvette will put air bubbles in the liquid. Wait 20–30 seconds for the bubbles to move out of the optical path and read the instrument scale. The reading at 20 seconds is a practical one; it is long enough to allow most of the air bubbles to clear but not long enough for appreciable settling of the bacteria to occur.

One way of preparing the calibration curve is to plot the instrument response (S) against relative concentrations (C) on rectangular coordinate paper. A second way is to plot S/C against C on rectangular coordinate paper and draw the best straight line through the points. The equation of the line is

$$S/C = \alpha - \beta^2 C. \tag{1}$$

The Y-coordinate of S/C for $C = 0$ is α. Compute β^2 from the value of S/C for some value of C between 50 and 100. Now put the equation in the form $S = \alpha C - \beta^2 C^2$ and compute S for assigned values of C. The computed values of S are plotted against C and the curve drawn. This curve is one of smoothed data and probably is better to use than the one constructed directly from the measurements.

The method of making a calibration curve described above involved the preparation of an exact concentration of bacteria which is not very easy to do with an uncalibrated instrument. A concentration of bacteria such that the measured optical density is near to but not exactly 1 can be used. Dilutions are prepared as described, measured, and the equation of the curve computed. The resulting curve is just as applicable to the problems as the more elegant one ($100C = 1.00$ optical density). It can be transformed mathematically into one such that $100C = 1.00$ optical density.

An example of the response of four different makes of instruments to the same set of suspensions of heat killed *S. aureus* is given in Fig. 2. Two of the instruments were self-balancing modifications of the Lumetron Model 402E. One instrument had a scale graduated so that $100 =$ optical density of 1.00. The other had a percentage transmittancy scale. The third was a Beckman Model B spectrophotometer. The fourth instrument was a Klett-Summerson Model 900-3 fitted with a self-emptying cuvette. The fifth was the Bausch & Lomb Spectronic 20. None of the instruments with an optical density type of scale gave a linear response. The Lumetron instruments gave smooth straight lines with slightly different slopes when S/C was plotted against C. The per cent T readings must be converted into optical density units before making the computations. The Klett-Summerson scale is such that 500 represents an optical density of 1.00.

A calibration curve is inconvenient to use if a large number of measurements are to be converted to relative concentrations. A table is more convenient than a curve and is prepared from a large scale graph computed from Eq. (1).

The calibration curve is used to convert instrument scale readings either into relative concentrations of bacteria or into absolute numbers. The relative numbers are sufficient for constructing growth curves, for computing log-probability response curves and for vitamin and amino acid assays. An absolute calibration for live bacteria is prepared by making a plate count on a suspension of the organism in the growth phase of interest. Bacteria in the log phase give very nearly the same total number and viable number. Since the bacteria are growing in a nutrient broth, the instrument is set to an optical density of 0.00 (100% T) with broth to compensate for broth absorption. The absolute calibration is not the same for all organisms. For one instrument, relative concentration of 100 represented 500 million of live *S. aureus* in the log

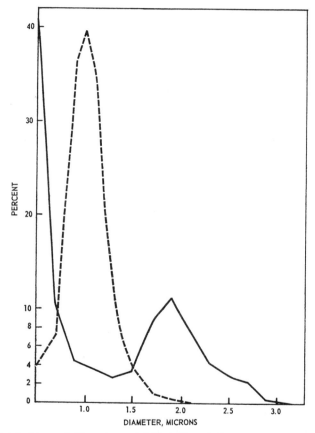

Fig. 3. Distribution of diameters of killed *Straphylococcus aureus* grown in the absence of penicillin (– – – –) and in the presence of 0.05 u./ml. of penicillin G (—). The counts were made with a Model A Coulter counter.

phase, 800 million of live *Klebsiella pneumoniae,* or 1300 million of living *Salmonella gallinarum* cells per milliliter.

In constructing growth curves or inhibition curves the *assumption* is made that all of the samples measured have organisms of the same size or same distribution of sizes. Otherwise the curves should be constructed in terms of cell mass rather than cell numbers. Ordinarily, concentrations of cells are preferred because we are used to thinking in terms of numbers not masses. There is no assurance that use of relative concentration in the figures of this chapter to mean relative concentrations of numbers of cells is correct in all details. Correct or not it will be so used as a matter of convenience and always with the implied reservation that relative cell mass may be the correct designation of the quantity.

The two curves in Fig. 3 show the distributions of sizes of *Staphylococcus aureus* grown in the absence (log phase) and presence of penicillin G. The concentration of the penicillin was sufficient to give a suspension with only 15% of the turbidity of the suspension grown without penicillin. The measurements were made with a Model A Coulter counter with a 19-μ diameter aperture. The distribution of *S. aureus* in the log phase is sharp with the peak at the expected size of 1μ. The other curve shows that most of the particles were either smaller or larger than 1μ. The small particulate material is unidentified. The bacterial peak in the penicillin containing culture is at about twice the diameter of bacteria grown without penicillin. There is a suggestion of another peak at about 3 times the expected diameter. These results suggest that clumps of bacteria as well as the enlarged bacteria obtained in the presence of penicillin are present. The "relative concentration" for this solution as determined by light scattering is subject to several errors. The small material (0.5μ) probably is not bacteria and makes the measured cell concentration high. The clumping makes the count low. There is another indeterminate error caused by the different instrument responses to equal numbers of particles of different sizes. Some of the curvature of the dose-response curves may be caused by this inherent error of the turbidimetric method. This possible source of error needs investigation.

IV. Inoculum

A. Preservation of Organisms

Lyophilization has great current popularity as a method of preserving stock cultures. Over a long period, a culture preserved this way may not be quite the same as the material lyophilized because differential death may reduce the proportions of desirable or undesirable individuals in the population.

Another method of culture preservation is by means of the frozen vegetative procedure of Squires and Hartsell [19] used by Tanguay.[20] The latter author kept eight strains of bacteria and a yeast for more than 1 year in a state suitable for immediate use after thawing by storing

[19] R. W. Squires and S. E. Hartsell, *Appl. Microbiol.* **3**, 40 (1955).
[20] A. E. Tanguay, *Appl. Microbiol.* **7**, 84 (1959).

washed cultures at −40°C. Two bacteria, *Streptococcus faelcalis* R (ATCC 8043) and *Sarcina lutea* (ATCC 9341) gave a somewhat greater growth response at the end of storage for 1 year than the original culture. Of the methods of preservation given here, only the last one can be used directly to inoculate a test; the others must go through an intermediate stage of propagation.

A good method of preparing a frozen vegetative culture is as follows: Propagate the organism for 24 hours in an appropriate medium at the proper incubation temperature. Centrifuge and wash the cells once with sterile phosphate buffer, volume for volume. Suspend the cells in 0.4 volume of glycerol-phosphate buffer, i.e., use 40 ml. of buffer for each 100 ml. of original culture solution. Adjust the cell concentration, if necessary, to 1 to $2 \cdot 10^9$ cells/ml. by adding glycerol-phosphate buffer. Add inoculum sufficient for 1 day's use to a 5- or 10-ml. sterile glass ampoules. Do not fill to more than one-half volume. Seal the ampoules by fusing the neck in a flame, slant, freeze in a −40°C. bath, and store at −40°C. until needed. Thaw the inoculum rapidly in a 25°C. water bath, shake the ampoule, break, and aseptically remove sufficient inoculum for the application.

Squires and Hartsell used $M/15$ Sorensen pH 7.0 phosphate buffer. Tanguay used a $M/15$ pH 7.0 potassium phosphate buffer of the following composition: KH_2PO_4 0.75 gm., K_2HPO_4 1.0 gm., distilled water 1 liter. The glycerol-phosphate buffer was made by dissolving the salts in water, adding 150 ml. of glycerol followed by dilution to 1 liter with distilled water.

Nutrient media give little protection to the frozen cells, or even to cells stored at 4°C. for a week.

Overlaying of a slant with 5 to 10 ml. of sterile white mineral oil preserves bacterial and fungal cultures for a long time. It is a particularly valuable technique for maintaining cultures of fungi which do not produce spores. Many species of *Basidiomycetes* have remained viable for more than 5 years under oil. Storage at low temperature usually is better than storage at incubation temperature. Many organisms continue to grow even at 4°C.

Agar slant and liquid cultures have been frozen very quickly in a "dry ice" bath and stored in a "dry ice" box. Such cultures must be very tightly sealed to prevent loss of water and killing as a consequence of absorption of carbon dioxide.

Another way to prepare a suspension of cells which can be used for a week to give quite reproducible antibacterial assay curves is to wash the cells from a large slant with pH 7.0 phosphate buffer. If the agar slant culture is in the log phase, it may stay in it during storage in the refrigerator. Whether or not the log phase is maintained during storage depends upon the species of bacterium, and should be determined if it is important. The suspension is standardized by measuring the total number of cells by means of a calibrated photometer or nephelometer.

If the bacteria used in assay of antibacterial substances are in the logarithmic stage of growth, then the lag phase of the bacteria in the inoculated test will be short or absent. A long lag phase prolongs the time of incubation of the test unnecessarily.

B. Propagation of Cultures

If microorganisms were stable as is tacitly assumed all too often, the method of propagation of a culture would not be important. Unfortunately for the microbiologist, the biological world is one of unceasing change. This fact must be kept in mind and considered in designing assay procedures. It is especially important in selecting the methods of preserving the assay organisms and in the development of the inoculum.

The simplest way to preserve a culture is in stabs or on agar slants with transfers at intervals of from 1 to 2 months. Certain cultures have been maintained in this manner since 1943 without essential change in sensitivity to penicillin and streptomycin.[21] The transfer should be from agar to agar and not from liquid to liquid. The latter procedure can lead to a uniform (the so-called laboratory cultures) or a balanced culture which, however, may not be the same in essential properties as the original culture.

Consider for a moment the sudden appearance of a more rapidly growing mutant in the population of bacteria in a freshly planted agar slant. The mutant multiplies until its growth is stopped by growth of the surrounding normal bacteria. The number of mutant cells is larger than the average number produced from a single normal cell but not vastly larger. Growth on a fixed spot on agar has limited the expression of the advantage accruing to the mutant. Subcultures from the slant will contain a very small proportion of the mutant. The proportion of the mutant to the total population could increase slightly or even decrease depending upon the proportion of the mutant in the area of the slant selected for transfer. A long time and many transfers might be required for the mutant to dominate the population.

The situation with liquid culture to liquid transfer may be quite different from the agar slant to agar slant transfers. With only a slight growth advantage, a mutant can change from an insignificant member of the population to the dominant one fairly quickly. This happens in practice fairly often to the consternation of the person who was depending upon his culture to remain unchanged. The example in Table II illustrates the small advantage in growth rate needed for a mutant to supplant the

[21] F. Kavanagh, *Bull. Torrey Botan. Club* **74**, 303 (1947).

normal form when it is free to divide without limit. Let the generation time (G) of the mutant be 30 minutes and of the normal be 30.6 minutes in one example and 32.0 minutes in another example and with no lag period. Start with a mixture containing one mutant per 100 normal cells. At what times will the concentration of mutant be equal to the normal, 10, 100, 1000 times as great as the normal? (See Table II.)

TABLE II

Time of Growth Required to Give the Indicated Ratio of Mutant to Normal (M/N) for Two Generation Times (G) of the Normal. The Mixture Contains One Mutant Per 100 Normal Cells at Zero Time

Ratio M/N	G (min.)	
	30.6	32
1	133 hr.	53.2 hr.
10	200	79.8
100	267	106.4
1000	334	133

Many common laboratory strains of bacteria will grow for about 6 hours at 37°C. in the A.C. inoculum broth (Difco), and *Staphylococcus aureus* will reach a terminal population of $2 \cdot 10^9$ viable cells per milliliter in that time. A total growth time of about 133 hours would be obtained in 22 subcultivations or in about one working month.

A population containing 1% of resistant mutants could be quite satisfactory for plate assay; the mutants would not interfere with measurement of zones of inhibition. A population which contained equal numbers of the two forms would be useless; zones of inhibition would not be apparent. This latter condition could be reached in just 22 transfers if the mutant grew 2% faster than the sensitive form. Obviously, if the mutant grew slower than the normal, it would disappear from the population and not be observed. The plate assay will tolerate many more resistant forms than a turbidimetric assay. The inoculum of 500 cells per tube in the serial dilution method must not contain even one resistant form (see Chapter 3). One resistant cell of *S. aureus* could become $2 \cdot 10^7$ in a 24-hour incubation period and cause an erroneous assay. This is the reason for testing a bacterial strain for naturally occurring cells resistant to the antibiotic substance being assayed, and for the use of a small inoculum. A bacterial strain which develops resistance very rapidly cannot be used in a serial dilution method but can be used, if necessary, in a plate method.

Resistant forms can interfere with the photometric assay too. The following discussion will be illustrated by data obtained during the development of an assay for erythromycin with *S. aureus*. Erythromycin reduced the growth rate of the organism but did not affect the lag phase. The total growth period was 3 hours. The initial concentration of bacteria was $13.5 \cdot 10^6$. The generation time was 40 minutes in the absence of erythromycin and 52 minutes in the presence of 0.05 μg./ml. of erythromycin free base. If the inoculum contained 0.43% of a contaminant which had a generation time of 40 minutes in the presence of the erythromycin, it would amount to 1% of the total population in 3 hours. This amount of contaminant would not affect the test by a detectable amount. Five times the amount of contaminant (2%) in the inoculum would make a detectable error in the calibration curve at the 0.05 μg./ml. level. The error in the curve would be greater at the high end. Appreciable contamination would cause the curve on a log-probability plot to depart from a straight line by curving upward as antibiotic concentration increased.

Contamination as used in this discussion means any organism which has a growth rate greater in the presence of the antibiotic than the growth rate of the general population under the same conditions. It could be a resistant mutant of the test organism or be a different organism.

These computations show the reason that nonsterile samples are assayed successfully by photometric methods but not by a serial dilution method. A sample which contained $1.35 \cdot 10^5$ bacteria per milliliter of the dilution assayed (as in the example above) would be from a source so obviously contaminated that no experienced analyst would attempt to assay it without first sterilizing the sample.

C. Preparation of Inoculum

The quality and quantity of inoculum have only second-order effects upon the response of an assay system, be it one of inhibition of growth (antibiotics) or stimulation of growth (vitamins). More important than the quality and quantity of inoculum is the suitability of the test organism for the purpose. This point will not be discussed here (see Chapter 3). Relatively few authors of papers on turbidimetric methods give more than a brief description of the routine followed in their own laboratories. For example, they rarely report the concentration of bacteria inoculated into an assay medium and age of the culture may be anywhere between 2 and 24 hours.

The inoculum usually is grown in a medium similar to or identical with the assay medium. The inoculation may be from a single colony picked from a plate, suspended in 100 ml. of broth and incubated overnight at 37°C. Green,[22] who did this, also checked the grown inoculum microscopically for purity, measured the turbidity in a calibrated photometer,

[22] C. A. Green, *J. Pathol. Bacteriol.* **58**, 559 (1946).

and inoculated the assay broth to give 5 million cells per milliliter. The usual procedure is to inoculate a tube of nutrient broth from a slant, incubate overnight, and dilute by a fixed amount. This procedure depends upon the limitations imposed by the medium to produce each time very nearly the same (generally limiting) population of bacteria. For example, *S. aureus* H grown in A.C. broth for 6 hours at 37°C. reaches a population of $2 \cdot 10^9$ cells/ml.

McMahan[3] used a heavy inoculum prepared from a heavy broth culture incubated for 14 to 18 hours. There are more reports of this procedure than any other. Grove and Randall [23] follow two general procedures. *Klebsiella pneumoniae* is grown overnight on freshly prepared agar surfaces, the surfaces washed with sterile distilled water, the suspension diluted to a standard concentration (concentration not given), refrigerated and used for not more than two weeks. *Staphylococcus aureus* is prepared daily by washing the growth from a freshly prepared agar slant (grown overnight at 32° to 35°C.) with nutrient broth and diluting to a standard light transmission (concentration of cells not given).

There are occasions when a single suspension of bacteria must be used for all of the assays during a given period. One circumstance which dictates this is the requirement that the dose-response curve as measured (not log-probability form) be nearly the same from day to day. To do this, everything from medium to inoculum must be held constant. The medium can be carefully prepared in several lots or a single large lot used. The inoculum as ordinarily prepared each day is not constant and contributes to the daily variation in the dose-response curve. A single lot of *S. aureus* cells washed from an agar surface with $M/15$ pH 7 phosphate buffer can be kept in the refrigerator for a week with a loss in number of about 10% and sensitivity to penicillin nearly unchanged. Cells washed off the agar with nutrient broth, die in the refrigerator too rapidly for broth to be a satisfactory suspending medium.

D. Concentration of Bacteria

The proper initial concentration of cells in the inoculated assay broth depends upon the length of time the assay is incubated. A light inoculum (100–1000 cells/tube) is used in an assay incubated 14–24 hours; and a heavy inoculum (5–10 million cells/ml.) is used in a short-time assay. Relatively few turbidimetric assays are now incubated for 14 to 24 hours. The serial dilution assay and its variants are incubated for 18 to 48 hours (Chapter 3).

Many authors have published curves showing the influence of size of

[23] D. C. Grove and W. A. Randall, "Assay Methods of Antibiotics," 238 pp. Medical Encyclopedia, New York, 1955.

inoculum on the response of the organism to antibacterial substances. A typical set was given by Osgood and Gamble[24] for penicillin and *S. aureus*. The larger the inoculum, the greater the M.R. (median response) and the less the slope of the log-probability lines computed from their data taken after either 3.5 hours or 30 hours of incubation.

The influence of inoculum size needs to be ascertained for each assay method, medium, test organism, and substance assayed. The inoculum for an assay should be taken from cultures in the log phase of growth. This will ensure that a population density measured by a photometric method will have about the same high proportion of viable cells each time the inoculum is prepared and that the lag phase of the assay will be short. A calibration curve in terms of living cells is used to standardize the inoculum for each assay. The necessity for standardizing the inoculum can be illustrated by citing a streptomycin assay in which the M.R. (see Section VIII.A for definition) decreased from 1.7 μg./ml for $22 \cdot 10^6$ cells/ml. of inoculated broth to 0.62 μg./ml. for $5 \cdot 10^6$ cells/ml. The cultures were allowed to reach the same end point of $500 \cdot 10^6$ cells/ml. in the tube without streptomycin (different incubation periods). The assay for erythromycin showed a similar but smaller increase in sensitivity of assay with decreasing size of inoculum. In one example the M.R. was 0.012 μg./ml. for an inoculum of $5 \cdot 10^6$ cells/ml. and 0.015 for $13 \cdot 10^6$ cells/ml.

TABLE III

M.I.C. VALUES FOR MINOMYCIN

Dilution of inoculum	M.I.C. (μg./ml.)
10^2	2
10^3	1
10^4	0.5
10^5	0.5
10^6	0.062
10^7	0.004
10^8	0.004

One of the widest ranges of inoculum sizes reported is that of Shimohira[25] for minomycin against *S. aureus* 209 P. The concentration of the original inoculum before diluting was not stated but possibly was $4 \cdot 10^8$. The minimum inhibitory concentrations (M.I.C.) were determined by a serial dilution assay (Table III).

[24] E. E. Osgood and B. Gamble, *J. Lab. Clin. Med.* **32,** 444 (1947).
[25] M. Shimohira, *Ann. Rept. Shionogi Research Lab.* p. 21 (1960).

V. Media

A. Composition

Growth rate in the absence of an added inhibitor is affected by composition of the medium, pH, temperature of incubation, and aeration as well as by the inherent synthetic capabilities of the organism. The usual assay medium is a mixture of peptones of natural origin, salts, and water. Sometimes glucose and buffers are added. The peptone may or may not furnish all of the nutrients in best proportion required for rapid growth of bacteria to high densities. Sugar may increase growth rate and extent of growth.

Donovick and Rake[26] reported that raising the concentration of tryptone in tryptone-water medium from 0.5 to 1.0% increased the minimum inhibitory concentration of streptomycin from 0.036 to 0.084 unit/ml. for *Klebsiella pneumoniae*. Addition of glucose also decreased the sensitivity of the assay somewhat. Addition of 0.5 mg./ml. sodium thioglycollate to the broth reduced the sensitivity by a factor of about 30.

Pope and Stevens[27] developed a 16-hour sharp end-point assay for penicillin. The assay depended upon a particular medium, a papain digest of horse meat, for its success. This medium would be an interesting one to test for use in a short-time assay.

Osgood and Gamble[24] preferred a tryptose-phosphate medium because their strain of *Staphylococcus aureus* (Oregon-J) grew rapidly to high densities in it. They wrote "The faster the growth, the more uniform its rate, and the higher the final turbidity, the better are the results with penicillin; and the slower the growth, the greater is the sensitivity to streptomycin. Still better media no doubt can be discovered."

How well growth of the test organism should be supported depends upon the incubation time available for the assay. If only a short time is available, a rich medium capable of providing nutrients sufficient for maximal growth is used. The ones used for most of the turbidimetric assays in this book contain peptone, meat extract, yeast extract, glucose, and phosphate buffer. The cell multiplication rate of *S. aureus* in this medium at 37°C. under the microaerophilic to anaerobic conditions obtaining in an assay tube is such that the generation time is about 0.7 hour.

The pH of the medium affects the rate of growth of bacteria as shown by Cohen and Clark.[28] They made plate counts of bacteria growing in peptone-phosphate buffer mixtures over a wide range of pH values. The

---

[26] R. Donovick and G. Rake, *Proc. Soc. Exptl. Biol. Med.* **61,** 224 (1946).

[27] C. G. Pope and M. F. Stevens, *League Nations Bull. Health Organisation,* **12,** 274 (1945/46).

[28] B. Cohen and W. M. Clark, *J. Bacteriol.* **4,** 409 (1919).

pH range for growth was rather sharply defined at each end of the range. Growth rate was not greatly influenced by pH in the region between the boundaries. Table IV gives the pH range for sensibly constant growth

TABLE IV

Organism	pH range	Generation time (hours)
Escherichia coli	5.0–8.1	0.40
Aerobacter aerogenes	4.7–8.4	0.38
Proteus vulgaris	6.5–8.8	0.65
Alcaligenes faecalis	6.9–9.7	1.1

rate and the generation time for 4 gram-negative bacteria. Sherman and Holm[29] showed that sodium chloride in a peptone medium reduced pH dependence of the growth rate of *E. coli*. Assay media usually are adjusted to a pH between 6.8 and 7.8 depending upon the assay organism and the sensitivity required of the assay.

Medium ingredients should not contribute substances which interfere with the action of the antibiotic substance. These substances could inhibit or enhance antibiotic action. Foley and Eagle[30] investigated the mode of action of actinomycin D on several species of *Lactobacilli*. The bacteria required an exogenous supply of pantothenic acid for growth, and the inhibition by actinomycin D could be reversed competitively only by pantothenic acid. Noncompetitive reversal of inhibition was obtained by adenine, L-methionine, orotic acid, and pyruvate. Slotnick[31] observed that a strain of *Saccharomyces cerevisiae* requiring β-alanine for growth was inhibited by actinomycin D only at suboptimal concentrations of β-alanine. These two observations show in a striking manner that the medium constituents interfering with the assay of a particular substance may be different for different assay organisms. The observations also show that the concentration of the interfering substance must be the same from time to time if the dose-response curve is to be reproducible. This may be difficult or impossible except in a synthetic assay medium. Synthetic media are rarely used.

Different lots of supposedly identical media may affect the response of the test organism to give different M.R.'s for the different lots. The M.R. of *Staphylococcus aureus* to erythromycin was 0.030 μg./ml. for a lot of medium made on one day and 0.036 μg./ml. in a lot made 3 days later from the same basic ingredients. In a test with 12 single colony isolates of *S. aureus* and a synthetic penicillin, the M.R. varied more between lots of

[29] J. M. Sherman and G. E. Holm, *J. Bacteriol.* **7,** 465 (1922).
[30] G. E. Foley and H. Eagle, *Cancer Research* **18,** 1012 (1958).
[31] I. J. Slotnick, *Ann. N. Y. Acad. Sci.* **89,** 342 (1960).

"identical" media than between isolates. The assay media in use are adaptations of standard bacteriological media and may not necessarily be the best for all assays. Much work needs to be done on media with proper regard paid to the purpose of the assay and the biological properties of the substance assayed.

Synthetic media should be used more in assaying for unknown substances. The sensitivity of the test may be considerably greater in synthetic than in complex media. If a minimal medium is employed and total growth is limited by restricting the concentrations of essential ingredients, hitherto unknown inhibitors may be discovered.

B. pH and Activity

Most of the antibiotics in current use are either acidic or basic substances and their activities are influenced by the hydrogen ion concentration of the medium. The relationship between pH and toxicity of organic compounds has been known for about 40 years, and its significance has not been fully appreciated for about the same length of time. Crane[32] investigated the influence of pH upon the toxicity of bases, including strychnine, for *Paramoecium*. She apparently was the first to attribute the toxicity to the free, undissociated base. She computed the concentration of the free base from the pH of the culture medium and the ionization constants of the base. Although Foster and Woodruff [33] showed that the activity of penicillin in a plate method of assay increased with a decrease in pH, Foster and Wilker[34] did not show the same for a photometric assay. Ample evidence[35] has accumulated since then to establish the pH dependency of the sensitivity of photometric assays of many acidic and basic antibiotics including penicillins. The activity of an acidic antibiotic (penicillin) increases and the activity of a basic antibiotic (streptomycin) decreases as the pH of the medium is lowered. This phenomenon can be used to change the M.R. of the tests to a considerable extent. The increases in M.R. of the tests for erythromycin and streptomycin are about fivefold for an increase in pH of 1 unit in the region from pH 6.5 to 7.5. The increase in M.R. of the penicillin assay is about 1.5 times for a decrease in pH of 1 unit. This great sensitivity of some assays to the pH of the medium causes much unexplained day to day variation of the 50% inhibition points. Obviously, if the pH of the broth changes greatly during the course of the assay, the apparent sensitivity of the bacteria also changes, and the assumption that the bacteria are growing in a constant environment is incorrect. The desired linear log-

[32] M. M. Crane, *J. Pharmacol. Exptl. Therap.* **18,** 319 (1921).
[33] J. W. Foster and H. B. Woodruff, *J. Bacteriol.* **46,** 187 (1943).
[34] J. W. Foster and B. L. Wilker, *J. Bacteriol.* **46,** 377 (1943).
[35] F. Kavanagh, *Advances in Appl. Microbiol.* **2,** 65 (1960).

probability line will not be obtained if the environments in the zero tubes (ones without antibiotics) and the antibiotic containing tubes differ enough in the early hours of incubation to affect growth rates appreciably.

The pH drops as growth proceeds in sugar-containing medium and may go as low as pH 4.9 in *Staphylococcus aureus* or *Escherichia coli* cultures. Enough buffer to hold the pH constant at some point between 6 and 7 for the entire 3–4 hours of an assay probably would be inhibitory. A compromise is made by adding enough buffer to keep the pH within 0.2 unit of the starting pH for a period of about 2 hours. The first 2 hours seem to be the important ones at least for penicillin or erythromycin acting against *Staphylococcus aureus*.

VI. Incubation

A. Temperature

Although the range of temperature specified in most procedures is 36–38°C., it may not be optimal for a particular test. Many bacteria will grow in complex media at temperatures higher or lower than 38°C. Lowering the temperature reduces the growth rate and prolongs the incubation time needed to obtain the standard population. Bond and Davies[36] reported experiments which showed that a deviation of 3° from 37°C., either above or below, approximately halved the growth rate of *Staphylococcus aureus*. The large temperature effect shows very clearly that to achieve high accuracy the temperature must be uniform during incubation. Uniformity of temperature in a water bath can be accomplished by efficient agitation.

An unexpected temperature effect was reported by Brown and Young[37] for an end-point type of assay for streptomycin with *Escherichia coli*. When the tests were incubated at 37°C., 11% of the tubes were "out of order" in that the *E. coli* had multiplied at concentrations of streptomycin higher than that showing inhibition. The number of tubes "out of order" was 3% when the test was incubated at 28°C., a temperature they then adopted. For a discussion of this problem and another solution of it, see Chapter 3.

Tubes in the center of a 6 × 10 compartment rack might not be subjected to exactly the same temperatures and rates of changes of temperature in the water bath and Arnold sterilizer as those on the outside. This possibility was tested by putting control tubes in alternate columns of the rack and tubes containing 0.3 μg. of erythromycin also in alternate

[36] C. R. Bond and O. L. Davies, *Analyst* **73**, 251 (1948).
[37] A. M. Brown and P. A. Young, *J. Gen. Microbiol.* **1**, 353 (1947).

columns. The rack contained, in effect, 5 columns of 6 tubes each and 6 rows of 5 tubes of each treatment. The inoculated tubes were incubated for a standard time in the water bath, then heated in the Arnold sterilizer to kill the bacteria. There were no significant differences between rows, columns, or rows and columns of the assay. This test shows that a random distribution of samples in a rack is not necessary to eliminate variations caused by position in a rack of tubes. This argument applies only to deviations attributable to the incubation and killing steps.

A position-related error will be caused under some conditions if the time between filling the first and last tubes is very long. This is the reason that two racks of tubes (80), standards and samples, constitute a complete test in one assay laboratory. Systematic differences were found in another laboratory between standards put in the first and in the last rack of large (300–400 tubes) assays for streptomycin. The differences disappeared when the practice of using chilled inoculated broth was restored.

B. Aeration

The tubes in the usual turbidimetric assay stand in racks in a water bath. Bacteria slowly settle to the bottom of the tube during the incubation period producing a gradient of organisms and a gradient in the environment as they metabolize. The conditions are microaerophilic at best (except on the surface) and anaerobic at the bottom.

There are several investigations of the effect of oxygen supply upon an assay. Foster and Wilker[34] selected *Bacillus adhaerans* as the test organism for a short-time (4–5 hours) assay because it was as sensitive to penicillin as *Staphylococcus aureus*, and grew faster. They used 10 ml. of medium in 50-ml. Erlenmeyer flasks and shook them to increase oxygenation and promote growth. They wanted maximum growth per unit of time. Lewis *et al.*[38] compared culture tubes shaken periodically during incubation with undisturbed tubes and showed that shaking reduced inhibition by 30% for a given level of subtilin. Kavanagh[35] felt that a uniform environment in the tube would be advantageous and could be obtained by gently shaking the tubes in an inclined position on a rotary shaker. The shaking kept the bacteria suspended and increased aeration increased growth rate considerably over that of the static tubes. The generation time in the tubes shaken at 33°C. was 27 minutes. In static tubes it was 2 hours at 33°C., and 38 minutes at 38°C. A slight increase in sensitivity occurred with some antibiotics in the shaken test. As is evident from the experience of Lewis *et al.*, each assay system (drug and organism) must be tested for effect of shaking. The time of incubation of

[38] J. C. Lewis, E. M. Humphreys, P. A. Thompson, K. P. Dimick, R. G. Benedict, A. F. Langlykke, and H. D. Lightbody, *Arch. Biochem.* **14**, 437 (1947).

many assays could be shortened substantially by the simple operation of gentle shaking of the tubes. A concomitant of the shaking is the extention of the linear portion of the log-probability plot to higher concentrations of antibiotic (to 10% growth).

C. Time

Incubation times range from less than 2 hours for photometric antibiotic assays using shaken tubes to 3 days for vitamin assays with *Lactobacillus sp.* The incubation period accounts for about one-half of the elapsed time of a rapid photometric assay. The principal advantage of the photometric method is speed.

During the early days of the penicillin program, overnight assays were impractical. Peak titers, as much as 160 u./ml., could drop by 50% in the following 4 hours, thus making time of harvest of the fermenter very important. When, several years later, fermenter yields reached 1000 u./ml., the decrease beyond the peak concentration was proportionately less significant and short-time chemical assays were now available. A 4-hour photometric assay also fitted into an 8-hour day with little wasted time for the operators. Thus, there is not much need for a still shorter assay unless procedures can be devised to permit two assays per day and occupy all of the operator time.

Time of incubation may affect an assay answer very little. The usual practice is to incubate the test until the zero tube attains an arbitrarily selected reading, which may be such as to represent an optical density of 0.7 to 0.8. The time required for a test to grow to this density is a function of such variables as composition of medium and its pH, temperature of incubation, size of inoculum, aeration, and strain of test organism. The time varies between 3.0 and 4.5 hours. If proper attention is paid to the important factors affecting time of incubation, it can be kept within narrow limits. The requirement for high bacterial density (large scale reading) is more apparent than real; the calibration curve increases in steepness of slope with increasing time of incubation and thus becomes easier to read in figuring the test. The scale of the photometer can be expanded electrically, or a different way of plotting the standard dose-response curve can be used to avoid the requirement for large cell density and thus reduce the incubation time.

A short-time incubation of an erythromycin assay changed the sensitivity but not the slope of the line in the log-probability plot. For example, the M.R. was 0.033 μg./ml. for an incubation time of 2.66 hours and 0.024 μg./ml. for 4 hours. In a penicillin V assay, the M.R. went from 0.0074 μg./ml. for 3.66 hours to 0.0060 μg./ml. for incubation of 4.25 hours. This variation of assay with incubation time means that the standard curve and all tubes to be estimated in terms of it must be treated exactly alike during incubation and the killing period. The exact time of incubation is not important so long as it is the same for all tubes in the test.

There is probably a minimum time of incubation for a satisfactory test. It probably is different for antibiotics which kill only dividing cells (penicillin) and those (streptomycin) which kill nondividing cells. The minimum time for penicillin assay should be at least three generation times, since two divisions occur before growth ceases. This is the rationale for the use of chilled inoculated broth in setting up an assay. If broth at room temperature is used, bacteria in the tubes filled first will start to grow and will be exposed to the antibiotic for an effective time longer than those in the latter part of the test. The same concentration of antibiotic in different parts of the test will give different inhibitions and compute to substantially different answers. All this causes concern to the

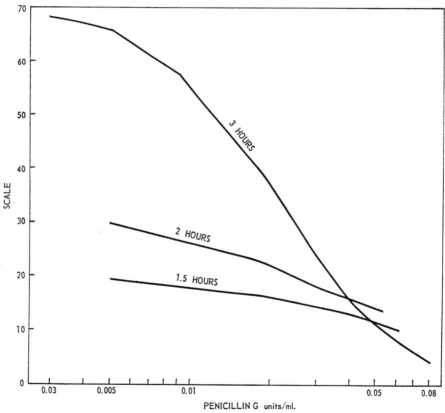

Fig. 4. Log-arithmetic plot of photometer scale reading for the same curves as those in Fig. 5. The curves were computed from the straight lines of Fig. 5 and thus represent smoothed data.

technicians and needless conflict between those who submit samples and the assayers.

Curves for three different times of incubation of a penicillin G assay are given in Figs. 4 and 5. The data are plotted in the usual way in Fig. 4 to show instrument scale reading as a function of the log-concentration of penicillin G in the tubes. The curves for 1.5 and 2 hours are too flat to be of much use in assaying. The 3-hour curve is the one normally used. It is apparent that lysis occurred between 2 and 3 hours at 0.05 u./ml. The same data are plotted in Fig. 5 on log-probability paper. The three tests

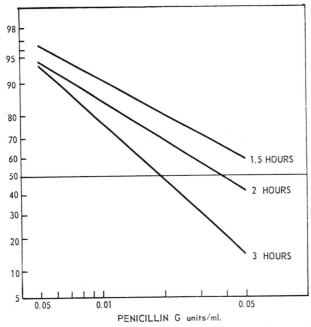

FIG. 5. Log-probability plot of turbidimetric standard curve for penicillin G after three incubation times at 38°C. Curve 1 is after 1.50 hours (2G), curve 2 is after 2 hours (2.66G), and curve 3 is for 3 hours (4G), where G is the generation time measured in the tubes without penicillin.

plot into straight lines with different slopes and M.R.'s. The slopes of the 2- and 3-hour tests indicate that the distribution of apparent susceptibilities changed in that 1-hour period; the distribution being smaller at 3 than at 2 hours. The advantages of the log-probability plot over the usual one for theoretical and practical purposes is readily apparent. Incubation of the test for more than 3 hours changes the sensitivity of the test, but not the slope of the line. In general, an increase of time of incubation beyond a minimum for a fixed level of inoculum increases the sensitivity of the test but not the slope of the log-probability line.

So far, attention has been placed on inhibition of growth in tests inoculated with high concentrations of bacteria. The incubation time of these assays can be expected to be short. What about vitamin and amino acid assay in which the inoculum is low? Here the minimum time used seems to be 18 hours. Such short incubation times should be regarded with suspicion until proof is obtained that limiting growth is attained in that time. Short incubation times can cause undue error. Ideally, the only factor limiting growth in a vitamin or amino acid assay is the substance assayed. In actual practice and in numerous published curves, the extent of growth was determined by time, pH, and exhaustion of nutrients as well as by the substance assayed. This is particularly true in the high concentration end of the assay curve. Sufficient time needs to be allowed for the growth of the organism to be limited by the amount of the substance assayed and not by time. The amount of vitamin per tube and not its concentration determines the total mass of cells produced. If a short-time assay is required, then the range of the assay is reduced to only that portion at the low end of the standard curve that does not change on incubation longer than the minimum time available for the assay. For example, three identical *Lactobacillus leichmannii* standard curves for assay of B_{12} harvested at 1, 2, and 3 days will show quite clearly the range for each time of incubation. The range at 3 days will be considerably greater than the range at 1 day.

VII. Response of Bacteria

A. Antibiotics

The apparent response of a population of bacteria to an antibiotic is a reduction in growth rate. The effect of a penicillin or erythromycin on *Staphylococcus aureus* is illustrated by Figs. 6 and 7. At low concentrations of the synthetic penicillin (0.05 μg./ml. of broth), the growth rate was depressed but the total population measured after 22 hours of incubation was large; lysis was not apparent. At a higher level of the penicillin, 0.2 μg./ml., the growth rate was greatly reduced and, finally, lysis occurred (Fig. 7). The curves show that after about two divisions the new growth rate is established. The curves change slope during this period from that of the uninhibited growth to that obtaining in the presence of the penicillin. The assumption is made that any change in size of the bacteria is too small to influence the growth curves. The shapes of the growth curves taken in conjunction with the mode of action of the penicillin indicates that part of the population of bacteria is killed and that the survivors grow at a reduced rate. If the only action was killing of part of the population, the growth curve after the first two or three divisions

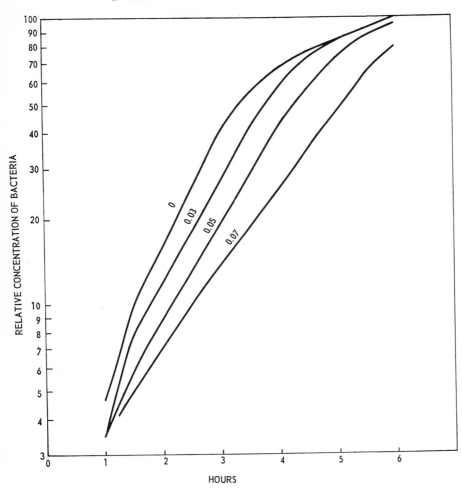

FIG. 6. Growth of *Staphylococcus aureus* H in the presence of the indicated concentrations of erythromycin in micrograms of base per milliliter of broth. The generation times for the straight line portions of the curves are: 0.73, 0.84, 0.88, and 1.11 hours for increasing concentrations of erythromycin.

would be displaced to the right of and would have the same slope as the zero concentration (control) curve. The curve is displaced as expected but the slope is less than the control, indicating that less than a fatal concentration of the penicillin causes an apparent reduction in growth rate. This was proved for a synthetic penicillin (Fig. 8) by adding a large amount of penicillinase at 2 hours to destroy the penicillin; the apparent growth rate rapidly increased to that of the medium without penicillin showing that no permanent harm was done to the bacteria by the penicillin.

A striking feature of growth of bacteria in the presence of a small

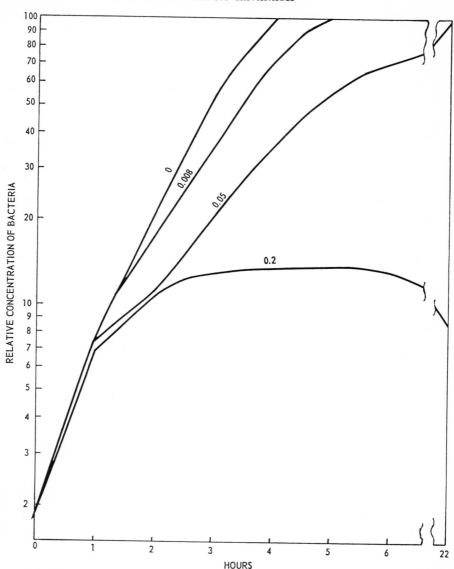

FIG. 7. Growth of *Staphylococcus aureus* H in the presence of the indicated concentrations of α,α'dimethylpenicillin V (phenoxyisopropylpenicillin).

amount of an antibiotic substance is the extended linear region of the growth curve. The straight part of the curves as measured by a photometric method extend to a higher cell population in the presence than in the absence of an antibiotic substance. This may be real or an artifact of the method of measurement of growth. If the cells in the control (no

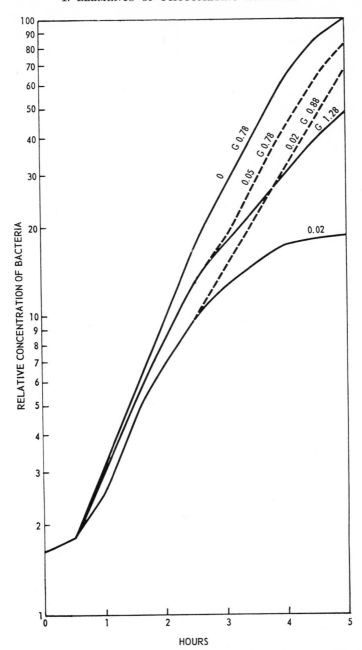

FIG. 8. Growth of *Staphylococcus aureus* H in the presence of the indicated concentrations of α,α′-dimethylpenicillin V. Penicillinase (500 u./ml.) was added at 2
hours. The uninhibited growth rate was re-established in the penicillinase treated
samples in about 0.5 hour.

antibiotic) are aging by 4 hours, their average sizes are decreasing and the growth curve has the proper shape (no artifact). Cells grown in the presence of the antibiotic may undergo the aging change in size at a later time and higher cell population than uninhibited cells. Direct measure of cell size and population density would answer the question.

B. Growth Substances

Organisms respond to vitamins and amino acids by an increase in mass. The increase in mass of bacteria is expressed by an increase in size and in number. If growth is limited by a single substance, the relation between the *amount* of the substance and the *amount* (cell mass of bacteria) of microorganism may be linear. Another way of stating the relation is that the *number* of cells produced is proportional to the *number* of molecules of the vitamin. This also means that the number of cells is independent of the concentration of the vitamin as can be shown by diluting the medium with an equal volume of water after a given amount of vitamin has been added, inoculating, and measuring the response in comparison with an undiluted control. The two different concentrations but equal amounts of vitamin will give very nearly the same response as measured by the number of cells or dry weight. Any difference will be a second-order effect caused by diluting the medium.

The implicit assumption is made that the test organism has an absolute deficiency[39] for the growth substance. If the deficiency is partial, a small amount of growth may ensue and complicate the problem considerably. Assay organisms which have absolute deficiencies are always chosen, if possible. Usually, the vitamin level in medium and inoculum can be reduced to an undetectable amount. Any assay in which the zero level tube shows appreciable growth should be regarded with suspicion and discarded unless the response is known to be normal for the system.

Hamilton[40] was asked at a symposium why he regarded microbiological assay methods as applicable only when the response was linearly proportion to the concentration of the substance to be estimated. He said, in reply, "that his credo was that a bio-assay was only good when the growth response was linearly proportional to the concentration of the substance to be estimated and that logarithmic plots of bio-assays were mathematical smoke screens calculated consciously or unconsciously to conceal a variety of variables that could not be measured. Patently, if one Euglena required 4800 molecules of vitamin B_{12}, two Euglenas required 9600 molecules, and three Euglenas, . . . , etc., then this uncomplicated relationship was linear. Any departure from linearity at the upper end of the response curve indicated intrusion of other variables and

[39] W. J. Robbins and V. Kavanagh, *Botan. Rev.* **8**, 411 (1942).
[40] L. D. Hamilton, S. H. Hutner, and L. Provasoli, *Analyst* **77**, 618 (1952).

at this point the validity of the assay ended. He believed this generalization to hold no matter what kind of response was used, e.g., acid production by lactobacilli; it was precisely at the point where acid production impeded growth or other limiting factors came into play that these assays lost validity. In certain amino-acid assays, it was true that unidentified factors bent the curves away from linearity, affecting both the curves for amino-acid standards and the assayed amino acids in simplified materials such as protein hydrolysates in approximately equal measure. But he was inclined to be more suspicious of the results when complex materials were assayed, where there was less reason to suppose that the variables encountered evenly affected both standard and assayed material. He was not satisfied that an assay was suitable for general use until the response curve was linear from the origin. The region near the origin was especially informative because not only did it indicate the extent of carry-over and background contamination, but it also clearly brought out imperfections in the basal medium. Lag phase phenomena and induction periods, showing up as concavities in the curve, were signs that the medium was not complete enough for a clean-cut response to a single metabolite. These inadequacies in the basal medium tended to be blotted out by higher concentrations of natural materials and might allow a misleading linearity of response. These remarks applied only to bio-assays in liquid media."

The composition of the medium should be such that growth is not limited by a lack of essential elements (including minerals) or by inhibitory substances including unfavorable amino acid ratios. Although this ideal situation probably never obtains, it can and should be approximated.

The time and temperature of incubation should be chosen to give a total growth limited by amount of growth substance being assayed and not by the time of incubation. In general, the shorter the time of incubation, the shorter the range of the assay; a relation not always recognized by assayers. The time and temperature of incubation selected should be such as to reduce to secondary importance the influence of associated materials added with the substance assayed. In too many procedures, exhaustion of medium, low pH, and short-time exert an undue influence in determining the dose-response curve especially at the high concentration end of the curve.

All of these requirements for accurate assay of vitamins and amino acids were given by Snell [41] in an article well worth reading by assayers of growth promoting compounds. He also pointed out, as is easily demonstrated, that the recommended procedures of washing inoculum to remove "adhering" vitamins is superfluous if the inoculum is grown in the

[41] E. E. Snell, *Wallerstein Labs. Communs.* **11**, 81 (1948).

presence of a limiting amount of the vitamin. Some organisms if grown in the presence of a large excess of an essential vitamin will accumulate enough of the vitamin to make nearly maximal growth when transferred to a vitamin-free medium. *Lactobacillus leichmannii* grown in the presence of a large excess of vitamin B_{12} will be useless as inoculum for B_{12} assay; it will carry so much B_{12} that it will not make a measurable response to the vitamin in the medium. Washing an inoculum may be ineffective and does increase the chances of contaminating the inoculum.

The statement made above that the growth response was linearly proportional to the concentration of the substance to be estimated seems contrary to current concepts and practices as judged by publications during the last decade. The most popular response is the one between concentration of the vitamin and optical density of the resulting bacterial growth. One who early recognized the linear relationship between cell mass and amount of growth substance was Snell.[42] He showed in 1948 that *Lactobacillus casei* gave a linear relation between dry weight and the riboflavin content of the tube. Several vitamins and amino acids give linear dosage-response curves with the appropriate organism.

The essay by Hutner *et al.*[43] discusses the many details of good procedures of assaying vitamins and amino acids and should be read by all who direct assay laboratories.

VIII. Modes of Expressing Response

A. Antibiotic Dosage-Response Curves

The photometer used to measure the response of an organism to a drug may have a scale graduated in percentage transmittancy or in optical density units. The type of instrument scale graduation influences the shape of the dose-response curve as is illustrated by the curves in Figs. 9 and 10. The data are from one curve measured on one instrument. Scale reading (not relative numbers of bacteria) is the dependent variable. The relationship between the concentration of penicillin G in the tubes and the instrument response to the consequent turbidity of the test organism (*Staphylococcus aureus*) is presented in three ways. Figure 9 shows the relation between logarithm of the concentration and optical density, and log of concentration and percentage transmittancy. Figure 10 gives the relation between concentration of the standard solutions and percentage transmittancy. All of the curves are satisfactory calibration curves in the range of 0.01 to 0.03 penicillin G units per milliliter of

---

[42] E. E. Snell, *in* "Vitamin Methods" (P. György, ed.), Vol. 1, pp. 327–505. Academic Press, New York, 1950.

[43] S. H. Hutner, A. Cury, and H. Baker, *Anal. Chem.* **30,** 849 (1958).

broth. The shapes of the curves in Figs. 9 and 10 are subject to considerable day to day variation.

Some assayers use only the straight line portion of the percentage transmittancy-concentration curve (Fig. 10). This procedure may restrict

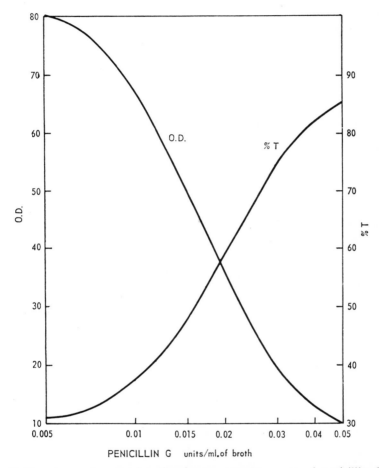

FIG. 9. Two ways of expressing the dosage-response curves of penicillin G and *Staphylococcus aureus* with scale reading (proportional to optical density), percentage transmittancy, and logarithm of concentration of penicillin in the broth as parameters. Data from a standard curve.

the range of the assay unduly for some antibiotics and assay laboratories. The straight line region of Fig. 10 extends from 0.01 to 0.025 u./ml. If a larger error in assay is acceptable, the "linear" range can be extended to include the 0.008 and 0.03 u./ml. points. This is sometimes done when computating assays by a digital computer to reduce the number of missed assays by reason of restricted range. When a fast computer is available, curvature of both standard and sample dosage-response lines can be taken in consideration. Curvature is inherent in this manner of representing the data. Statistical variation is superimposed upon the inherent curvature.

The relation between dose of antibiotic substance and measured response may be expressed in many ways. The three ways of relating turbidity to drug concentrations given in Figs. 9 and 10 are all sigmoid.

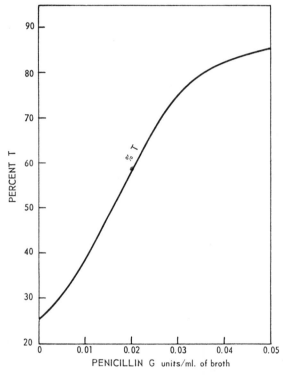

Fig. 10. Same data as Fig. 9 plotted as percentage transmittancy against concentration of penicillin.

Small daily differences in pH of medium, medium "quality," and concentration of bacteria contribute to a nonreproducibility of the curves that is rarely mentioned in publications. Considerable improvement in constancy of position and shape of the dose-response curve can be obtained by referring the response in the presence of the antibiotic to that of a control (no antibiotic) as was done by Lewis et al.[38] and Joslyn and Galbraith.[44] These authors computed the ratio of the turbidity (optical

[44] D. A. Joslyn and M. Galbraith, J. Bacteriol. **59**, 711 (1950).

density of the suspension) of the sample to that of the control. Joslyn and Galbraith obtained a slightly sigmoid standard curve upon plotting per cent of growth (turbidity) of the control against the corresponding concentration of chloramphenicol on a logarithmic scale. Lewis *et al.* rectified the sigmoid curves obtained in the assay of subtilin by plotting either on linear probability paper or logarithmic-probability paper depending upon the test organism.

The use of the probability plot to straighten the calibration curve apparently went unnoticed until Treffers[45] called attention to the several advantages of the method a decade later. "The basic requirements for application of the method is a knowledge of the percentages of the organisms in the culture which grow at certain specified drug concentration." The percentages are computed relative to that of a control (no antibiotic). The numbers of cells per milliliter can be determined by a plate count or, better for assaying, with some photometric device calibrated in terms of cell concentration.

The few authors who related growth in the presence of a drug to a control grown in the absence of the drug took the optical density as a measure of the concentration of bacteria. To make their comparisons valid they needed to correct their photometers for deviations from linearity as was done by Alper and Sterne,[9] Longsworth,[8] and, more recently by Toennies and Gallant.[18] Nonetheless, the procedure of relating response to a control was a big advance over the usual one of reporting response as a galvanometer deflection, optical density, or transmittancy In antibiotic assays, the preferred response is the per cent inhibition. The response as a percentage is obtained by multiplying the population densities in the tubes with antibiotic by 100 and dividing the product by the population density in the zero tube. The population density obtained in the zero tube (the one without any antibiotic) is not important so long as the cells are in the log-phase and the time is long enough for complete action of the antibiotic. If the cells are grown for such a long time that they are no longer in the log phase, the population density will be less than it should have been and the log-probability line will curve upward in the range of low antibiotic concentrations; at high concentrations of drug only slight effect will be observed.

In vitamin and amino acid assays, the response is the cell density. As indicated earlier (Sections II.A and III.D), the relative concentration of cells as measured by a calibrated photometer is as useful in assaying as absolute values of cell concentration.

Many assay systems, antibiotic and test organism, give approximately straight line dosage-response curves when the responses are plotted on

[45] H. P. Treffers, *J. Bacteriol.* **72**, 108 (1956).

log-probability paper.[46] The logarithm of the dosage is used instead of the dosage because it is the function of the concentration of the drug that gives the straight line (compare Fig. 11 with Figs. 9 and 10). An occasional system, such as that of subtilin with *Micrococcus conglomeratus* (MY) or *Staphylococcus aureus* H as test organism, gives the straight line when the plot is made on linear-probability paper.

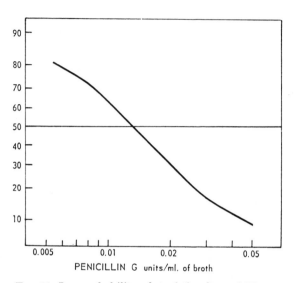

Fig. 11. Log-probability plot of the data of Fig. 9.

A straight line is the easiest one to fit to a set of points. The best fit can be computed by any standard method, or it can be drawn to give the best apparent fit. A straight line relation is most economical of tubes; the tubes of the standard could be concentrated at the two ends of the straight portion of the dose-response curve. However, a standard curve should not be set up with only the end points established; a mid-point should be included as a check on curvature and position of the ends. Usually, at least one point at each end beyond the straight portion of the curve would be included.

Two valuable parameters, the M.R. (median response) and slope of the line, characterize the inhibition of a particular organism by a particular inhibitor and are obtained from the dosage-response curve. The M.R. is the concentration of the drug that permits attainment of a population density one-half as great as that obtained in the same growth period in

[46] Codex Book Co., Norwood, Massachusetts, paper 31.376, 32.376, or 3228, or Keuffel and Esser, New York, N. Y. paper no. 358-22. This writer used 3228. All of the papers are poorly designed for microbiological assaying.

the absence of the drug (zero tube of the standard curve). The slope of the curve is the change in the normal deviate units for a tenfold change in dosage. The slope is inversely proportional to the standard deviation of a distribution represented by the line.

The greater the slope, the steeper the dosage-response and the shorter the range of the assay. The curves (Fig. 15) for erythromycin and vancomycin illustrate this clearly. The curve for vancomycin is so steep that it is useless for assay of solutions of unknown concentrations. If the concentration is known accurately enough to bring it within the range of the turbidimetric assay, then it is known accurately enough for most purposes other than control. The erythromycin curve is typical of dosage-response curves suitable for assaying of unknowns.

The slope of the response curves is determined by the characteristics of both drug and organism. The response of *Klebsiella pneumoniae* and *Staphylococcus aureus* to dihydrostreptomycin and streptomycin are good illustrations of this (Fig. 12). The curves obtained with *Klebsiella pneumoniae* are too steep to be good standard curves. The *Staphylococcus aureus*, a strain selected for the purpose, gives a curve well suited to the assay of streptomycin.

The slope of the dosage-response curve (log-probability form) of a particular combination of organism and antibiotic seems to be influenced little by small changes in pH of the medium, by lot to lot differences in the medium, by concentration of organism in the inoculum by length of incubation (beyond a minimum time?), and temperature of incubation. These factors should be tested for effect on any new combination of organism and drug. Occasionally, a change in slope does occur without any apparent cause.

The other parameter, the M.R., is a measure of the sensitivity of the organism to the drug. The M.R. is affected by pH of the medium, concentration of organism in the inoculum, lot to lot variations in the medium, the incubation period, temperature of incubation, and the quality (richness) of the assay medium. Although these environmental influences cause measurable changes in the M.R., the influence on the dosage-response curve is much less in the log-probability form than in the optical density-log concentration form or the other representations given earlier.

The M.R.'s for dihydrostreptomycin (1.10 μg./ml. of broth), and streptomycin (1.18 μg./ml.), against *S. aureus* are the same within the experimental error of the assay (Fig. 12). The M.R.'s for dihydrostreptomycin (0.96 μg./ml.), streptomycin (1.7 μg./ml.) acting against *Klebsiella pneumoniae* are significantly different. The same medium and same sets of standards were used to obtain the two sets of curves. All were done on the same day. The two *K. pneumoniae* curves can be rectified by plotting on arithmetic-probability paper instead of log-probability paper. The streptomycin slope on arithmetic paper is considerably less than the dihydrostreptomycin slope; the M.R.'s are not changed by the different ways of plotting the data. The difference in slopes as well as the different shapes of the dose-response curves indicate a significant difference in

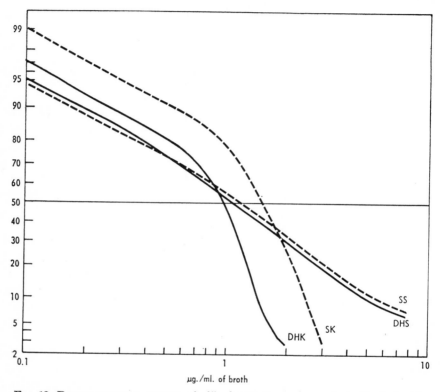

FIG. 12. Dosage-response curves of dihydrostreptomycin against *Staphylococcus aureus* (DHS) and *Klebsiella pneumoniae* (DHK), and streptomycin against *S. aureus* H (SS) and *K. pneumoniae* (SK). The strains of bacteria were *S. aureus* H (ATCC 9996) and *K. pneumoniae* (ATCC 10031).

response of the two organisms. The reason for the differences in responses between these two typical gram-positive and gram-negative bacteria would be worth investigating.

The equivalence of dihydrostreptomycin and streptomycin in inhibiting *Staphylococcus aureus* is apparent from the curves (Fig. 12); there is no difference between either their M.R.'s or their slopes. A derivative may also have the same slope but an M.R. different from the parent com-

[47] B. W. Coursen and H. D. Sisler, *Am. J. Botany* **47**, 541 (1960).

pound as, for example, cycloheximide[47] and its semicarbazone inhibiting *Saccharomyces pastorianus* Hansen. The M.R. was 0.018 μg./ml. for cycloheximide and 0.37 μg./ml. for the semicarbazone, and the dose-response curves were parallel. The oxime had an M.R. of 12 μg./ml. and a curve parallel to the others. A chromatographic investigation of the three compounds showed that the active compound in the oxime was indistinguishable from cycloheximide. This experience shows that identity of slope suggests identity of action and possibility of identity of active compound. Such curves cannot be used to prove identity of compounds. They, as chromatography, are better used to indicate lack of identity of active compounds.

Dosage-response curves are available on several sets of related antibiotic substances. Erythromycin B has the same slope as erythromycin but a larger M.R. when the activity is measured against *Staphylococcus aureus*. The largest group of related antibacterial substances are the penicillins. Hundreds have been made biosynthetically and thousands by chemical synthesis from 6-aminopenicillanic acid. Some of the synthetic penicillins have dose-response curves different from those of the biosynthetic penicillins as will be shown. The number of dosage-response curves available to this writer are few and are only those done personally. The

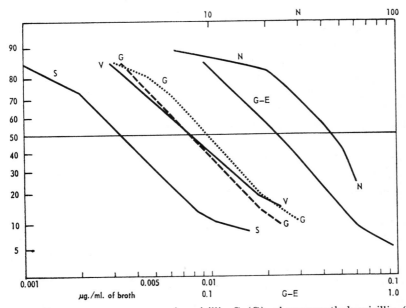

Fig. 13. Dosage-response curves of penicillin G (G), phenoxymethylpenicillin (V), and phenylmercaptomethylpenicillin (S) against *Staphylococcus aureus* H (0.001–0.1 concentration scale); 6-aminopenicillanic acid (N) against *S. aureus* H (upper scale); penicillin G (G-E) against *Escherichia coli* ATCC 4157 (0.1–1.0 concentration scale)

first group discussed are representative of the biosynthetic penicillins. The dosage-response curves of the penicillins against *S. aureus* are given in Fig. 13. All of the biosynthetic penicillins have about the same slope. The M.R.'s for penicillins G and V are about the same. The phenylmercaptomethylpenicillin is about twice as active against *S. aureus* as either penicillin G or V. The 6-aminopenicillanic acid curve is somewhat different from the others. This compound is the penicillin nucleus from which synthetic penicillins are made. It is more active against some gram-negative bacteria than against *S. aureus*.

The synthetic penicillins are represented by α-methylpenicillin V (phenoxyethylpenicillin, Syncillin Bristol), α,α'-dimethylpenicillin V (phenoxyisopropylpenicillin), and dimethoxyphenylpenicillin (Staphcillin, Bristol). The three synthetic penicillins are more active than penicillin G against penicillinase producing ("penicillin resistant") staphylococcii. There are other striking differences between these nonbiosynthetic penicillins and a typical biosynthetic penicillin (penicillin G). Substitution of one hydrogen on the α-carbon of the phenoxyacetic acid side chain by a methyl group did not change the M.R. or the general shape of the curve for inhibition of *S. aureus*. Substitution of two methyl groups for the two hydrogen atoms on the α-carbon to give α,α'-dimethylpenicillin V caused such a profound change in the dose-response curve taken with *S. aureus* that at first glance it no longer resembles a penicillin curve (Fig. 14). The M.R. is slightly more than the M.R. of penicillin G. However, the comparison has little meaning because the great difference in slopes of the two dose-response lines makes the ratio of activities dependent upon the value of the response selected to be the comparison point. The dimethoxyphenylpenicillin has little activity against *S. aureus*. If the comparison is made in the 10–50% response region, it is about 3% as active as penicillin G. The 10–50% region has a slope characteristic of penicillins.

The dose-response curves of the three synthetic penicillins against a "penicillin-resistant" *Staphylococcus* is most illuminating and illustrates the utility of the log-probability plot in research on drug sensitivity. The resistant *Staphylococcus* produced so much penicillinase that an "inhibition curve" with penicillin G extended over more than three decades of concentration. The dimethoxyphenyl penicillin was one-fourth as active against the resistant *Staphylococcus* as against the sensitive strain and more important, the slope was that of a penicillin. The concentration at the 90% inhibition point was 3 times that of the 50% point, which is the same ratio as for penicillin G and the sensitive *S. aureus*. The other two synthetic penicillins have very flat dose-response curves. The concentrations at the 90% inhibition point is 100–200 times that of the 50% point. The dose-response curves of these synthetic penicillins against the resis-

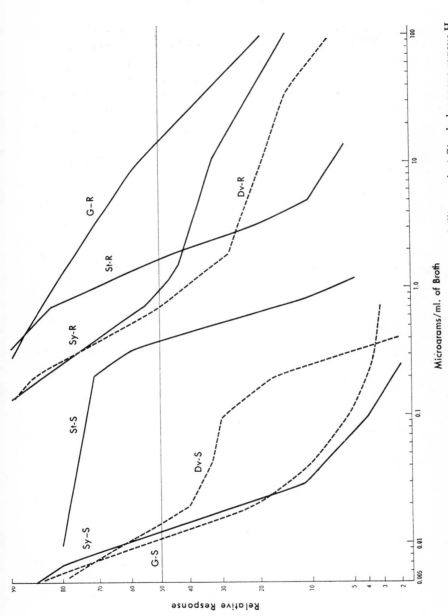

FIG. 14. Log-probability plot of dose-response curves for four penicillins against *Staphylococcus aureus* H (S), and a penicillin resistant (penicillinase producing) strain of *S. aureus* (R). The penicillins are: penicillin G (G), α-methylpenicillin V (S), α,α′-dimethylpenicillin V (DV), and dimethoxyphenylpenicillin (St).

tant *Staphylococcus* may give an indication of their therapeutic potentiality.

Mixtures of biosynthetic penicillins give typical dose-response curves against *S. aureus*. A mixture of biosynthetic and synthetic penicillins will give a dose-response curve which depends upon the relative activities and concentrations of the penicillins. For example, a 1:1 by weight mixture of phenylmercaptomethylpenicillin and α,α'-dimethylpenicillin V had a dose-response curve characteristic of the phenylmercaptomethylpenicillin, the more active member. The curve was plotted assuming phenylmercaptomethylpenicillin to be the only active substance in the mixture. The M.R. of the mixture was 10% less than that of phenylmercaptomethylpenicillin alone, a rather small increase in activity to obtain from such a large amount the second penicillin.

Dose-response curves for eight antibacterial substances are given in Fig. 15. The data used in computing the curves were obtained over a 4-year period. The values of M.R. are influenced by the composition of the media, pH, inoculum, temperature of incubation, etc., none of which was standardized. The curves are given solely to illustrate the kinds encountered and are not "standard" curves to which all subsequent curves should correspond.

An advantage of the log-probability plot is that related antibiotics give dose-response curves with the same slope. Consequently, mixtures can be assayed in terms of one of the pure components. This rule has been tested with pure compounds and mixtures of: biosynthetic penicillins, streptomycin and streptomycin B, the tylosins, and erythromycin and erythromycin B. Mixtures of dissimilar antibiotics may give slopes less than that of the antibiotic with the lesser slope or slopes approaching that of the dominantly active antibiotic depending upon the ratio of the antibiotics.

Usually the dose-response curve is concave to the concentration axis at high responses (90%) and convex to the axis at the low response (20%) ends. The reasons are unknown. If the size, shape, and degree of clumping of the bacteria are different at the different concentrations of the antibiotic substance as it is for *S. aureus* grown in the presence of penicillin, than the shape of the dose-response curve will depart from linearity because of the effect upon the photometric measurement of bacterial concentration.

The dosage-response curves in this chapter are all uncorrected for the turbidity of the inoculum which amounts to about 1.5 to 2%. The concentration of bacteria in a penicillin assay should at least double, even in the concentration ranges that eventually kill all of the bacteria. The correction to the curve for the turbidity of the inoculum is small and does not affect use of the curve in assaying. Correction for growth of the

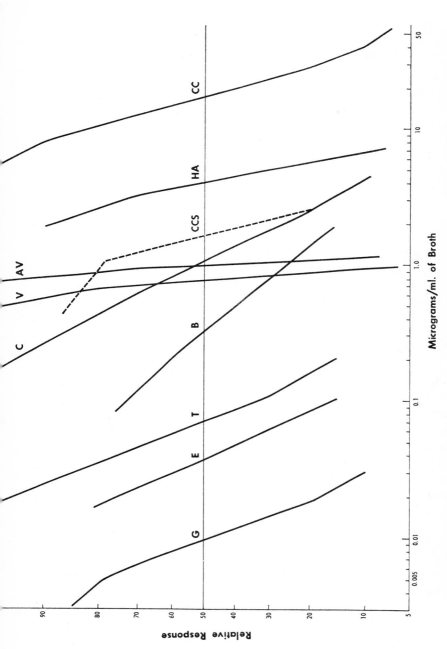

FIG. 15. Log-probability plot of dose-response curves for: *Staphylococcus aureus* H and penicillin G (G), erythromycin (E), tylosin (T), bacitracin (B), vancomycin (V), aglucovancomycin (AV), chloramphenicol (C), and cephalosporin C (CC); *Salmonella gallinarum* and cephalosporin C (CCS); *Klebsiella pneumoniae* and hygromycin A (HA). The concentration of the bacitracin is given in units per milliliter.

inoculum in the presence of a concentration of penicillin sufficient to kill all of the bacteria can be obtained by setting the photometer to its zero point (no growth) with tubes containing 10 times the concentration of penicillin required to produce a response of 3 to 5%. As judged from the curves in Fig. 14, this correction would be impossible with penicillinase-producing *Staphylococci* and penicillinase sensitive penicillins. The correction when applicable extends, somewhat, the linear range of the assay toward the region of high concentration of antibiotics. The whole problem of turbidity and its measurement in relation to assaying needs a thorough theoretical and experimental investigation by modern techniques.

The dosage-response curve is plotted on log-probability paper because the log of the dose usually gives a straight line response. The log scale also allows easy expansion of the concentration scale by decades as was done for the synthetic penicillins. Occasionally one encounters a system that gives a straight line on linear-probability paper. Streptomycin, dihydrostreptomycin, *Klebsiella pneumoniae* (ATCC 10031) (Fig. 12), and subtilin acting against either *Micrococcus conglomeratus* or *Staphylococcus aureus* are examples. The linear concentration scale would be used in constructing a standard curve for these combinations of antibiotics and organisms.

Although the log-probability graph seems to be the most generally useful for assaying, it is inconvenient to use because of the extra operation of converting photometer reading to concentration of bacteria. This inconvenience can be eliminated by a linear turbidimeter which can also compute the required ratios of dose tube to zero tube. The ratios are then plotted directly on the log-probability paper. A digital computer can be programed to perform the conversions and computations.

A log-probability dosage-response curve can be constructed without a standard preparation by using dilutions of the solution under investigation. The M.R. will be obtained in arbitrary units and other samples can be assayed in terms of it. For example, a set of M.R.'s of a new antibiotic can be taken for several bacteria and the relative activities computed from the M.R.'s.

The percentage scale on the probability paper is symmetrical about the 50% point. The spacing between units of percentage is a minimum at the 50% point and increases rapidly beyond 10 and 90%. A small constant error in measurement in these two regions will be magnified when plotted. This must be considered when drawing the curves, and the data at the extremes should be given less weight than those between 20 and 80%.

Reference was made earlier to shaking the tubes of an assay to speed growth and shorten the time of an assay. Shaking may or may not change the shape of the dosage-response curve. Shaking increased the sensitivity but did not change the shape of the dosage-response curves for vancomycin, erythromycin, and penicillin assayed with *S. aureus*. Shaking the tubes decreased the sensitivity and made a minor change in the shape of the α-methyl penicillin V curve. When changes in shape of the response curve are investigated, the range of concentrations of antibiotic should be such that the responses range from less than 10 to more than 90%.

B. Vitamins and Amino Acids

The dose-response line of vitamins and amino acids, unlike those of the antibiotics, should be straight lines passing through zero when the total number of cells is plotted against the amount of vitamin in the tube. The basis for this statement is the assumption that a definite and invariant number of molecules of a vitamin gives rise to one individual when that vitamin is the only factor limiting cell division. There is good experimental basis for the assumption as will be shown.

The calibration curve will pass through zero growth at zero vitamin level if the basal medium and inoculum are free from the vitamin. Although the amount of vitamin determines total growth, concentration units can be used in practical assaying because the total volume per tube is the same for standards and samples.

Wood [48] showed in 1946 that the acid produced by *Lactobacillus helveticus* in his medium was a linear function of the amount of riboflavin in the range 0.03–0.2 μg. per tube. Snell obtained a linear relation between dry weight of *Lactobacillus casei* and the amount of riboflavin in the tube. The linear range for B_{12} turbidimetric assay by *Lactobacillus leichmannii* ATCC 7830 is up to about 70 μμg. per tube when the optical density reading are converted to concentration of bacteria before plotting.

A linear dose-response curve has many advantages over the per cent transmittancy-concentration curve. Deviation from a linear relation is apparent immediately; any abnormality of response, out-of-line tube, or curvatures caused by inadequate basal media are not concealed in the gently changing form of the usual nonlinear calibration curve. The linear relation is so much more useful than any other that it should be used if at all possible. Wood showed that the linear dosage-response curve leads to the most efficient design of the assay.

Sometimes, the dose-response curve is linear to a point near zero and then falls off more rapidly than it should. Wood observed in a riboflavin assay employing *Lactobacillus helveticus* that the response (acid production) below 0.025 μg. riboflavin was uncertain and not on the linear part of the dose-response curve. He added an amount of riboflavin to the basal medium equivalent to 0.03 μg. per tube to bring his "blank" on the linear

[48] E. C. Wood, *Analyst* **71**, 1 (1946).

portion of the curve. This procedure has general utility in making a "blank" reproducible. A particular lot of prepared medium for assaying for vitamin B_{12} by *L. leichmannii* showed a drop below 0.01 mμg./tube and could have been corrected by adding 0.01 mμg. B_{12} to each tube (Chapter 7.11). The loss in B_{12} activity was caused by a slightly inadequate amount of reducing agent; the curve was linearized to the zero point by addition of reducing agents. The expedient of adding the substance assayed to linearize the assay curve should be done only if a medium that permits a linear curve is not available.

IX. Design and Validity

A. Antibiotics

1. *Validity*

There is no simple test of validity of an assay for an antibiotic substance. If the dosage-response curves of standard and sample have the same slope, the test is assumed to be valid. The slopes must lie within the boundaries set by the range of daily variation of the slope of the standard curve. If the slope of the sample is much different from that of the standard, a valid assay is impossible and the sample should be investigated.

2. *Design*

The minimum number of points needed to determine a standard curve depends upon the mathematical expression of the dose-response relationship and whether it is one of inhibition or growth (see Section VIII). If a straight line representation is valid, then only the two end points need be located. Even if 8 tubes were used per point, fewer tubes would be used than in establishing the usual multipoint curve. If a straight line relation is not demonstrable, the points on the concentration curve should be selected so that adjacent points differ by 10 to 15% in response. A preliminary assay of an antibiotic substance at concentration intervals represented by a geometric series (such as 0.1, 0.2, 0.4, 0.8, 1.6 u./ml. for a penicillin G solution would determine a log-probability dosage-response curve with *Staphylococcus aureus* accurately enough so that it in turn could be used to estimate concentrations needed to give responses differing by 10 to 15%. A penicillin curve, for example, is straight over the 20–80% response portion (log-probability form). The operator soon learns to space the concentrations to obtain a satisfactory curve. A set of concentrations once decided upon need not be changed so long as all other factors determining response are not changed.

If economy of operation is important, a technique leading to a straight line dose-response plot must be used. As an example of poor economy of tubes and labor, consider a penicillin curve made by plotting optical density against logarithm of concentration (Fig. 9). No portion of the curve is straight, although the portion between 0.1 and 0.25 u./ml. of sample could be considered to be straight with little loss of accuracy. A total of 10 points (including zero) are used and all are needed. Five points would have sufficed had the log-probability representation been used (Fig. 11). If only the straight portion of the latter curve were used, then only the two end points need to be established. The two-point design is a little too economical; a mid-point should be included as a check of linearity. Assay curves for many antibiotics plotted in this manner are linear over the most important portion, as is shown in Figs. 14 and 15.

B. Vitamins and Amino Acids

1. Nonlinear Assay Lines

Most of the assay curves for vitamins and amino acids reported in the literature are in a nonlinear form. Part of the nonlinearity stems from the mode of reporting the measurements of turbidity, part from the design of the test and part seems to be inherent in the test. As examples of amino assay lines both straight and curved see Chapter 8.

If the standard curve is not a straight line, then points at every 10 to 20% of the concentration range will be needed to establish it. The assay is read from a curve.

2. Linear Assay Lines

The assay curves for many of the vitamins and amino acids are straight lines when concentration of cells of the test organism is plotted against amount of growth promoting substance in the tube. This relationship seems to be relatively unknown as judged by its absence in the pertinent literature. Probably more straight line assay curves would be used if the utility of the linear representation were better known. The straight line representation reveals deficiencies of medium and procedure usually concealed in the gently flowing curves of the customary dose-response line. Curvature of the line in the linear representation is a danger signal alerting the assayer to deficiencies in the assay. These concealed deficiencies are responsible for much of the large variation in assays. Replication in an effort to obtain valid answers is a laborious and unsatisfactory substitute for good design and operations. Accuracy comes from meticulous attention to the details of correct design, not from statistical legerdemain—from microbiology, not mathematics.

One of the most useful linear representations of assay lines is the efficient common-zero five-point assay first introduced by Wood[48] for the acidimetric assay for riboflavin. The method has the further advantage that the potency of the sample can be obtained from a computed ratio of slopes of standard and sample lines and the hypothesis of similarity of sample and standard can be tested easily. The similarity of sample to standard should be tested for all samples until there is no doubt of the similarity. Samples known to be similar or identical to the standard need be assayed at only one level near the upper end of the curve. In vitamin B_{12} assays, a third level (10 $\mu\mu$g./ml.) of the standard near the lower end should be included as a check of the medium. It would not be used in constructing the standard curve. A medium deficient in reducing agent will give less response at 10 $\mu\mu$g. (*Lactobacillus leichmannii* assay) than it should (Fig. 1, Chapter 7.11). The higher levels are not affected so much. When this loss of response occurs, the assay should be repeated with a better medium. A small amount of vitamin B_{12} in the medium will make the response at the zero level greater than zero but will not interfere with the assay or with the computation of the ratio of the slopes of sample and standard curves. A zero-tube growth of more than 10% of that of the highest standard is unnecessary and should not be tolerated. Such growth indicates a poor medium or dirty glassware and can cast doubt on the validity of the entire assay.

Wood and Finney[49] discuss the basic statistics of the assay and the reasons they selected the particular distributions of tubes in the assay. They preferred the fully symmetrical arrangement where the number of tubes in each of the five groups was the same. The standard tubes and those of each sample may be thought of as a complete assay and treated as if no other samples were in the assay. The design assigns one-fifth of the tubes to each of the following groups: zero, mean standard, high standard, mean sample, and high sample. The mean is one-half of the high.

Application of the common-zero five-point assay will be illustrated by a vitamin B_{12} assay. The turbidities of the tubes are read in a photometer and converted into relative concentrations or, better read in a linear nephelometer. Let N be the total number of observations, or 20 if 4 tubes are in each group as in the example. Let y_0, y_m, y_h represent the relative concentrations of bacteria at zero, mean standard, and high standard, respectively. Let x_0, x_m, x_h represent the relative concentrations of bacteria in the zero tubes, mean test, and high test, respectively. $x_0 = y_0$ by definition. Test the linearity of each regression line by forming the quantities below using the mean values of each response:

$$L_s = y_0 + y_h - 2\,y_m \qquad (1)$$

$$L_t = y_0 + x_h - 2\,x_m. \qquad (2)$$

If the test is perfect and the lines not curved, $L_s = L_t = 0$.

[49] E. C. Wood and D. J. Finney, *Quart. J. Pharm. and Pharmacol.* **19,** 112 (1946).

The equations of the two regression lines are obtained by calculating the three parameters a, b_s, and b_t:

$$a = y_0 - (L_s + L_t)/7 \tag{3}$$

$$b_s = y_h - y_0 + (6L_t - L_s)/35 \tag{4}$$

$$b_t = x_h - y_0 + (6L_x - L_t)/35. \tag{5}$$

The terms involving L_s and L_t are corrections for the fact that unless L_s and L_t are both zero, the points do not lie on two straight lines, and therefore the best lines that can be drawn pass close to, but not through, the five points.

The mean estimated potency ratio is $R = b_t/b_s$ where b_s and b_t are the slopes of the two lines defined in Eqs. (4) and (5).

If the validity of the assay is in question or if the standard error is needed for some purpose, the following calculations are made. The standard error of a single observation is

$$s = \sqrt{\frac{\sum (y - y_p)^2}{N - 5}} \tag{6}$$

were y_p represents the appropriate group mean to be subtracted from any given observation y, $(y - y_p)$ is the deviation from the mean, and N is the total number of observation (20 in this example). The summation includes the deviations of the standards and of one sample.

The linearity of the lines, and hence the validity of the assay, may now be checked. The standard error of either L_s or L_t is given by

$$s_L = s \sqrt{30/N}, \tag{7}$$

so that for a 20-tube assay $s_L = 1.225s$. "If L_s and L_t are each less than twice s_L, the assay is valid. If either or both is over 2.5 times s_L, it most certainly is not."

"If the regression for the standard preparation is satisfactorily linear while that for the test preparation, over the same range of response, shows clearly a departure from linearity (or vice versa), the hypothesis of similarity is contradicted and the assay is invalid. But it is quite possible for both regressions to be nonlinear in a valid assay. . . ."[48] If the assay is valid, the standard error of the potency ratio R is calculated from

$$s_R = \frac{s}{b_s} \sqrt{\frac{8}{7N} (8R^2 - 9R + 8)} \tag{8}$$

which simplifies when $N = 20$ to

$$s_R = \frac{0.239s}{b_s} \sqrt{8R^2 - 9R + 8}. \tag{9}$$

"It is now possible to state the answer in the form of its 'fiducial limits,' that is to say, two figures, one higher and one lower than the mean result, such that the true answer is unlikely to lie outside the range covered by these limits. More accurately, the odds are 20 to 1 against the possibility that the observed mean estimate of the results could have arisen from a true value lying outside these limits." They are given accurately enough for all practical purposes by $R \pm t\ s_R$ where t has its usual statistical meaning and is 2.131 for a 20-tube assay.

These equations will now be applied to the assay of two samples for vitamin B_{12} by the *Lactobacillus leichmannii* method. Two variants of the assay were used. In one, the assay was incubated for 48 hours, and the potencies of two levels (0.3 and 0.5 ml. of sample) read from a calibration curve. In the first method the curve was constructed by plotting scale readings (optical density) against the amount of vitamin B_{12} added to the tubes. In the second method, the five-point common-zero assay was applied and the computations made according to Eqs. (1) to (9). In run no. 1, the assay was read after incubation periods of 48 to 72 hours. The validity of the assays was tested for two responses, scale readings (optical density), and the relative concentrations computed from them. The scale readings give a curved calibration line as would be expected from the discussion in Section II of this chapter and the curvature may be sufficient to make a valid assay into an invalid assay when the statistical procedure of Wood and Finney for the five-point common-zero assay is applied.

The details of the computations are given for the 72-hour, relative concentration response of run no. 1 (Table V). This run differed from all others in that the photometer scale readings were made 1 minute after the cuvette was filled and not at various times approximating 20 seconds after filling. This in accordance with the principle set forth in Section III.C.

TABLE V

RELATIVE CONCENTRATIONS

Standards	Individual tubes				Mean	Deviation from mean				$\Sigma\ (y - y_p)^2$
0	0	0	0	0	0	0				
30 $\mu\mu$g.	26.9	27.9	25.1	26.4	26.6	0.3	1.3	−1.5	−0.2	4.07
60 $\mu\mu$g.	54.0	50.5	50.5	52.5	51.9	2.1	−1.4	−1.4	0.6	8.69
Sample 1										
0.3 ml.	25.1	25.1	25.6	21.5	24.3	0.8	0.8	1.3	−2.8	10.81
0.6 ml.	47.1	48.6	49.1	44.7	47.4	−0.3	1.2	1.7	−2.7	11.69
Sample 2										
0.3 ml.	22.9	21.5	23.3	20.7	22.1	0.8	−0.6	1.2	−1.4	4.40
0.6 ml.	49.1	44.2	45.1	47.6	46.5	2.6	−2.3	−1.4	1.1	16.63

Sample 1:

$$L_s = 0 + 51.9 - 2(26.6) = -1.3$$
$$L_t = 0 + 47.4 - 2(24.3) = -1.2$$
$$b_s = 51.9 - 0 + [6(-1.2) - (-1.3)]/35 = 51.7$$
$$b_t = 47.4 - 0 + [6(-1.3) - (-1.2)]/35 = 47.2$$
$$R = 47.2/51.7 = 0.913$$
$$s = \sqrt{35.26/15} = 1.35$$
$$s_L = 1.255s = 1.88$$

L_s and L_t are each less than twice s_L; therefore, the test is valid. The standard error of the potency ratio is given by

$$s_R = \frac{0.239\,(1.53)}{51.7}\sqrt{8(0.913)^2 - 9(0.913) + 8} = 0.0179.$$

The 95% confidence limits of R are given by 0.913 ± 0.038. R is the ratio of the highest sample (0.6 ml.) to the highest standard (60 $\mu\mu$g.); therefore, the sample as measured contained 0.913 (60)/0.6 or 91.3 $\mu\mu$g./ml. of B_{12}.

The sample was diluted 10,000 for assay, therefore, the potency of the sample as submitted was 0.913 ± 0.038 μg./ml.

Sample 2:

$$L_s = 0 + 51.9 - 2(26.6) = -1.3$$
$$L_t = 0 + 46.5 - 2(22.1) = 2.3$$
$$b_s = 51.9 - 0 + [6(2.3) - (-1.3)]/35 = 52.3$$
$$b_t = 46.5 - 0 + [6(-1.3) - (2.3)]/35 = 46.2$$
$$R = 46.2/52.3 = 0.883$$
$$s = 1.50$$
$$s_L = 1.83$$

L_s and L_t are each less than twice s_L, therefore, the assay is valid. The standard error of R is

$$s_R = \frac{1.50\,(0.239)}{52.3}\sqrt{8(0.883)^2 - 9(0.883) + 8} = 0.0211.$$

R and its 95% confidence limits was 0.883 ± 0.049 and the potency of the undiluted sample was 0.883 ± 0.049 μg./ml. Similar computations were made for run 1 at 48 hours to illustrate that the assay was not valid when the assumption was made that the relationship between optical density of the suspension and amount of vitamin B_{12} in the tube was linear. (The validity of sample 2 was the same as that of sample 1). The 72-hour assay was valid when the relative concentration of the bacteria was one of the parameters.

The relationship between turbidity and relative concentration of bacteria in run 1 was linear up to 40 $\mu\mu$g./tube for the 48-hour assay and up to 60 $\mu\mu$g./tube for the 72-hour assay. In run 2 in contrast to run 1, the highest standard (60 $\mu\mu$g./tube) was on the linear portion of the calibration line at 48 hours and the assays were statistically valid. The range of linear response in 48 hours varies with the lot of prepared medium and has ranged from 40 to 70 $\mu\mu$g./tube. As a practical matter, the computational part of the five-point assay should be done only when the response (concentration of bacteria) is linear at least to the level of the high standard (60 $\mu\mu$g./tube in the design of run 1). Otherwise, the basic assumptions of the statistical design are not valid. The 48-hour assay of run 1 was not only of questionable validity ($L_s > 2.5s_L$, $L_t < 2s_L$) in the five-point common-zero assay but probably was inaccurate because of the short incubation time. The questionable validity came from the application of a statistical model which obviously did not fit it. The potencies obtained from a calibration curve ranged from 0.81 to 0.96 μg./ml. for sample no. 1 and from 0.83 to 0.96 μg./ml. for sample no. 2, and are given in Table VI. Samples 1 and 2 were from

TABLE VI

ASSAY OF TWO SAMPLES FOR VITAMIN B_{12} BY TWO MODIFICATIONS OF THE
L. leichmannii METHOD

Run no.	Response	Assay valid?	Sample no.	
			1	2
1	48 hr., rel. conc.	?	0.842 ± 0.056	—
	72 hr., scale	—	0.915 ± 0.939	—
	72 hr., rel. conc.	Yes	0.913 ± 0.038	0.883 ± 0.049
2	48 hr., rel. conc.	Yes	0.905 ± 0.049	0.895 ± 0.064
1	48 hr., scale, curve	—	0.96	0.96
1	72 hr., rel. conc., curve	—	0.92	0.90
3	48 hr., scale, curve	—	0.84	0.84
4	48 hr., scale, curve	—	0.81	0.83
5	48 hr., scale, curve	—	0.83	0.86
6	48 hr., scale, curve	—	0.83	0.91

similar but not identical sources. The two valid five-point common-zero assays gave essentially identical answers noticeably different from the potencies obtained from the 48-hour assays read from calibration curves. There are two objections to the 48-hour assays: (1) there was no proof that the incubation period was long enough to obtain limiting growth by both standards and samples, and (2) the speed of measuring the turbidities was such as to introduce considerable variation in the measurements (Section III.C).

The five-point common-zero design should be considered in selecting the assay procedures for vitamin B_{12}, other growth promoting substances, and certain amino acids. Its successful application requires that the following conditions obtain:

1. The response be obtained as protoplasmic mass and not as some instrument response nonlinearly related to it.

2. The dose level be restricted to the region of linear response.

3. The incubation period be long enough to obtain growth limited by the amount of growth substance present. The required incubation period depends upon the quality of the medium, the dose, the inherent characteristics of the test organism, the contaminating substances accompanying the test material, and the physical conditions of incubation. The smaller the highest dose, the shorter the time of incubation—within limits.

The standard error of the potency ratio will be small only if scrupulous attention is paid to the details of preparing the assay and if the turbidity is measured properly (Section III). If conditions 1–3 cannot be met, use a calibration curve and replicate if a statistical treatment is required.

The principles of turbidimetric assaying given in this chapter were applied to the problem of reducing the variation in the vitamin B_{12} assay. A careful operator was selected and trained. He worked by himself in a separate room with equipment used by no one else. He did the final steps in the washing of all glassware. He prepared the assay medium from commercial dehydrated media (B.B.L., Baltimore Biological Laboratory, Inc.) and distilled (not deionized) water. He prepared the standard solutions, dilutions of samples, and performed all other operations of the assay. Before he measured the turbidity he shook the tubes and allowed them to stand for 15 minutes to permit air bubbles to clear from the medium. In one assay he even measured the turbidity 1 minute after pouring the contents of a tube into the cuvette in an effort to mimimize the influence of flow birefringence upon measurement of turbidity. (A waiting time this long is impractical in any except a small special assay. A turbidimeter can be designed so that the effect from flow birefringence is insignificant and it should be used in any precise assay for vitamin B_{12}.) The use of the proper response, i.e., cell mass, made possible the linear relationship required for the common-zero five-point design. The results of assays of two samples are given in Table VI.

The growth period of 3 days was necessary in some of the tests to obtain the linear relationship. Even this time may not be long enough in some tests if linearity to the 60 $\mu\mu$g./ml. point is required. Small uncontrolled variation of unknown origin was still present as was indicated by different lots of broth prepared from commercial medium of the same lot number giving a linear response to 60 $\mu\mu$g./ml. in 48 hours in one test and in 72 hours but not in 48 hours in the subsequent test.

An operator working full time can assay about 20 samples per day using an average of 10 tubes per sample. The same laboratory technician can assay 28 samples per day by a multipoint (9 levels) standard curve and multipoint (6 levels) of sample but will not prepare either the glassware or the medium. The assays will require an average of 16 tubes per sample. The poor design of the test procedure makes replication necessary to obtain acceptable standard errors.

This experience suggests that the most efficient procedure for assaying vitamin B_{12} samples of fairly accurately known potency would be to use the common-zero five-point design with all that its use implies and assay each sample once. For control purposes or where certainty of answer is demanded, or for samples of imprecisely known potency, assay each sample twice and take the average if the two assays agree within statistically acceptable limits. The second assay should be done on a different day or by a second operator if on the same day. Start with the material submitted for assay each time and prepare the second set of dilution independently of the first set.

This section may be ended appropriately with two quotations from Wood.[48]

"I must here insert, however, a word of warning. While the result of any statistically invalid assay must be for that reason in error, it does not follow that the result of a statistically valid assay is not in error. If, for example, in a riboflavine assay the Test Preparation should happen to contain not only riboflavine, but also some other growth-stimulating factor, and if this other factor stimulated growth proportionally to the dosage at all dosage-levels, no statistical test and no method of calculating the result could possible detect anything suspicious in the result obtained. The combined riboflavine and other factor would be estimated as riboflavine.

"I hope it will not be thought presumptious if I conclude with an exhortation to all analysts, no matter what their speciality may be. It is not making the best use of a statistician, nor is it fair to him, to call him in at the end of an experiment to extract therefrom information the analyst cannot obtain for himself; indeed, if the experiment was not properly designed the statistician will probably fail also. He should be consulted when the experiment is being designed. The result will be peace of mind for both analyst and statistician; a greater return of information per unit of experimental labour; and the satisfaction that comes from knowing that, whatever the results may be, no critic can infuriate the analyst by proclaiming pontifically 'this assay is statistically un-sound.' "

An excellent example of the difficulties caused a statistician by data inadequate both microbiologically and in mode of presentation is the report by Bliss[50] on a collaborative A.O.A.C. study of a B_{12} assay using *Lactobacillus leichmannii*.

X. Sensitivity

The sensitivity of photometric assays is so dependent upon specific details that only general statements can be made. Sensitivity, for the purposes here may be defined as the smallest change in concentration that can be detected with certainty. It is influenced to a minor degree by such environmental factors as pH, temperature of incubation, composition of medium, concentration of cells in the inoculated broth, and aeration and to a major degree by the reciprocal properties of test organism and substance assayed. For example, changes in the environment make small changes in the M.R. of the penicillin assay whereas a change from the usual test organism (*Staphylococcus aureus*) to *Escherichia coli* causes a large decrease in sensitivity. Sensitivity can be increased by decreasing the volume of broth in a tube; the 10-ml. volume is not sacrosanct. The sensitivity of a penicillin assay can be varied by a factor from 1 to 5 or more by changes in volume and environment. A range of variation of more than 5 is possible in several of the basic antibiotics-assay systems.

A comparison of three methods is given in Table VII. The test or-

TABLE VII

SENSITIVITY OF THREE METHODS OF ASSAY
FOR PENICILLIN G IN UNITS/ML.
OF SAMPLE

Serial dilution	Turbidimetric	Plate
0.02	0.007	0.09

ganism was *S. aureus*. The value for the serial dilution method was taken from Table I in Chapter 3.

The assay procedures given in Chapter 6 for specific methods of assay for antibiotics are neither the most nor least sensitive, they are the ones being used. The sensitivity can be changed to suit the need. The

[50] C. I. Bliss, *J. Assoc. Offic. Agr. Chemists* **39**, 816 (1956).

methods for vitamins and amino acids are of practical sensitivity. Some find the method for vitamin B_{12} to be embarrassingly sensitive.

Procedures for changing sensitivity while maintaining reliability need to be developed for many of the common assays.

XI. Meaning of a Photometric Response

Let us suppose that the principles of turbidimetric assaying as given in this chapter have been followed and an answer obtained. Perhaps confidence or even lack of confidence limits, depending upon the sophistication of the assayer, have been placed upon the answer. Now just what does the answer mean? Unfortunately, for our peace of mind this question does not have a universal answer because microbiological assays do not measure quantities of substances; they measure responses. The response must then be converted into quantity of active substance with the aid of a standard curve (responses of a "standard" preparation). An assay can give on a sample of unknown composition (most samples) its activity in terms of the standard and nothing more. More information than is obtained from an assay is necessary to determine the *concentration* of the putative active substance. Usually, definite information is lacking, and two assumptions are made as a substitute for knowledge. They are that the standard and sample each contains one and only one active principle and that the two active substances are identical. The two assumptions are correct guides to interpretation of assays of pure materials which, however, are a minority of the samples assayed. Most samples are mixtures. The mixtures may be those of closely related substances, as, for example, a penicillin fermentation sample, or a mixture of unrelated substances. The dose response curves may or may not be modified by the contaminating substances depending upon the character and quantity of the contaminants. The assayer needs always to be upon his guard against overenthusiastic interpretation of a response as indicating a definite quantity of substance. It may or may not.

Nothing can be learned about interfering substances from an assay at one concentration. Additional information can be obtained from assays at two or more (dose-response curve) levels or from a chromatogram of the sample. If the dose-response curve of preparation and standard differ appreciably in slope, the assayer is confronted with a problem to be solved before he can report his assays in terms of the standard. Chromatography is an invaluable guide to interpretation of assays of samples of unknown purity. As an example of an assay without definite meaning consider the assay of a sample of unknown origin for vitamin

B_{12}. The assay gives the growth promoting activity of the sample for a species of *Lactobacillus* in terms of cyanocobalamin, but tells nothing of the relative concentrations of the individual compounds that make up the vitamin B_{12} complex or the activity of the preparation when administered to test animals. A chromatogram might identify the active substance and thus give definite meaning to the assay.

There is a tendency to measure reliability of a microbiological method by comparing it with a chemical method. This measure of reliability fails when applied to impure preparations because the two classes of methods do not measure the same thing. Perfect agreement between the chemical and microbiological methods is more than fortuitous only on the assay of pure materials. The methods should be expected not to agree because of the presence of substances (a second or third antibiotic, for example) which interferes with each assay to different degrees. Usually, but not invariably, the answer obtained by a chemical method will be higher than the microbiological assay. Neither is right, neither is wrong, they measure incommensurable. Differences between chemical and microbiological assays of impure preparations should be expected and should be considered as a lead to be investigated and not as indication of poor work by one or both groups of assayers.

Most of the chemical methods are not specific for the microbiologically active groups of the active compound. All too often the chemical method consists of measuring the amount of a sugar or an ultraviolet absorption and converting the measurements to an equivalent of the active substance by means of a calibration curve prepared from pure compound. Such chemical methods have little specificity when applied to crude preparations. However, there is a chemical assay which has the same molecular specificity as the microbiological assay. Both the microbiological assay and the hydroxamic acid assay for penicillins measure the intact β-lactam group. The chemical method obtains its specificity not from the chemical reactions, which are not specific, but from the enzyme, penicillinase, used in preparing the blank. Since the chemical method will measure penicillins which have little if any antimicrobial activity, the answers may be higher than those obtained in the microbiological assay. Or again, the chemical assay may be lower than the microbiological as, for example, assays on a 1:1 mixture of penicillins G and K assayed in terms of penicillin G.

The point of this discussion is not that a definite answer cannot be given, but that the meaning may not always be that attributed to it. Obviously, an assay group must supply answers; otherwise it will soon cease to exist. The practical procedure is to assign potencies as indicated by the assay but always to be ready to reinterpret the answer and change method of assays should significant systematic errors be indicated. The interpretations are made by the assayer in the light of his experience accumulated over a number of years in assaying many kinds of samples. This experience is invaluable in the day-to-day operations of the assay laboratory and cannot be replaced by an "electronic brain," however sophisticated, or by routine statistical operations.

XII. Variations and Errors

Many factors cause variation and error in turbidimetric assays. Some of these, their causes and preventions will be considered. Errors caused by obtaining an unrepresentative sample or errors in preparing the sample for assay are considered under the specific methods of assay and discussion of them will not be duplicated here.

A. Glassware Cleaning

An ample supply of *clean* glassware is essential to successful assaying. Dish washing is one of the lowest rated jobs in an organization and, to an assay group, one of the most important. Poorly cleaned glassware is a source of endless annoyance, confusion, and expense. Unless the cleanliness of the glassware is beyond question, sources of error in an assay are difficult to find. A residue of chromic and sulfuric acids from a chromic acid cleaning solution can be as effective in reducing growth as an antibiotic. This difficulty can be avoided by proper rinsing of the glassware, preferably by properly operated automatic equipment. Although convincing the operator that he should do his own dish washing might be difficult, it should reduce the total amount of work by improving duplication of such sensitive assays as those for vitamin B_{12} and folic acid. Glassware cleaned in a washroom should be good enough for most other assays.

Another and better way to avoid the difficulties associated with chromic acid cleaning mixtures is not to use it; other cleaning methods usually are at least as effective and do not have its disadvantage of a toxic residue and corrosiveness. A residue of detergent can be just as damaging to an assay as an acidic residue. Any cleaning agent can be removed by proper rinsing, preferably by machine because hand washing seems to be a lost art. Proper rinsing requires *clean distilled* water as the final rinse. Distilled water containing oil (steam condensate) or deionized water are not "just as good but cheaper" than clean distilled water. Clean glass and also glass contaminated with detergents drain with a continuous film of water. Modern industrial dish washing machines are used extensively throughout the pharmaceutical industry for washing all of the glassware used in assay laboratories. Even pipets may be washed by the machines. In many laboratories pipets are soaked in detergent solution and rinsed in an automatic washer. The automatic flushing-type of pipet washer may do a poor job of washing if the water flow is too great. Pipets requiring more vigorous cleaning than can be obtained with the usual alkaline detergents are best cleaned by filling with a nitro-sulfuric acid mixture (95 ml. conc. H_2SO_4 + 3 to 5 ml. conc. HNO_3) and rinsing with distilled water after a contact time of 1 to 48 hours. This seems to be a more effective mixture than chromic acid and avoids possible trouble with chromic and chromous ions. Sintered glass filters should be cleaned with the nitro-sulfuric acid mixture; never with chromic acid mixture.

B. Volumetric Equipment

The range of dilutions required to prepare the many kinds of samples submitted to a large assay laboratory extends over many orders of magnitude. The dilutions may range from 10^0 as for some blood samples of antibiotics to 10^4 for other samples of antibiotics. The most extreme range is found in the samples of vitamin B_{12} where the range is from <100 $\mu\mu$g./ml. in serum of patients with pernicious anemia to 4×10^{10} $\mu\mu$g./ml. or more in concentrates of the vitamin.

Errors creep into the dilutions at two points, the pipetting and the measurement of the diluent. Both transfer and measuring pipets are used in making dilutions. A 1-ml. transfer pipet with intact tip will measure a volume with an error of about ±0.01 ml. and should be used for making the decimal steps in the dilution. A good quality measuring pipet (Mohr type with intact tip) should be used for the nondecimal dilutions. The size of the measuring pipet should be chosen so that not less than one-half of its total volume is delivered; otherwise, the measuring error may be more than 1%. The pipet must be clean and carefully used to achieve such accuracy. If an accuracy of 3 to 5% is sufficient, measuring pipets could be used for all dilutions. Nonaqueous or viscous solutions should be measured by a pipet calibrated to contain (wash-out type). The practice of using a 1-ml. serological pipeted as a wash-out pipet can lead to a positive error of 3% caused by design (contained volume is greater than delivered volume) of the pipet or to negative errors of unknown size depending upon completeness of washing. Viscous solutions may be weighed if wash-out pipets are unavailable.

Volumetric flasks or dilution bottles are used in making large dilutions. Dilution bottles, filled by a good quality (preferably a direct read ing type) automatic pipetting machine, are satisfactory for most work especially if the total dilution is less than 5000–10,000. A test on a direct reading pipetting machine (Filamatic) showed that it would repeat to give a total volume, for 10 consecutive samples of 990.3 ml., when set for 99.0 ml. It will not do this if there are leaks in the system or if bits of cotton are in the valve chambers. One milliliter of sample is added to a 99.0-ml. blank, the bottle stoppered and shaken vigorously two or three times. Further dilution can be made in the same way if necessary. The bottle and its closure must be washed, sterilized, and handled so as not to contaminate it with the substance being assayed. The prevention of contamination is most important for equipment used in folic acid, biotin, and vitamin B_{12} assays. Volumetric flasks (20–200 ml.) are used when more accurate dilutions than are obtainable with bottles are wanted. Many volumetric flasks, including those made to M.C.A. style, have so little volume above the graduation that complete mixing is difficult to achieve without more care than too many operators exercise.

What are the possible diluting errors in preparing the 40 mg./ml. vitamin B_{12} sample mentioned above? To dilute the sample to the assay range of 40 $\mu\mu$g./ml. requires a total dilution of 10^9. This can be done in many ways, two of which are illustrated. To dilute a 40 mg./ml. sample to 40 $\mu\mu$g./ml. in three steps accurately requires three 1-ml. transfer pipets and three 1-liter volumetric flasks and could have a maximum dilution error of 3% caused by errors of calibration of the measuring pipets and flasks. To make the dilution in five steps (four of 100 and one of 10) instead of three gives a maximum error of about 4% and would require more (but less expensive) glassware and only an eighth as much diluent. Most of the error results from the uncertainty of the calibration of the pipets. These estimates of diluting errors are minimal, they can very well be much larger and, unless the dilution procedure were duplicated, could go unrecognized.

After the dilutions are prepared, accurately measured samples are added to test tubes. In too many assay groups, the samples are added to the sample tubes with a capillary pipet or a 1-ml. serological pipet. Volumes from 0.02 to 0.1 ml. are measured with the former, and from 0.1 to 1.0 ml. with the latter. Obviously, pipets used in this way can contribute an error of 10% or more in measuring the volume of sample. However, the convenience of the capillary pipet is more important in some survey work than an accurate answer. For those who need as accurate answer as possible, e.g., control groups, the assay should be so designed that a fixed volume of sample, 0.5 ml. is added to each tube. Then use carefully made Mohr-type pipets with graduations (and calibration) at 0, 0.5, 1.0, 1.5, and 2.0 ml. Fewer errors in measuring will be made with such pipets with their widely spaced graduations than with pipets graduated in 1/100 ml. steps. (If such special pipets are available, there will be no reason to use 5-ml. measuring pipets to measure 0.5-ml. samples as I have seen done.) Other graduation intervals, of course, can be had depending upon the details of the assay. These special pipets cost no more in quantity than the commercially available ones and could be of higher quality if each graduation also is a calibration point.

Other assay schedules call for volumes ranging from 0.5 to 4.5 ml. of sample per tube and customarily are added with a 5-ml. measuring pipet. A change in these designs to reduce the total number of levels assayed and to use more accurate pipets would improve the tests.

Making many large dilutions is time consuming, uses large amounts of glassware, and becomes a boring operation. Better mechanization of the operation would be welcomed. Many times, the necessity for making large dilutions in several steps could be avoided with considerable saving in glassware and time by using a less sensitive test organism. Ideally, the dilution in any assay should be just large enough to eliminate the inhibitory and stimulatory effects of the substances which may accompany any sample of natural origin. The size of the dilution depends upon the substance assayed, the test system, and the test organism. Some indication of the minimum dilution, where it is known, is given in the detailed instruction for performing each assay. The possibility of avoiding large dilutions by using a relatively insensitive assay organism has not been adequately investigated. If a method seemed to "work," it was used with a minimum of investigation because the people who had to assay an ever-increasing number of samples each day did not have the time to make a careful study of the assay methods. They made do with what they had.

Details of a dilution scheme can introduce unexpected error as will be illustrated by an unusual error encountered in the assay of Tylosin. The samples were diluted for routine assay to a concentration of 2.0 μg./ml. by several schemes using buffer solutions in soft-glass bottles. The pH 7 buffer blanks were measured by automatic pipetting machines for the large volumes (19, 49, 99, and 149 ml.) and by pipets for the small volumes (2, 9, and 14 ml.) with errors in measuring the volumes of less than 0.1 ml. The antibiotic was known to be stable in the buffer for several months in the refrigerator. Residue of detergent and bottle washing were eliminated as causes of the erroneous answers which were always low. Dilutions of a test sample prepared by different operators occasionally gave quite different answers as did dilutions prepared by the same operator on different days. Then a relation was observed between the diluting scheme and a discordance in answers much larger than possible errors of diluting. For example, a sample known to contain 2 μg./ml. assayed 1.34 μg./ml. by one scheme and 2.00 by another in the same assay. The conclusion that the antibiotic was not in the solution when assayed was inescapable. Then where was it? Inspection of the dilution scheme revealed that the low answers were obtained only when a small amount (10 or 15 ml.) of the 2 μg./ml. solution was prepared in a soft-glass bottle. A test in which pyrex vessels were used in preparing the dilutions suggested the hypothesis that the antibiotic (a large basic molecule) was adsorbed on the glass surface. Dilute solutions of Tylosin should be handled only in Pyrex glass vessels. This soft-glass adsorption effect should be considered when assaying other basic antibiotics.

C. Medium

Small variations in composition of medium, including those caused by day-to-day differences in sterilization, will cause small changes in shape and position of the dose-response curve but should not cause variation in answers read from the daily standard curves. A composite curve should never be used except to obtain an approximate answer, which then would be used as a guide in preparing the dilutions for the accurate assay. A composite curve in the form of a log-probability plot has less variation than a curve constructed from instrument response-log concentration of antibiotic. An assay (antibiotics, amino acids, vitamins) to be accurate must be read from a standard curve prepared at the same time with the same medium, inoculum, etc.

D. Organism

The usual test organism is a stable laboratory stain and rarely gives trouble. A contaminated culture may give erroneous answers as would be expected. A fastidious organism is more difficult to maintain and use than a nonfastidious one and should be used only when there is no substitute.

E. Incubation

After the samples are put in the tubes, and broth and inoculum added, they must be incubated at a selected temperature for a given time period which may range from 3 hours to 3 days. A water bath is always used for incubating the tubes. The bath should be stirred to ensure uniform temperature and the same rate of heating of tubes at the center of a rack as at the periphery. This is especially important when chilled inoculated broth is put into the tubes as in certain antibiotic assays. One size of test tube should be used throughout a test also to ensure identical rates of heating of the contents of the tubes. A tube with a temperature schedule different from the standards is in reality part of a different assay and may give erroneous answers. If a bath is suspected of nonuniform temperature, a set of standard tubes (about 50% inhibition) in each rack will soon show if the heating schedules of the racks are different.

All racks of tubes of a test should be in the water bath the same length of time. This is an unobtainable ideal that can only be approximated. However, it can be approximated very closely by keeping the size of the test to an easily handled number (4–6 racks). In some laboratories, a complete test is only 2 racks (80 tubes total). In others, a test may be 10 racks (600 tubes)! Growth rate studies show that at the higher levels of growth, a time interval as short as 2 minutes will give a detectable change in turbidity. After the racks are removed at the end of the incubation period, the bacteria are killed by steaming or by addition of an antiseptic. If the latter, then the killing agent must be added to all of the tubes simultaneously as by a peristaltic pump with as many delivery tubes as there are test tubes in a rack and followed by a holding time long enough to ensure complete inhibition of growth. Steaming usually is a more convenient way of killing than by a chemical agent. Steaming should not cause coagulation of material (including the bacteria) or formation of color in some tubes and not in others. Again the procedures for handling the racks must be the same for all racks.

F. Mechanical Errors

Certain errors may be listed as mechanical and include those caused by not shaking the tube sufficiently, not allowing the air bubbles to escape from the shaken broth before measuring turbidity, reading too soon after filling the cuvette (both cocci and rods will give this error but it is much larger for rods), using an inherently unstable photometer, reading the scale with less than maximum accuracy (for example to only the nearest ±1 division), using a photometer with an inadequate scale (the scale of the Spectronic 20 is much too short for good assays), and making errors in recording the readings.

The several kinds of errors in reading the photometer can be avoided with possible introduction of other kinds of errors by employing electromechanical means to effect the recording. The form of the record may be holes in a punched card or tape, or record of pulses on a magnetic tape. Records in these forms can be further processed by computers.

G. Errors of Measurement

Errors of measurements are of two types, those inherent in a calibration curve and those stemming from the variation in turbidity of the several tubes in a set. The error caused by inability to distinguish small differences on a graph (on 9 × 12 inch paper) depends somewhat upon the type of plot used. Typical curves for penicillin G, bacitracin, and vancomycin were examined and the error in concentration, C, caused by a small error in response (0.5 division) was estimated at high (80%) and low (20%) response (Table VIII).

The penicillin curve has a slope typical of most antibiotics. The bacitracin curve is much flatter than the penicillin curve and the vancomycin curve is much steeper. No type of calibration curve is best at the higher

TABLE VIII

	Type of calibration curve			
Antibiotic	Log-probability (%)	T-log C (%)	O.D.-log C (%)	T-C (%)
Penicillin G	3–1.8	8–1.6	6–1.9	8–1.9
Bacitracin	5–3	—	—	—
Vancomycin	0.7–0.5	—	—	—

concentrations of antibiotic. The log-probability curve is the best one at low concentrations.

The scale reading error can be reduced below those given in the tabulation by increasing the size of the graph, a procedure that may accomplish little with respect to accuracy, as will be explained. All of the points in the straight line portion of a log-probability plot are used to determine a best line fitted to the points by eye. A point obviously in error is given little weight in drawing the line. The three other representations of calibration curves are curved or have empirically determined straight line portions. The customary practice is to connect the points on the curve by straight lines or to draw the curve with a French curve. All points are given equal weight and an error in position is not obvious and affects a region on each side of the erroneous points by unknown and variable amounts. A large scale reduces the inherent error in reading a calibration curve but does not affect in any way the frequently much larger error caused by error in the position of a single point. Drawing a smooth curve through or near points may or may not have the effect of achieving a valid smoothing of the curve. This writer knows of no way of being certain that the "best" curve has been drawn unless the log-probability form of plot is used.

The second type of error is that caused by the different turbidities of the different tubes of a set. The differences in the tubes are the end products of several variations, each of which may be small. There is the difference caused by the measuring instrument; it should be imperceptible. If the instrument will not duplicate readings of the same uniform suspension, it should be improved or replaced. Individual techniques of preparing the suspension, such as not shaking the tubes sufficiently or not allowing the air bubbles to clear from the suspension before reading, contribute to the tube-to-tube differences and can cause rather large and avoidable errors. Errors of pipetting samples and inoculated broth also contribute their part to the variations within a set of tubes. The sample pipetting error can be kept to less than 2% if 0.5-ml. volumes or larger are carefully measured with good quality pipets. The error of pipetting the inoculated broth should be well under 1% if a good automatic pipetting machine is properly used. Dirty tubes (residual acid from an acid bath) can contribute large and random errors to both antibiotic and vitamin assays. Equipment, medium, and glassware contaminated with the growth substance assayed can cause large errors and is especially important in the assay for B_{12} which is the most sensitive assay. The preceding is one way of saying that accurate assaying requires careful work by properly trained operators who understand and practice the principles of microbiological assaying. No mention has been made of the contribution of microbiological variation to assay errors; this author has never seen a significant one. What is popularly called microbiologically variation is found to have a macrobiological origin when investigated.

The turbidity of the individual tubes of a set should measure within 2% of the average of the set of 3 or 4 tubes of an antibiotic assay. Frequently the measurements of the tubes within a set will deviate by less than 1% from the average which indicates either that the estimates of errors to be expected are too high or that there are compensating errors. Table IX

TABLE IX

1000 µg./ml. erythromycin standard

990 }	1000	960 }	1000	1020 }
980 }		990 }		980 }
	1000 }	960 }	1000 }	
1000	1000 }		980 }	980
		1010 }	980 {	
990 }	1010 }	1010 }	970 }	1010 }
990 }	990 {			1010 }
	970 {	970 }	1010 }	
980 }	1020 }	970 }	1020 }	1020
990 }				

lists the activities of 1000 µg./ml. erythromycin standard solutions found in 17 consecutive assays obtained during a 4-week period. Four sets of stock standard solutions were prepared. The same 1000 µg./ml. stock solution from which the standards were prepared was also diluted to 2 µg./ml. for assay. Each figure represents a single determination (3 tubes) of a separate dilution from the stock solution. The figures enclosed in a bracket were from the same assay.

The figures were taken from routine assays in which the best technique was not followed. The calibration curve was read to the nearest 1% of the concentration. Hence the unit interval of the answers was 10 µg./ml. Presumably, if carefully done, the samples with unknown potency could be measured with the same accuracy as the standards in so far as the mechanics of the assay are concerned.

The inherent errors of a turbidimetric assay are less than is commonly believed. The sources of most of the errors are ignorance of proper procedure or indifference to their practice, especially if good practice is inconvenient or requires a change of habits. This is as true of the supervisors as of the technicians who carry out the operations.

TABLE X

TYLOSIN STANDARD CURVE BY GOOD TECHNIQUE

Conc. of standard solution (µg./ml.)	Photometer readings, individual tubes				Avg.	Rel. conc.	Response (%)
0	73.5	74.0	73.5	74.0	73.8	68.1	100
0.6	66.0	64.5	65.0	64.0	64.9	58.1	86
0.8	58.5	57.5	58.5	58.0	58.1	51.6	75.8
1.0	50.5	51.0	50.0	50.0	50.4	44.0	64.6
1.4	39.0	39.0	38.5	39.0	38.9	33.0	48.5
2.0	25.5	25.5	26.0	25.5	25.6	21.2	31.1
3.0	15.0	15.5	15.5	15.0	15.3	12.4	18.2

The data in Table X are given to show the quality of measurements to be expected from a careful worker and a favorable system. The individual photometer readings are reported for the tubes (in quadruplicate) which had stood for 15 minutes after being shaken vigorously. The average values were converted into relative concentrations of *Staphylococcus aureus* by a conversion table prepared according to Section III.D.

H. Inherent Errors

The inherent errors are those caused by unavoidable variations in the operations. Consequently, the greater the number of operations the greater the variation in the resultant. This is as true of mechanisms and electronic assemblages as of microbiological procedures. Simplicity of design is conducive to speed and accuracy.

A photometric assay requires at least 10 steps which can be further subdivided into 25 operations; a plate assay has at least as many operations. Since there are at least 25 operations, there are at least 25 places where a variation can be introduced or a mistake made. Certain errors, the use of wrong organism, for example, would be fatal to the assay and immediately apparent; other errors would cause only minor deviation from the normal pattern of response and would be evident only by an increase in variability.

Accuracy of a simple chemical assay often is the standard used to judge a microbiological assay without realizing that the variation in the "standard" curve in a microbiological assay is inherently much greater than the variation in the standard curve for a chemical assay. The reason for this is a good one; the number of factors causing large second-order variation is much greater in the bio-assay than in the chemical assay and are subject to less definite control. When a chemical method has more than one critical step, exactness of control decreases and variation increases just as it does for the bio-assay. The superior precision of the precise chemical method comes not from an inherent superiority but from simplicity relative to microbiological methods. Precision departs with simplicity. Production of a precisely determined standard curve for a microbiological assay imposes an impractical degree of control over the variables. A group "standard" curve is not used in work of high precision.

Solutions to the problems presented by inherent variation are being sought in two areas. In one area, the design is simplified and the number of operations reduced to a minimum. The design is chosen to obtain maximum value from each tube. The common-zero five-point assay for vitamin B_{12} is an example of a simple design. It is efficient and accurate when and only when the microbiological part of the assay is done properly. That is when the variations in the manual operations are so small as not to affect the assay and when growth of the test organism is limited only by the amount of vitamin B_{12}. The difficulty a human being has in doing the same thing twice in exactly the same way leads to the second area where an answer is sought. Machines can be constructed to do certain repetitive operations with a variation at least an order of magnitude less than that of hand operation. These machines are just now being introduced into assaying and their evaluation will take several years. Even the best machine requires constant care and knowledgeable supervision. Although the machine may be faster and even more convenient than hand operations, to obtain the most out of it may require a quality of operators considerably higher than those it replaces. No machine so far proposed eliminates the need for human intelligence, knowledge, and judgment in designing and performing assays.

XIII. Work Load and Personnel

A team of two operators can prepare 100 samples, make appropriate dilutions, pipet the standard curve and the 100 samples, inoculate, incubate, kill, read, and compute the answers in less than an elapsed time of 8 hours. During the incubation period (3–4 hours), dilution bottles are filled, and other preparation for the work of the next day occupies their time. The assay computations are checked and recorded by another technician. If pressed, the team can do 150 samples, but errors increase and blunders occur. A constant load of about 100 samples is conducive to the best work and should not be exceeded unless urgency justifies the lessened accuracy and the possible mistakes.

Simpson (Chapter 2) reports that 10 operators using large plates can put on 100 samples in duplicate (200 samples total) to achieve a standard error of about 5%. The same number of operators could do 500 samples of an antibiotic substance in a photometric assay with approximately equivalent error but with an occasional blunder because each sample would be in only one assay.

The selection and training of the people who do the assaying is a very important subject as neglected in theory as in practice. It is important because successful assays are dependent upon careful work by intelligent, trained people. The operators in a large group may be put into one of two categories. In one are those who are made unhappy by any departure from familiar, well-established procedures; they are suited for routine assays. In the other are those who are bored by routine and are eager to try the new; they should be selected to assist in developing new procedures and investigating application of old ones. One task of the supervisors is to keep the latter group happy with daily routine. If he does not recognize the existence of the two groups or misplaces his operators, efficiency decreases and his personnel problems increase.

The operators need to be aware of the seriousness of possible errors and be trained to recognize them. The supervisor must by constant monitoring of operations prevent, insofar as possible, those small deviations from a standard operating procedure which naturally occur in any human undertaking.

In some assay laboratories each step in the assay is done by a different person as on an assembly line. In other laboratories, one person does all of the work from sample preparation to computation of answers. In the former, an operator may have little interest in the assay or even in what he is doing. In the latter, he has an intense personal interest in his work and knows exactly where to place responsibility for errors that make repetition of the assay necessary. When one operator does most of the steps of the assay, the number of replications needed to obtain a given standard error is smaller than if he is responsible for only one step. The same experience is reported for plate assaying (Chapter 2).

An operator who cannot see clearly the lines on a graduated pipet cannot do assay of high precision. A visit to an assay laboratory reveals that from several to many of the operators have difficulty in seeing the graduations on volumetric glassware. Either they need spectacles or are using bifocal spectacles of the same form as those of their inventor, Benjamin Franklin. Ordinary reading bifocals are not suited to laboratory work. The near vision is the one used constantly and should be in the place most comfortable to use, which is the center of the lens. Such bifocal glasses would decrease fatigue to neck and back and would improve accuracy of operations.

CHAPTER 5

Automation of Microbiological Assays

THOMAS A. HANEY, JOHN R. GERKE, AND JOSEPH F. PAGANO *

The Squibb Institute for Medical Research, New Brunswick, New Jersey

I. Introduction

The mechanical and biological principles for automating microbiological assays have been known for many years. It is only recently, however, that instrumental systems have been applied to the automation of these assays. Some systems are automatons simulating the analyst's actions; others depart so far from human actions that the methods would not give meaningful answers if done manually. Regardless of the system employed, the objectives of automation are to increase accuracy of answers and sample handling capacity of the laboratory.

The advantages of automation for a microbiological assay laboratory were first discussed by Hucke and Roche.[1] They described a digital recording system for photometric assays, automatic measurement of the sizes of zones in petri dish diffusion assays, and automatic data processing. Gerke et al.[2] introduced the concept of using continuous flow analysis, such as the AutoAnalyzer instrumental system, to automate microbiological assays and described both turbidimetric and respirometric

---

* Present address, Sterling-Winthrop Research Institute, Rensselaer, New York.

[1] D. M. Hucke and C. H. Roche, Antibiotics Ann. **1959–60**, 556.

[2] J. R. Gerke, T. A. Haney, J. F. Pagano, and A. Ferrari, Ann. N. Y. Acad. Sci. **87**, 782 (1960).

methods. In systems operating according to the continuous flow principle, the inhibitory effects of antibiotic solutions are sequentially impressed upon a continuous stream of metabolizing microbes. Subsequently, Haney et al.[3] improved the instrumentation of the respirometric method so that it became practical for routine analyses. Pagano et al.[4] described a modification of the respirometric method whereby a sample can be continuously diluted and the resultant response curve monitored on a strip chart recorder.

The systems for automatic analysis described in this chapter are mechanized manual methods and continuous flow methods. Detailed respirometric methods employing the continuous flow principle are presented for the antibiotics amphotericin B, nystatin, and the tetracyclines.

II. Mechanization of Manual Methods

A. Manual Aids

Repetitive manual manipulations in an assay method are frequently time consuming and may be the cause of error. They may be classified in the following broad categories: preparing standards and samples, combining these with other reagents in the method to produce biological responses, measuring the responses, and calculating the potencies of the sample using, as a reference, the responses of the standards. To increase the speed of analysis and reduce error, automatic instruments are used to carry out these repetitive manipulations.

Instruments used in sample preparation are: shaking machines to aid in extraction; blenders to reduce particle size and to aid in solubilizing and extracting; centrifuges to hasten sedimentation of solids and breaking of emulsions; and automatic pipetting machines (such as those sold by the Baltimore Biological Laboratories, Inc., and National Instrument Company to deliver preset volumes of diluents and extractants. A recent development in automatic pipetting machines are those that dispense predetermined volumes of sample and diluent in one operation. Instruments employing this principle are manufactured by several companies [Scientific Products (Auto Dilutor), Fisher Scientific Company (Dilumat) and National Instrument Company (Filamatic)].

The operation of combining the prepared sample with the reagents can be mechanized in a number of ways. For the agar diffusion methods, molten agar is measured and dispersed into plates by automatic pipetting

[3] T. A. Haney, J. R. Gerke, M. E. Madigan, J. F. Pagano, and A. Ferrari, Ann. N. Y. Acad. Sci. 93, 627 (1962)

[4] J. F. Pagano, T. A. Haney and J. R. Gerke, Ann. N. Y. Acad. Sci. 93, 644 (1962).

machines, the agar solidifies as the plates are carried on a continuous belt through a cooling tunnel (Lampson Conveyor), and cups are positioned on the surface of the agar with the aid of an automatic cup setter. For turbidimetric methods, inoculated medium is measured and dispensed by a pipetting machine into assay tubes.

The type of automatic measuring device employed for measuring responses is dependent upon the assay method. For the photometric method a colorimeter, spectrophotometer, or nephelometer is used to measure the response. To decrease the time required for reading optical densities of microbial cultures, a solenoid-controlled vacuum exhaust system is connected to the cuvette of the photometer. This device, by facilitating repetitive readings in the same cuvette, increases speed and eliminates the need for large numbers of optically matched tubes. Manual aids to measurement of zone diameters are projectors (slightly modified slide projectors, see Chapter 6, Part II, Nystatin, Large Plate Method) and the Fisher-Lilly Antibiotic Zone Reader. For methods that require counting, three instruments are in general use. The Electronic Colony Counter (New Brunswick Scientific Company) is used to register automatically the number of colonies touched with a sensing probe. To preserve the integrity of the colonies, counting may also be done on the outside of the bottom of the petri dish. A similar instrument is the Klett Colony Marker and Tally with which colonies may be counted at the rate of 160 colonies per minute. The Coulter electronic particle counter (Coulter Electronics, Inc.) measures bacterial number, and may also be used to obtain the distribution[5] of particle sizes in growing cultures. Manually operated electric calculators are used for computations

Recently, instruments which perform several operations have become available. Such instruments are the Automatic Zone Comparator (Technical Controls, Inc.), the Bristol Zone[6] Comparator, the Turbidimetric Digital Recording System (Datex Corporation), and the Robot Chemist (Research Specialties), all of which can be used in conjunction with electronic data processing systems.

B. Semiautomatic Measurements of Responses

1. TCI Zone Comparator

The TCI (Technical Controls Inc.) Zone Comparator automatically measures and records the diameters of zones of inhibition developed by standard techniques of the agar diffusion assay. It may be used with commercially available digital devices (such as those manufactured by

---

[5] G. Toennies, L. Izard, N. B. Rogers and G. D. Shockman, *J. Bacteriol.* **82,** 857 (1961).

[6] Anon., *Instruments* (January 1949).

Datex Corporation) to record the zone diameters automatically on tape or cards. The system contributes precision and objectivity to agar diffusion assays. Also, there is the advantage of entering data directly into the computor by way of a punched tape or a card system. Reproducible measurements are obtained, even when the zone edges are not clearly defined. When compared to visual plate reading, the Zone Comparator

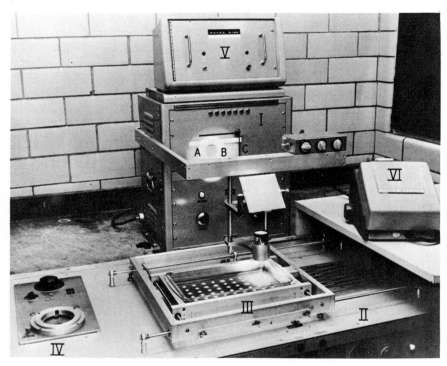

Fig. 1. Automatic Zone Comparator with a large plate in the carriage.

requires more contrast between zone and background and a greater uniformity of background density. Both of these requirements are satisfied by suitable changes in inoculum size, incubation time, medium composition, or by addition of an indicator to the media. For example, the colorless 2,3,5-triphenyltetrazolium chloride (oxidized form) is reduced to a red, insoluble formazan by the dehydrogenase enzyme systems of many species of microorganisms. The contrast will be increased by incorporating this indicator into the assay agar medium. The Zone Comparator offers no advantage if only qualitative results are required or overlapping zones are obtained.

The basic units of the automatic zone comparator shown in Fig. 1 are: the vertical cabinet or zone reader (I), the horizontal movable bench

(II) with which the operator can position a unit holding the baking dish (III) or petri dish (IV) in the optical system of the zone reader, the control chassis (V), and the automatic digital print-out unit (VI) which records the zone diameters.

The dish holding unit containing the dish with its zones of inhibition is positioned over a stationary light source. A fourfold magnification of the zone image is projected on a translucent frosted glass screen. The frosted glass enables the operator to see the zone image for the purpose of focusing and adjusting the sensitivity of the phototubes. When the dish holding unit is put into operation, the following events occur. The zone image travels from right-to-left across the screen. When the sharp difference in light intensity at the zone edge strikes the phototube positioned behind the screen (A of Fig. 1), movement of the dish stops. At the same time another signal activates the motor of a drive shaft to move the mobile phototube from position C toward the right edge of the zone (B of Fig. 1). Movement of this tube can be visually observed for adjusting the sensitivity of the phototube because the indicator bar (C of Fig. 1) mounted in front of the screen, moves with the phototube. When the mobile phototube detects the right hand edge of the zone, it stops and an encoder attached to the phototube drive shaft transmits a signal to the control chassis where it is transformed into a digital code. The encoder measures the shaft position. The code is relayed to the printer and the zone diameter is recorded on paper tape. An electric signal from the control chassis instructs the mobile phototube to return to its starting position and the dish holding unit to move to the next zone. The entire process is repeated until all zones have been measured.

2. Photometers

Manual recording of photometric assay responses is a time consuming, monotonous, and sometimes subjective operation. Coupling of a digital device, such as that described for the Zone Comparator, to the photometer obviates these problems. Such a system was first reported by Hucke and Roche.[1]

Figure 2 is a photograph of a twin channel system used by the authors. It consists of a mixer (I), two spectrophotometers (A and B) each with its respective start button (AA and BB), a Bristol twin channel self-balancing potentiometer (II), a control chassis (III), and a paper tape printer (IV). A Spectronic 20 spectrophotometer generates a signal proportional to the percent of light transmitted by the suspension. The signal is converted to a shaft position by the self-balancing potentiometer of the recorder (II). The shaft position is sensed by the attached Datex encoder and subsequently transformed on command, via the control chassis (III), to a digital print-out on paper tape (IV).

A homogeneous suspension of cells with a minimal entrapment of air in the medium is assured by dispersing the microorganisms in the assay tube with a Vortex Mixer (Scientific Industries Inc.) prior to pouring the contents of the tube into the cuvette of the spectrophotometer. The technician depresses the start button of the timer after pouring the suspension into the cuvette. The timer is adjusted for either a 4- or 6-second delay depending on the nature of the microbial suspension. During the delay period air bubbles rise out of the culture medium and movement

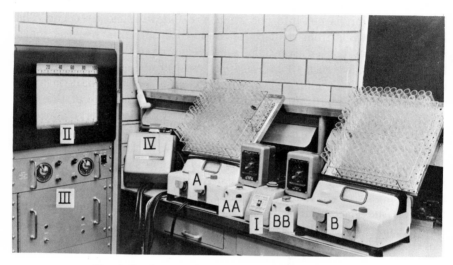

Fig. 2. Dual system for recording photometric assays.

of the liquid ceases. At the end of the time cycle, the printer is commanded to print-out the percent transmission; the solenoid valve connected to the bottom of the cuvette opens; a 2-second delay timer starts and the cell suspension is removed from the cuvette by a vacuum-operated exhaust system. After being open for 2 seconds, the solenoid valve closes and the system is ready to receive the cell suspension from another tube.

C. Automatic Systems of Photometric Assay

An automatic photometric bioassay instrument manufactured by Research Specialties Company to meet the specifications of the authors is currently under investigation at the Squibb Institute for Medical Research. The instrument automatically measures a preset amount of antibiotic solution, adds inoculum and nutrient medium, incubates the mixture for 100 minutes, and transfers the incubated mixture to a photometer where the results are automatically read and recorded. Fifty complete assays are carried out during each hour of operation. Before the instrument can operate by itself, inoculum and reagents must be prepared and test solutions placed in a predesigned order on a sample table. The start-up time is between 2 and 3 hours.

FIG. 3. Robot photometric assay system.

A photograph of the system is shown in Fig. 3. The sample table (I) holds 100 tubes of sample and standard solutions and is mounted on a drive mechanism. Each time the sample table advances to the next tube the following events occur.

(1) One milliliter of inoculum from the refrigerated reservoir (II) is added to a clean tube in the incubation turn-table (III).

(2) A probe on the transfer unit (IV) aspirates 1 ml. of a test solution from a tube in the sample table. The transfer unit then turns 90° and delivers the sample to the tube containing the inoculum on the incubation turntable. After the sample has been delivered, the transfer unit moves to two more positions before aspirating the next sample. In the first position, rinse water is aspirated from a reservoir to wash the probe and in the second position the water is discarded.

(3) Eight milliliters of nutrient medium from the reservoir (V) is added to the test solution and inoculum already in the tube.

(4) After incubation, 1 ml. of formaldehyde from its reservoir (VI) is added to the tubes to prevent further growth.

(5) The killed cells are aspirated through a transfer line to a single channel version of the Spectronic 20—Datex measuring and recording system described previously.

(6) Rinse water is added to clean the tube, transfer line, and cuvette in preparation for the next cycle. The rinse water is aspirated from the tube, through the transfer line, and through the cuvette without being registered by the system.

III. Continuous Flow Methods

A. Principles of the AutoAnalyzer

The basic principle of the AutoAnalyzer is analysis via a continuously flowing reagent system. This suggested a means of assaying antibiotics by measuring their inhibitory effect on a stream of metabolizing microbial cells.

The AutoAnalyzer consists of specific units that can be assembled in a variety of ways to fit the requirements of an analysis. The units that pertain to microbiological assays are: a proportioning pump, a sampler, a dialyzer, an incubation bath, a colorimeter, and a recorder.

The heart of the AutoAnalyzer is the proportioning pump. It consists of two parallel stainless steel roller chains with spaced roller bars that press continuously against a spring-loaded pump platen. Across this platen lies a "manifold"—a set of flexible tubes whose different sized lumens determine the rate of flow through each. The separate tubes are uniformly, simultaneously, and completely occluded in concert so that proportioned delivery is accurate and reproducible from test to test. The rate of flow of each material entering into the test can be established by pumping it through a tube with a specified lumen.

Samples are usually delivered to the system by the sampler which consists of a synchronously driven circular sample plate holding 40 sample cups. As the loaded sampling plate rotates (at a choice of speeds representing 20, 40, and 60 cups per hour), a hinged pick-up crook dips into each cup in succession, aspirates its contents for a given time interval and feeds it into the system. At the proper instant, the pick-up crook automatically lifts out, the sampling plate rotates and the pick-up crook dips into the next sample cup, and so on, repeating this in-and-out action until all the samples on the plate have entered the instrument.

Although the samples are introduced intermittently, the reagents are continuously pumped through the system. Mixing of two or more converging streams is accomplished by constant inversion through horizontal glass helices (mixing coils).

The streams flowing through the system are segmented with air. This action is extremely important and serves a dual purpose: it aids in cleansing the tubes by the mechanical wiping action of the segments of air and it provides a number of barriers between successive samples in the timed sequence as they flow through the system.

The dialyzer unit selectively transfers a substance from one stream to another. It comprises a pair of plates whose mating surfaces have grooves (arranged as mirror images) to provide a continuous channel when the plates are brought into contact. A cellophane membrane is sandwiched between them and the plates are clamped together leaving the two pathways separated only by a thin semipermeable membrane. The plates are immersed in a constant temperature water bath to prevent the effects of changing temperatures on dialysis.

The incubation bath, thermostatically controlled at 37°C., contains glass or polyethylene tubing. Incubation time is determined by the length and inside diameter of the tubing as well as by the flow rate of the segmented stream.

The stream issuing from the system passes through a trap to remove the air used for segmentation and enters the cuvette of the colorimeter. The colorimeter continuously monitors the concentration level of the samples flowing through it.

Using the continuous flow principle of the instrumental system described, antibiotic activities have been measured by two separate methods: cellular growth (photometric method) and cellular respiration (respirometric method).

B. Photometric Methods

The instrumental design used in the photometric method is illustrated in the schematic flow diagram of Fig. 4. The proportioning pump mixes and conveys two streams: one composed of sample with nutrient medium as diluent and segmented with air; the second composed of bacterial suspension and nutrient medium and also segmented with air. The two streams pass through opposite sides of a double dialyzer (two dialyzers in tandem) to permit the antibiotic to dialyze into the stream containing the bacteria. The stream containing the spent sample is discarded and the other stream, containing bacteria and sample, is pumped through an incubation coil consisting of 80 feet of 0.125-inch-diameter polyethyene tubing. As the stream emerges from the incubation coil, it is dialyzed against formalin to kill the bacterial cells.

Forty minutes after entering the system, the portion of the stream that has been influenced by the antibiotic enters the flow cuvette. The resulting turbidity is recorded as absorption of light at about 540 mμ. During this short incubation period, the concentration of bacteria increases by 70%, an amount insufficient to give a full-scale deflection of the recording system. To improve readability of the chart, the amplitude of the deflection is increased fourfold with the range expander.

Several problems remain to be solved before this method can be used for a routine analysis. The major problems are poor efficiency of dialysis,

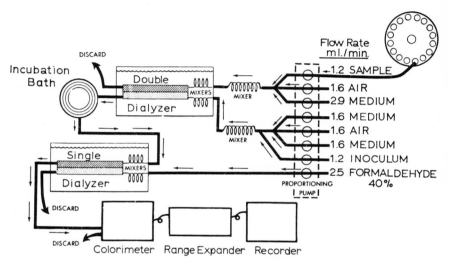

FIG. 4. Schematic of photometric assays using the AutoAnalyzer. Automatic microbiological determination of antibiotics via measurement of growth density. By permission: *Ann. N. Y. Acad. Sci.* **87**, 782 (1960).

high internal pressure, maintenance of the cells in uniform suspension, and inadequate separation between samples.

Dialysis, the method used to add the antibiotic to the nutrient-inoculum stream, presents a problem in that most antibiotics, due to their large molecular weight, dialyze poorly through a cellophane membrane. Usually the small amount of antibiotic passing through the membrane limits the use of this system to concentrated antibiotic solutions. A possible solution to this problem is to devise a method in which dialysis is not required. Under the conditions described for the photometric method, dialysis was necessary because bypassing the double dialyzer resulted in erratic responses.

The great resistance to flow offered by a long length of smaller diameter coil, or a faster flow rate through the coil causes high pressures which in turn are responsible for erratic flow rates and frequent ruptures of joints in the tubing. Slower flow rates or larger diameter tubing make difficult maintenance throughout the system of the uniform bubble pattern required for good separation of samples. The 80 feet of 0.125-inch tubing gave a constant flow rate and a uniform bubble pattern.

Microbial cells tend to settle and adhere to the surface of the tubing, thus causing drift within the system. This problem is greater with the large yeast cells than with the small bacterial cells. Inadequate separation between samples is obtained in the system described. The influence of a sample upon the following sample is particularly noticeable when the concentrations are widely different. Perhaps the use of formaldehyde as described in the following respirometric method would minimize the effect of cell deposits and improve the separation between samples. Although tetracycline, neomycin, and streptomycin were accurately assayed using *Klebsiella pneumoniae,* the above problems must be solved before an efficient photometric method can be realized.

C. Respirometric Methods

1. Introduction

The respirometric method is based upon the principle that antibiotics affect the respiratory process of susceptible microorganisms. Antibiotic dose-response curves can be obtained by feeding into the AutoAnalyzer a continuous stream of inoculum with nutrient medium and sequentially introducing therein different levels of antibiotic, and by continuously measuring the carbon dioxide produced.

The instrumental system used in this method is illustrated in Fig. 5. Each of the proportioning pumps serves a specific function: one dilutes and combines the sample (or formaldehyde) streams with a continuous stream of water; the second takes an aliquot of the diluted sample (or formaldehyde) stream, combines it with inoculum and nutrient medium, and introduces air for segmenting the combined stream. The stream flows through an incubation coil. After emerging from the incubation coil, the stream is acidified to release dissolved metabolic carbon dioxide. As the stream enters the liquid-gas separator, an aliquot of the gas phase is continuously aspirated by the pump and used to segment a stream of alkaline buffered phenolphthalein solution. The carbon dioxide dissolved in the indicator solution decreases the pH and decolorizes the indicator. The color intensity as measured at 555 mμ is inversely proportional to the amount of carbon dioxide produced. A detailed method for the assay of antibacterial and antifungal antibiotics is described at the end of this chapter.

Accuracy of the results obtained in the respirometric method was compared with the accuracy of the routine photometric and agar diffusion methods. Column 1 of Table I lists a group of typical samples assayed.

FIG. 5. Instruments used in the AutoAnalyzer respirometric method of assay. By permission: *Ann. N. Y. Acad. Sci.* **93,** 628 (1962).

TABLE I*

CORRELATION OF AUTOANALYZER METHOD WITH PHOTOMETRIC AND
AGAR DIFFUSION METHODS

Sample type	Sample code	AutoAnalyzer	Photometric	Ratio
Amphotericin B	A	71	74	0.96
Fermentation broths	B	67	71	0.94
	C	62	64	0.97
	D	69	64	1.08
				Ave. 0.99
Amphotericin B	A	97	91	1.06
Formulations	B	91	87	1.05
	C	87	85	1.02
	D	45	46	0.98
				Ave. 1.03
Tetracycline	A	25	25	1.00
Formulations	B	19	18	1.05
	C	15	14	1.07
	D	92	90	1.02
				Ave. 1.03
		Agar Diffusion		
Tetracycline	E	24	25	0.96
Formulations	F	15	16	0.94
	G	39	38	1.03
	H	18	19	0.95
				Ave. 0.97

* By Permission: *Ann. N. Y. Acad. Sci.* **93**, 638 (1962).

Columns 2 and 3 show the results obtained by the different methods. The last column shows the ratios of the AutoAnalyzer to the photometric or agar diffusion results. In looking at the averages of these ratios it is evident that there are no significant differences between methods.

The efficiency of the AutoAnalyzer was obtained by comparing its productivity to that of the photometric and agar diffusion methods. One technician operating two AutoAnalyzers can obtain 240 responses per day as illustrated in Table II. This number is $\frac{2}{3}$ as many responses as by the photometric method and $\frac{1}{2}$ as many as by the agar diffusion method. However, because the number of responses required to obtain equal precision is much less for the AutoAnalyzer, it is twice as efficient as the photometric method and three times as efficient as the agar diffusion method.

TABLE II*

EFFICIENCY OF AUTOANALYZER INSTRUMENTAL SYSTEM

Method	Responses per day per person[a]	Response required for standard deviation 5%	Relative efficiencies (%)
AutoAnalyzer	240	2	100
Photometric	360	6	50
Agar diffusion	480	12	33

* By permission: *Ann. N. Y. Acad. Sci.* **93**, 639 (1962).

[a] Excluding sample preparation and calculation.

To obtain the efficiency as described for the AutoAnalyzer, the technician operating the instruments must be well versed in its principle of operation, the methods used, the troubles that can occur, and remedies for these troubles (see Trouble Shooting Guide, Section III. C. 4). This means that the quality of the results, therefore, is proportional to the skill of the operator.

2. Respirometric Assay for Amphotericin B, Nystatin, and the Tetracyclines

a. *Instrumentation.* The basic components of the AutoAnalyzer instrumental system, as illustrated in Fig. 5, consists of: an inoculum ice bath, an automatic sample table (A), two proportioning pumps (B and C), a 37°C. incubation bath (D), a colorimeter (E), and a strip chart recorder (F).

The inoculum ice bath maintains the cells in a static state. The cells are contained in a 1-liter flask surrounded with ice in a 2-gallon polyethylene container modified to maintain a constant water level of 2 to 3 inches deep. Constant agitation with a magnetic stirrer prevents settling of the cells.

As shown in Fig. 6, the sample table is modified in three ways. A metal housing with a red transparent Lucite cover placed around the sample pick-up plate assembly protects light-sensitive polyene antibiotics. A formaldehyde reservoir is attached to the side of the sample table to provide a formaldehyde flush between successive samples. The formaldehyde pick-up arm is attached to the back of the sample pick-up arm. The sample and formaldehyde lines are exactly equal in length. These modifications permit (in a 3-minute cycle) alternating deliveries of sample for 2 minutes and formaldehyde for 1 minute.

The manifold of the proportioning pump used for dilutions is designed to segment the streams of formaldehyde or sample with air and to combine and mix this stream with a water diluent (Fig. 7). High concentra-

FIG. 6. Detail of sample table with formaldehyde reservoir. By permission:
Ann. N. Y. Acad. Sci. **93,** 629 (1962).

tions of water insoluble antibiotics, such as the polyenes, contained in
water miscible solvents are uniformly diluted and subjected to analysis
before precipitation occurs. The air-segmented stream passes through a
mixing coil to a "T" joint where a small portion of either the diluted
sample or formaldehyde is aspirated by the second pump. Because of the
difference in flow rates, excess liquid and all air are discarded from the
system at the "T" joint. Elimination of air from the stream is a necessary

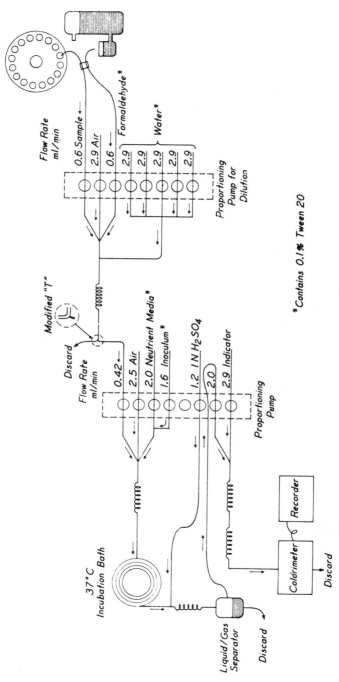

Fig. 7. Schematic of respirometric assays using the AutoAnalyzer. Automatic microbiological analysis of antibiotics via measurement of respiratory CO_2. By permission: *Ann. N. Y. Acad. Sci.* **93**, 630 (1962).

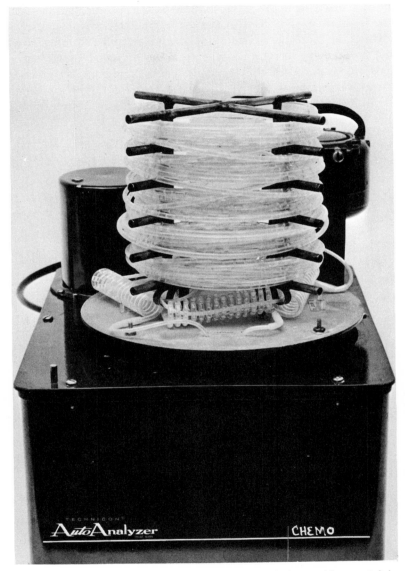

Fig. 8. Incubation bath assembly showing the polyethylene tubing containing the incubating samples. By permission: *Ann. N. Y. Acad. Sci.* **93**, 631 (1962).

condition to the maintenance of a uniform flow rate in the incubation bath.

Streams of inoculum and nutrient medium introduced through the second pump are combined either with the sample or the formaldehyde, segmented with air, and mixed. The stream passes through polyethylene tubing immersed in a 37°C. incubation bath. The incubation bath assembly is shown in Fig. 8, with the incubation coil raised and inverted. Eighty feet of 0.125-inch inside diameter polyethylene tubing is wound around a stainless steel frame. The frame is attached to a transparent (for inspection purposes) Lucite cover. The mixing coils are mounted on the cover, to maintain them at a constant temperature by immersion in the bath. All connections between polyvinylchloride (Tygon) tubing and

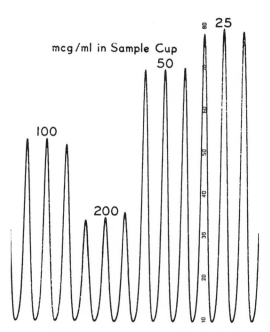

FIG. 9. Instrument response showing inhibition of respiration of CO_2 of *Escherichia coli* by tetracycline HCl. By permission: *Ann. N. Y. Acad. Sci.* **93,** 632 (1962).

glass tubing or polyethylene tubing are made outside the bath, because these connections will separate if immersed in water. A sample moves through the incubation coil in 30 minutes, a time ample to obtain the effect of the antibiotic upon respiration of the metabolizing cells.

Sulfuric acid (1 N) is injected into the stream as it emerges from the incubation bath. The stream then passes through a mixing coil where the carbon dioxide released by the acid enters the air segments. The stream then goes to a liquid-gas separator where 80% of the carbon dioxide rich gaseous phase is continuously aspirated by the second pump (see Fig. 7) and the remaining gas and all the liquid are eliminated from the system. The carbon dioxide enriched stream is used to segment a weakly alkaline buffered stream that contains phenolphthalein. Two mixing coils in a series are used to equilibrate the gas with the indicator solution.

The colorimeter measures the color intensity of the indicator stream (at a wavelength of 555 mμ) as it flows through a 10-mm. flow cuvette. The color intensity, recorded on a strip chart recorder, is inversely proportional to the amount of metabolic carbon dioxide. The amount of metabolic carbon dioxide is dependent upon the concentration of antibiotic as is shown by the response of *Escherichia coli* to tetracycline (Fig. 9). As the concentration of antibiotic increases there is less carbon dioxide available to decolorize the indicator. Since the formaldehyde prevents the cells from respiring between samples, the color intensity of the indicator is not affected and the recorder pen returns to the base line obtained with the formaldehyde solution (bottom of figure) between samples.

b. Stock Culture. The test organisms are subcultured weekly on yeast beef (Difco) agar slants. *Candida tropicalis* (Squibb 1647) is used for the amphotericin B and the nystatin assays, and *Escherichia coli* (Squibb 1559) is used for the tetracyclines assays. The subcultures are incubated at 37°C.

c. Inoculum. Amphotericin B and Nystatin Assays. Candida tropicalis is transferred from the refrigerated stock slant to 500 ml. of medium A (Table III) contained in a 2-liter Square-Pak flask (American Steri-

TABLE III

MEDIA (gm./100 ml.)

			Medium			
	A^a	B^b	C^a	D^a	E^b	F^a
Glucose	1.10		1.10	2.00	1.00	0.20
Yeast extract	0.65		0.65	0.50		1.30
Tryptone	1.00		1.00			2.00
Beef extract	0.15		0.15			0.30
Peptone	0.50		0.50			1.00
NaH$_2$PO$_4$·H$_2$O		2.65			2.65	
Na$_2$HPO$_4$·7H$_2$O		0.225			0.225	
Sodium citrate		1.00		1.00	1.00	
KH$_2$PO$_4$			0.132	0.10		0.264
K$_2$HPO$_4$			0.368	0.10		0.736
Casitone				0.90		
Tween 20^c		0.1				

^{a} Sterilize for 20 minutes at 121°C.

^{b} These need not be sterilized if used on the day prepared.

^{c} If the medium is to be sterilized, Tween 20 is added after sterilization. Tween 20 is manufactured by the Atlas Powder Company, Wilmington, Delaware.

lizer Company). The plastic cap and rubber seal of the Square-Pak flask are preferred over cotton plugs because strands of cotton can lodge within the system and cause the bubble pattern to break up. The use of cotton plugs is avoided in all media, reagents, and antibiotic solutions for the same reason. To increase the cell population, the flask is incubated at 37°C. for 18 hours on a mechanical shaker. The flask of inoculum is held at 5°C. for about 24 hours to allow the cells to settle. Four hundred milliliters of the clear supernatant is removed and a 0.3% (v./v.) suspension of the yeast prepared by diluting the residue to 1500 ml. with cold medium B. One and a half milliliters of Tween 20 is added.

d. *Inoculum. Tetracycline Assay. Escherichia coli* is transferred from a refrigerated stock slant to 500 ml. of medium *C* contained in a 2-liter Square-Pak flask. Each flask opening is covered with a sterile gauze pad secured with a rubber band. Incubation at 30°C. is carried out for 18 hours on a mechanical shaker. To prepare a 0.25% (v./v.) suspension of the microorganism, the inoculum is diluted fivefold with cold sterile water containing 0.1% Tween 20.

e. *Preparation of Standards and Samples.* The standards are prepared in triplicate (weighings 1, 2, and 3) to ensure assay accuracy.

Both standards and samples are dissolved in dimethyl sulfoxide to contain about 1 mg. of amphotericin B activity or 3000 units of nystatin per milliliter. To approximate the center of the dose-response curve, a further dilution is made in a mixture composed of (by volume) 3 parts dimethyl sulfoxide, 4 parts methanol, and 3 parts water. The center of the curve for amphotericin is 1 mg./ml. and for nystatin 120 u./ml. See Antifungal Assays (Chapter 6, Section II) for further information about handling and preparation of standards and samples.

Tetracycline standards and samples are dissolved in 0.1 N HCl to contain about 1 mg. of activity per milliliter. To approximate the center of the dose-response curve, further dilutions are made in pH 4.5 phosphate buffer (0.1 M). This low pH minimizes conversion of the tetracyclines to the less active eip-isomers. The center of the curve is about 100 μg./ml. for tetracycline, 40 μg./ml. for 7-chloro-tetracycline, 110 μg./ml. for 5-hydroxytetracycline, and 55 μg./ml. for 7-chloro-6-demethoxytetracycline.

The above are sample cup concentrations for the medium (M) standards. High (H) and low (L) standards, $H = 1.5M$ and $L = 0.66M$ are also prepared. The over-all dilution of samples is 241-fold for the system shown in the schematic diagram (Fig. 7). This dilution may, of course, be altered to suit specific requirements.

f. *Design.* During the course of an assay, successive responses to any one level of antibiotic may drift. Drift is characterized by a progressive increase or decrease in the responses. For example, if the responses to successive samples are 50, 51, 52, etc., per cent transmission, the responses are drifting. The causes of drift are discussed in the Trouble Shooting Guide, Section III. C. 4. Since drift introduces an error, an assay design and method of calculation were developed to compensate for it. An example of this is illustrated in Fig. 10. In essence, a standard is repeated

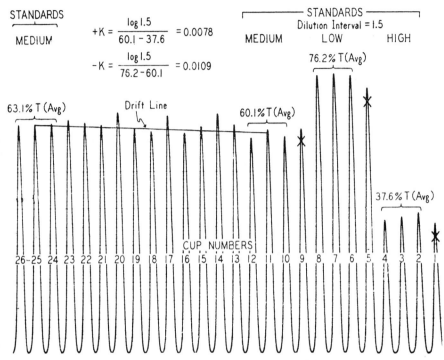

FIG. 10. Slope constants ($\pm K$) and drift line.

periodically throughout the run. The successive responses to this standard are connected graphically with a series of straight lines to form the drift line. All other responses are expressed as differences from the drift line. Other features of the design are a method for determining slope of the dose-response curve and independent replication of all standards.

Forty AutoAnalyzer cups containing standards and samples are positioned on the sample table according to the sequences given in Table IV. If the concentrations of antibiotic in adjacent cups (adjacent peaks on recorder) differ markedly, then carry-over from the leading cup may bias the response indicated by the trailing peak. For this reason, the second peak of a pair is considered to be more representative of the sample than the first peak and duplicates of a sample are put in adjacent cups.

TABLE IV

ORDER OF STANDARDS AND SAMPLES

Position	Sample
1, 2	High standard, weighing 1
3	High standard, weighing 2
4	High standard, weighing 3
5, 6	Low standard, weighing 1
7	Low standard, weighing 2
8	Low standard, weighing 3
9, 10	Medium standard, weighing 1
11	Medium standard, weighing 2
12	Medium standard, weighing 3
13–23	Samples in duplicate
24	Medium standard, weighing 1
25	Medium standard, weighing 2
26	Medium standard, weighing 3
27–37	Samples in duplicate
38	Medium standard, weighing 1
39	Medium standard, weighing 2
40	Medium standard, weighing 3

g. *Operation.* Components of the system are assembled according to the schematic diagram of Fig. 7. Water diluent (0.1% Tween 20 in water) is pumped through all lines except air and buffered indicator line until a smooth flow and a uniform bubble pattern is established (about $\frac{1}{2}$ hour). The inside of the gas separator is coated with silicone stopcock grease to prevent foaming.

The buffered-phenolphthalein solution is prepared by adding 12 ml. of carbonate-bicarbonate buffer (56 gm. $NaHCO_3$ and 35 gm. Na_2CO_3 dissolved to make 1 liter of solution) and 21 ml. of 1% (w./v.) methanolic phenolphthalein solution to three liters of distilled water. Flow of this buffered indicator solution is started through the system.

The angle of the bubblers (a junction of two or more tubes where air (gas) is introduced into the system) is adjusted to deliver from 30 to 35 bubbles per minute to the stream going to the incubator bath and from 100 to 135 bubbles per minute in the indicator stream.

The formaldehyde reservoir is filled with 10% formaldehyde solution (25% formalin) containing 0.1% Tween 20. Adjustment of the sample line in the pick-up mechanism is made so that its tip is just above the bottom of the sample cup. When the pick-up mechanism is in the sampling position, the formaldehyde line is adjusted so that its tip is $\frac{1}{4}$ inch above the formaldehyde surface.

Ten sample cups are filled with the M (medium concentration) standard and placed on the sample table. These standards are used for adjusting the instrument. Subsequent standards and samples are placed in the sample table according to the design presented. The sample table is started and the flow of $1 N$ H_2SO_4 containing 0.1% Dow-Corning Antiform B, inoculum, and assay medium in their respective lines. Medium E is used for amphotericin B and nystatin, and medium F for the assay of the tetracyclines (Table III). About 35 minutes after starting, when the responses begin to appear on the strip chart, the response to the M standard is observed. If the response is greater than 50% T, the concentration of the buffer in the alkaline-buffered indicator is increased by adding more buffer; if lower than 45% T, it is diluted with a phenolphthalein solution of the composition of 7 ml. of methanolic 1% phenolphthalein per liter of water.

After the last sample has entered the system at the end of the working day, all lines are flushed with $2 N$ NaOH for 5 minutes except the acid, air, and indicator lines. The caustic wash is followed with water-diluent (0.1% Tween 80 in water) until the system is rinsed (about 20 minutes). After the last sample is recorded, the indicator lines are flushed as outlined above.

The trouble shooting guide (Section III. C. 4) aids in the diagnosis and correction of many of the most frequently encountered difficulties that might arise during operation of the instrumental system.

h. Calculations of Potency. The method for calculating potency is based on the assumption that a straight-line relation exists between the height or peaks and logarithm of concentration of antibiotic in the sample cups. Since the calibration line may be slightly curved, the straight-line relation is approximated by dividing the curve into two portions each with a different slope. One limb of the curve is that between medium and high standard, the other between medium and low standard (Fig. 10). The slopes are computed as follows: The average peak heights of the medium standard (cups 10, 11, 12) are subtracted from the peak heights of the high standard (cups 2, 3, 4) to obtain $+T$. The average of the peak heights of the low standards (cups 6, 7, 8) is subtracted from the average of the peak heights of the medium standards (cups 10, 11, 12) to obtain $-T$. Each difference is divided by the logarithm of the dilution interval ($\log 1.5 = 0.176$) to obtain a positive slope constant ($+K$) or a negative slope constant ($-K$).

The average peak heights of the medium standards are used to construct the drift lines. The first drift line (for samples 13–23) is constructed by connecting the average peak heights of 10, 11, and 12 with the average of 24, 25, and 26 (Fig. 10). The second drift line (samples 27–37) connects 24–26 and 38–40. Concentrations of samples as put in the cups is computed from dilution of the sample and concentration of the medium standard. To obtain the concentration of a sample relative to the medium standard, the difference from the drift line is multiplied by the appropriate slope constant. For example, $+K$ is used if the sample peak height is below the drift line and $-K$ if above. The log total potency is calculated by adding the sum of log (dilution) and log (concentration of medium standard) to the log (relative concentration).

3. Continuous Dilution Modification of the Respirometric Method

In estimating the concentration of an antibiotic in solution by one of several analytical methods (chemical or biological), it is customary to

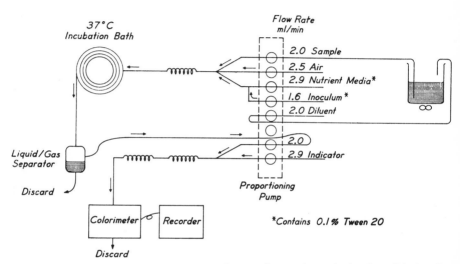

FIG. 11. Schematic of continuous dilution respirometric analysis of antibiotics. By permission: *Ann. N. Y. Acad. Sci.* **93**, 645 (1962).

prepare preselected dilutions of the sample calculated to produce varying inhibitory values in the range of the linear portion of the standard dose-response curve. Usually the readings for two or three doses of the unknown are plotted and a curve drawn through them. The dose-response line of the sample is compared with that of the standard. The analyst is at a disadvantage because he must estimate the concentrations of antibiotic that, when diluted and assayed, would yield responses on the linear portion of the dose-response curve. A second disadvantage is that a curve produced by graphing the values for several doses of the unknown solution gives only an approximation of its shape since all other points are determined by interpolation or extrapolation.

A procedure more satisfactory than that described is to obtain all points on the dose-response curve. A full description of dose-response curves can be obtained by substituting a continuous dilution device (Fig. 11), for the sample table in the respirometric instrumental system.

A similar application of this principle was recently described by Menzies[7] for preparing dilutions of soil suspension for making bacterial counts. He also presented a review of the mathematics. Figure 11 shows a vessel containing a solution of antibiotic agitated continuously by a magnetic stirrer. Two polyvinylchloride tubes of equal diameter are inserted into the solution; as one tube withdraws a quantity of sample the other

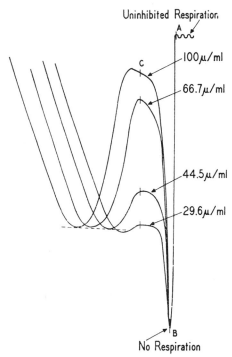

FIG. 12. Example of continuous dilution analysis of nystatin via measurement of respiratory CO_2. Constant volume: 12 ml.; input/output: 2 ml./minute. By permission: *Ann. N. Y. Acad. Sci.* **93**, 647 (1962).

simultaneously restores an equal quantity of diluent. Since the solution in the vessel is maintained at a constant volume while it is simultaneously being sampled and diluted, the concentration of solutes in solution will decrease logarithmically as a linear function of time.

As an example of an application of the continuous dilution technique the authors compared the dose-response for 4 concentrations of nystatin. The results are shown in Fig. 12. In the figure, time increases from right to left, a characteristic of most strip chart recorders. At point A, a 0.2% formaldehyde solution was put into the stream and this interrupted the respiration of the cells. After one minute (point B) the formaldehyde

[7] J. D. Menzies, *Can. J. Microbiol.* **6**, 583 (1960).

treatment was discontinued and the antibiotic was introduced into the stream. At point C, continuous dilution was started. As the nystatin was continuously diluted from a high concentration, the amount of respiratory carbon dioxide decreased to a minimum and then increased with continued dilution. The stimulatory effect of high concentrations of nystatin, shown in Fig. 12, occurs when medium D is used (Table III) but does not occur with medium E.

The continuous dilution method may find application in the study of biochemical and nutritional processes in microorganisms. It is also a new means of characterizing and quantitating antibiotics.

4. Trouble Shooting Guide

This guide was prepared to aid the technician in locating and correcting common troubles that arise and to understand the causes of these troubles.

1. Erratic pen movement resulting in obliteration of the normal peak pattern. This trouble is caused by foam or liquid entering the gas line from the liquid-gas separator. To rectify, disconnect gas line from the separator and coat the inside wall of the separator with silicone grease.

2. Smooth movement of pen followed repeatedly by short, quick movements. Gas bubbles that are too large or breaking up in the indicator stream can cause this. The bubble size can be made smaller by angling the bubbler upward. The bubbler should be adjusted and firmly secured so that gas bubbles form at a rate of from 100 to 135 per minute. Breaking up of the bubbles in the stream is caused by foreign matter on the inside walls of the indicator line. To clean the indicator line flush with $2 N$ NaOH for 5 minutes followed by distilled water for five minutes. If there is no improvement in the bubble pattern, replace the line. Another cause is foreign matter in the flow cuvette. If this is suspected, remove and clean the cuvette.

3. Formaldehyde base line does not return to the same per cent transmission between successive sample peaks. This is caused by a leak in the formaldehyde line, partial blockage in the line, or nonuniformity of delivery. Either change the location of the tube in the manifold or, if this fails, replace the tube.

4. Different responses to identical samples. This trouble is caused by: failure to fill the sample cups to a uniform level, a leak in the sample line, partial blockage in the sample line or incubation coil as evidenced by pulsation of the stream emerging from the incubation bath, or a nonuniform bubble pattern emerging from the incubation bath. A partial blockage in the sample line (or incubation coil) can be removed sometimes by flushing $2 N$ NaOH through the blocked line (or coil) for 5 minutes followed by distilled water for 5 minutes. If this fails to remove the foreign object, replace the line (or coil). A uniform bubble pattern

can be obtained by adjusting the angle of the bubbler so that the number of air segments entering the bath is between 30 and 35 per minute. If the bubble pattern does not improve, flush all lines with $2 N$ NaOH as described above.

5. Recording pen not moving across chart. The cause is frequently a sticking sample pick-up mechanism, a burned out colorimeter lamp, formaldehyde not entering the system, formaldehyde entering the system continuously, a too strongly alkaline-buffered indicator solution as shown by a near straight line response in the 0–10% transmission region, or a too weakly alkaline-buffered indicator solution as shown by a near straight line response in the 90–100% transmission region. When the formaldehyde solution is either not entering the system or is entering continuously, adjust the top of the formaldehyde line so that it is $\frac{1}{4}$-inch from the formaldehyde surface when the sample pick-up mechanism is in the sampling position. If the concentration of the alkaline-buffered indicator is the cause of the trouble, adjust it according to the procedure outlined in the text.

6. Drift. Drift can be caused by a gradual deposit of cells on the inside wall of the incubation coil, microbial contamination of reagents, a rise in the temperature of inoculum above 5°C., or a settling of inoculum. Cells may deposit on the inside wall of the incubation coil because of a poor bubble pattern entering the incubator or a break-up of the pattern in the incubation coil. As mentioned before, a uniform bubble pattern can be obtained by adjusting the angle of the bubbler. If the break-up of the bubble pattern in the incubation coil is due to foreign matter in the coil, the material usually can be removed by flushing the system with $2 N$ NaOH as previously described. If a reagent is turbid or has sediment, it is probably contaminated. Flush the lines of the contaminated reagents with $2 N$ NaOH as previously described and replace reagent bottles with their contents. To maintain the temperature below 50°C., add more ice to the bath and to prevent the cells from settling, adjust the stirrer.

7. Slope of the dose-response curve is less than usual. This trouble is caused by an alkaline-buffered indicator that is too weak or strong, a set of standards improperly prepared, or a change in the susceptibility of the inoculum.

IV. Computations

The main objectives of automating bioanalytical data processing are to provide accurate answers and to enable effective utilization of statistical procedures. The information commonly required of analytical calculations are: (1) potency of the sample with its confidence limits and significance; (2) detecting out-of-control situations; (3) evaluating the skill of technicians; and (4) a means of evaluating new analytical methods.

The methods of automating bioanalytical computations are essentially the general methods of electronic data processing. The basic instrument is the electronic computor such as the IBM 650 used by the authors. It

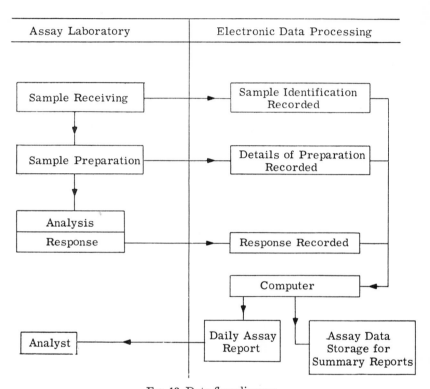

Fig. 13. Data flow diagram.

must be instructed in minute detail for each operation to be carried out. These instructions, called a program, take the form of specially wired circuits, pin boards, punched cards, punched tape or magnetic tape, etc. The system of calculation routines embodied in the program is called a logic. It determines the usefulness of the computer.

Writing a logic is an exacting task because the logic must instruct the computor to handle most situations that can arise in the assay. When computations are made by hand, a number of decisions are made and alternate actions are taken by the analyst. The logic of an electronic data processing program must provide for making these decisions, taking action, and recording the decisions in the print-out. For example, if a potency is calculated from data containing a large number of missing values, a statement to this effect in the print-out would prevent the potency from being given weight equal to that of a potency calculated from a complete set of data.

The methods section of this book shows that several different assay designs are used depending upon the agent (penicillin, vitamin B_{12}, etc.) to be assayed, the method (diffusion, turbidimetric, etc.) used, and the purpose of the assay. Each assay design requires a specific logic.

Data for the automatic computers are usually handled according to the flow diagram of Fig. 13. The sample is received and the identification information recorded. After the sample is prepared for analysis, the details of preparation are also recorded. Details of preparation can be included on the same card (or tape) as sample identification and would include such information as method of extraction, weight of sample, total dilution of sample, etc. After the sample is analyzed, the response is recorded. All recordings of all standards and samples are entered into the computer system where the data are sorted and analyzed. The computer system then produces, in printed form, an assay report. The report can contain, in a form that is easily assimilated by the analyst, such information as: sample identification, potency of the samples, parallelism of the sample responses to those of the standards, a comparison of different preparations of the standard or samples, and shape of the standard curve. In addition to the daily assay report the computer stores the assay data on cards (or tape). These cards (or tape) are used for summary reports. When processed through the computer they can report, also in printed form, the analytical results of long range stability studies, charge account numbers, etc. The practice of automating bioanalytical computations is still young and we can expect basic changes and simplifications in the future. The design and application of improved automated assay machines should increase the accuracy and precision of individual assay values and thereby reduce the number of responses that must be digested to achieve the desired confidence limits. Mechanical and electronic analog devices suited to direct coupling with bioanalytical response measuring devices will probably be developed to transform responses to potency automatically and instantaneously. The analog devices will therefore replace many steps now carried out by digital computers.

Antibiotic Substances
Part I   Antibacterial Assays

6.1 Introduction

FREDERICK KAVANAGH

Eli Lilly and Company, Indianapolis, Indiana

I. Plan of the Chapter

Enough examples of specific methods of assay for antibiotic substances are given in this chapter to illustrate the principles of assaying and to indicate current practices. The antibiotic substances are chosen from the older, well-established ones of commercial importance. It will be noticed that the practices may not always represent the best principles. Historically, practice developed and became well established before the principles were elaborated. Change is slow. None of the methods described is the "best" one. All are useable and could be improved in one respect or another. The best way of achieving a particular result is rare, if it exists at all in biology. The essence of the biological world is the multiplicity of ways of obtaining the same end product; the alternate route permits adaptation and survival.

The methods for antibiotics are given in alphabetical order beginning with bacitracin and ending with vancomycin. The details of turbidimetric and plate assaying will be described for the assay of penicillin G (see also chloramphenicol plate method) as the model. Large plate methods are illustrated by an assay for penicillin G in Chapter 2, for cephalosporin

C in Chapter 6.3, Section III, and for nystatin in Part II of Chapter 6. The assay for cephalosporin C has a method of plating that greatly simplifies the design and the statistical treatment of large plate assays. Methods for antifungal substances are given in Part II of this chapter. Respirometric methods in simple form are described in Part II and in complex form in Chapter 5. Other procedures will be detailed only as they differ from the penicillin model thus avoiding needless repetition of details common to all methods. No attempt is made to describe preparation of all possible types of samples; there are entirely too many. Grove and Randall [1] give the details for many pharmaceutical preparations.

The concentrations of standards and samples will be given in micrograms (or units for impure preparations) per milliliter for both turbidimetric and plate assays. The concentrations are those of the solutions pipetted and are not those attained in the broth as in Chapter 4, Section VIII. Most of the turbidimetric assays are designed for 0.5-ml. samples and 10 ml. inoculated broth. Since the same volumes of samples and standards are pipetted into the tubes, the exact volumes are not important as long as they are the same in all tubes. The 0.5-ml. volume was selected because it is large enough to pipet accurately and small enough to obtain four aliquots from one 2-ml. pipet.

The methods given here are easily modified for application to a new antibiotic substance. An example is the assay for tylosin which is essentially the same as the assay for erythromycin.

A minimum of references to the literature will be given.

The compositions of generally used media and buffers are collected in Chapter 6.19. The Grove and Randall media are designated by the number assigned them in Chapter 21 of "Assay Methods of Antibiotics." This book contains much material not duplicated in this chapter and should be in every laboratory where antibiotics are assayed. The seventh edition of "The Merck Index" [2] is a valuable reference work for an assay laboratory and should be available.

II. Sources of Standard Preparations

New and commercially unimportant substances usually can be obtained in small amounts from the discoverer of the substance. The most important of those sold as drugs may be obtained from the U.S.P. or N.F.

The following U.S.P. reference standards for assays given in this chap-

---

[1] D. C. Grove and W. A. Randall, "Assay Methods of Antibiotics: A Laboratory Manual," 238 pp. Medical Encyclopedia, New York, 1955.

[2] P. G. Stecher, ed., "The Merck Index," 1641 pp. Merck and Co., Rahway, New Jersey, 1960.

ter are obtainable from U.S.P. Reference Standards, 46 Park Avenue, New York 16, New York: bacitracin, chloramphenicol, dihydrostreptomycin, erythromycin, neomycin sulfate, novobiocin, nystatin, phenoxymethylpenicillin, penicillin G, polymyxin B sulfate, streptomycin sulfate, and tetracycline hydrochloride.

The following reference standards, N.F., of antibiotic substances may be obtained from Director of N.F. Revision, American Pharmaceutical Association, 2215 Constitution Avenue N. W. Washington 7, D.C.: chlortetracycline hydrochloride, gramicidin, oleandomycin chloroform adduct, oxytetracycline, triacetyloleandomycin, and tryothricin.

International standards and reference preparations for antibiotics are listed in Table I. The activities are given as international units per milli-

TABLE I

ACTIVITIES OF INTERNATIONAL PREPARATIONS

	i.u./mg.	1 i.u. equivalent to 1 "American" μg. of:
International standards for antibiotics[a]		
Bacitracin	55	
Chlortetracycline (hydrochloride)	1000	Hydrochloride
Dihydrostreptomycin (sulfate)	760	Base
Erythromycin (base)	950	Base
Oxytetracycline (base dehydrate)	900	Anhy. base
Penicillin G (sodium salt)	1670	
Phenoxymethylpenicillin (free acid)	1695	
Polymyxin B	7874	
Streptomycin (sulfate)	780	Base
Tetracycline (hydrochloride)	990	Hydrochloride
International reference preparations for antibiotics		
Amphotericin B	960	
Kanamycin (sulfate)	812	Free base
Neomycin (sulfate)	680	Free base
Novobiocin (sodium salt)	835	Free acid
Oleandomycin (chloroform adduct)	845	Free base
Vancomycin	1007	
Viomycin (sulfate)	730	Free base

[a] J. W. Lightbown, *Analyst* **86**, 216 (1961).

gram (i.u./mg.). In the U.S., many of the activities are expressed as micrograms of acid, base, or salt per milligram of preparation. Table I shows that 1 i.u. is equivalent to 1 μg. of the appropriate form as given in the last column.

III. Interferences

The combination of two or more antibiotics in a preparation frequently creates problems in the quantitative assay of each component by microbial procedures. Arret et al.[3] considered the problems and prepared tables that can be used to obtain an approximation to an answer. The interferences and sensitivities in their tables may not be identical with those obtained in other laboratories. Similar tables should be prepared in each laboratory for the combinations and the methods of interest. Their tables and comments follow.

"If the test organism used for the assay of one antibiotic (A) is not affected by a second antibiotic (B) the analytic problem is uncomplicated; such preparations are assayed as if they contained only A. However, if the test organism is affected by B, erroneously high or low values for A are obtained and methods must be developed to eliminate the effect of B. Such methods could involve: (1) inactivating B; (2) using a different test organism, which is sensitive to A and relatively resistant to B; (3) separating the antibiotics by differential solubility techniques; or (4) compensating for the presence of B by adding it to every solution of A used for the standard response curve. Numerous applications of these possibilities are discussed in the laboratory manual by Grove and Randall." [1]

Arret et al. regarded the specific antibiotic as A. They prepared several solutions containing the reference concentration of A and different concentrations of antibiotic B (Table II). Several solutions containing different concentrations of B alone were made in the diluent used in the assay of A. The two series of solutions were assayed using the reference concentration of A alone as the standard of comparison.

The lowest concentration of B, which, in combination with the reference concentration of A, gave a relative potency (as compared to the reference concentration of A) of greater than 110% or less than 90% was termed the "interference threshold."

The lowest concentration of B alone which caused a measurable response in the assay procedure for A was established as the "sensitivity threshold."

The interference thresholds and the sensitivity thresholds for the antibiotics and methods tested are given in Tables II and III,[3a] respectively.

The data in Table II indicate the conditions under which the second antibiotic interferes with the assay of the first. To find how bacitracin interferes with the *Sarcina lutea* assay for chloramphenicol, for example, find "bacitracin" in the left-hand column headed "Second antibiotic pres-

[3] B. Arret, M. R. Woodard, D. M. Wintermere, and A. Kirshbaum, *Antibiotics & Chemotherapy* 7, 545 (1957).

[3a] Tables II and III are taken from Arret et al.[3]

TABLE II

INTERFERENCE THRESHOLDS: CONCENTRATION OF ANTIBIOTIC B INTERFERING IN GIVEN ASSAY METHOD WHEN MIXED WITH THE REFERENCE CONCENTRATION OF ANTIBIOTIC A

Antibiotic A: Assay organism used in each method[a] and reference concentration of A present

Second antibiotic (B) present	M. aureus, 1 u./ml. penicillin	M. flavus, 1 u./ml. bacitracin	S. lutea, 50 μg./ml. chloramphenicol	M. albus, 1 μg./ml. neomycin	M. aureus, 10 μg./ml. neomycin	S. lutea, 1 μg./ml. erythromycin	S. lutea, 1 μg./ml. carbomycin	Br. bronchiseptica, 10 u./ml. polymyxin	B. cereus, 0.20 μg./ml. tetracycline	B. subtilis, 1 μg./ml. streptomycin[b]	K. pneumoniae, 30 μg./ml. streptomycin[b]	M. aureus, 0.24 μg./ml. tetracycline
Penicillin, u./ml.	—	0.2	0.03	6.0	0.1	0.04	0.03	>500.0	2.0	0.3	1000.0	0.05
Bacitracin, u./ml.	15.0	—	3.0	50.0	10.0	1.0	4.0	17.0	2.0	500.0	160.0	0.5
Chloramphenicol, μg./ml.	70.0	8.0	—	15.0	32.0	6.0	5.0	25.0	3.0	2.0[c]	1.2	3.0
Neomycin, μg./ml.	80.0[c]	>500	120.0	—	—	140.0	18.0	160.0	370.0	3.0	2.0	0.5
Erythromycin, μg./ml.	40.0[c]	4.2	0.04	0.5	1.0	—	0.2	350.0	2.0	0.6	2.0	0.02
Carbomycin, μg./ml.	20.0[c]	2.0	1.0	3.0	4.0	0.3	—	300.0	3.0	1.0	2.0	0.2
Polymyxin, u./ml.	>5000.0	>5000.0	>5000.0	110.0	1000.0	>5000.0	4000.0	—	>5000.0	1500.0	5.0	30.0
Tetracycline, μg./ml.	6.0	2.0	6.0	70.0	2.0	30.0	6.0	4.0	—	4.0	0.2	—
Chloretetracycline, μg./ml.	2.0	0.5	1.0	60.0	3.0	40.0	7.0	0.9	—	0.7	0.08	—
Oxytetracycline, μg./ml.	6.0	3.0	7.0	70.0	2.0	30.0	10.0	4.0	—	3.0	0.1	—
Streptomycin,[b] μg./ml.	90.0	9.0	40.0	100.0[c]	2.0	60.0	10.0	>500.0	12.0	—	3.0	3.0

[a] All assays are plate methods, except the K. pneumoniae streptomycin and M. pyogenes var. aureus tetracycline turbidimetric methods.
[b] Strepomycin and dihydrostreptomycin may be considered as acting identically.
[c] Indicates a depression, rather than an enhancement of activity.

TABLE III

Sensitivity Thresholds: Concentration of Antibiotic Alone Producing Minimum Measurable Growth Inhibition in Given Assay Method

Antibiotic tested	Penicillin,[a] M. aureus	Bacitracin, M. flavus	Chloramphenicol, S. lutea	Neomycin, M. albus	Neomycin, M. aureus	Erythromycin, S. lutea	Carbomycin, S. lutea	Polymyxin, Br. bronchiseptica	Tetracycline, B. cereus	Streptomycin, B. subtilis	Streptomycin, K. pneumoniae	Tetracycline, M. aureus
Penicillin, u./ml.	0.1	0.1	0.02	8.0	0.03	0.02	0.02	>1000.0	0.9	0.06	>1000.0	0.05
Bacitracin, u./ml.	15.0	0.1	0.4	8.0	6.0	1.5	1.5	20.0	1.0	250.0	160.0	0.5
Chloramphenicol, µg./ml.	3.0	6.0	5.0	4.0	8.0	8.0	8.0	10.0	2.5	12.0	0.6	3.0
Neomycin, µg./ml.	20.0	0.05	80.0	0.1	1.0	4.0	4.0	70.0	6.0	2.0	4.0	0.5
Erythromycin, µg./ml.	5.0	1.0	0.06	0.2	0.08	0.1	0.06	70.0	1.0	0.07	2.0	0.05
Carbomycin, µg./ml.	0.1	0.5	0.2	0.5	0.5	0.2	0.1	9.0	1.0	1.0	2.0	0.2
Polymyxin, u./ml.	1100.0	200.0	2000.0	80.0	90.0	300.0	300.0	1.0	1000.0	500.0	15.0	70.0
Tetracycline, µg./ml.	0.08	0.7	2.0	700.0	0.8	8.0	8.0	1.0	0.025	3.0	0.3	0.04
Chlortetracycline, µg./ml.	0.03	0.05	1.0	500.0	0.8	20.0	20.0	0.2	0.015	0.7	0.08	0.01
Oxytetracycline, µg./ml.	0.3	0.9	2.0	1000.0	1.0	10.0	10.0	2.0	0.025	0.7	0.3	0.04
Streptomycin,[b] µg./ml.	9.0	1.0	7.0	>1000.0	0.4	4.0	4.0	70.0	1.0	0.1	5.0	3.0

[a] All assays are plate methods, except the K. pneumoniae streptomycin and S. aureus tetracycline turbidimetric methods.

[b] Streptomycin and dihydrostreptomycin may be considered as acting identically.

ent," follow that row horizontally across the column head "*S. lutea* 50 µg./ ml. chloramphenicol." The figure 3.0 u./ml. indicates that samples containing bacitracin:chloramphenicol ratios of less than 3:50 (u:µg.) could be assayed accurately for chloramphenicol by this method.

Table III shows what effects different antibiotics alone have in the different assay procedures, as judged by the concentration which is just sufficient to produce a measurable zone of inhibition. This information can be used as a guide to the extent of interference to be expected from different antibiotics in a given assay procedure. For example, to find the amount of bacitracin alone which produces an inhibitory zone on *S. lutea* chloramphenicol plates, find "bacitracin" in the left-hand column headed "Antibiotic tested," follow that row horizontally across to the column headed "Chloramphenicol method, *S. lutea*." The figure 0.4 indicates that 0.4 u./ ml. of bacitracin produces a minimum measurable inhibitory zone. The data may even suggest alternate assay organisms for various antibiotics and combinations of them. Caution must be exercised in this respect, however, since many atypical responses occur, that is, the zones are not as clear and as well defined as those usually obtained with *A*. The data in Table III can also be used as a guide to the specificity of the given assay procedure.

Frequently *B* alone produced inhibition in the assay for *A* at a much lower concentration than that which caused significantly different assays when combined with the reference concentration of *A*. This lack of sensitivity to *B* may be caused by the assay conditions being so much more favorable to *A* than *B* that the effects of *B* are almost completely masked. For this reason, these data should not be construed to indicate any synergistic or antagonistic relationships.

The work of Arret *et al.* was directed toward the simplification and elucidation of problems of assaying combinations of antibiotics. Combinations of two antibiotics were specifically considered; however, the data can be used, in a general way, as guides for assaying combinations of three or more antibiotics.

IV. Solubilities

Assay of an antibiotic mixture may require separation of the active components or groups of components. Water immiscible organic solvents can be used to separate such antibiotics as penicillins, erythromycin, tetracyclines, and bacitracin, to list a few, from organic solvent insoluble compounds such as streptomycin, vancomycin, polymyxin, ristocetin, etc. Weiss and his co-workers[4] measured the solubility of 54 salts of 32 anti-

---

[4] P. J. Weiss, M. L. Andrew, and W. W. Wright, *Antibiotics & Chemotherapy* **7**, 374 (1957); M. L. Andrew and P. J. Weiss, *Antibiotics & Chemotherapy* **9**, 277 (1959).

TABLE IV

SOLUBILITIES OF ANTIBIOTICS, mg./ml.*

Antibiotic	Water	Methanol	Ethanol	Isopropanol	Isoamyl alcohol	Cyclohexene	Benzene	Toluene	Petroleum ether	Isooctane	Carbon tetrachloride
Tetracycline phosphate†	15.90	10.9	2.20	0.40	0.30	0.042	0.017	0.065	0	0	0.015
Oleandomycin Phosphate†	>20	>20	>20	11.20	6.40	0.082	1.60	1.06	0	0	0.515
Triacetyl-oleandomycin	0.25	>20	>20	>20	>20	1.41	>20	>20	0.492	0.170	>20
Potassium penicillin V†	>20	>20	1.35	0.105	0.250	0.060	0.020	0.030	0	0	0.040
Dihydronovobio-cin calcium salt	2.20	>20	13.65	3.60	2.80	0.042	0.025	0.030	0.002	0.022	0.020
Dihydronovo-biocin	0.180	>20	>20	>20	14.75	0.087	0.590	0.345	0.022	0	0.030
Novobiocin acid calcium salt	2.85	>20	>20	4.25	1.90	0.047	0.127	0.027	0.005	0	0.005
Anisomycin	6.55	>20	>20	16.80	>20	0.117	3.39	1.52	0.050	0.010	0.687
Cycloserine	>20	1.95	0.45	0.50	0.31	0.027	0.047	0.030	0.020	0	0.002
Bryamycin	0.24	0.55	0.65	0.75	1.30	0.032	0.58	0.040	0.025	0.002	0.012
Kanamycin base	>20	0.95	0.027	0.067	0.157	0.020	0.050	0.25	0.010	0.002	0
Kanamycin sulfate	>20	0.028	0.017	0.06	0.157	0.022	0.020	0.022	0.075	0	0.010
Calcium amphomycin	5.35	12.60	1.65	1.45	1.70	0.017	0.022	0	0.005	0	0.007
Soframycin	>20	0.080	0.030	0.40	0.130	0.022	0.022	0.012	0	0.005	0.015
6-Demethyl 7-chlortetracycline hydrochloride	>20	>20	4.60	0.50	0.50	0.030	0.032	0.022	0.012	0.007	18.88
6-Demethyl-tetracycline hydrochloride	4.75	>20	17.70	8.25	7.10	0.025	0.035	0.030	0.007	0	0.032
Neomycin-tetra-cycline salt	13.75	2.35	0.50	3.20	1.00	0.297	0.030	0	0.442	0	0.042
Tyrocidine hydrochloride	0.80	>20	17.55	5.45	1.20	0.027	0.025	0.042	0.037	0.002	0.012
Neomycin palmitate	0.35	>20	17.15	14.10	>20	2.71	6.45	4.64	2.28	1.31	3.05
Neomycin pelargonate	13.40	>20	>20	15.00	11.70	0.032	0.050	0.057	0.045	0	0.007
Neomycin picrate	1.70	4.50	2.90	0.140	0.49	0.015	0.035	0.037	0.035	0	0.010
Neomycin sorbate	>20	11.85	0.85	0.35	2.85	0.032	1.13	0.702	0.042	0.347	0.125
Neomycin undecylenate	0.95	>20	>20	>20	>20	1.92	4.73	3.42	2.62	1.11	7.95
Bacitracin methylene disalicylate spray dried	1.85	12.10	9.40	4.60	3.85	1.26	1.11	1.51	1.18	1.13	0.59
Bacitracin methylene disalicylate sodium salt	18.55	17.30	2.05	0.12	0.50	0.090	0.125	0.047	0.017	0.085	0.042
Bacitracin methylene disalicylate potassium	18.30	13.85	2.20	0.11	0.50	0.032	0.022	0	0	0.002	0.017
Erythromycin ethyl carbonate	0.30	>20	>20	>20	>20	0.410	>20	>20	0.370	0.060	5.09
Erythromycin glucoheptonate	>20	>20	>20	11.35	8.30	0.170	1.54	1.06	0	0.140	0.620
Hygromycin A	>20	16.15	16.10	11.05	9.65	0.037	0.030	0	0.017	0	0.007
Hygromycin B	>20	13.45	3.80	0.60	0.33	0.027	0.010	0.005	0.012	0.032	0
Vancomycin hydrochloride	>20	4.90	0.60	0.130	0.130	0.032	0.012	0	0.007	0	0.007
Vancomycin acetate	19.55	1.70	0.35	0.087	0.35	0.030	0.015	0	0.015	0	0
Vancomycin phosphate	>20	1.30	0.150	0.105	0.55	0.015	0.027	0.005	0.012	0.007	0
Vancomycin sulfate	17.30	0.190	0.35	0.075	0.165	0.022	0.035	0	0.017	0.012	0
Vancomycin gluconate	>20	1.35	3.00	0.047	0.30	0.020	0.035	0	0.007	0.005	0
Amphotericin B	0.75	1.60	0.50	0.107	1.05	0.022	0.060	0	0.010	0	0.002
Gramicidin	0.140	>20	>20	>20	14.10	0.020	0.192	0.040	0.007	0.005	0.047
Ristocetin	>20	0.60	0.050	0.102	0.090	0.025	0.52	0	0.015	0.002	0.020
Solvent blank	0.020	0.015	0.017	0	0.037	0.017	0	0	0	0.005	0.012

* All values uncorrected for solvent blank; nonitalicized values ± 0.05 mg.; italicized values ± 0.0025 mg.

† Pooled commercial preparations.

Ethyl acetate	Iso-amyl acetate	Acetone	Methyl ethyl ketone	Diethyl ether	Ethylene chloride	1,4-Dioxane	Chloroform	Carbon disulfide	Pyridine	Formamide	Ethylene glycol monomethyl ether	Benzyl alcohol
0.275	0.30	0.095	0.40	0.470	0.167	0.30	0.232	0.055	8.25	>20	>20	2.70
3.95	5.60	>20	10.75	1.265	0.132	>20	19.83	0.65	>20	>20	>20	>20
>20	>20	>20	>20	>20	>20	>20	>20	>20	>20	>20	>20	>20
0.65	1.10	0.222	0.60	0.840	0.40	0.565	0.45	0.075	0.130	2.55	>20	2.65
0.50	0.85	>20	11.25	0.535	0.30	14.15	0.140	0.35	>20	>20	>20	14.85
>20	>20	>20	>20	>20	4.45	>20	>20	0.095	>20	>20	>20	>20
0.35	1.50	9.95	1.15	0.660	0.122	14.25	0.147	0	>20	>20	>20	4.20
15.45	8.50	>20	>20	4.300	>20	>20	>20	0.30	>20	>20	>20	>20
0.35	0.070	0.85	0.45	0.457	0.060	0.45	0.067	0	0.90	1.60	1.50	1.00
0.65	0.85	0.75	1.30	0.513	0.30	8.20	0.027	0.30	>20	>20	>20	>20
0.022	0.027	0.042	0.057	0.360	0.037	0.30	0.027	0	0.020	0.062	0.23	0.720
0.035	0.030	0.070	0.082	0.470	0.037	0.322	0.045	0	0.040	0.105	0.062	0.525
0	0.80	0.065	0.30	0.357	0.080	0.590	0.025	0	1.25	2.10	3.70	5.25
0.012	0.025	0.177	0.077	0.390	0.020	0.385	0.045	0.30	0.087	0.265	0	2.70
0.055	0.102	0.155	0.135	0.478	0.032	0.75	0.062	0.015	>20	>20	>20	3.80
0.50	0.90	3.35	1.80	0.381	0.195	6.65	0.75	0	>20	>20	>20	>20
0.110	0.082	0.167	0.165	0.413	1.65	0.340	0.180	0.045	0.65	1.15	2.80	2.95
0.017	0.40	0.40	0.45	0.437	0.022	1.55	0.055	0.040	18.00	>20	>20	14.90
4.85	7.85	7.40	6.35	17.028	1.25	9.15	>20	13.20	>20	9.35	>20	17.55
0.027	0.30	3.90	0.30	0.527	0.080	0.40	0.055	0.30	3.45	5.70	>20	16.00
0.65	0.30	>20	>20	0.431	0.050	0.35	0.040	0.060	>20	>20	>20	7.60
0	3.30	0.025	1.65	0.458	0.050	5.45	0.230	0.130	0.55	0.45	7.50	17.35
4.70	7.40	16.00	5.85	8.271	0.80	6.70	>20	9.40	>20	>20	>20	>20
4.15	4.15	5.80	4.15	7.157	0.95	4.65	1.20	1.30	14.00	>20	18.25	19.20
0.027	0.30	0.040	0.146	0.598	0.072	0.30	0.085	0.090	0.55	10.45	15.70	0.40
0	0.30	0.202	0.092	0.440	0.070	0.315	0.035	0.015	0.40	9.65	>20	0.75
>20	>20	>20	>20	17.907	>20	>20	>20	3.60	>20	>20	>20	18.85
3.95	2.15	>20	9.35	1.820	6.20	>20	12.25	0.95	>20	>20	>20	19.70
0.80	0.40	7.20	3.60	0.492	0.032	9.90	0.050	0.120	>20	>20	>20	9.00
0	0.067	0.30	0.075	0.450	0.035	0.367	0.065	0.165	7.75	12.90	12.95	7.65
0	0.090	0.057	0.087	0.437	0.062	0.30	0.035	0.120	4.10	10.15	2.10	1.35
0	0.050	0.50	0.090	0.482	0.042	0.317	0.050	0.055	4.70	17.10	1.55	0.50
0.30	0.072	0.090	0.062	0.410	0.047	0.255	0.040	0.30	1.80	5.15	0.35	0.30
1.55	0.052	0.147	0.062	0.473	0.032	0.395	0.025	0.045	0.85	4.55	0.127	0.30
0	0.032	0.122	0.085	0.376	0.015	0.265	0.035	0.030	6.00	>20	1.0	0.30
0.30	0.30	0.155	0.505	0.55	>20	0.082	0.235	1.75	6.40	>20	2.60	0.75
11.90	>20	18.80	18.10	10.701	2.15	>20	>20	0.100	>20	>20	0.45	0.30
0	0.042	0.097	0.055	0.492	0.040	0.35	0.075	0.180	0.95	3.50	0	0.30
0	0.040	0.005	0.055	0.282	0.030	0.215	0.017	0	0.010	0.030	0	0.305

TABLE V

Solubility of Antibiotics, mg./ml.*

Antibiotics	Water	Meth-anol	Eth-anol	Iso-prop-anol	Isoamyl alcohol	Cyclo-hexane	Ben-zene	Tol-uene	Petro-leum ether	Iso-octane	Carbon tetra-chloride
Benzathine penicillin G	0.315*	16.9	15.4	3.65	0.60	0.315	0.45	0.72	0.99	0.09	0.57
Benzathine penicillin V	0.321	>20	14.6	7.15	1.6	0.47	0.32	0.67	0.505	0.20	0.73
Chloroprocaine penicillin O	9.2	>20	>20	>20	6.05	0.29	0.465	0.52	0.645	0.125	0.60
1-Ephenamine penicillin G	1.2	19.5	2.5	0.45	1.0	0.085	0.12	0.085	0.0	0.032	0.122
Hydrabamine penicillin G	0.075	7.3	5.2	1.7	3.1	0.115	0.60	0.39	0.0	0.055	0.50
Hydrabamine penicillin V	0.05	11.05	5.8	1.75	6.85	0.12	1.4	1.07	0.06	0.065	3.30
Sodium penicillin G	>20	>20	10.0	0.75	2.1	0.105	0.047	0.02	0.0	0.032	0.042
Procaine penicillin G	6.8	>20	>20	6.5	2.6	0.075	0.075	1.05	0.12	0.0	0.12
Penicillin V acid	0.90	>20	>20	>20	16.95	0.08	0.45	0.26	0.245	0.037	0.097
Bacitracin	>20	>20	9.1	1.85	1.65	0.075	0.025	0.015	0.035	0.055	0.018
Zinc bacitracin	5.1	6.55	2.0	0.16	2.6	0.06	0.065	0.02	0.025	0.015	0.12
Tyrothricin	2.1	>20	>20	5.6	2.4	0.20	0.30	0.15	0.275	0.042	0.455
Polymyxin B sulfate	>20	0.30	0.115	0.007	0.175	0.06	0.045	0.015	0.0	0.022	0.037
Viomycin sulfate	7.8	0.35	0.052	0.40	0.065	0.025	0.16	0.0	0.0	0.027	0.035
Dihydrostrepto-mycin sulfate	>20	0.35	0.10	0.35	2.85	0.07	0.035	0.05	0.025	0.030	0.032
Streptomycin hydrochloride	>20	>20	0.90	0.12	0.117	0.10	0.05	0.0	0.02	0.017	0.042
Streptomycin sulfate	>20	0.85	0.30	0.01	0.30	0.04	0.027	0.03	0.015	0.015	0.035
Streptohydrazide	>20	1.6	0.115	0.30	1.3	0.055	0.045	0.0	0.0	0.010	0.025
Neomycin B hydrochloride	15.0	5.7	0.65	0.05	0.33	0.06	0.03	0.0	0.0	0.08	0.01
Neomycin B sulfate	6.3	0.225	0.095	0.082	0.247	0.08	0.05	0.0	0.005	0.027	0.092
Carbomycin	0.295	>20	>20	4.65	8.1	0.44	18.6	3.78	0.095	0.065	13.53
Chloramphenicol	4.4	>20	>20	>20	17.3	0.13	0.26	0.145	0.085	0.022	0.295
Chloramphenicol palmitate	1.05	>20	>20	>20	>20	0.335	10.7	13.52	0.225	0.085	2.97
Erythromycin	2.1	>20	>20	>20	9.65	>20	>20	>20	4.69	0.477	>20
Novobiocin monosodium	15.6	>20	>20	6.45	4.4	0.06	0.10	0.005	0.0	0.044	0.07
Novobiocin acid	0.05	>20	>20	16.1	1.05	0.055	0.10	0.005	0.03	0.047	0.056
Nystatin	4.0	11.2	1.2	1.2	2.4	0.505	0.28	0.285	0.16	0.030	1.23
Chlortetracycline hydrochloride	8.6	17.4	1.7	0.45	0.172	0.045	0.09	0.03	0.005	0.010	0.132
Oxytetracycline	0.60	18.5	8.1	0.30	0.087	0.055	0.037	0.005	0.0	0.027	0.055
Oxytetracycline hydrochloride	6.9	16.35	11.95	7.3	7.45	0.055	0.027	0.0	0.01	0.025	0.072
Tetracycline	1.7	>20	>20	16.1	14.2	0.095	1.05	0.595	0.005	0.027	0.315
Tetracycline hydrochloride	10.9	>20	7.9	1.15	1.4	0.075	0.25	0.21	0.0	0.027	0.10
Solvent blank	0.02	0.015	0.04	0.0	0.045	0.005	0.025	0.0	0.0	0.05	0.017

* All values uncorrected for solvent blank; nonitalicized values ± 0.05 mg.; italicized values ± 0.0025 mg.

biotics in 24 solvents. Their results are given in Tables IV and V.[4a] They did not identify the material only slightly soluble in certain of the solvents (isooctane, for example). It might not be the antibiotic being measured but an impurity, since none of the antibiotics were chemically pure. The tables should be used with this reservation in mind.

[4a] Tables IV and V are taken from Weiss et al.[4]

Ethyl acetate	Isoamyl acetate	Acetone	Methyl ethyl ketone	Diethyl ether	Ethylene chloride	1.4-Dioxane	Chloroform	Carbon disulfide	Pyridine	Formamide	Ethylene glycol	Benzyl alcohol
1.2	0.55	3.0	3.2	0.40	0.90	2.4	2.1	0.50	>20	>20	>20	12.45
9.0	0.55	>20	>20	1.2	1.9	>20	>20	0.55	>20	>20	>20	>20
11.5	2.9	>20	>20	1.45	6.4	>20	>20	0.45	>20	>20	>20	>20
0.80	0.26	0.75	0.85	0.485	0.75	4.55	1.56	0.07	>20	>20	9.0	9.95
1.65	1.4	3.4	3.65	0.70	>20	14.65	>20	1.4	>20	>20	>20	>20
4.0	4.9	10.2	13.7	0.095	>20	7.5	>20	1.3	>20	>20	>20	>20
0.40	0.22	0.19	0.147	0.06	0.30	1.9	0.05	0.083	1.15	>20	>20	11.2
3.35	1.2	14.95	13.7	0.60	2.0	9.8	>20	0.51	>20	>20	>20	>20
>20	>20	>20	>20	11.75	12.65	12.60	>20	0.30	>20	>20	>20	>20
0.047	0.09	0.75	0.20	0.065	0.025	0.70	0.0	0.30	9.15	19.9	>20	>20
1.3	0.45	1.0	0.85	0.02	1.1	0.49	0.01	0.30	4.05	>20	7.95	10.35
2.65	2.9	6.8	12.3	3.25	1.3	11.1	1.6	0.65	>20	>20	>20	>20
0.025	0.09	0.11	0.052	0.012	0.025	0.90	0.0	0.076	0.45	0.105	1.55	0.45
0.25	0.06	0.03	0.097	0.07	0.45	1.2	0.0	0.06	0.055	0.075	0.045	0.85
0.105	0.06	0.01	0.35	0.24	0.055	0.43	0.022	0.01	0.50	0.50	0.04	9.0
0.30	0.10	0.015	0.07	0.01	0.08	0.80	0.0	0.056	0.14	>20	>20	0.90
0.30	0.10	0.00	0.05	0.035	0.30	0.60	0.0	0.25	0.195	0.107	0.25	5.55
0.25	0.10	0.08	0.055	0.31	0.055	0.80	0.01	0.056	0.15	0.55	0.102	0.30
0.047	0.13	0.06	0.115	1.55	0.08	0.70	0.06	0.05	0.35	0.45	3.85	1.0
0.045	0.13	0.17	0.062	0.135	0.045	0.70	0.0	0.037	0.95	0.35	0.045	12.35
>20	>20	>20	>20	13.95	>20	>20	>20	0.65	>20	>20	>20	>20
>20	>20	>20	>20	>20	2.3	>20	1.95	0.35	>20	>20	>20	14.6
>20	>20	>20	>20	>20	>20	>20	>20	2.05	>20	>20	>20	>20
>20	>20	>20	>20	>20	>20	>20	>20	5.05	>20	>20	>20	>20
0.60	1.45	5.9	1.65	0.22	0.25	4.9	0.0	0.04	>20	>20	>20	>20
10.57	1.35	3.5	>20	0.30	0.30	2.9	1.35	0.04	>20	>20	>20	>20
0.75	0.55	0.390	0.75	0.30	0.45	2.1	0.48	0.40	>20	>20	8.75	2.65
0.35	0.12	0.12	0.18	0.085	0.25	1.45	0.02	0.023	>20	5.9	3.0	1.8
0.85	0.15	1.6	1.35	0.13	0.25	4.1	0.0	0.066	>20	>20	>20	0.70
2.05	1.0	10.8	4.4	0.135	0.35	6.3	0.40	0.063	>20	>20	>20	>20
17.3	11.6	17.4	>20	3.7	11.25	14.6	13.8	0.50	>20	12.75	>20	14.35
0.75	0.35	0.75	0.70	0.60	0.80	7.7	2.85	0.35	>20	>20	17.75	10.8
0.017	0.07	0.02	0.057	0.065	0.075	0.56	0.00	0.063	0.027	0.0	0.012	—

6.2 Bacitracin

L. J. DENNIN

Eli Lilly and Company, Indianapolis, Indiana

I. Introduction

Two test organisms have been employed in the cylinder-plate assay for bacitracin. Either of the cultures (*Micrococcus flavus*, ATCC 10240, or *Sarcina subflava*, ATCC 7468) may be used in a conventional assay or in one whose sensitivity is increased by the use of a thinner "base" layer and a lighter inoculum. The *S. subflava* assay has come into general usage because of the somewhat greater reproducibility of the method as compared to the *Micrococcus flavus* assay.

261

II. Test Organisms

Maintain stock cultures of *Micrococcus flavus* or *Sarcina subflava* by weekly transfers to fresh sterile slants of G. & R. No. 1. Incubate the freshly prepared slants at 32° to 35°C. for 16 to 24 hours and store at 4° to 6°C. until ready for use in the preparation of the inoculum suspension.

III. Standard Solutions

Dry a quantity of the working standard in a tared weighing bottle with capillary tube for 3 to 4 hours at 60°C. and a residual pressure of 5 mm. Hg or less. Weigh sufficient dried standard to make a solution containing 100 units of bacitracin activity per milliliter, transfer to a suitable container, and dilute to volume with pH 6 phosphate buffer. Store the stock standard solution at 4° to 6°C. for a period not to exceed 3 weeks.

IV. Sample Preparation

Most bacitracin samples present little difficulty to the assayist from the standpoint of sample preparation. The antibiotic itself is readily soluble in pH 6 phosphate buffer. The zinc salt is insoluble in water and is dissolved by adding several drops of concentrated hydrochloric acid to the suspension to lower the pH to 2 to 3. The methylene disalicylate derivative is insoluble in water and is dissolved in 2% sodium bicarbonate solution immediately before diluting to the assay level of approximately 1.0 unit/ml. Ointments are extracted by first dispersing in ethyl ether and then washing repeatedly with pH 6 phosphate buffer.

V. Mechanics of the Assay

A. Design

Refer to Penicillin, Chapter 6.10, Section III.E.1.

B. Media

The assays under consideration (*Micrococcus flavus* and *Sarcina subflava*) make use of the two-layer agar system. If assay sensitivity is not an important factor, use a 21-ml. "base" layer of G. & R. No. 2 and a 4-ml. "seed" layer of G. & R. No. 1. To increase the sensitivity of either assay, reduce the above "base" layer to 10 ml.

C. Inocula

Prepare suspensions of either of the test organisms in the following manner.

Prepare an agar slant of the culture as described in Section II. Wash the growth on this slant from the agar surface using 2–3 ml. of sterile 0.85% sodium chloride solution. Transfer the suspension to a Roux bottle containing approximately 300 ml. of sterile G. & R. agar medium No. 1 and distribute it evenly over the surface of this agar with the aid of sterile glass beads. Incubate the inoculated Roux bottle at 32° to 35°C. for 16 to 24 hours. Harvest the resultant growth from the Roux bottle by washing the agar surface with 25 ml. of sterile 0.85% sodium chloride solution.

If the test organism is *Micrococcus flavus,* adjust the stock inoculum suspension prepared above so that a 1:50 dilution in 0.85% sodium chloride solution will give 75% transmittance at 6500 A (Lumetron). This adjusted stock suspension will contain approximately 5×10^9 viable cells per milliliter. If *Sarcina subflava* is used, adjust the stock inoculum suspension so that a 1:50 dilution will give 50% transmittance under the same conditions as with *Micrococcus flavus.* The adjusted stock suspension wil contain approximately 5×10^9 viable cells per milliliter. The adjusted stock suspension, not the 1:50 dilution, is used in inoculating the "seed" agar. Inoculate the "seed" agar by adding from 0.1 to 0.4 ml. of the stock inoculum suspension per 100 ml. of liquefied and cooled (48°C.) agar and pour the plates immediately. The inoculum concentration will depend upon the assay sensitivity required. The stock suspensions may be stored at 4° to 6°C. for a period of 2 weeks.

D. Standards

Dilute samples of the stock standard solution in sterile pH 6 phosphate buffer immediately prior to use on the day of the assay. The choice of standard levels will, of course, determine to what extent these aliquots are diluted. In my experience, both organisms yield a linear response in the range from 0.5 to 4.0 u./ml. with the 21-ml. "base" layer and from 0.05 to 1.0 u./ml. with the 10-ml. "base" layer and its somewhat lower level of inoculum. The general considerations regarding diluents which were discussed under Penicillin, Section III.D, also hold true for this antibiotic.

E. Samples

Refer to Penicillin, Section III.E.5.

F. Incubation

Incubation of the tests described is carried out at 32° to 35°C. for a period of 16 to 18 hours.

VI. Measuring the Response

See Penicillin, Section III.F.

VII. Computation of Answers

See Penicillin, Section III.G.

6.3 Cephalosporin C

FREDERICK KAVANAGH

Eli Lilly and Company, Indianapolis, Indiana

I. Introduction

Cephalosporin C is an antibiotic[1] produced by mutant strains of the fungus *Cephalosporium acremonium*. It is weakly active against gram-positive bacteria and somewhat more active against certain gram-negative bacteria. It is of interest because it has the β-lactam ring characteristic of penicillins but is resistant to penicillinases. It has the same side chain,[2,3] D-α-aminoadipic acid, as cephalosporin N and is considerably less active against gram-negative bacteria than cephalosporin N. Since the two anti-

---

[1] G. G. F. Newton and E. P. Abraham, *Biochem. J.* **62**, 651 (1956).
[2] E. P. Abraham and G. G. F. Newton, *Biochem. J.* **62**, 658 (1956).
[3] E. P. Abraham and G. G. F. Newton, *Biochem. J.* **79**, 377 (1961).

biotics occur together in fermentation broths and are active against the same bacteria, cephalosporin N is destroyed by treatment of the broths with penicillinase at pH 6.5–7.0 or by incubating at pH 2 and 37°C. for 2 hours before assaying. Cephalosporin N can be assayed by the hydroxamic acid method used for other penicillins. Although cephalosporin C gives the hydroxamic acid reaction, specific chemical assay for cephalosporin C is not possible because of the lack of a β-lactamidase active against the β-lactam group in cephalosporin C. If the penicillinase used to inactivate cephalosporin N contains an esterase active against the acetyl group in cephalosporin C, the acetyl group will be removed and desacetyl[4] cephalosporin C formed. Desacetyl cephalosporin C is less active against the test organism than cephalosporin C and its formation causes a low microbiological assay but may not affect the hydroxamic acid assay greatly.

II. Plate Assay

A. Introduction

The plate assay is the method of choice because of its wide range, 10–100 μg./ml., in contrast to the turbidimetric assay, which has a range from 20 to 40 μg./ml. using the same test organism. Derivatives of the nucleus of cephalosporin C which are active against gram-positive bacteria are assayed by methods used in assaying penicillins.

Bond et al.[5] described a large plate assay that was about 10 times as sensitive as the plate methods given here. They employed a strain of *Vibrio cholera* 1077 obtained from Dr. I. N. Asheshov who had used it for many years as a host organism for cholera phage. The *V. cholera* was of attenuated pathogenicity and would grow on the surface but not when deep-seeded in agar media. The papain digest medium of Asheshov[6] was used for growing the organism. Since the conditions of the assay were critical, they were described in detail in the publication by Bond et al.

B. Test Organism

Salmonella gallinarum is maintained by the procedure described under Penicillin, Chapter 6.10, Section III.B. A gram-negative test organism is used because fermentation broths contain cephalosporin P, an antibiotic active against gram-positive but not against gram-negative bacteria.

[4] J. D'A. Jeffery, E. P. Abraham, and G. G. F. Newton, *Biochem. J.* **81**, 591 (1961).
[5] J. M. Bond, R. W. Brimblecombe, and R. C. Codner, *J. Gen. Microbiol.* **27**, 11 (1962).
[6] I. N. Asheshov, *Can. J. Pub. Health* **32**, 468 (1941).

C. Standard Solutions

Dry the standard for 4 hours at 70°C. in a weighing bottle (fitted with a capillary tube) in a vacuum of 5 mm. Hg or lower. Prepare a stock solution of 1 mg./ml. in pH 6.0 buffer. Make four standard solutions, 10, 20, 30, and 50 μg./ml., in pH 6.0 buffer. The stock solution may be used for a period of 7 days.

D. Preparation of Samples

Filter the fermentation samples and dilute with pH 6 buffer to an estimated concentration of 20 μg./ml. Dissolve the solid preparations in buffer and dilute to 20 μg./ml.

E. Inoculum

Transfer a loop full of growth from a slant to nutrient broth and incubate at 37°C. until the cell population reaches about 800 million per milliliter. If growth under static conditions is too slow, shake to aerate and accelerate growth.

F. Mechanics of the Assay

1. Design

The design is the simple one of one level of standard (20 μg./ml.) and one of sample per petri dish. Two or more dishes are used for each sample depending upon the precision needed. Five plates of standard curve (10, 20, 30, and 50 μg./ml.) are prepared. (See Penicillin, Section III.E.1.) The paper disks, 0.5 inch in diameter, are dipped in the solutions and then placed on the agar.

2. Medium

One layer of agar is used. It has the following composition:

Peptone, Bacto	6 gm.	Beef extract	1.5 gm.
N-Z case	4 gm.	Agar	20 gm.
Yeast extract	8 gm.	Water, distilled	1 liter

Adjust pH of the medium to 5.8 with sulfuric acid before sterilizing.

3. Inoculum

Add 5 ml. of inoculum (Section II.E) to each bottle (350 ml.) of melted and cooled (48°C) agar, mix, and pour 7-ml. layers into 100-mm. plastic or glass petri dishes.

4. Incubation

Incubate the plates in the 37°C. incubator over night (Penicillin, Section III.E.6).

5. Measuring the Response

See Penicillin, Section III.F.

6. Computation of Answers

See Penicillin, Section III.G. The dose response line is linear over the range of standards given.

III. Large Plate Assay

A. Introduction

Two versions of large plate assays (Chapter 2 and Nystatin, Chapter 6, Part II) are given in this book. Elaborate statistical designs are used in an effort to compensate for the bias introduced by the large time difference between the applications of the first and the last sample on the plate. Only in a frankly inaccurate assay design are the standards as few as 13% of the total number. The statistical design requires application of standards and samples at random. Assembly of the replicates of each sample (or standard) from a random distribution is time consuming unless it is done by a computer. A procedure which eliminated the necessity for a randomized design would be more convenient and more efficient than present procedures.

Using paper disks to contain the samples and placing them on the agar all at once eliminates time dependent bias. This can be done for both petri dishes and large plates by placing the disks on a loading plate as they are dipped and then transferring the disks by inverting the agar plate and moving it into contact with the disks to pick them up on the agar surface. Test of this procedure showed that size of an inhibition zone was independent of location on the plate. The zones were placed so far apart that one zone did not interact with adjacent zones in the direction of the diameter measured. The agar plates were similar to those described in Chapter 2 and were prepared as described there. The loading plate was made from $\frac{1}{4}$-inch plate glass cut about $\frac{1}{4}$ inch smaller in size than the inside dimensions of the agar plate. A thin coating of silicone grease was rubbed on the glass plate to prevent adhesion of the wet disks to it. A pattern for locating the disks was placed under the loading plate. When viewed from the top, the first row of disks on the plate is the bottom row on the loading plate. Since I prefer to read the zones in increasing numerical order beginning with 1 in the upper left-hand corner of the plate, the pattern was made with zone 1 in the lower left-hand corner. A minimum of 9 standards (3 low, 3 medium, 3 high concentrations) are put on each plate. Certain samples, shake flask investigation, for example, may be represented by a single zone on each of two plates. Although this seems to be inadequate replication of a sample, it is statistically much superior to the present practice of having the two zones both on one petri dish. More important samples, large fermenters and processing, for example, may be replicated on each of the plates put on in a day.

Zone diameters on the large plate can be measured with an error of ±0.1 mm. with needle-point calipers or, preferably, with the projection device described in Nystatin, Large Plate Method (Chapter 6, Part II). Use of the latter device permits simple and rapid method for obtaining potency directly from zone diameters as they are measured. The scale on which the image of the zone is projected is calibrated in units of potency, not millimeters. Thus potencies, not diameters, are reported for the samples. By reading zone diameters directly in potency, time is saved and chance of error from the operation of interpolation from a graph is avoided. This simplified procedure is possible only because of the absence of significant variation in zone diameter correlated with position.

B. Mechanics of the Assay

1. Design

The range of the assay is wide. Satisfactory zones are obtained for standards in the concentration range from 5 to 800 μg./ml. depending upon composition of the assay medium. The medium in Section II.F.2 prepared without yeast extract gives such zones that the range of the assay is from 6.25μg./ml. (14 mm. zone) to 200μg./ml. (26 mm. zone). Omission of the N-Z case reduces the zone size so that the range extends from 25 μg./ml. to 400 μg./ml. The medium used in a particular assay is selected according to the sensitivity (and range) needed.

The greater thickness (2 mm.) of the agar reduces the sensitivity below that of the petri dish assay made with the same medium.

2. Medium

Adjust the medium (Section II.F.2) to pH 5.8 with sulfuric acid before sterilizing. Put sufficient agar in each bottle to form a layer 2.0 mm. deep when poured into a large plate.

3. Inoculum

Use the same proportion as in Section II.F.3.

4. Incubation

See Section II.F.4.

5. Measuring the Response

Measure the diameter by any appropriate means.

6. Computation of Answers

The type of plot depends upon the range of the assay. Use the one appropriate to the design.

6.4 Chloramphenicol

ROBERT HANS, MARGARET GALBRAITH, AND WILLIAM C. ALEGNANI

Research Division, Parke, Davis & Company, Detroit, Michigan

I. Introduction

Chloramphenicol [1] is a neutral, bitter, colorless, crystalline antibiotic which can be prepared both by biochemical and synthetic processes. The empirical formula of crystalline chloramphenicol is $C_{11}H_{12}Cl_2N_2O_5$ (mol. wt. 323) and it is known chemically as D-*threo*-1-*p*-nitrophenyl-2-dichloro-acetamido-1,3-propanediol. It is somewhat soluble in water (2.5 mg./ml.

[1] The trade name of Parke, Davis & Company for chloramphenicol is Chloromycetin.

at 25°C.), very soluble in methanol, ethanol, butanol, acetone, ethyl ether, ethyl acetate, and propylene glycol, and insoluble in benzene and petroleum ether. Aqueous solutions are highly stable even on boiling and are stable in the dark at room temperature over the pH range 2–9 for more than 24 hours.

Chloramphenicol is a broad spectrum antibiotic which has been reported to be effective against gram-positive and gram-negative bacteria, rickettsiae, and viruses. Some of the diseases against which the drug has proved to be effective include typhoid fever, *Hemophilus influenzae* meningitis, brucellosis, tularemia, plague, bartonellosis, bacterial and viral pneumonia, various ophthalmic and otic infections and urologic and enteric infections, psittacosis, typhus, spotted fever, and lymphogranuloma. It has also proved to be effective in the treatment of systemic infections caused by penicillin-resistant staphylococci.

A number of chemical and microbiological assay methods have been proposed for the determination of chloramphenicol in fermentation and fractionation samples, body fluids, and pharmaceutical products. Those methods making use of chemical and physical properties of the drug, such as the spectrophotometric,[2] polarographic,[3] titrametric,[4] and countercurrent[5] procedures, are useful for the assay of relatively pure preparations. The limitation of each of these methods is obvious, however, when one remembers that a physical or chemical characteristic is being used as the parameter instead of the inherent antimicrobial activity of the intact chloramphenicol molecule. For this reason, microbiological assay methods are preferred for any pharmaceutical formulations in which spurious activity might be detected, for body fluids specimens, and for materials on long-term stability study or those which have been subjected to adverse storage conditions.

There are two basic microbiological assay procedures for determining chloramphenicol potency, viz., the cylinder plate assay method and the turbidimetric method. These will be discussed below.

II. Cylinder Plate Assay Method

For pharmaceutical forms, we employ the cylinder plate assay described in the F.D.A. regulations[6] using *Sarcina lutea* (ATCC 9341) as

[2] A. J. Glazko, L. M. Wolf, and W. A. Dill, *Arch. Biochem.* **23**, 411 (1949).

[3] G. B. Hess, *Anal. Chem.* **22**, 649 (1950).

[4] R. Truhaut, *Ann. pharm. franç.* **9**, 347 (1951).

[5] A. Brunzell, *J. Pharm. and Pharmacol.* **8**, 329 (1956).

[6] U.S. Food and Drug Administration. Compilation of Regulations for Tests and Methods of Assay and Certification of Antibiotic Drugs. Washington, D.C., 1951. Vol. I, Part 141d.

the test organism. The range of the standard curve is from 32.0 to 78.0 μg. chloramphenicol/ml. with the reference point at 50 μg./ml. A cylinder plate procedure employing *Bacillus subtilis* (Parke, Davis culture No. 04969) for the assay of body fluids and tissue extracts has also been described.[7]

A. Cylinders (Cups)

Stainless steel cylinders having an outside diameter of 8 mm. ($\pm$0.1 mm.), an inside diameter of 6 mm. ($\pm$0.1 mm.), and a length of 10 mm. ($\pm$0.1 mm.) are used for this assay. During use, cylinders may become encrusted with fats, oils, proteins, etc., which, if not thoroughly removed from the cylinders, could significantly affect the size and shape of the zones produced. Therefore, as soon as the cylinders are removed from the assay plates, they are treated in the following manner:

(1) They are placed in a stainless steel container, covered with an alcoholic-sodium hydroxide solution (12% denatured alcohol and 4% sodium hydroxide), brought to a boil and allowed to sit for 15 minutes.

(2) The cleaning solution is then decanted, and the cylinders are rinsed six times with tap water and three times with distilled water.

(3) They are covered with distilled water and boiled an additional 15 minutes. After boiling, the rinse water is cooled and the pH is checked with Brom Thymol blue. The color should be green or yellow. If blue, rewash with distilled water until the desired color is obtained.

(4) The distilled water is then removed, and the cylinders are rinsed with ethyl alcohol, dried on a cloth towel, placed in petri dishes and sterilized in a hot air oven.

Cylinders treated in the above manner have been used in these laboratories for years without evidence of etching or deterioration.

B. Culture Media

Both the nutrient broth and nutrient agar used in preparing the seed and base layers and for carrying the test organism may be prepared from basic ingredients or purchased as dehydrated culture media from commercial laboratories.[8,9] The nutrient broth contains peptone, 5.0 gm.; yeast extract, 1.5 gm.; beef extract, 1.5 gm.; sodium chloride, 3.5 gm.; dextrose, 1.0 gm.; monobasic potassium phosphate, 1.32 gm.; dibasic potassium phosphate, 3.68 gm.; and distilled water to make 1000 ml. The pH after sterilization is 7.0. The nutrient agar contains peptone, 6.0 gm.; pancreatic digest of casein, 4.0 gm.; yeast extract, 3.0 gm.; beef extract,

---

[7] D. G. Smith, C. B. Landers, and J. Forjacs, *J. Lab. and Clin. Med.* **36**, 1 (1950).

[8] Baltimore Biological Laboratory, Inc.: Antibiotic Assay Broth and Seed Agar (U.S.P. Peptone Casein Agar).

[9] Difco Laboratories, Inc.: Penassay Broth and Penassay Seed Agar.

1.5 gm.; agar, 15.0 gm.; and distilled water to make 1000 ml. The pH after sterilization is 6.5 to 6.6.

If numerous assays are being performed routinely, it pays to prepare the agar medium in large quantities (10–20 liters). The medium is then dispensed into appropriate containers (1000-, 500-, or 250-ml. florence flasks), sterilized, and stored under refrigeration at 5°C. Medium kept for 1 month under these conditions has produced satisfactory results.

Flowing steam is used to remelt the agar. For this purpose, place the flasks in an Arnold sterilizer (approximately 20 minutes) or in an autoclave, the valves of which have been adjusted so that the pressure does not rise and the temperature, therefore, approaches but does not exceed 100°C. Melted medium for pouring base layers may be used immediately or cooled to suit the needs of the operator. Place the medium used for the seed layer in a 48°C. water bath and allow it to reach thermoequilibrium before adding the test organism. The time required to reach thermoequilibrium will depend on the size of the flask used, the amount of the medium in the flask, and the frequency of shaking.

C. Test Organism

Maintain the test organism by transferring once a week on nutrient agar slants and incubating for 24 hours at 26°C. Inoculate Roux bottles containing 300 ml. of nutrient agar with the growth from a 24-hour slant washed off in 3 ml. of nutrient broth. Glass beads aid in the distribution of the suspension over the entire surface of the agar. Two to three dozen beads of approximately $\frac{1}{4}$ inch diameter are dry heat sterilized at one time in a small flask. The entire contents of the flask are usually added to the Roux bottles, where they remain during incubation and aid in breaking up the surface growth when preparing the cell suspension.

After incubation (24 hours at 26°C.), wash the growth from the surface of the agar with about 20 ml. of nutrient broth. Adjust the bulk suspension so that a 1:10 dilution in nutrient broth gives a 10% light transmission in a Lumetron 400-A photoelectric colorimeter equipped with a filter having a wavelength of 650 mμ. In our laboratories a 1:50 dilution was found to give a similar reading (10% transmittance) when measured on a Bausch & Lomb Spectronic 20 set at a wavelength of 650 mμ. Decant the adjusted cell suspension aseptically into a sterile rubber stoppered centrifuge bottle (250 ml.) containing approximately 2 dozen glass beads. This suspension may be used up to 1 month when kept at 5°C. Shake the material thoroughly before removing an aliquot for preparation of the seed layer. The amount of bulk suspension needed to produce suitable zones of inhibition with good sensitivity will vary. However, 1.0 to 1.5 ml. of the adjusted bulk suspension when added to 100 ml. of nutrient agar which has been melted and cooled to 48°C. should yield good results.

D. Working Standards

Keep the chloramphenicol working standard (1000 μg./mg.) in a tightly stoppered vial in a desiccator at 5°C. Weigh an appropriate amount of the standard, dissolve in ethanol (about 5 ml.), and dilute in 1% phosphate buffer at pH 6.0 (2.0 gm. K_2HPO_4 and 8.0 gm. KH_2PO_4 in 1000.0 ml. distilled water) to give a solution containing 1000 μg. chloramphenicol/ml. This stock solution, when stored at 5°C., may be used for as long as 1 month.

Weighing of chloramphenicol standards, samples, and other antibiotic preparations which are being assayed in the laboratory should be done in a room separate from that in which the assay plates are poured. Air currents in a room are capable of whipping up microscopic particles of spilled material and depositing them on the assay plates, causing irregular zones and spotting of the plates.

E. Preparation of Plates

Prior to the introduction of plastic labware, flat bottom, glass petri dishes (20 by 100 mm.) were used routinely. Disposable nonsterile plastic petri dishes (20 by 95 mm.) have been used over the past year without evidence of interfering contamination. With glass plates, porcelain covers glazed only on the outside, or aluminum petri dish tops with absorbent disks[10] may be used to cover the plates. With plastic dishes, special covers have been found unnecessary. Condensation on the tops is minimal and presents no problems. The cost of disposable plastic labware is offset by the breakage incurred with regular glassware, cost of washing and sterilizing. A 20-mm. depth in the petri dish is critical in order that a 10-mm. stainless steel cylinder, when placed on the agar surface, will not touch the cover of the petri dish.

1. Base Layer

Add 21 ml. of nutrient agar to each petri dish, distribute the agar evenly over the surface, and place the dishes on a flat, level surface to harden. Put the hardened plates in a 37°C. incubator for approximately 1 hour. Warming of these plates facilitates spreading of the seed layer. When a large number of plates are to be poured, the use of dispensing syringes or a Brewer automatic pipetter[11] will speed up the operation considerably.

[10] Baltimore Biological Laboratory, Inc.: Brewer Petri Open Type Metal Tops with Absorbent Disks.

[11] Baltimore Biological Laboratories, Inc.: Brewer Automatic Pipetting Machine.

2. Seed Layer

Cool the melted nutrient agar to 48°C., seed (1.0–1.5 ml. of adjusted bulk suspension per 100 ml. of agar), mix thoroughly, and hold in a 48°C. water bath throughout the seeding period. Add 4 ml. of the seed agar to each plate and spread evenly over the agar surface by gently tilting the plate back and forth with a rotary motion. Place the plates on a flat, level surface until the seed layer has hardened. By placing the plates on a level surface, an even thickness of agar is obtained throughout the plate. Differences in the thickness of the seed and/or base layers will produce irregular zones.

3. Dispensing of Cylinders

Several methods are available for dispensing the cylinders onto the seeded agar plates. A plastic template with holes bored of the proper size and interval, a Behmer cylinder dispensing machine,[12] or a piece of cardboard appropriately marked may be used for this purpose.

In the chloramphenicol assay, six cylinders are placed on the inoculated agar surface so that they are at approximately 60° intervals on a 2.8-cm. radius. The plates are allowed to stand ½ hour at room temperature to ensure proper seating of the cylinder on the agar surface. Cylinders are then filled with the appropriate concentration of standard and sample, respectively. It is not necessary to pipet an exact amount of material into each cylinder. If the cylinders are filled to the top, an excess of material is available which allows sufficient antibiotic to diffuse into the surrounding seed agar to produce suitable and reproducible zones. When adding material to the cylinders, care should be taken to avoid trapping air bubbles in the cylinders. Trapped air would prevent the antibiotic from coming into contact with the agar surface.

F. Standard Curve and Assay Procedure

Prepare dilutions of the following concentrations in 1% phosphate buffer, pH 6.0: 32.0, 40.0, 50.0, 62.5, and 78.1 μg./ml. to be used for constructing the standard curve. Twelve plates are required for the preparation of the standard curve, three plates for each concentration except the 50.0 μg./ml. concentration. The 50.0 μg./ml. concentration is used as the reference or correction point and is included on each plate. Fill, on each of three plates, three of the six cylinders (alternately) with the 50.0 μg./ml. reference standard and the other three cylinders with one of the standard concentrations. Thus, 36 determinations of the 50.0 μg./ml. reference point and 9 determinations for each of the other points on the curve are obtained. Incubate the plates for 16 to 18 hours at 32° to 35°C. and measure

[12] J. L. Behmer, Inc., Philadelphia, Pennsylvania.

the zones of inhibition with a Fisher-Lily zone reader. Other equipment such as a caliper, millimeter rule, or a projection instrument may be used.

For each set of three plates at each concentration, obtain the average zone diameter for the 50 μg./ml. reference concentration and for the standard concentration being tested. Also average the 36 readings of the 50 μg./ml. reference standard for all 12 plates. Add to or subtract from each standard concentration average zone diameter the difference by which the 50 μg./ml. average zone diameter associated with that standard concentration falls short of or exceeds the average of all 36 readings of the 50 μg./ml. concentration. If, for example, in the case of the three plates bearing the 40 μg./ml. concentration, the average zone diameter of the 50 μg./ml. concentration (3 zones on each of 3 plates) is 18.3 mm. and the average zone diameter of the 50 μg./ml. concentration on all 12 plates is 18.0 mm., 0.3 mm. is subtracted from the average zone diameter of the 40 μg./ml. concentration. If the average zone diameter of the 40 μg./ml. concentration is 16.4 mm., the corrected value would be 16.1 mm. Plot the corrected value for the 32.0, 40.0, 62.5 and 78.1 μg./ml. concentrations, and the average of the 50 μg./ml. concentration on two-cycle semilog paper. The concentrations are plotted on the ordinate (log scale) and the zone diameters on the abscissa. Draw the standard curve through these points, or as we prefer to do, draw it through two zone diameter values determined by the following equations:[13]

$$L = \frac{3a + 2b + c - e}{5}, \qquad H = \frac{3e + 2d + c - a}{5},$$

where

L = calculated zone diameter for the lowest concentration,

H = calculated zone diameter for the highest concentration,

c = average zone diameter of 36 readings of the 50 μg./ml. standard and,

a, b, d, e = corrected average values for the 32.0, 40.0, 62.5 and 78.1 μg./ml. standard solutions, respectively.

G. Determination of Sample Potency

Sample treatment will vary according to the nature of the product being assayed. Extractions with various organic solvents are required in some cases (ointments, creams, etc.). Refer to the F.D.A. regulations for exact and official procedures for handling such forms. Most powders and liquids need no special treatment, however, and can be easily dissolved and diluted with buffer. Final dilution of the sample is made in 1% phosphate buffer, pH 6.0, to an estimated concentration of 50.0 μg./ml. Three plates are required for each sample. As above, three cylinders on each

[13] J. Deutschberger and A. Kirshbaum, *Antibiotics and Chemotherapy* **9,** 752 (1959).

plate are filled with the 50.0 μg./ml. standard and three cylinders with the sample. The plates are incubated for 16 to 18 hours at 32° to 35°C. and the zones of inhibition measured and recorded. The zone diameters for the sample and for the 50 μg./ml. standard are each averaged. If the average diameter for the sample is larger than for the 50 μg./ml. standard concentration, the difference is added to the zone diameter read from the standard curve at the 50 μg./ml. level. If the average diameter for the sample is smaller than for the 50 μg./ml. standard concentration, the difference is subtracted from the zone diameter read for the 50 μg./ml. level. This latter value, in either case, is then read from the standard curve to obtain the corresponding chloramphenicol concentration (μg./ml.) in the diluted sample. This concentration is multiplied by the appropriate dilution factor to obtain the potency of the original sample.

H. Esters of Chloramphenicol

Chloramphenicol palmitate and chloramphenicol succinate are microbiologically inactive *in vitro*. The potency of these preparations is determined by chemical methods.

III. Turbidimetric Assay Method

The turbidimetric assay for chloramphenicol generally employs *Shigella sonnei* ATCC 11060 as the test organism.[14] When circumstances preclude the use of this organism, as in some children's hospitals, *Agrobacterium tumefaciens* (Parke, Davis culture No. 05057) can be used. Although less sensitive to chloramphenicol than the recommended *Shigella sonnei* culture, *Agrobacterium tumefaciens* is nonpathogenic for humans. Directions for use of the latter culture, where different from those for *Shigella sonnei*, are included in brackets [ ] in the following description. *Escherichia coli* has also been used.[15]

A. Inoculum

The stock culture is maintained by weekly transfer on freshly slanted nutrient agar, incubation at 37°C. for 24 hours and subsequent storage in the refrigerator. A rapidly growing culture is used for assay, and preparation of inoculum must be started a day prior to use. Do this by suspending a loopful of organisms from the stock culture slant in a tube containing 10.0 ml. of brain heart infusion broth,[16,17] inoculating a second tube of

[14] D. A. Joslyn and M. Galbraith, *J. Bacteriol.* **59,** 711 (1950).

[15] D. M. Wintermere, W. H. Eisenberg, and A. Kirshbaum, *Antibiotics and Chemotherapy* **7,** 189 (1956).

[16] Baltimore Biological Laboratories, Inc.: Brain Heart Infusion.

[17] Difco Laboratories, Inc.: Brain Heart Infusion.

brain heart infusion broth with 0.3 ml. of the suspension, and incubating the second tube overnight at 37°C. Transfer 0.4–0.6 ml. [*Agrobacterium tumefaciens:* 1.0 ml.] of the resulting broth culture to a second tube containing 10.0 ml. of brain heart infusion broth and incubate the second tube 1.5–2.5 hours [*A. tumefaciens:* 2.5–3.5 hours] at 37°C. until the culture has 73% light transmission on the Coleman junior spectrophotometer or 34–36% on the Lumetron Model 402E at a wavelength of 575 mμ, when the machines are adjusted to 100% transmission with an uninoculated tube of broth. This culture will contain approximately 2.9×10^8 viable cells [*A. tumefaciens:* 2.4×10^7]. Dilute the resultant culture 1:8 with additional brain heart infusion broth and store it in an ice bath until needed to inoculate the test.

B. Standard

Chloramphenicol is assayed gravimetrically. Use any pure chloramphenicol powder (1000 μg./mg.) as a working standard, storing it over a desiccant at 5°C. Prepare a stock solution by dissolving an accurately weighed portion of the standard in a small amount of methanol or ethanol and adding sufficient sterile distilled water to give a final solution containing exactly 50 μg./ml. [*A. tumefaciens:* 200 μg./ml.]. The resultant solution can be kept in the refrigerator or frozen for at least a month without detectable loss of potency.

C. Sample Preparation

The turbidimetric assay has proved successful in the assay of fermentation beers, fractionation samples, pharmaceutical preparations such as capsules, ointments and creams, ophthalmic and otic solutions, and suppositories, and clinical specimens such as blood, spinal fluid, urine, fecal suspensions, and tissue extracts. Fermentation samples, fecal suspensions, and other heavily contaminated specimens should be sterilized by Seitz filtration prior to assay. Most other samples need not be sterilized so long as they are free of gross contamination and visible particulate matter.

Most pharmaceutical forms are easily dissolved in water. Where difficulty is encountered, dissolve the sample in a small amount of methanol or ethanol and dilute with distilled water. Dissolve forms with petrolatum-type bases in a small amount of petroleum ether and recover the chloramphenicol by shaking with one or more small volumes of sterile distilled water. Refer to the F.D.A. regulations for exact and official procedures for handling of specific pharmaceutical preparations.

Microbiologically active product forms of chloramphenicol are assayed antimicrobially. Microbiologically inactive product forms, such as the palmitate and succinate esters, are generally assayed by physical chemical methods.

D. Standard and Test Solutions

Dilute the 50 μg./ml. working standard solution 1:10, 15, 20, 30, and 40 in sterile distilled water [A. *tumefaciens:* 200 μg./ml. working standard solution 1:20, 30, 40, 60, 80, and 100] and pipet 1.0-ml. volumes of each into single tubes in each of two rows of five matched test tubes.[18] Pipet 1.0 ml. of sterile distilled water into a sixth matched test tube in each of the duplicate rows. These latter tubes serve as culture controls.

Make similar dilutions of the unknown sample, calculated from the estimated potency, and pipet 1.0-ml. volumes of these dilutions into a single series of five (to nine) matched test tubes. Again, pipet 1.0 ml. of sterile distilled water into a sixth (or last) tube, the culture control.

E. Assay Procedure

Perform the assay in brain heart infusion broth. Add 8.8 ml. of 1.1 strength medium to each of the 12 tubes of standard and the 6 tubes representing each sample. Inoculate the test by pipetting 0.2 ml. of the test culture, prepared as described above, into each of the standard and sample tubes of the test including culture control tubes. As each row is inoculated, transfer it to a 37°C. water bath to incubate about $3\frac{1}{2}$ hours [A. *tumefaciens:* about $2\frac{1}{2}$ hours] until its culture control attains 35–39% light transmission in the Lumetron or 69–72% in the Coleman junior spectrophotometer. When the culture control of each reaches this stage, take the entire row from the water bath, dry the tubes on a towel, and shake them well. A Vortex Junior Mixer (Scientific Industries, Inc.) is very effective and convenient for shaking the tubes but manual shaking is equally satisfactory. Measure and record the per cent light transmission of each tube. Throughout the test the colorimeter or spectrophotometer is set to read 100% light transmission through a matched tube containing 9.0 ml. of culture media plus 1.0 ml. of water (or of sample, if sample is colored or turbid). Each tube of broth, the light transmission of which is now being measured, must be aligned in the machine exactly as it was aligned at the time of its original selection for use in the assay. Hence the need for etching or otherwise marking the tubes when they were selected.

Convert the light transmission values to optical density values by means of a table. Calculate the per cent growth of each tube of standard and

[18] Matched test tubes are regular test tubes which, when filled with brain heart infusion broth, will give a uniform per cent transmission (usually 50 ±0.5%) in the spectrophotometer or colorimeter when the tube is rotated to some suitable angle in the light beam. When such tubes are found, a vertical line is drawn at the top with a wax pencil and the tubes are later etched at that point with a file, grinder, or diamond pencil. The etched line indicates the position in which the tube is to be inserted in the colorimeter during the assay. On the average, approximately 75% of all new test tubes checked will meet this requirement.

unknown by dividing the optical density of the test dilution by the optical density of the culture control and multiplying this quotient by 100. Average the per cent growth values of duplicate tubes of standard and plot these averages on a linear scale against the corresponding dilutions of standard on a logarithmic scale. Draw a smooth curve to connect these points. Use this standard curve to convert growth values in the sample tubes to equivalent dilutions of the 50 μg./ml. [*A. tumefaciens:* 200 μg./ ml.] chloramphenicol standard. Divide the value read from the curve by the actual dilution of the unknown and convert this quotient to its logarithm. Determine this latter value for as many dilutions of the unknown as can be read from the standard curve and average these logarithmic values for each sample. Subtract the average logarithmic value for each sample from the logarithm of 50 (standard solution = 50 μg./ml.) [*A. tumefaciens:* Since standard solution = 200 μg./ml., subtract average logarithmic value for each sample from the logarithm of 200]. The antilog of the difference is the assayed potency of the unknown solution in micrograms per milliliter. Replicate assays usually agree within 5% of their mean.

6.5 Dihydrostreptomycin

FREDERICK KAVANAGH AND L. J. DENNIN

Eli Lilly and Company, Indianapolis, Indiana

I. Introduction

Dihydrostreptomycin is the product obtained by reducing the aldehyde of the sugar in streptomycin to a primary alcohol group by catalytic hydrogenation. It is somewhat more stable than streptomycin and is more easily purified. Its chemical and antibacterial properties are similar to those of streptomycin.

Dihydrostreptomycin is soluble in water and nearly insoluble in the usual organic solvents. The usual salts of commerce are: $C_{21}H_{41}N_7$-

$O_{12} \cdot 3HCl$ (mol. wt. 693) and $(C_{21}H_{41}N_7O_{12})_2 \cdot 3H_2SO_4$ (mol. wt. 1461.4). The assay is in terms of the base (mol. wt. 583.6) as 1000 $\mu g./mg.$ On this basis the theoretical assay of the chloride is 842 $\mu g./mg.$ and of the sulfate is 799 $\mu g./mg.$

The range of the *Bacillus subtilis* plate assay is from 0.1 to 4 $\mu g./ml.$, for the *Staphylococcus aureus* turbidimetric assay from 3 to 40 $\mu g./ml.$, and for the *Klebsiella pneumoniae* assay from 7 to 14 $\mu g./ml.$ of sample.

II. Turbidimetric Assay

A. Test Organism

1. Stock Culture

Klebsiella pneumoniae (ATCC 10031) cultivated and treated as is *Staphylococcus aureus* in the penicillin assay (Chapter 6.10).

Staphylococcus aureus (ATCC 9996) may also be used with advantage in reading the test.

2. Standard Inoculum

Prepare as for the penicillin assay using either organism.

B. Standard Solutions

Dihydrostreptomycin is assayed in terms of the free base. Weigh enough of the standard to give 25 mg. of free base into a 25-ml. volumetric flask, dilute to volume with pH 6 buffer. The stock solution is stored in the refrigerator and may be used for a period not to exceed 1 month.

C. Preparation of Samples

Dissolve weighed sample of solids in pH 6 buffer and dilute to an estimated 15 $\mu g./ml.$ for the *S. aureus* assay. Dilute liquid samples with pH 6 buffer to a concentration of 15 $\mu g./ml.$ (or 11 $\mu g./ml.$ for *Klebsiella pneumoniae*).

Weigh about 1 gm. of a sample of an ointment containing several antibiotics (dihydrostreptomycin, erythromycin, and penicillin, for example) into a separatory funnel, add 20 ml. ethyl ether and shake until the suspension is homogeneous (the dihydrostreptomycin is insoluble in ether). Extract the ether suspension once or twice with 50 ml. of pH 6 buffer, acidify the extract to pH 2.0 with HCl, and let stand at room temperature for 1 hour to destroy the erythromycin. If the penicillin is not an acid stable one, it too will be destroyed. Neutralize the solutions to about pH 6.0, dilute to 100 ml. with pH 6 buffer. Dilute to an estimated concentration of 15 $\mu g./ml.$ If the penicillin is acid stable, treat with penicillinase to destroy it and then assay.

D. Mechanics of the Assay

1. Design

Prepare standard solutions with concentrations of 0, 6, 10, 14, 20, and 30 μg./ml., or more, depending upon the test organism and the mode of measuring and expressing the response of the test organism (see Chapter 4). Pipet 0.5 ml. of standards or prepared samples into the tubes. Proceed as for the penicillin assay.

2. Medium and Inoculum

Inoculate nutrient broth with 0.5% by volume of *Klebsiella pneumoniae* or 3% *Staphylococcus aureus* inoculum prepared as in the penicillin assay.

3. Incubation

Same as the Penicillin assay.

E. Response and Answers

Proceed as for Penicillin. The results of the assay are obtained in terms of the free base.

F. Remarks

Samples may be frozen and held for weeks before assay. Although there are fewer interferences by other antibiotics when *Klebsiella pneumoniae* is the test organism, the *Staphylococcus aureus* assay is easier to read because it does not show streaming birefringence which interferes with rapid reading. Sensitivity of the assay can be increased by increasing the pH of the medium.

III. Plate Assay for Dihydrostreptomycin and Streptomycin

A. Introduction

These antibiotics are assayed under identical conditions, except that appropriate standards must be used; i.e., streptomycin standard for the assay of streptomycin, and dihydrostreptomycin standard for the assay of dihydrostreptomycin. As with penicillin, two plate methods are available; one for the assay of the pharmaceutical dosage forms and another slightly more sensitive method for the determination of the antibiotics in biological fluids. The greater sensitivity of the latter method is achieved merely by reducing the thickness of the "base" agar layer and by a slight reduction in the inoculum level.

B. Test Organism

Maintain stock cultures of *Bacillus subtilis* (ATCC 6633) by monthly transfers to fresh sterile slants of G. & R. No. 1. Incubate the freshly prepared slants at 37°C. for 16 to 24 hours and store at 4 to 6°C. until used in the preparation of the inoculum spore suspension.

C. Standard Solutions

Dry the appropriate working standard for 3 hours at 60°C. and a residual pressure of 5 mm. Hg or less in a tared weighing bottle with attached capillary tube. Weigh a quantity of the dried standard sufficient to make a stock solution containing 1000 μg./ml. of streptomycin (or dihydrostreptomycin) activity (base). Transfer this quantity to a suitable container and bring to appropriate volume with sterile pH 8 phosphate buffer. Store the stock standard solution at 4 to 6°C. for a period not to exceed 30 days.

D. Sample Preparation

Preparation of pharmaceutical samples generally presents no difficulties to the assayist. The antibiotic is readily soluble in water or pH 8 phosphate buffer, and is only occasionally found in combination with other antibiotics which would interfere in this assay, generally penicillin, which is readily inactivated by the enzyme penicillinase.

In the assay of blood serum the considerations discussed under Penicillin, Section III.D are applicable. The assay of either streptomycin or dihydrostreptomycin in milk, however, presents an interesting phenomenon. Levels as high as 0.5 μg./ml. of streptomycin or dihydrostreptomycin often fail to demonstrate measurable activity when diluted in a control milk. A twofold predilution of this control milk in pH 8 phosphate buffer usually will permit detection of antibiotic concentrations as low as 0.1 μg./ml. Because of this "binding" property, milk must be diluted before being used to prepare the standard response curve. Unknown samples of milk are similarly diluted prior to assay.

E. Mechanics of the Assay

1. Design

See Penicillin, Section III.E.1.

2. Medium

The two plate methods commonly employed for these antibiotics use G. & R. No. 5 for both "base" and "seed" layers. In the assay of pharmaceutical preparations, a "base" layer of 21 ml. and a "seed" layer of 4 ml. are used. In the assay of body fluids and tissues, the "base" layer is reduced to 10 ml.

3. Inoculum

Prepare an agar slant of the culture as described in Section III.B. Wash the growth on this slant from the agar surface using 2–3 ml. of sterile distilled water. Transfer the suspension to a Roux bottle containing approximately 300 ml. of sterile G. & R. agar medium No. 1 and distribute it evenly over the surface of this agar with the aid of sterile glass beads. Incubate the inoculated Roux bottle at 37°C. for 7 days. Harvest the resultant growth from the Roux bottle by washing the agar surface with 50 to 75 ml. of sterile distilled water. Heat the bacterial suspension for 30 minutes at 65°C., centrifuge, wash the cells three times with sterile distilled water, and resuspend in sterile distilled water. Repeat the heat treatment and again centrifuge, wash, and resuspend the resultant spores in 100 ml. of sterile distilled water. This spore suspension may be stored at 4° to 6°C. for at least 1 month.

Determine the concentration of inoculum required for "seeding" of assay plates by running test plates each time a new spore suspension is prepared. In my experience, this is generally in the range 0.3–0.6 ml. per 100 ml. of liquefied and cooled (48°C.) agar; the lower concentration being better suited to the serum assay.

4. Standards

Dilute samples of the stock standard solution in sterile pH 8 phosphate buffer (or appropriate biological diluents) immediately prior to use on the day of the assay. The choice of standard levels will, of course, determine the extent of dilution. The conventional assay (21 ml. "base" layer) yields a linear response in the range 0.5–4.0 μg./ml. The body fluid assay (10 ml. "base" layer) appears linear, in most cases, from about 0.1 to 4.0 μg./ml.

5. Sample

See Penicillin, Section III.E.5

6. Incubation

Incubate both assays at 37°C. for 16 to 24 hours. Also see Penicillin, Section III.E.6.

F. Measuring the Response

See Penicillin, Section III.F

G. Computation of Answers

See Penicillin, Section III.G

6.6 Erythromycin

FREDERICK KAVANAGH AND L. J. DENNIN

Eli Lilly and Company, Indianapolis, Indiana

I. Introduction

Erythromycin is the one of the group of macrolide antibiotics. It is a basic substance with a molecular weight of **734**. The free base is insoluble in water and soluble in such organic solvents as amyl acetate and chloroform. Erythromycin forms water soluble salts with weak and strong acids. It is unstable in highly acidic or alkaline solution and has its maximum stability in the range of pH 6 to 9.5. Its aqueous or alcoholic solutions lose considerable antibacterial activity upon storage even in the refrigerator for less than 1 week. It forms esters, for example, erythromycin ethyl carbonate and erythromycin propionate, which have little if any antibacterial activity until hydrolyzed.

Erythromycin decomposes rapidly at 25°C. in strongly acid solution as is shown by the following tabulation interpolated from that of Korecká.[1] $t_{1/2}$ is the half-life in minutes:

pH	2.0	2.5	3.0	3.5	4.0
$t_{1/2}$ (min.)	1.65	5.0	14.5	45	127

This sensitivity to acid is a useful property when relatively acid stable antibiotics, tylosin and streptomycin, for example, must be assayed in mixtures with erythromycin.

The antibacterial activity is against gram-positive bacteria and is usually assayed by means of its inhibition of *Staphylococcus aureus*. Either a turbidimetric (range 0.3–2.0 µg./ml.) or a plate method (range 0.5–2.0 µg./ml.) is used for assay.

II. Turbidimetric Method

A. Test Organism

Staphylococcus aureus, see Penicillin, Chapter 6.10, Section II.B.

B. Standard Solutions

Dry the standard sample at 60°C. and 5 mm. pressure for 3 hours. Weigh a 100-mg. sample of the standard as free base into a 100-ml. volumetric flask, add 20 ml. methyl alcohol to dissolve the erythromycin, and then dilute the solution to volume with pH 7 phosphate buffer. This stock solution should be stored in the refrigerator and used for not more than 1 week. Erythromycin is not very stable in aqueous or alcoholic solutions. If the solution must be kept for long periods, dissolve the sample in dry acetone and keep in the refrigerator. Always warm the acetone solution to room temperature before taking an accurately measured sample.

C. Activity and pH

Erythromycin is a basic substance; consequently, its antibacterial activity increases with increase in pH. The activity of erythromycin against *S. aureus* in the assay increases about fivefold for an increase of 1 unit of pH in the range from 6.5 to 7.5.

[1] E. Korecká, *Proc. Symposia Antibiotics, Praha, 1959* p. 355 (1960).

D. Preparation of Samples for Assay

1. Treatment of Samples

a. *Fermentation beers.* Filter or centrifuge and dilute with sterile pH 7 buffer to an estimated concentration of 0.7 μg./ml.

b. *Powders.* Dissolve the sample in methyl alcohol and dilute with sterile pH 7 buffer. Grind tablets (1–5) in a mortar and then wash into a virtis homogenizer with 100 ml. methyl alcohol and blend for 20 minutes. Make further dilutions with pH 7 buffer.

c. *Ointments.* Dissolve approximately 2 gm. of an ointment or one suppository in 50 ml. of petroleum ether and then extract with four 20-ml. lots of 80% v./v. aqueous methanol. Dilute the combined extracts to 100 ml. Dilute samples of this solution to 0.7 μg./ml. with pH 7 buffer solution.

d. *Erythromycin ethyl carbonate and propionate.* These two forms of erythromycin are assayed after hydrolysis. Weigh a sample large enough to contain about 100 mg. of the free base into a 100-ml. volumetric flask, dissolve in 40 ml. methyl alcohol, dilute to about 90 ml. with sterile water (if ethyl carbonate) or pH 8 buffer (if lauryl sulfate), place in a 60°C. water bath for 3 hours (or let stand at 25°C. for 48 hours) to achieve hydrolysis, cool, and dilute to 100 ml. with water. Make subsequent dilutions with pH 7 buffer to an estimated concentration of 0.7 μg./ml.

2. Samples Containing Other Antibiotics

a. *Penicillin.* Dissolve the finely ground sample in methanol, dilute with buffer to an estimated concentration of 7 μg./ml., place 1 ml. of this diluted solution in a 10-ml. volumetric flask, add about 2 ml. of buffer followed by enough penicillinase to inactivate 1000 units of penicillin G, let stand at room temperature for 15 to 30 minutes, and then dilute to volume with buffer.

b. *Other antibiotics.* No general procedure can be given. Advantage can be taken of the solubility of erythromycin base in organic solvents (amyl acetate, chloroform) at pH 9.5 to separate it from such solvent insoluble antibiotics as streptomycin.

E. Inoculum

Prepare as for Penicillin assay. Add 15 ml. of the standardized inoculum to each liter of erythromycin assay broth.

F. Conditions of the Assay

The range of the test is from 0.3 to 2.0 μg./ml. of erythromycin free base (0.5-ml. sample per assay tube). Standard solutions with a concentration of 0, 0.3, 0.4, 0.6, 0.8, 1.2, 1.6, and 2.0 should be enough to define the curve for any instrument. If the log-probability plot is used, then the minimum number of concentrations would be 0, 0.4, 0.75, 1.6, and 2.0 μg./ml. All final dilutions of standards and samples are made in pH 7 buffer.

An incubation time of 3 hours in the 36°–37°C. water bath should be enough to give a highly turbid O tube which would measure in the upper range of the turbidimeter.

G. Computation of Answers

The answers obtained by the method given here are in terms of erythromycin free base. The concentrations or amounts of other forms of erythromycin are obtained by multiplying the answers by a factor which is the ratio of the molecular weight of the other form to the molecular weight of erythromycin free base. The factors for some common derivatives of erythromycin are given in Table I.

TABLE I

THEORETICAL FACTORS FOR CONVERTING WEIGHT OF
ERYTHROMYCIN BASE INTO COMMON DERIVATIVES

	Mol. wt.	Factor
Erythromycin free base	733.92	1.000
Erythromycin ethyl carbonate	806.02	1.098
Erythromycin glucoheptonate	960.14	1.308
Erythromycin lactobionate	1092.26	1.487
Erythromycin propionate	791	1.077
Erythromycin propionate lauryl sulfate salt	1056	1.439

III. Plate Assay

A. Introduction

As with many antibiotics, two cylinder plate methods are in common usage for the assay of erythromycin. Both methods use the same test organism and the same plating media. The method used for serum or body fluid assays achieves somewhat greater sensitivity by the use of a thinner "base" agar layer and a slightly lower incubation temperature.

B. Test Organism

Maintain stock cultures of *Sarcina lutea* (ATCC 9341) by weekly transfers to fresh sterile slants of G. & R. agar medium No. 1. Incubate the freshly prepared slants at 26°C. for 16 to 24 hours and store at 4° to 6°C. until ready for use in the preparation of the inoculum suspension.

C. Standard Solutions

Dry the working standard at 60°C. and residual pressure of 5 mm. Hg or less for a period of 3 to 4 hours in a tared weighing bottle fitted with a capillary tube. Accurately weigh a quantity of the dried standard sufficient to make a stock solution containing 1000 μg. of erythromycin activity (base) per milliliter. Transfer the weighed portion to a suitable container, add sufficient reagent grade methyl alcohol to dissolve the erythromycin (approximate 1 ml./10 mg.), and bring to volume with pH 8 phosphate buffer. Store the solution at 4° to 6°C. for a period not to exceed 1 week.

D. Sample Preparation

Most salts and esters of erythromycin may be readily dissolved in small quantities of methyl alcohol and diluted to volume with pH 8 phosphate buffer. Esters such as the ethylcarbonate and propionyl lauryl sulfate must be hydrolyzed to erythromycin base brior to assay as described in Section II.D.1.d. Make subsequent dilutions with pH 8 phosphate buffer so that each milliliter of the final dilution will contain approximately 1.0 μg. of erythromycin activity.

In the assay of serum or biological materials, the standard samples are diluted in appropriate diluents (see discussion, Penicillin, Section III.D) to an estimated potency of 0.2 μg./ml.

Erythromycin may be extracted from animal feeds by blending in methanol. The extracts are filtered and the filtrate further diluted with pH 8 phosphate buffer. Care must be exercised to make the methanol concentration in the final dilution of the sample equal to that used in preparing the standard response curve.

E. Mechanics of the Assay

1. Design

Refer to Penicillin, Section III.E.1.

2. Medium

Both of the assays use the two-layer system. The conventional assay uses a "base" layer of 21 ml. of G. & R. agar medium No. 11 and a "seed" layer of 4 ml. of this same agar. The somewhat more sensitive serum assay uses a "base" layer of 10 ml. and a "seed" layer of 4 ml. of G. & R. agar medium No. 11.

3. Inoculum

Prepare an agar slant of *S. lutea* as described in Section III.B. Wash the growth from the slant with 3 to 4 ml .of G. & R. broth medium No. 3 and use this suspension to inoculate the surface of a Roux bottle containing 300 ml. of G. & R. agar medium No. 1. Distribute the inoculum evenly over the agar surface with the aid of sterile glass beads. Incubate the Roux bottle for 24 hours at 26°C. Wash the growth from the Roux bottle with 20 ml. of G. & R. broth medium No. 3 and adjust so that it contains approximately 5×10^8 viable cells per milliliter. The adjusted suspension constitutes the stock inoculum suspension. This stock inoculum suspension may be stored at 4° to 6°C. for a period of 2 weeks. To prepare the "seed" layer, add 0.3 to 0.5 ml. of the inoculum suspension to each 100 ml. of liquefied and cooled (48°C.) agar, and pour the plates immediately.

4. Standards

Immediately prior to use, dilute samples of the stock standard solution in sterile pH 8 phosphate buffer, in a mixture of pH 8 buffer and methanol, or in appropriate biological diluents, depending on the type of samples being run. The choice of standard levels will, of course, determine to what extent these samples are diluted. The conventional assay (21- and 4-ml. layers) yields a linear response in the range 0.5–2.0 µg./ml. and the serum assay (10- and 4-ml. layers) in most cases is linear over the range 0.02–1.0 µg./ml. The linear range will depend, to a great extent, on the quality of the diluent as was pointed out in Penicillin, Section III.D. Bovine albumin appears to exhibit excessive binding of this antibiotic; and, therefore, pooled human serum is the diluent of choice in the assay of samples of human serum. Doses should be evenly spaced on the logarithmic scale.

5. Samples

Refer to Penicillin, Section III.E.5 and to Section III.D of this method.

6. Incubation

The conventional assay is incubated at 32° to 35°C. and the serum assay at 26°C. Incubation times should be from 16 to 24 hours. Refer also to the discussion of incubation under Penicillin, Section III.E.6.

F. Measuring the Response

Refer to Penicillin, Section III.F.

G. Computation of Answers

See Penicillin, Section III.G.

6.7 Fumagillin

ROLAND L. GIROLAMI

Abbott Laboratories, North Chicago, Illinois

I. Introduction

Fumagillin is an antibiotic produced by certain strains of *Aspergillus fumigatus.* It was discovered as a result of its inhibitory action on *Staphylococcus* bacteriophage and described first by Asheshov et al.[1,2] who named it phagopedin sigma and later by Hanson and Eble[3,4] who called it fumagillin. The similarity of the two compounds was established subse-

---

[1] I. N. Asheshov, F. Strelitz, and E. A. Hall, *Can. J. Public Health* **39**, 75 (1948).
[2] I. N. Asheshov, F. Strelitz, and E. A. Hall, *Brit. J. Exptl. Pathol.* **30**, 175 (1949).
[3] F. R. Hanson and T. E. Eble, *J. Bacteriol.* **58**, 527 (1949).
[4] T. E. Eble and F. R. Hanson, *Antibiotics & Chemotherapy* **1**, 54 (1951).

quently.[5] The antibiotic has limited antibacterial or antifungal activity and little or no effect against several common animal viruses. It is of interest mainly because of its amebicidal action and has been used in the treatment of *Endamoeba histolytica* infections.

The antibiotic has the emperical formula $C_{26-27}H_{34-37}O_7$ and has been shown to be a monoester of decatetraenedioic acid $HOOC—(C=CH)_4—COOH$ and fumagillol $C_{16-17}H_{25-26}O_3$. The antibiotic is subject to oxidation and photolytic degradation so care must be taken to protect the compound from light and air if biological activity is to be maintained. It is poorly soluble in water, slightly soluble in ethanol and amyl acetate and soluble in chloroform, acetone and ether.

There are two assays for the antibiotic, a spectrophotometric method which utilizes the characteristic absorption of the compound in the ultraviolet region and a plate assay method which is based on the ability of the antibiotic to inhibit a bacteriophage in a mixture of cells plus specific, lytic phage. The response differs from the normal antibiotic plate assay result in that exhibition rather than inhibition zones are obtained.

In fairly pure materials the method of choice is the spectrophotometric assay because of its inherent reproducibility and reliability. Comparisons of the two methods on materials of varying potency have given good correlations. In the assay of complex materials such as fermentation beers, or of samples which contain low levels of the antibiotic together with UV absorbing materials, the plate assay method is preferred.

II. Plate Method

The biological system used in the plate assay of fumagillin is the bacterium *Staphylococcus aureus* ATCC 6538P (FDA 209P) and its specific phage ATCC 6538B (FDA 209).

A. Culture

Maintain the culture by biweekly serial transfer to slants of Antibiotic Medium No. 1 (Difco). Incubate for 24 hours at 37°C. Stock cultures are maintained in lyophilized form also and are referred to occasionally as a check on assay performance. Experience has shown that the culture is very stable; serial transfer over a period of several years has not altered the response of this strain.

B. Preparation of Test Organism

Inoculate a quantity of Antibiotic Assay Broth [BBL (Baltimore Biological Laboratories, Inc.)] from a stock slant and incubate on a rotary

[5] I. N. Asheshov, F. Strelitz, and E. A. Hall, *Antibiotics & Chemotherapy* **2**, 361 (1952).

shaker (150 r.p.m.) for 18 hours at 37°C. The cell concentration under these conditions will be about 0.5 to 1.0 × 10⁹ per milliliter. The culture can be refrigerated and used over several days.

C. Preparation of Phage

Inoculate several tubes of Heart Infusion Broth (Difco) to which has been added 0.005 M Ca⁺⁺ as the chloride with *S. aureus* 6538P and incubate for 8 hours at 37°C. Inoculate the culture with approximately 2% of a previously prepared suspension of phage 6538B and continue to incubate overnight. The culture tubes should be clear; if not, repeat the procedure using the final tubes as phage inoculum for the next cycle. Use the clear lysate at a level of 20% as inoculum for a freshly prepared 8-hour culture of *S. aureus* inoculated from a slant and continue incubation overnight. Transfer the phage suspension to small screw capped containers and store under refrigeration.

Phage suspensions are stable and have given an acceptable assay response after storage for 1 year under these conditions. The titer of the suspension should be established to aid determination of seeding levels.

D. Phage Titer

Prepare an 18-hour Heart Infusion Broth (Difco) culture of *S. aureus*. Add 1 ml. of a 1 to 10 dilution of this culture to petri plates containing Heart Infusion Agar (Difco) plus 0.005 M Ca⁺⁺. Remove excess moisture by pipet and dry the plates for 1 hour at 37°C. inverted with lids ajar. Prepare serial tenfold dilutions of the phage suspension in Heart Infusion Broth. Place one loopful (0.01 ml.) of the diluted suspension in a designated area of the dried plates. Up to 10 or 12 different dilutions or replicas are tested on a single plate. The phage forms a very small plaque, about 0.5 mm., so test areas containing up to 40 plaques are counted easily. The titer should be 1 × 10⁹ per milliliter or greater.

E. Standards

Fumagillin has been assigned a potency of 1000 μg./mg. Potency of the standard is established by the spectrophotometric assay (see Section III). Care must be taken to protect the standard from deterioration. Stock standards are sealed under nitrogen and stored in the refrigerator away from light. Working standards are kept in the refrigerator within tightly sealed brown bottles containing a desiccant. The potency of working standards decreases under these storage conditions but generally remains satisfactory for 2 to 3 months.

To prepare standard solutions, accurately weigh a sample of fumagillin standard and dilute to 2000 μg./ml. with acetone. Make further dilutions in 1% pH 7.5 phosphate buffer (see Table I) to give solutions containing

2.5, 5.0 (reference point), 10.0, and 20.0 μg./ml. Protect the solutions from light by using nonactinic glassware or by covering the vessels with aluminum foil or a dark cloth. Prepare fresh solutions for each assay.

TABLE I

Buffer

        Phosphate buffer pH 7.5 (1%)
           KH$_2$PO$_4$ 0.9 gm.
           K$_2$HPO$_4$ 9.1 gm.
        Distilled water to 1000 ml.

3-A Alcohol

        100 gal. 190° proof ethyl alcohol +5 gal.
          commercially pure methyl alcohol.

Media

        Antibiotic Medium No. 1 (Difco)
        Antibiotic Assay Broth (BBL)
        Heart Infusion Broth (Difco)
        Heart Infusion Agar (Difco)

F. Sample Preparation

There are only a limited number of pharmaceutical forms of the antibiotic. It is available in either powder or tablet form as the pure acid, a salt such as bicyclohexylamine or a mixture of acid plus a protective agent. None contains substances which interfere with the assay.

1. Powders

Handle powders the same as the standard. Dissolve in acetone and dilute in buffer to a theoretical potency of 5 μg./ml.

2. Tablets

Crush four 10-mg. tablets in a tube type tissue grinder. Add 20 ml. of acetone and stir until dissolved. Further dilute with phosphate buffer pH 7.5 to give a fumagillin concentration of 5 μg./ml.

3. Fermentation Beers

Centrifuge fermentation samples and dilute an aliquot of the supernatant in phosphate buffer pH 7.5 to a theory of 5 μg./ml.

4. Feeds

Animal feeds or other complex mixtures which contain as little as 0.01% of the antibiotic may be assayed by the plate method.

Add to the cup of a high-speed mechanical blender 5 gm. of sample together with 100 ml. of phosphate buffer pH 7.5. Blend for 3 minutes. Allow to settle, filter if necessary, and dilute an aliquot to a theoretical potency of 5 μg./ml. Samples which contain more than 0.1% fumagillin require a primary extraction in alcohol before dilution because of the low solubility of the antibiotic in buffer.

G. Mechanics of Assay

Prepare a sufficient quantity of Heart Infusion Agar (Difco) and after sterilization add a sterile solution of $CaCl_2$ to give a final concentration of 0.005 M Ca^{++}. Cool to 48°C.

Mix a quantity of *S. aureus* suspension (Section B) together with sufficient bacteriophage suspension to give a phage to cell ratio between 0.3 and 10, usually 1 to 1. Add approximately 2% by volume of the inoculum mixture to the cooled agar and distribute evenly to flat-bottomed petri plates, 5 ml. per plate. Allow plates to cool, invert and store in a refrigerator until used. Plates should be used the same day they are prepared.

If the phage to cell ratio is too low poor cell lysis will occur and zone edges will be difficult to read. If too high the fumagillin will not inactivate all the phage as it diffuses into the agar and no zones will be obtained. Slight variations in the seeding level may be required as judged by assay performance.

H. Standard Curve and Sample Assay

1. Standard Curve

The plate assay is run using filter paper disks (Schleicher and Schuell Company, No. 740-E, 12.7 mm.) to which is added 0.08 ml. of the sample to be assayed. Place one disk of each concentration of the standard solutions, 2.5, 5.0, 10.0, and 20.0 μg./ml., on each of 10 plates. Invert the plates and incubate for 16 to 18 hours at 37°C protected from light. Measure the size of exhibition zones with a suitable reader. Construct a standard curve with the average zone diameters plotted arithmetically against the standard concentrations plotted logarithmically on semilogarithmic paper. The slope of this curve, that is, the increase in zone diameter per doubling of antibiotic concentration, should be 2.0–2.4 mm.

2. Unknowns

Use four plates for each sample. Place two disks of the sample (theory, 5 μg./ml.) and two disks of the 5.0 μg./ml. reference standard alternately on each plate. Invert the plates and incubate for 16 to 18 hours at 37°C. Measure the zone size and average the zone diameter for the two standards and for the two samples. Correct for plate variation: if the plate standard gives a larger average zone size than that corresponding to the 5.0 μg./ml. reference point of the standard curve subtract this difference from the average sample diameter, if the plate standard gives a smaller average zone size add the difference to the average sample zone size. Use the corrected sample zone size and read the antibiotic concentration directly from the standard curve. Multiply by appropriate dilution factors to obtain the potency of the sample.

III. Spectrophotometric Method

Pure fumagillin exhibits a characteristic absorption spectrum in the ultraviolet with peaks at 335 and 351 mμ. The slightly greater absorbance at 335 mμ is used for the quantitative assay of this antibiotic. This method, a modification of the procedure of Garrett and Eble,[6] is used to establish potency of standards and of other materials which contain a relatively high level of antibiotic in relation to extraneous materials which absorb in this region of the spectrum.

Add approximately 100 mg. of sample, accurately weighed, to a 100-ml. volumetric flask and dissolve in 15 ml. of reagent grade chloroform. Dilute the sample to volume with alkali distilled 3-A alcohol. Dilute a 3-ml. aliquot of this solution to 1000 ml. with alkali distilled ethanol or denatured alcohol 3-A to obtain a solution containing approximately 3 μg. of fumagillin per milliliter. Determine the absorbance of this solution in a Beckman DU spectrophotometer in a 1-cm. cell at 335 mμ using alkali distilled 3-A alcohol (see Table I) to set the instrument to 100% transmittance. Protect against light destruction of the sample during preparation.

Calculation:

$$\frac{\text{Absorbance at 335 m}\mu}{\text{Sample weight (mg.)} \times 0.000003} \times \frac{1000}{1560}$$

$$= \mu\text{g. of fumagillin per milligram}$$

$$E_{1\,\text{cm.}}^{1\%} = 1560 \text{ at } 335 \text{ m}\mu.$$

IV. Other Methods

There is no turbidimetric method for determining fumagillin activity; however, a tube dilution procedure against *Endamoeba histolytica* can be used to demonstrate low levels of the antibiotic. In this method serial tenfold dilutions of the antibiotic are added to an egg yolk, rice powder medium. The tubes are inoculated with *E. histolytica* and incubated at

[6] E. A. Garrett and T. E. Eble, *J. Am. Pharm. Assoc. Sci. Ed.* **43,** 385 (1954).

37°C for 48 hours. The assay end point (highest dilution which inhibits growth) must be read microscopically because of the turbidity caused by growth of the associated bacteria. The end point varies with the strain of *E. histolytica* used and with the inoculum level but inhibition can be demonstrated at fumagillin dilutions over 1:1,000,000. Further refinements in this assay are necessary before it can be recommended as an analytical procedure.

6.8 Hygromycin B

L. J. DENNIN

Eli Lilly and Company, Indianapolis, Indiana

I. Introduction

Hygromycin B is assayed by a two-layer cylinder plate method using *Bacillus subtilis* as the test organism.

II. Test Organism

Maintain stock cultures of *B. subtilis* (ATCC **6633**) by monthly transfers to fresh sterile slants of G. & R. agar medium No. 1. Incubate the freshly prepared slants at 37°C. for 16 to 18 hours and store at 4° to 6°C. until ready for use in the preparation of the inoculum suspension.

III. Standard Solutions

Dry the hygromycin B working standard for 3 hours at 60°C. and a residual pressure of 5 mm. Hg or less in a tared weighing bottle fitted with capillary tube. Accurately weigh a quantity of the dried standard to make a solution containing 1000 units/ml. of hygromycin B activity (base). Transfer this quantity of standard to a suitable container and bring to volume with pH 7 phosphate buffer. Store the stock standard solution at 4° to 6°C. for a period not to exceed 2 weeks.

IV. Sample Preparation

A. General

The antibiotic hygromycin B will generally be seen by the analyst in the form of a finished animal feed or a feed premix. The premixes are extracted by blending in pH 7 phosphate buffer and do not normally present an assay problem. Because of the relatively low order of anti-microbial activity of the antibiotic, the assay of feeds presents two problems to the analyst: the extraction of the active ingredient, and its concentration to a level at which it can be measured. The complete procedure for accomplishment of the foregoing tasks is described below.

B. Animal Feeds

Accurately weigh a quantity of the feed to be assayed (for exact quantity recommended see Table I) and place in a suitable blendor jar. Add an

TABLE I

FEED CLASSIFICATION CHART

Sample potency range (units/lb.)	Sample weight (gm.)	Buffer volume for extraction (ml.)	Sample size for column (ml.)	Final volume (ml.)
6,000–12,000	50	300	100	10
18,000–24,000	30	500	100	10
30,000–36,000	20	500	100	10
42,000 or greater	20	500	75	10[a]

[a] Or a volume which will result in a final concentration of approximately 25 units/ml.

accurately measured volume of pH 7 phosphate buffer (for exact quantity see Table I) and blend at high speed for a minimum of 5 minutes. Transfer the resultant slurry to two 250-ml. centrifuge bottles and centrifuge at 2600 r.p.m. for 10 minutes. Pool the supernatants and transfer to a suitable container. Adjust the pH of the supernatant to 5.0 with concentrated hydrochloric acid. Transfer to two 250-ml. centrifuge bottles, each containing 50 ml. of chloroform, stopper, and shake thoroughly. Centrifuge at 2600 r.p.m. for 10 minutes. Draw off and pool the aqueous phases in a suitable container. Adjust the pH of the aqueous phase to 7.0 using 40% sodium hydroxide solution. Transfer a measured aliquot (for recommended quantity see Table I) of the neutral solution to the reservoir of an ion-exchange column prepared as directed below. Allow the sample to flow through the column at the rate of 40 drops per minute. Wash the column with four 20-ml. aliquots of sterile distilled water. Discard the effluent and washings. Elute the antibiotic from the resin with 50 ml. of a 3% ammonia solution (1:10 dilution of concentrated ammonium hydroxide). Collect the eluate in a Pyrex beaker and place on a hot plate until all the ammonia has been evolved and the volume is reduced to approximately 3–5 ml. Adjust the pH of the sample to 7.0 with $1\,N$ hydrochloric acid and wash into an appropriate size of volumetric flask (see Table I for recommended volume). Bring to volume with pH 7 phosphate buffer. Make further dilutions of the sample, if necessary, with pH 7 phosphate buffer.

1. Preparation of Ion-Exchange Resin

The ion-exchange resin used, IRC-50 (Rohm and Haas Company), is regenerated in the following manner. Slurry the resin in a $1\,N$ sulfuric acid solution for 3 hours. Wash the resin with distilled water until the pH of the water supernatant is above 5.0. Add slowly, and while stirring, solid lithium hydroxide until the pH remains constant and between 7 and 8. Allow to stand overnight. Wash the resin at least 5 times with fresh distilled water. Adjust the pH to 7.0 with $1\,N$ phosphoric acid. Store under distilled water in a tightly closed container until ready for use.

2. Preparation of Ion-Exchange Column

Select an ion-exchange column of the following dimensions: I.D., 6 mm.; length, 140 mm.; solution reservoir, approximately 50 ml. The column should be fitted with a small glass wool plug and suitable valve or screw clamp to control the flow rate. Fill the column with freshly distilled water and add sufficient resin to just fill the column when the resin has settled. Drain the water from the column until the water level is approximately 5 mm. above the resin surface. Immediately prior to use, wash the resin with approximately 25 ml. of freshly distilled water.

3. Special Considerations

It should be noted that in the preparation of samples of feeds, all pH adjustments, except the final one, are accomplished with concentrated reagents. This is done to minimize to volume changes involved which generally may be disregarded in calculating the dilutions made. A typical dilution, as recommended, would be that of a 6000-u./lb. feed; i.e., $50:300 \times 100:10 = 1:0.6$, which, of course, is actually a concentration of the antibiotic rather than a dilution. This is necessary in order to reach the approximate test level (about 25 u./ml.) with samples of such low potency.

As with most ion-exchange techniques, there may be some loss of the antibiotic. It is good practice to check each newly purchased or regenerated batch of resin to determine whether significant column loss may be anticipated in its use. For additional confidence, a standard should be run through the column system and assayed along with the samples each day. If the column loss is significant, and if it can be shown to be constant for a given batch of resin, one may have justification for the use of a correction factor.

In my experience, the assay for hygromycin in feeds is a workable method only when *all* details are strictly controlled. If difficulties are experienced in the conduct of the assay, check each step carefully for compliance with the method.

V. Mechanics of the Assay

A. Design

See Penicillin, Chapter 6.10, Section III.E.1.

B. Medium

The "base" agar layer consists of 10 ml. of G. & R. agar medium No. 5, while the "seed" layer requires 4 ml. of this same agar.

C. Inoculum

Prepare an agar slant of the culture as described in Section II. Wash the growth on this slant from the agar surface using 10 ml. of sterile Trypticase Soy broth [BBL (Baltimore Biological Laboratory)] made to contain 4 p.p.m. manganese chloride. Transfer this washing to a flask containing 100 ml. of sterile medium of this same composition and incubate for 48 hours on a mechanical shaker at 37°C. After this incubation period, transfer a 5.0-ml. aliquot of the culture to a Roux bottle containing approximately 300 ml. of sterile modified Trypticase Soy agar. This medium is identical to the above-mentioned broth except that it contains 2.0% agar. Distribute the inoculum evenly over the surface of the agar with the aid of sterile glass beads. Incubate the inoculated Roux bottle at 37°C. for 7 days. Harvest the resultant growth from the Roux bottle by washing the agar surface with approximately 50 ml. of sterile distilled water. Heat the bacterial suspension for 20 minutes at 65°C., centrifuge, decant, and resuspend the cells in 50 ml. of sterile distilled water. Repeat the heating, centrifugation, and washing steps two additional times to ensure a well-prepared spore suspension. This spore suspension may be stored at 4° to 6°C. for at least 1 month.

Determine the concentration of inoculum required for "seeding" of assay plates by running test plates each time a new spore suspension is prepared. In my experience, this is generally in the range 0.1–0.3 ml. of a 1:10 dilution of the stock suspension per 100 ml. of liquefied and cooled (48°C.) agar.

D. Standards

Dilute samples of the stock standard solution in sterile pH 7 phosphate buffer immediately prior to use on the day of the assay. The choice of standard levels will, of course, determine to what extent these samples are diluted. This assay yields a linear response in the range 20–100 u./ml. Standard levels should be evenly spaced on the logarithmic scale.

E. Samples

See Penicillin, Section III.E.5.

F. Incubation

Incubate the assay plates for 16 to 18 hours at 37°C. Also refer to Penicillin, Section III.E.6.

VI. Measuring the Response

See Penicillin, Section III.F.

VII. Computation of Answers

See Penicillin, Section III.G.

6.9 Neomycin

L. J. DENNIN

Eli Lilly and Company, Indianapolis, Indiana

I. Introduction

Two organisms may be used for the plate assay of neomycin. The *Staphylococcus aureus* method is commonly used for the assay of most pharmaceutical dosage forms although a few require the tenfold greater sensitivity afforded by the *Staphylococcus epidermidis* method. Sensitivity of the latter method may be increased by a slight reduction in both the thickness of the "base" agar layer and the size of the inoculum. Under the last-mentioned conditions, the assay is suitable for the measurement of neomycin in serum and other body fluids.

II. Test Organisms

Maintain stock cultures of *S. aureus* (ATCC 6538P) and of *S. epidermidis* (ATCC 12228) by weekly transfers to fresh sterile slants of G. & R. agar medium No. 1. Incubate the freshly prepared slants at 32° to 35°C. for 16 to 24 hours and store at 4 to 6°C. until ready for use in the preparation of the inoculum suspension.

III. Standard Solutions

Dry the neomycin working standard for 3 hours at 60°C. at a residual pressure of 5 mm. Hg or less in a tared weighing bottle fitted with a capillary tube. Accurately weigh a quantity of the dried standard to make a solution containing 1000 µg./ml. of neomycin activity (base). Transfer this portion to a suitable container and bring to volume with pH 8 phosphate buffer. Store the stock standard solution in a refrigerator at 4° to 6°C. and use for a period not to exceed 30 days.

IV. Sample Preparation

Since neomycin sulfate (probably the most frequently encountered form of the antibiotic) is quite soluble in water and relatively insoluble in petroleum ether, diethyl ether, and chloroform, few problems are encountered in preparing samples for assay. The antibiotic may be found in combination with other antibiotics, such as penicillin, streptomycin, or dihydrostreptomycin, which would interfere in the *S. aureus* assay. These can usually be inactivated enzymatically, chemically, or merely by dilution. The last-named is often possible if the more sensitive *S. epidermidis* assay is employed.

V. Mechanics of the Assay

A. Design

See Penicillin, Chapter 6.10, Section III.E.1.

B. Medium

Both the *S. aureus* and *S. epidermidis* assays use a two-layer agar system. The "base" agar layer consists of 21 ml. of G. & R. agar medium No. 11, while the "seed" layer requires 4 ml. of this same agar. To increase the sensitivity of the *S. epidermidis* method, reduce the "base" layer to 10 ml. of G. & R. No. 11.

C. Inoculum

1. Staphylococcus aureus

Prepare an agar slant of the culture as described in Section II. Wash the growth on this slant from the agar surface using 2–3 ml. of sterile 0.85% sodium chloride solution. Transfer the suspension to a Roux bottle containing approximately 300 ml. of sterile G. & R. agar medium No. 1, and distribute it evenly over the surface of this agar with the aid of sterile glass beads. Incubate the inoculated Roux bottle at 32° to 35°C. for 24 hours. Harvest the resultant growth from the Roux bottle by washing the agar surface with approximately 50 ml. of sterile 0.85% sodium chloride solution. Adjust the inoculum suspension to contain approximately 1×10^8 viable cells/ml. and store at 4° to 6°C. for a period not to exceed 2 weeks. Inoculate the "seed" agar by adding 0.3–0.5 ml. of this suspension to each bottle of 100 ml. of liquefied and cooled (48°C.) agar and pour into plates immediately.

2. Staphylococcus epidermidis

Proceed exactly as for preparation of *S. aureus* except harvest the growth from the Roux bottle with only 30–35 ml. of sterile 0.85% sodium chloride solution. Adjust the inoculum suspension to contain approximately 1×10^7 viable cells/ml. and store at 4° to 6°C. for a period not to exceed 2 weeks. Inoculate the "seed" agar by adding from 0.3 to 0.5 or 1.0 to 1.5 ml. of this suspension per 100 ml. of liquefied and cooled (48°C.) agar. The lighter inoculum range mentioned above must be used for the serum assay which also employs somewhat lower standard levels of neomycin than does the conventional assay.

D. Standards

Dilute samples of the stock standard solution in sterile pH 8 phosphate buffer immediately prior to use on the day of the assay. The choice of standard levels will, of course, determine to what extent these samples are diluted. The *S. aureus* assay yields a linear response (semilog plot) in the range 5–20 µg./ml. and the *S. epidermidis* method in the range 0.5–2.0 µg./ml. The serum assay using *S. epidermidis* normally has a linear response in the range 0.02–1.0 µg./ml. assuming a careful selection of the "control" diluent. Standard levels should be evenly spaced on the logarithmic scale.

Bovine albumin is an unsatisfactory diluent of serum samples because of its excessive "binding" of the antibiotic. Pooled human serum is the diluent of choice when assaying samples of human serum. In general, serum samples should be diluted with serum of the same species.

Bovine albumin is a convenient diluent but it may destroy the value of the assay.

E. Samples

See Penicillin, Section III.E.5.

F. Incubation

Incubate the assay plates for 16 to 18 hours at 32° to 35°C. Also refer to Penicillin, Section III.E.6.

VI. Measuring the Response

See Penicillin, Section III.F.

VII. Computation of Answers

See Penicillin, Section III.G.

6.10 Penicillins

FREDERICK KAVANAGH AND L. J. DENNIN

Eli Lilly and Company, Indianapolis, Indiana

I. Introduction

Penicillins may be thought of as N-acyl derivatives of 6-aminopenicillanic acid. The acyl group of naturally occurring penicillins are derivatives of acetic acid. Of the very many biosynthetic penicillins, two are of primary importance in medical practice. One has phenylacetic acid; the other, phenoxyacetic acid, as the acyl group. Recently, acyl groups not derived from acetic acid have been attached to 6-aminopenicillanic acid by standard chemical methods for acylating an amino group. The number of commercially important penicillins produced by chemical synthesis will increase yearly. The 6-aminopenicillanic acid portion of the penicillin molecule contains the β-lactam ring which gives the molecule its instability (and bioactivity?) and the free carboxyl group (pK_a 2.8). The β-lactam ring is opened by acids, alkalies, and penicillinase to form the biologically inactive penicilloic acids. The structure of the side chain affects the rate of hydroylsis of the β-lactam structure by acid or penicillinase. For example, phenoxymethylpenicillin is more stable in acid solution than penicillin G and some of the synthetic penicillins are hydrolyzed much more slowly by penicillinase than are naturally occurring penicillins. Penicillin solutions always need protecting against acids, alkalies, heavy metals, primary alcohols, and penicillinase-producing bacterial contaminants. Penicillin solutions in primary alcohols rapidly lose biological activity on standing at room temperature by opening of the β-lactam ring to form α-esters of d-α-penicilloic acid. Catalytic amounts of zinc, tin, or copper promote the reaction which can be prevented by small amounts of dimercaptopropanol [1,2] (BAL). Standard solutions of penicillins are made in phosphate or citrate buffers which act as stabilizing agents. The free acids of most penicillins are readily soluble in alcohols, ketones, esters, and ethers, and insoluble in aliphatic hydrocarbons. Penicillins may be extracted from aqeous solution at pH 2 with amyl acetate and concentrated by subsequent extraction with the proper buffer. These extractions must be made rapidly and at low temperature (0°–5°C.) if substantial loss is to be avoided.

Penicillins are active against many gram-positive and a few gram-negative bacteria at low concentration and against many gram-negative bacteria at high concentration. The two test organisms used routinely in assaying for penicillins are gram-positive bacteria.

Antibacterial activity of penicillins increases with decrease in pH. The

[1] H. W. Florey, E. Chain, N. G. Heatley, M. A. Jennings, A. G. Sanders, E. P. Abraham, and M. E. Florey, "Antibiotics," 2 vols., 1774 pp. Oxford Univ. Press, New York and London, 1949.

[2] A. B. Segelman and N. R. Farnsworth [*J. Pharm. Sci.* 59, 726 (1970)] showed that solutions of potassium penicillin G in 0.5% benzyl alcohol or 40 or 70% aqueous ethanol were as stable as aqueous solutions for a period of 10 days at room temperature.

pH limits are set by the tolerance of the bacteria and the stability of the penicillin. The former is the more important. Lowering the initial pH of the medium from 7.2 causes an increase in sensitivity[2] of the photometric assay at the rate of 1.5 times per unit of pH. The sensitivity to pH is so great that attention should be paid to the pH and buffer capacity of the standard solutions and samples. Usually, dilutions of standards and samples in the phosphate buffer will be great enough to eliminate any differences in the pH and buffer capacities of the two classes of solutions.

Assays for penicillin G will serve to illustrate general principles and methods. One photometric procedure and two plate methods will be discussed. A simple serial dilution type of turbidimetric assay is given in Chapter 3. Chemical determination of penicillins will be illustrated by the hydroxamic acid method of Boxer and Everett. The photometric method has a range from 0.2 to 0.6 u./ml. of penicillin G solution. The method is applicable to samples not grossly contaminated with bacteria, and free from penicillinase. The solutions should be clear and uncolored when diluted to the assay concentration of 0.3 u./ml. Bacteria and suspended solids but not penicillinase can, of course, be removed by Seitz filtration. The method is capable of good accuracy and the answer can be obtained within 5 hours after the sample is submitted for assay. It is not suitable in its present form for assay of blood because of interference by blood proteins. The plate method is more difficult to do accurately and is more widely used than the photometric method. It is applicable to non-sterile samples (free from penicillinase), to bloods, and to samples containing slowly diffusing substances which would interfere with the photometric method. The plate method is thus applicable to those samples submitted to hospital laboratories for assay. The range of the plate assay is from 0.05 to 2 u./ml. when *Sarcina lutea* is the test organism and from 0.5 to 4 u./ml. when *Staphylococcus aureus* is the test organism. The time interval between submission of sample and computation of answer is about 24 hours. The plate method seems to be favored by managers of control laboratories perhaps because of its deceptively simple mechanics, the large number of replicates which of necessity must be used and its well worked-out statistics. Plate, photometric, and chemical methods will not necessarily give the same answer when mixtures of penicillins are assayed.

An excellent discussion of the early (up to 1948) work on the production, chemistry, antibacterial activities, and pharmacology of penicillin is given in Volume 2 of "Antibiotics" by Florey *et al.*[1] General methods of assay for antibiotics are given by Heatley[1] in Volume 1.

---

[2] F. Kavanagh, *Advances in Appl. Microbiol.* **2**, 65 (1960).

TABLE I
POTENCIES OF SALTS OF PENICILLIN G

Salt of penicillin G	Mol. wt.	Potency (u./mg.)
Benzathine·4 H$_2$O	981.18	1211
Calcium	706.83	1681
Potassium	372.47	1595
Procaine	588.71	1009
Sodium	356.38	(1667)

Theoretical potencies of several common forms of penicillin G are given in Table I. The potencies are computed from the molecular weights and the assigned potency of 1667 u./mg. for the sodium salt.

II. Photometric Assay

A. Introduction

The earliest turbidimetric methods were based upon a serial dilution type of assay. The interval between tubes could be geometric (twofold dilutions) or arithmetic in steps of 5 or 10%. The tests were performed with light inoculum and overnight incubation. Details of twofold serial dilution method are given in Chapter 3. Ingenious attempts were made to shorten the time of assay. One of the shortest tests was that of Rake and Jones,[3] who employed a hemolytic streptococcus in a penicillin assay requiring 55 to 90 minutes. The end point in this serial dilution assay was absence of hemolysis in the presence of sufficient penicillin to inhibit growth of the streptococcus. Technical difficulties prevented wide use of the assay. The end points of the above tests were estimated visually; the tubes were classified as being either turbid with growth of bacteria or not turbid. The uncertainty of the end point was one tube and could be 5–50% depending upon the design of the test. Foster[4] disliked end-point assays and measured turbidity with a photometer after incubation of 16 hours. He later abandoned turbidimetric assay in favor of plate methods. Josyln[5] and McMahan[6] described 4-hour photometric assays which, with slight modifications, became the standard turbidimetric antibiotic assays.

Details of the 4-hour assay for penicillin G are given here as a general

---

[3] G. Rake and H. Jones, *Proc. Soc. Exptl. Biol. Med.* **54**, 189 (1943).
[4] J. W. Foster, *J. Biol. Chem.* **144**, 285 (1942).
[5] D. A. Joslyn, *Science* **99**, 21 (1944).
[6] J. R. McMahan, *J. Biol. Chem.* **153**, 249 (1944).

method. In describing assays for other antibiotics, only details of procedure different from those of the penicillin G assay will be given.

B. Test Organism

1. Stock Cultures

The assay organism for penicillin is a strain of *Staphylococcus aureus*. It forms a good suspension without lumps and does not give a precipitate of extraneous material when the suspensions are heated at the end of the incubation period.

The particular strains of *S. aureus* (ATCC 9144, a Heatley strain) used in the test described here gives a nearly linear calibration curve (log-probability) in the range 0.16 to 0.6 unit of penicillin G per milliliter of standard solution. The strain has been cultivated in our laboratory for a number of years and may not be identical with a recent culture of that number obtained from the American Type Culture Collection.

The culture is maintained in the form of lyophilized pellets. Prepare the pellets as follows: To one slant incubated overnight at 36°C., add 10 ml. sterile beef serum Difco), suspend the bacteria in it, transfer 0.05 ml. of suspension to sterile 7-mm. lyophile tubes, place on the lyophilizer, freeze at −38°C., and dry under high vacuum ($<100 \mu$) as the bath warms to 0°C. Remove tubes from the bath and continue drying for an hour and seal.

Prepare fresh slant cultures each week. Place a pellet from a lyophile tube in a tube of nutrient broth, incubate the tube for 24 hours at 36° to 38°C., plant slants on appropriate agar medium from the suspension of bacteria, and incubate for 24 hours at 36°C. Keep the slants in a refrigerator and use from the set for a week. Freshly slanted tubes of AC-agar or the Penassay seed agar are satisfactory. *Staphylococcus aureus* can also be maintained with its sensitivity to penicillin relatively constant by weekly slant-to-slant transfers. The transfers should never be from liquid to liquid for reasons given earlier (Chapter 4, Section IV.B).

2. Standard Inoculum

The inoculum can be prepared in several ways. The important point is to have an adequate concentration of the bacteria in the log phase when needed. The procedure to be described here has been satisfactory for more than a decade. Transfer a loop full of bacteria from a slant to 500 ml. of nutrient broth in a 1-liter flask. Incubate the inoculated flask for 12 hours at 34° to 36°C. without shaking. Little growth is apparent at this time. Shake the flask by hand to aerate and then place in a 36°C. water bath. Shake it at about ½-hour intervals and incubate until the population reaches the required density of 350 million cells per milliliter, then chill in an ice bath, and store in a refrigerator until needed to inoculate the assay broth. Add 40 ml. of this inoculum to each liter of prechilled assay broth just before the broth is added to the tubes. Measure the population density of the inoculum in a calibrated photometer (Chapter 4, Section III.D) or nephelometer. Calibration can be made accurately enough for this purpose by making plate counts on suspensions of bacteria in the log phase. The inoculum need not be grown to the density suggested above as long as the concentration of cells is known and they are in the log phase. The concentration of cells in inoculated assay broth should be the same from day to day. It is $14 \cdot 10^6$/ml. in the example above.

Daily preparation of inoculum may be neither necessary nor advantageous. A suspension stable for a week can be made by washing the bacteria from a large agar surface (flat bottle) with $M/20$ pH 7 phosphate buffer. To grow the bacteria, inoculate an agar surface with a heavy suspension of bacteria, grow for from 4 to 6 hours at 37°C., and wash off with buffer. Chill the suspension and keep it in the refrigerator where the viability may decrease by only 10% in a week. The bacteria die more rapidly in nutrient broth than in buffer.

C. Standard Solutions

Prepare the standard solutions from a secondary standard which has been carefully standardized with reference to F.D.A. standard penicillin G. The secondary standard should be free from more than traces of other penicillins and should have a well established purity. The pure sodium salt (mol. wt. 365.38) has an assigned activity of 1667 u./mg. and the pure potassium salt (mol. wt. 372.47) an activity of 1595 u./mg. U.S.P. reference standard is received dried and is stored over a desiccant. Secondary or house standards can be dried *in vacuo* at 60°C for 3 to 4 hours. The dried preparation should be stored over a desiccant and used with all of the precautions applied to a primary standard. An alternate procedure is to determine moisture by appropriate means and then correct the weight of the compound taken.

Make the primary stock solution in sterile pH 6 phosphate buffer. Dissolve a sample of about 25 mg. in buffer and dilute to 25 ml. in a volumetric flask. Compute the concentration in terms of units per milliliter and make further dilutions as indicated later. Some prefer to make the stock solution of exactly 1000 u./mg. by weighing 62.7 mg. of the potassium salt and diluting to 100 ml. with buffer.

The stock solution of penicillin G is fairly stable if prepared without gross contamination and kept in the refrigerator at 5°C. The loss in activity in a week may be no more than 5%. It is good practice to prepare the stock solution at least once each week (usually Monday morning). The large concentration of phosphate in the stock solution does not prevent an independent check of the correctness of its concentration by the hydroxamic acid method.

The working standards are made by diluting the stock solution with buffer to the concentrations needed for preparing the standard curve (Section E.2).

D. Preparation of Samples for Assay

1. Treatment of Samples

Samples submitted for assay may be liquids, solids, or semisolids such as tissues. The liquids may be nearly pure solutions, whole beers from a fermentation area, or process samples with widely varying purities. The solids may be pure salts, mixes, tablets from a manufacturing area, or agricultural grade mixes which may contain one or more antibacterial substances in addition to penicillin. The problem is to separate the penicillin from the contaminants without substantial loss or, better, to extract the penicillin and then dilute so that the nonpenicillin materials do not affect the assay. Examples of several classes of samples will be considered.

a. Fermentation beers. Filter the samples through filter paper (Whatman No. 1) to remove mycelium, calcium carbonate, and oils. Dilute the slightly turbid or clear filtrate with sterile pH 6.0 phosphate buffer to approximately 0.30 u./ml. These filtered samples are not sterile and will become grossly contaminated if allowed to incubate at room temperature for several hours. The lack of strict sterility is of no disadvantage so long as the dilutions are kept cool and are assayed promptly. Should the samples be contaminated with a penicillinase-producing bacterium, appreciable loss of penicillin would occur rapidly.

b. Powders. Dissolve about 50 mg. of penicillin powder in pH 6.0 buffer and dilute to 50 ml. in a volumetric flask. Make further dilutions in buffer as needed for the particular assay. If the sample contains the procaine salt of penicillin G, dissolve it in 5 ml. of absolute methanol and then dilute immediately with buffer. Remove starch and other methanol and water insoluble materials by filtration.

c. Ointments. Dissolve about 2 gm. of the ointment in 50 ml. anhydrous ether, transfer to a 250 ml. separatory funnel, extract with a total of 150 ml. of pH 6 buffer (three extractions with 50 ml. each), and dilute with buffer to a suitable concentration.

2. Samples Containing Other Antibiotics

The number of combinations of antibiotics is so large that no general procedure can be given for preparing all samples. Often considerable ingenuity must be used to exploit to the fullest the different chemical, physical, and antibacterial properties of the several antibiotics in a mixture

In many mixtures, the penicillin is so much more active than the other antibiotics that only its effect remains after dilution has been made to 0.3 u./ml. Dilution is tried first in assaying mixtures. If the dose-response line of the diluted sample is different from that of the penicillin standard, then the dilution was insufficient to remove effect of the other antibiotic substances. If the dose-response lines are the same, no valid conclusion is possible without knowledge of the identity and quantity of the other antibiotics. Usually the *assumption* is made that identity of dose-response curves indicates identity of active substance. This assumption may not be true. Penicillin can be removed (inactivated) from a mixture by treating the solution with penicillinase at pH 6 to 7.5. Most antibiotics are insoluble in isooctane and petroleum ether, which may dissolve an ointment base and leave the antibiotics on the filter as a dry powder. The solubilities of the important antibiotics in 24 solvents given earier in this chapter are a great aid in devising ways of separating these compounds.

Penicillin G is sensitive to acid, alkali, high temperature, and penicillinase and needs to be handled rapidly at low temperatures if marked loss is to be avoided. These sensitivities of penicillins must be kept in mind when devising extraction procedures. Penicillin can be concentrated with little loss as is done in a procedure given in the section on chemical assay.

E. Mechanics of Assay

1. Design

The range of the assay is from 0.15 to about 1 unit of penicillin G per milliliter of test solution; the best part of calibration curve lies between 0.2 and 0.6 u./ml. (The concentration of penicillin G in the assay broth is about 1/21 of these concentrations.) Dilute the unknowns to an estimated concentration of 0.3 u./ml. The number of dilutions needed to obtain concentrations within the range of 0.2 and 0.6 u./ml. depends upon the accuracy of the estimation of the potency of the unknown. Usually the potency of the sample will be known to lie between rather narrow limits. If not, then estimate the minimum concentration, dilute to the estimated 0.3 u./ml., and prepare further dilutions in steps of 4 (4, 16, 64, 256). At least one of the dilutions will fall on a usable portion of the calibration curve.

The influence of extraneous materials upon the penicillin assay will not be of concern because the dilution is usually large enough to eliminate such influences. Low potency preparations are assayed by plate methods. If the concentration should be so low that only small dilutions (50) are possible, then assay the sample at several dilutions (0.2, 0.3, 0.45 u./ml.). The concentration in the undiluted sample as estimated from the two extreme concentrations should be within 10% of that computed from the 0.3

u./ml. dilution and there should be no trend to the figures. A substantial trend indicates an assay of questionable validity and the cause should be sought if the assay is important. A plate method might give a valid assay.

The number of replications of each sample depends upon the accuracy with which the preassay concentration is known (assuming that a favorable portion of the calibration curve is used) and the purpose for which the answer is required. The control laboratory in one large pharmaceutical company uses two tubes and two levels for each sample. An assay laboratory in another firm considers three tubes at one level to be sufficient for their samples, most of which are fermentation samples from development laboratories. The control laboratory of the same company puts on three tubes at three levels and repeats the complete assay on several days. If the choice is between a large number of replicates on one assay or a smaller number on several, choose the latter, and always start each time with the material submitted. This procedure reduces the chances that a gross error in weighing or diluting will go undetected. The assumption is made that the sample submitted for assay is stable. If it is not, then several assays on one day may be the best compromise.

The suggested design given above of one concentration of unknown at an estimated concentration of 0.3 u./ml. will not be acceptable to all assayists. The single-point design can lead to unrecognized errors and should be used only to assay samples known to be free from interfering substances when diluted to the assay level. Many prefer to assay at more than one level as may be required by regulatory agencies. A slight change in the design permits this. If a three-point curve is needed to test for parallelism and curvature, assay at 0.15, 0.3, and 0.45 u./ml. Dilute sample to 0.3 u./ml. and use 0.25, 0.50, and 0.75 ml. amounts. Add buffer to bring the total volume to 0.75 or 1.00 ml., whichever is more convenient. The standards must have the same total volume as the unknown. If the sample is diluted several thousandfold to bring it to the assay range and the three levels give answers which show a trend, investigate the details of the assay. A trend indicates asymmetry of technique or a sample grossly contaminated with interfering substance.

Some assayists prefer to add large volumes of samples and standards (say 1, 1.5, 2, 3, 4, 5 ml.) to the tubes, dilute to a final volume to 5.0 ml. and then add 5.0 ml. of inoculated double strength assay medium. Volumes of samples added to the tubes is unimportant so long as they are measured with the required accuracy. We prefer 0.5 ml. as the basic volume and add it from a special 2-ml. pipet graduated in 0.5-ml. intervals.

2. Standard Curve

Prepare the standard response curve by pipetting 0.50 ml. of standards of the following concentrations, 0 (buffer), 0.20, 0.30, 0.40, and 0.60 u./ml. into clean, sterile, 18 × 150 mm. culture tubes. Make at least 8 of the 0 tubes, and four of each of the standard solutions. Pipet 0.5-ml. portions of each dilution of the samples into three or four culture tubes.

When all of the samples have been pipetted, add 10 ml. of chilled inoculated penicillin assay broth rapidly with a pipetting machine. As soon as all the tubes have received broth, place the racks in a well stirred 36°–38°C. water bath. The exact volume of broth added is not important as long as the same volume is added to each tube. Incubate the test for at least 3 hours. At this time, measure the turbidity of the 0 tubes, and continue to do so every 15 minutes until the population density reaches about 350 million cells/ml. At this point transfer the racks *quickly* to a steam cabinet and heat with flowing steam for 10 minutes to kill the bacteria. Put the racks of hot tubes into a water bath at about 20°C. (tap water) to cool the solutions to room temperature. The test is now ready to read.

The incubation period should be at least 3 hours. Incubation time should be long enough so that additional incubation changes the sensitivity of the assay but not the slope of the dose-response line. The 3-hour incubation period is also used when the inoculum is light and the assay is read with a nephelometer.

F. Responses and Answers

1. *Measurement of Response*

The turbidity of the solutions is measured with a photometer or turbidimeter. All of the commonly used instruments show a nonlinear relation between scale reading and concentration of bacteria above a rather small concentration as is discussed in Chapter 4. The instrument should be fitted with a quick-emptying cuvette.

Shake the tubes vigorously by hand (thumb over the end) or with a mechanical shaker for 5 to 10 seconds to resuspend bacteria that settled on the bottom of the tube during the incubation and killing periods. Shake the tubes at least 15 minutes before the test is read. Set the instrument to 0 (or 100% transmittancy) with uninoculated broth. Read all tubes of the standard curve in order of decreasing penicillin concentration. Then measure the sets of sample tubes. Readings made sooner than 10 to 20 seconds after filling the cuvette of the photometer may be in error by unknown and variable amounts because of the presence of air bubbles. When adjacent sets of tubes differ greatly in turbidity, the first tube read will be slightly different from its companions because of carry-over of small but appreciable amounts of the previous sample in the cuvette. The remaining two or three tubes should agree within 1% of the full scale (as represented by the reading of the 0 tube). If the potencies have been estimated accurately, the tubes of the samples will be nearly alike and the carry-over inappreciable. Carry-over can be eliminated by rinsing the cuvette with 3 to 4 ml. of the contents of the first tube of a group to be measured.

Characteristics of photometers are discussed in Chapter 4, Section III.A and a method of calibrating the instrument scale in terms of concentration of bacteria is given in Chapter 4, Section III.D. All of the commercial instruments could be improved for the purpose of accurate measurement of turbidity of bacterial suspensions. All of them respond too rapidly for accurate routine measurement when operated manually. A measurement can be made in a few seconds, a time much too short to permit air bubbles to clear from the medium (Chapter 4, Section III.B). The cylindrical cuvette used in most instruments has a circulation time so long that accurate measurement of suspensions of rod-shaped organisms is impractical (see Chapter 4, Section III.C). A rectangular cuvette has a short circulation time, but its use reduces sensitivity of photometers when the lesser dimension is in the direction of propagation of light. (A forward scattering nephelometer is less sensitive to thickness of liquid layer in the cuvette than a photometer.) A rectangular cuvette could be used with certain spectrophotometers with the light passing through the larger of the two dimensions and thus avoiding loss of sensitivity. The instruments may not be mechanically as rigid or electrically as stable as is needed for long time measurements in assay work.

Certain of the difficulties can be eliminated by operating the measuring instrument semiautomatically so that the operator fills the cuvette and then presses a button to initiate the sequence of events leading to measurement of turbidity. The relatively simple arrangement used at Eli Lilly & Company has proved satisfactory. A shaft encoder (Datex Model C-104) was installed in the Leeds & Northrup Type H indicators of the self-balancing photometers. The encoder signal passed through a Datex Model K154 Translator and then to an IBM 526 card punch. Two self-balancing photometers, a control box, a translator, and card punch form a unit. The operator fills the cuvette, presses the button to start the 15-second time switch which goes through the following sequence of operations: at 9 seconds the balancing motor (in the Type H indicator) starts and operates for 4 seconds; a solenoid valve opens to empty the cuvette; the position of the shaft encoder prints out on the IBM 526; the timer stops at 15 seconds completing the cycle. The circuits are arranged so that the second photometer can be started only after the first one is in operation and can print-out only after the first one has printed. By connecting the two encoders in parallel through a diode switch to one translator and card punch, a single-pole double-throw relay is sufficient to control which encoder is read. The relay is actuated by a contract on the first time switch. The number one photometer is assigned the odd-numbered columns in the IBM card; and the second photometer, the even-numbered columns. Provision is made to use only the number one photometer if only one instrument is needed. Two tests are read simultaneously, and the one with the larger number of tubes is always assigned to the number one photometer. If two photometers with identical responses were available, then one test could be read on two instruments. Since the instruments are not identical, the arrangement described is used. The cards are then fed to an IBM 1620 or similar fast computer to compute the two assays. One technician operates two photometers and processes 400 tubes per hour. He spends as much time per tube as he did when he read the test manually, but now there is sufficient time available for most of the air bubbles to move out of the light path. Slowing the operations improved accuracy. The automatic read-out eliminates human errors of reading and recording the indicator scale but makes possible introduction of errors of machine origin.

Both this system and the one designed for reading zones on petri dishes can be operated manually without any re-conversion should translator, card punch, or computer fail.

2. Computation of Potency of the Samples

Draw the standard curve through the points obtained by averaging the several readings for each concentration. Average the readings for the tubes of a sample and obtain the concentration from the calibration curve. Multiply the concentration so obtained by the dilution of the sample to obtain the concentration of penicillin in the undiluted sample.

The above procedure is the one followed in most laboratories. I prefer to put the standard curve into the log-probability form as described in Chapter 4, Section VIII.A. The considerable computations required when response is measured with the usual photometer can be done most easily with a digital computer. A linear nephelometer, which has a simple analog computer built in, gives the per cent response of the samples directly and makes the log-probability representation as simple and as easy to use as the customary semilog plot.

The turbidities of individual replicates should be within 2 to 3% of the average of the group. Occasionally a tube will be in obvious error and will not be included in the average. Standard curves obtained on different days fall within a small region on the log-probability paper, consequently a greatly erroneous reading is obvious.

Two samples, one at two dilutions, were assayed with the curve in Table II. A log-probability plot shows the curve to be straight between 0.2 and 0.6 u./ml. Each of the two assays of the sample are reported as 7200 u./ml. because differences less than 100 u./ml. have little if any significance.

TABLE II

STANDARD CURVE FOR PENICILLIN G ASSAY WITH *S. aureus* H

Standard (u./ml.)	Photometer reading					Relative conc.	Response (%)
	Individual tubes				Avg.		
0	70.5	70.5	70.5	70.5	70.5	64.5	100
0.1	53.0	54.0	53.5	53.0	53.4	47.0	73
0.15	50.5	50.0	50.0	50.0	50.1	43.6	67.6
0.2	44.5	45.0	46.0	46.0	45.5	39.2	61
0.3	29.0	29.0	29.5	30.0	29.5	24.7	38
0.4	17.7	18.1	18.6	18.3	18.2	14.9	23
0.5	12.3	13.3	13.0	12.9	12.9	10.5	16
0.6	9.0	9.1	9.2	9.0	9.1	7.4	11.5
0.8	6.5	6.7	6.9	6.5	6.6	5.4	8.4
1.0	5.0	5.5	5.5	5.1	5.3	4.4	6.8

Sample	Dilution				Avg.			(u./ml.)	Total (u./ml.)
1	20,000	24.4	23.5	22.0	—	23.5	19.4	30	0.36	7200
1	33,300	43.5	41.4	42.5	—	42.5	36.4	56.4	0.215	7170
2	20,000	40.0	41.0	39.5	—	40.2	34.0	52.8	0.23	4600

3. Meaning of the Assay

The answers are obtained in terms of units of penicillin G activity. The meaning of the answer depends upon the composition of the sample because a bioassay measures activity not quantity. If only penicillin G is present, then the answer is in units of penicillin G. If other penicillins are present (F, dihydro F, and K as in fermentation samples), the answer is in terms of penicillin G activity which cannot be precisely computed into total units of penicillin because the several penicillins are present in unknown amounts and each has a specific activity against *S. aureus* different from that of penicillin G. If the sample contains much K, then the assay in terms of G will be higher than it would have been had the K been replaced by an molecular equivalent of G.

Since the slope of the log-probability curves are about the same for the penicillins produced by fermentation, any one penicillin could be assayed in terms of penicillin G as the standard. This practice is not recommended because the relative sensitivities may not be invariant and an emperical conversion factor must be used.

The quantity of penicillin is easily and accurately measured in certain samples by the hydroxamic acid method. When the chemical assay in terms of G and the bioassay of the same sample are compared, the two assays may differ by much more than either group of assayers considers to be the probable error of its method. However, agreement should be expected only when a purified penicillin is assayed. Such agreement is obtained. The chemical method applied to fermentation samples measures, in addition to the active penicillins, 6-aminopenicillanic acid and a cephalosporin-N-like substance, which have inappreciable activity in the assay against $S.$ $aureus$. Thus, in a penicillin fermentation made without added precursor, the chemical assay can be more than six times as great as the antibacterial assay in terms of G because of the presence of large amounts of 6-aminopenicillanic acid and cephalosporin-N-like substance. The chemical assay may give answers higher than the true G content even on fermentations made in the presence of adequate precursor. The argument made here with respect to penicillin G applies to the assay of other penicillins produced by fermentation. Agreement of basically different assays should not be expected, except on those samples which contain a single penicillin. Tube and plate methods may not give the same answers on mixtures of penicillins.

G. Remarks

Some kind of turbidimetric method for penicillin has been used for nearly 20 years to assay millions of samples. Despite the seeming importance of the method, many of the factors that affect the assay have been incompletely reported. A careful systematic study of the principles of the assay is long overdue and should be made.

Other test bacteria preferably cocci, are needed. The Heatley strain of $S.$ $aureus$ has a sensitivity (50% response on log-probability plot, M.R.) of 0.3 u./ml. (sample concentration). A strain with only 1% of this sensitivity would permit the assay of high potency samples with considerable saving in time and equipment for preparing dilution. A strain of bacteria with more than 10 times the sensitivity of $S.$ $aureus$ is needed to assay low potency samples and those which require great dilution to eliminate the influence of contaminating substances affecting the dosage-response curve.

The composition of the assay medium needs investigation to find one that will support considerable growth of the test bacteria in the log phase.

The pH of the medium should not change very much during growth. How much is very much and during what time period is the change important?

Shaking the tubes during the incubation period increases the growth rate of the bacteria greatly, shortens the incubation period, and increases the sensitivity slightly. What is the optimum shaking and how is it related to medium, pH, strain of organism, concentration of bacteria in the inoculum, and time of incubation? How closely must the environmental condition be controlled to obtain reproducible assay curves?

Variants in procedure are best tested by their effect upon the standard curve in a log-probability form. This way of plotting the data is preferred to others because it is affected less by day-to-day variations than a plot of instrument response. The meanings of change in the family of nearly straight lines so obtained are easier to interpret than the usual collection of apparently unrelated curves of different shapes which results from the usual ways of presenting the data.

Some of the methods of preparing samples seem needlessly involved. For example, what proportion of the penicillin G is removed in the first extraction performed in Section D.1.c? Are the other extractions needed?

H. Special Equipment

No equipment not ordinarily found in an assay laboratory is required for this assay. The 0.50-ml. volumes of sample and standards would be more conveniently measured in a special pipet graduated to deliver at 0, 0.50, 1.00, 1.50, and 2.00 ml. than in a serological pipet. Gross errors in measurement caused by mistaking graduations would be improbable with such a pipet which has only 4 calibration marks (rings). Measurement of volumes with an accuracy better than 2% would then be possible; it is not when serological pipets are used to measure the volumes as is customary.

A linear turbidimeter with its associated analog computer makes convenient the use of the theoretically preferable log-probability plot of the standard curve.

III. Petri Plate Assay

A. Introduction

Although a number of different organisms have on occasion been utilized by various workers for the bioassay of penicillin by the cylinder-plate technique, only two have achieved widespread recognition and acceptance. The first from the standpoint of usage in the U.S. is *Staphylococcus aureus* (ATCC 6538P) which is particularly suited to the assay of penicillin in most pharmaceutical dosage forms or in other high-potency samples. The second commonly used organism is *Sarcina lutea* (ATCC 9341). This organism is somewhat more sensitive to penicillin and has found particular application in the assay of biological fluids, animal tissues, feedstuffs, and the like. In these materials the antibiotic is frequently present in microgram or even submicrogram amounts, and is often accompanied by active nonpenicillin factors. These extraneous materials, whether inhibitory or stimulatory to the test organism, must be removed or adequately diluted to obtain an accurate estimation of the penicillin content of the material being assayed. The greater susceptibility of *S. lutea* to penicillin will, in most cases, permit dilution of these factors to such an extent as to eliminate the need for their removal.

The methodology of typical assays employing each of the organisms will be described. These methods will be presented in some detail as being classical examples of antibiotic plate assays. Other antibiotic assays will be described as modifications of these methods. It must be emphasized, however, that many alterations may be made in assay techniques, depending upon their particular application. Somewhat more elaborate systems may be desirable if one seeks the ultimate in precision, accuracy, or sensitivity. An excellent discussion of the two-dose cylinder-plate method and the estimation of the standard error of the assay is contained in a publication by Grove and Randall [7] which should be consulted by those not familiar with these procedures. For those desiring even greater statistical insight into the performance of their assay (potency, error variance, validity tests and 95% confidence intervals), the methods outlined in the U.S.P.[8] and applied by Kirshbaum et al.[9] should be consulted.

Should the assayist be confronted with the need for further increasing the sensitivity of the methods outlined, or if he encounters poorly defined zones or significant departure from linearity, he is referred to Chapter 1 which is devoted to the theory of diffusion methods for a possible explanation of the mechanism involved. It is sufficient here to acknowledge that many factors such as pH, temperature, time, agar density, and diffusion constants play vital roles in the successful conduct of these assays (see Chapter 1). The device (paper disk, cylinder, etc.) for applying the test solutions may affect the diffusion of the antibiotic and thus alter the linear range of the dose-response curve.

B. Test Organisms

Maintain stock cultures of *Staphylococcus aureus* (ATCC 6538P) and of *Sarcina lutea* (ATCC 9341) by weekly transfers to fresh sterile slants of Grove and Randall agar medium No. 1. Incubate the freshly prepared slants at 32° to 35°C. for 16 to 24 hours and store under refrigeration at 4° to 6°C. until ready for use in the preparation of the inoculum suspension.

C. Standard Solutions

Prepare a stock solution of the appropriate penicillin working standard

[7] D. C. Grove and W. A. Randall, "Assay Methods of Antibiotics: A Laboratory Manual," 238 pp. Medical Encyclopedia, New York, 1955.

[8] Anonymous, *in* "United States Pharmacopeia," Vol. XVI, pp. 873-885. Mack Publ., Easton, Pennsylvania, 1960.

[9] A. Kirshbaum, B. Arret, and J. Kramer, *Antibiotics & Chemotherapy* **6,** 660 (1956).

by careful weighing of the dried standard with subsequent dilution to a convenient concentration (usually 100 u./ml.). Keep the solid penicillin working standard in tightly stoppered vials over a desiccant, such as silica gel, in a desiccator. Weigh the standard in an area with controlled humidity, preferably a relative humidity of 50% or less. Dilute the working standard with sterile pH 6 phosphate buffer. A small quantity of solvent, such as formamide, may be needed to solubilize some penicillins (not G) before diluting in buffer. Store the solutions in the refrigerator. Not all penicillins or penicillin derivatives exhibit identical dose-response curves even under identical assay conditions, and therefore it is extremely important that one uses a standard identical to the compound being assayed. Use the stock solutions for a maximum of 2 days.

D. Sample Preparation

In view of the extremely long list of diverse materials which may confront the assayist, it is impossible to describe here all of the techniques of preparing the samples for assay. An excellent guide to the extraction of penicillin from various pharmaceutical preparations and animal feeds, and to the assay of penicillin in the presence of biological fluids, will be found in the publication by Grove and Randall. The solubility tables given earlier in this chapter help one to develop a scheme for separation of penicillin from combinations with other antibiotics. Completely quantitative separations are rare.

In general, penicillin solutions and bulk powders may be diluted directly with pH 6.0 phosphate buffer prior to assay. If the particular compound is not sufficiently water soluble, a small quantity of solvent such as formamide or methanol may be required to solubilize the sample prior to dilution in buffer. Preparations such as ointments can usually be dispersed in anhydrous ether in which the base is soluble and the antibiotic insoluble, or at least relatively so. Repeated "shake-outs" or washings of the ether with phosphate buffer pH 6.0 should effect quantitative extraction of the antibiotic from the ether into the buffer phase. Tablets, troches, animal feeds, and the like should be finely divided either by grinding with mortar and pestle or by blending in a suitable high speed homogenizer to facilitate dispersion and solution of the antibiotic in the diluent. A solvent composed of 1 volume of acetone and 3 volumes of pH 6 buffer may extract penicillins from feeds much better than the buffer alone.

Should one be unable to effect quantitative separation of penicillin from an antibiotic combination, the success of the assay will depend upon the "interference threshold" of the extraneous materials. Arret *et al.*[10] discussed this problem and determined both "interference thresholds" and

[10] B. Arret, M. R. Woodard, D. M. Wintermere, and A. Kirshbaum, *Antibiotics & Chemotherapy* **7**, 545 (1957).

"sensitivity thresholds" for 11 different antibiotics in each of 12 procedures. Their tables are given in Chapter 6.1, Tables II and III.

Although space does not permit a detailed description of various sample preparation techniques, it does appear that a few precautions should be mentioned. Since the analyst may use a variety of organic solvents in preparing samples, he must be aware of the possible effect of these solvents on the assay system. If this point is not well established in his laboratory, he should use the same concentration of the solvent in the preparation of the standard curve (a compensated standard). If the samples vary widely in pH, adjust the pH to that of the standard curve even though dilutions are to be made in buffer. This is particularly important if subsequent dilutions are small in magnitude (<1:5) as the capacity of the buffer may not be sufficient to achieve the desired pH. In the assay of biological fluids such as serum and urine the problems become even more perplexing. The logic of diluting samples and standards in "normal" urine or "normal" serum is readily understood, but the definition of "normal" is anything but clear (or possible of attainment?). In actual practice the usual approach is to use a pool of these fluids from several apparently healthy individuals known not to be undergoing antibiotic therapy. This is probably the best that can be done in the average assay laboratory. In the performance of assays on human serum, a widely accepted alternate to human serum is bovine albumin (fraction V). Although this diluent has been widely used, it too may have shortcomings. Variation in antibiotic binding power of different lots of bovine albumin has made it necessary, with the acquisition of each new lot of the albumin, to determine the dilution which in our experience seems best to equate to our concept of "normal" serum. Solutions ranging from 4 to 9% of albumin (fraction V) have been accepted on this basis. Regardless of the choice of diluents, the assayist should be certain of one thing—their sterility. It is generally accepted that samples need not be sterile for the diffusion techniques. Indeed, this is often quoted as one of the advantages of these methods, but there are many times when this belief can cost the analyst his entire assay. This is particularly true when one uses human serum contaminated with penicillinase producing organisms. Usually difficulties of this nature will be made evident by the peculiar behavior of the lower concentrations on the dose-response curve. These levels will produce zones smaller than anticipated, or perhaps no zones at all. One further caution regarding the use of "biological" diluents—in addition to their abilities to "bind" antibiotics, these fluids often have antimicrobial activity. This presents a real problem. Samples and extracts of blood, urine, feces, tissues, feedstuffs, etc., have all been known to contain antimicrobial substances with great quantitative variation from individual to individual sample as well as from species to species. Obviously, selection of a diluent of this nature for the standard should entail reasonable proof that the particular batch being used does not, of itself, exhibit activity. If the analyst has no choice but to use such materials, he can at least temper his interpretation of the results obtained by the knowledge of the limitations of his system.

E. Mechanics of the Assay

1. *Design*

Like other facets of the assay, the design itself may be varied somewhat to fulfill the needs of the assayist. Rather elaborate statistical models may be devised to answer specific questions regarding the reproducibility of the method, the effect of environmental changes, or the homogeneity of the materials to be assayed. In routine practice, however, the common approach seems to be to accept the slightly greater variability of the simpler designs and to attempt to achieve the desired precision by means of replication. The economic situation in which the assayist finds himself, coupled with his judgement regarding the purpose of assay, will most likely be the factors dictating the degree of precision he seeks to attain.

Some laboratories make use of large flat glass plates or Pyrex baking dishes which permit extensive randomization and replication of both standards and samples in what may be considered a single, uniform environment. The reduction of variables in such a system favors the attainment of a high precision; however, the method does not appear to have gained widespread use in the routine laboratory in the U.S. The majority of laboratories here still use the 100-mm. flat-bottom petri plate accomodating only four to six cylinders or disks per plate.

The practice in my laboratory is to use the petri dish and, whenever possible, to limit the standard dose-response curve to four or five dose levels. In this way the entire curve may be accommodated on a single plate. If one must run more than six doses per standard curve, do it by running two sets of plates, each with one-half the levels, and tie the sets together with one standard level common to each set. A third method is to run two levels of standard per set repeating one of these levels within each set. In our laboratory, where the first method is used, the scheme is replicated on ten plates and the arithmetic mean response for each dose level is used to construct the dose-response curve. The sample or test plates are replicated five times, each plate containing four cylinders. Two alternate cylinders on each sample plate are used for the application of the unknown solution and the remaining two cylinders are used for the application of one of the concentrations used in the dose-response curve itself. Arbitrarily, this "reference" concentration is the median dose in a five-level curve, or the second lowest concentration in a four-level curve. The behavior of this "reference" concentration is used to compensate for the effect of plate-to-plate variables such as small differentials in incubation temperature, variation in agar thickness, variation in inoculum concentration, etc. The interpretation of the resultant "reference" zones will be described later in this discussion.

Perhaps the main drawback in the above design is that it represents a one-dose assay; that is, only one concentration of the unknown is measured against the response curve. Obviously, reading at a single point on a line or curve does not enable one to make any predictions regarding the parallelism of the sample with that line or curve. With samples likely to contain contaminants with microbiological activity, either stimulatory or inhibitory, at least two dilutions of the sample should be assayed. If such is done, and parallelism demonstrated, the assayist may have considerably more faith in the validity of the results.

2. Media

Both of the assays presently under consideration (*Staphylococcus aureus* and *Sarcina lutea*) make use of the two-layer agar system. In this system a layer of melted agar is poured into the plate and allowed to harden. After this "base" layer has solidified, a second layer of agar which was previously melted, cooled to approximately 48°C., and inoculated with the appropriate test organism, is applied to each plate. The reasons for the use of the two-layer system are to improve the definition of the resulting zones of inhibition and to provide the optical contrast necessary for their precise measurement.

The selection and application of agars in this phase of the assay is of utmost importance to the success of the method. In addition to the nutritive components of the media, there are many other considerations such as pH, agar content, and depth of agar, which have a direct bearing on the size and quality of the zones of inhibition. One of the most important steps in the entire assay is the pouring of agar layers of uniform depth and great care should be exercised in insuring level surfaces for this operation.

The *Staphylococcus aureus* penicillin method is performed on plates containing a "base" layer of G. & R. agar medium No. 2 and a "seed" layer of G. & R. agar medium No. 1. The *Sarcina lutea* method utilizes a "base" layer of G. & R. agar medium No. 1 and a "seed" layer of G. & R. agar medium No. 4. Although the exact quantities of these media required in the assay may be somewhat dependent upon the sensitivity needed, for the majority of applications the recommendations contained in the Code of Federal Regulations[11] and in the outline of assays by Kirshbaum and Arret[12] will suffice; that is, 21-ml. "base" and 4-ml. "seed" layers for the

[11] Anonymous, 21 CFR:141; Compilation of Regulations for Test and Methods of Assay and Certification of Antibiotic and Antibiotic-Containing Drugs. Federal Register, Washington, D. C.

[12] A. Kirshbaum and B. Arret, *Antibiotics & Chemotherapy* **9**, 613 (1959).

Staphylococcus aureus method, and 10-ml. "base" and 4-ml. "seed" layers for the *Sarcina lutea* method. These quantities are for the 100-mm. assay petri plates.

3. Inoculum

Prepare the suspensions of the test organisms for use in the seeding of the assay plates in the following manner:

a. *Staphylococcus aureus*. Prepare an agar slant of the culture as described in Section III.B. Wash the growth on this slant from the agar surface using 2–3 ml. of sterile 0.85% sodium chloride solution. Transfer the suspension to a Roux bottle containing approximately 300 ml. of sterile G. & R. agar medium No. 1 and distribute it evenly over the surface of this agar with the aid of sterile glass beads. Incubate the inoculated Roux bottle at 32°C. for 24 hours. Harvest the resultant growth from the Roux bottle by washing the agar surface with approximately 50 ml. of sterile 0.85% sodium chloride solution. The bacterial suspension thus obtained may be stored under refrigeration for at least 2 weeks with little or no effect on the response to the antibiotic.

The day-to-day variation in assays of this type is reduced by the standardization of the daily inoculum by one means or another. This may be done by actual viable count on the inoculum suspension just described, or by some approximation of bacterial population such as optical density. Standardization of the inoculum suspension may also be achieved by running test plates with a range of dilutions of the suspension and selection of the dilution yielding the desired response, although this is somewhat more time consuming. A workable method for this assay is dilution of an aliquot of the stock inoculum suspension to yield a suspension of approximately $1 \cdot 10^8$ viable cells per milliliter. Inoculate the "seed" agar by adding 1.0–1.5 ml. of this suspension per 100 ml. of liquefied and cooled (48°C.) agar. Pour the inoculated agar immediately after inoculating to prevent undue death of the bacteria at the temperature of the liquefied agar.

Other standardization methods have been used with equally satisfying results and were dependent upon the needs of, and equipment available in, the various laboratories. The main objective being sought is the recurrence of well-defined zones of approximately equivalent diameter for a given concentration of antibiotic (constancy of slope). The actual zone diameters are of little consequence, within limits, and will be a function of the concentrations of antibiotic chosen, the inoculum used in "seeding" the plates, and other factors discussed under diffusion theories elsewhere in this text. Generally, an attempt is made to establish that the net result of these interacting factors permits a linear response when the logarithm of the dose is plotted against the zone diameters or some function thereof, and that the slope is neither too steep nor too flat. This makes for ease of calculation and will increase the precision of the assay.

b. *Sarcina lutea.* Prepare an agar slant of the culture as directed in Section III.B. After the initial incubation, harvest the growth from the slant by washing with 2 to 3 ml. of sterile G. & R. broth medium No. 3. Transfer the suspension to a Roux bottle containing 300 ml. of sterile G. & R. agar No. 1 and distribute evenly over the surface of the agar with the aid of sterile glass beads. Incubate the inoculated Roux bottle at 30°C. for 16 to 18 hours. Harvest the resultant growth from the Roux bottle by washing the agar surface with approximately 25 ml. of sterile G. & R. broth medium No. 3. Adjust the suspension thus obtained to approximately $5 \cdot 10^8$ viable cells per milliliter. The adjusted suspension constitutes the stock inoculum suspension and is used for preparation of the inoculated media. This stock inoculum suspension may be stored under refrigeration for a period of at least two weeks. Prepare the "seed" layer by adding from 0.3 to 0.5 ml. of this inoculum suspension to each 100 ml. of liquefied and cooled (48°C.) agar and pour into plates immediately. Both layers must be uniform in thickness.

4. Standards

Dilute samples of the stock standard solution in sterile pH 6 phosphate buffer immediately prior to use on the day of the assay. The choice of standard levels will, of course, determine to what extent these aliquots are diluted. In this investigator's experience, the *Staphylococcus aureus* test yields a linear response in the range from 0.5 to 4.0 u./ml. and the *Sarcina lutea* method in the range from 0.05 to 2.0 u./ml. The latter range also appears to hold true when dilutions are made in serum or bovine albumin, assuming that the previously mentioned criteria for selection of these diluents are met. These concentration ranges refer to penicillin G and may vary considerably if one is working with other penicillins or penicillin derivatives.

The number and spacing of doses in the standard response curve is a matter which must be largely decided upon by the analyst in light of his knowledge of the materials which he is assaying. Select doses that are evenly spaced on the logarithmic scale (the concentrations form a geometric progression); this makes for simplicity of calculation of the line of best fit, and reduces possibilities of calculation errors.

It should be remembered in preparing the standard response curve that the buffer diluent mentioned above only applies when one can similarly dilute the samples being assayed. If one is measuring the activity present in biological fluids, organic solvents, etc., the standards must be so diluted as to contain equivalent concentrations of these materials or the assay may well be invalid. All too frequently the erroneous assumption that

5–10% extraneous materials would not affect the assay has caused considerable embarrassment to the analyst.

5. Samples

After having overcome any extraction problems, such as were discussed earlier, the analyst has only to dilute his sample to the range of his standard curve; remembering, of course, the importance of the diluent he chooses. Selections of the proper tools for this task are discussed elsewhere in this text. Assuming proper selection of pipets, flasks, diluents, etc., and assuming the adherence to accepted analytical techniques, one should have little difficulty in the performance of this portion of the assay. Although given little space here, all of these steps are extremely important to the success of the method. Without being facetious, this investigator believes that unless microbiological assays are built around accurate, easily read, and scrupulously clean glassware, high quality reagents, and intelligent and conscientious people, they should not be run. On the other hand, having all of these attributes is not a guarantee of success.

6. Plating

Standards and samples apparently have been successfully applied to the inoculated agar surface in a number of ways. The use of holes punched in the agar for this purpose is described in Chapter 2, and the use of stainless steel cups or cylinders is described under Chloramphenicol, Chapter 6.4, Section II.A. Paper disks of good quality with diameters of $\frac{1}{4}$, $\frac{3}{8}$, and $\frac{1}{2}$ inch are commercially available (as penicillin disks) and have been widely accepted in this country although some workers have experienced difficulty in their use when assaying samples of bacitracin. Loading of the paper disks may be accomplished by use of an especially designed self-filling pipet (0.09 ml. for $\frac{1}{2}$ inch disk), or by holding the disk with forceps and dipping an edge in the solution to be applied. If the dip method is used, care should be exercised to remove excess liquid adhering to the surface of the disk. This may be accomplished with a quick flip of the wrist or by touching an edge of the disk to the dry wall of the sample container. Other devices such as a fish-spine bead and a microsyringe have also been used for this purpose. With the latter device a microdrop of the solution is placed directly on the agar surface.

Possibly of greater importance than the device used for application of the solutions is the constancy of the technique employed. If the device used does not afford reasonable control of the volume of sample applied to the agar surface, it is imperative that the same individual apply all samples and standards within a given assay or test. If reasonable tolerances can be maintained, group effort with its inherent advantage of decreasing time effects on the assay may be used.

The influence of elapsed time on the assay is mentioned above since, in many laboratories, the assay design may not compensate for this effect. Within certain limits, this influence is merely a reduction in zone diameters without alteration of the slope of the dose-response line and a single reference level on each plate serves to compensate for it. It has been this investigator's unhappy experience, however, to encounter significant slope alterations in certain assays even in relatively short periods of time (10–20 minutes). In these cases, group effort by decreasing the time may obviate the need for an elaborate statistical design, an excessive number of standards, or several reference levels on the sample plates. In any event, it is probably advisable to hold all assay plates under refrigeration until immediately prior to use and then to apply all solutions as rapidly as is consistent with good technique.

Penicillin standards and unknowns give relative responses independent of mode application (hole, cup, disk). Other antibiotics and certain samples of them may not.

7. Incubation

For both of the tests described, an incubation time of 16 to 18 hours should be sufficient to attain inhibition zones of a desirable size and definition. The general practice is to incubate the *Staphylococcus aureus* plates at 32–35°C. and the *Sarcina lutea* plates at 26° to 28°C. One has some leeway in this incubation temperature, and in our laboratory it is common to incubate either assay at 28° to 30°C. Probably of greater significance, within limits, is the uniformity of temperature within the incubator chosen. The incubator should be tightly sealed and should have some means of providing constant temperatures to all shelf areas. One must be cognizant of the capabilities of the incubator, and must resist the temptation to overload or overstack within it. Stacks of petri plates, whether glass or plastic, are relatively poor conductors of heat and considerable plate-to-plate variation will be evidenced if care is not exercised in this step of the technique. Ideally, the plates should be incubated in a layer one plate deep until incubation temperature is attained. Then they can be stacked.

Obviously, if one uses the cylinder-plate, considerable care must be taken in transporting plates to the incubator, and in placing the plates on the shelves. Spillage from the cylinders causes irregularly shaped zones which cannot be interpreted. Various trays and holding devices have been developed to overcome this problem, but many of these present heat distribution problems that reduce their practicality.

F. Measuring the Response

Response in these assays is the appearance of circular translucent areas or zones surrounding the cylinder or disk containing the antibiotic. The zones are bounded at their periphery by an intensely opaque area of bacterial growth. Any device which will accurately measure the diameter of these circular zones will suffice in this phase of the assay. An assortment of calipers, millimeter rules, scale projectors, and automatic and semi-automatic measuring devices has been utilized for this purpose. The selection of this equipment will undoubtedly be governed by the number of assays being run, the degree of precision needed, and the allowable cost.

The use of "zone readers" specifically designed for this purpose normally requires the careful and symmetrical placement of cylinders or disks on the plates in the earlier stages of the assay. This is to ensure that, upon rotation, the true diameter of the zone will be presented to the instrument. Again, there are several commercially available devices to assist in accomplishing this task. In a small laboratory doing only an occasional assay, simple templates should serve equally well. With anything short of fully automatic equipment the principal error will be that of the human eye in distinguishing the boundary between the contrasting areas. In fact, even a fully automated system is not capable of perfect constancy in this regard. The error involved here is usually rather small except in those assays where there is very poor definition of zones, or where a "halo" effect is seen. Many times these undesirable effects are the result of trying to extend the lower limit of sensitivity of the assay. The semiquantitativeness of the assay is acceptable only because none better can be obtained. A valid assay is possible only when standard and sample give the same type of zones.

Considerable emphasis has been placed recently on automatic or semi-automatic read-out of assay responses and several devices to accomplish this goal have made their appearances. One of these, the TCI Zone Comparator, is described elsewhere in this text (Chapter 5). The approach at Eli Lilly & Company was to install a Datex Shaft Position Encoder (same as that used in recording photometric assays) on the conventional Fisher-Lilly Antibiotic Zone Reader along with a single push-button switch. The operator zeroes the instrument and traverses the zone of inhibition in the conventional manner. Then she presses the button which activates an IBM 526 card punch by way of a Datex Translator and Datex Junction Box. A number proportional to the observed zone diameter is automatically punched into a card system suitable for automatic calculation of assay results and their statistical evaluation. The shaft encoder has about three times as great a resolution as the scale of the instrument. This system has advantages in eliminating errors caused by manual recording of data and calculations. The system appears to be rapid (approximately 200 four-zone plates/hour), precise, and relatively free from mechanical defects.

G. Computation of Answers

1. Standard Curve

Construct the standard dose-response curve by plotting the average response value (millimeters) for each of the standard levels against the logarithm of the respective doses, and by joining these plotted points. Normally this is simplified somewhat by plotting this curve on semi-logarithmic graph paper with the doses represented by the log scale. Since this plot should be truly linear, in the ranges previously discussed, it is probably somewhat more accurate to draw the "line of best fit" for the observed data than it is to merely connect the observed points with a series of straight lines. This tends to "smooth out" the errors inherent in the measurement of each of the individual points, and gives one a con-figuration capable of somewhat simpler mathematical interpretation. It is impossible to overemphasize the need for satisfying the criterion of "linearity" in one's own laboratory. In spite of the most conscientious ad-herence to published detail, one often is unable to duplicate the conditions prevailing in other laboratories. The linear ranges given above and in the other plate assays should be taken as guides, not as doctrine.

If one has selected dose levels evenly spaced on the logarithmic scale, the calculation of the "line of best fit" becomes extremely simple as may be seen from the article by Deutschberger and Kirshbaum.[13] However, even if one has not fulfilled this latter criterion, the calculation of these lines by the method of least squares is not an unwieldy technique. In plot-ting response lines by this method it is probably well to plot the observed points as well as the calculated line. This will give the analyst some idea of the precision of his method (how well the observed and computed points agree) and will enable him to assure himself that no single point had an undue influence on the slope of this line. For instance, if the higher levels of standard tend to "plateau" it may be that, under the conditions of the particular assay involved, one is not really in the linear range. Such a visual check is especially necessary if no statistical tests are to be made on the data to prove that there is no regression from a straight line. In abbreviated assays, it is well to make a quick check on the "slope" of the response line, as this should remain fairly constant from day to day. Any wide deviation in slope should alert the analyst to check all facets of his test.

2. Potency of Samples

To determine the potency of the samples assayed, add to or subtract

---

[13] J. Deutschberger and A. Kirshbaum, *Antibiotics & Chemotherapy* **9**, 752 (1959).

from the sample-zone mean the difference between the mean of the standard level run on the sample plates (the so-called reference point) and the theoretical value for this concentration—i.e., the intercept of the ordinate for this same concentration with the plotted standard response line. In other words, if we assume that the standard response line describes the true response for each value of standard, then some environmental condition must have altered the response of the standard on the sample plates unless their mean is identical to that of the theoretical. The additional assumption is made that if some effect has depressed or enhanced the standard response on the sample plates by an amount "X" that this effect has also caused a similar "X" enhancement or depression on the size of the sample zones on those same plates. Therefore, whatever quantity (in millimeters) is needed to make the sample-plate standard-zone equal to the theoretical zone diameter is also applied to the sample zones on those same plates. To further clarify the matter let us consider an exampe: if the theoretical value for a 0.5-unit standard is 20.0 mm., and the 0.5-unit reference point averages 20.5 mm. on a given set of sample plates, then one would have to assume that the mean of the sample-zone diameters observed on that same set of plates is also 0.5 mm. larger than it should be. To correct this situation, subtract 0.5 mm. from the observed mean of the sample zones before reading the response from the standard dose-response line.

Once the "adjusted" sample response is determined, read this value from the standard curve and multiply the reading obtained by the dilution factor for the particular sample to obtain the potency of that sample.

3. Validity of the Assay

The degree to which one wishes to pursue this concept will probably hinge upon the purpose of assay, the time and equipment available for computation, and the statistical talents of the analyst. With only slight modification, the one-dose assay method described can be converted to a three-dose technique which allows for ascertaining the parallelism of the standard and sample response lines and curvature of the sample lines. Three-dose levels should always be used when assaying very complex materials or when a new sample type is presented to the analyst. If assay conditions are altered, or if a new diluent is proposed, it would certainly be well to demonstrate that the response line is truly linear over the range of doses being considered and that standard and sample dosage response lines have the same slope. Many other applications of statistical methods present themselves in these methods, but are beyond the scope of this discussion. Those interested in pursuing this matter further will find the literature in the field of biometrics ample to satiate their curiosity.

IV. Large Plate Assay

Mr. Simpson in Chapter 2 illustrates application of large plate methods to antibiotic assay by giving the details of a penicillin G assay. Standards range from 0.5 to 4 u./ml.

The petri plate method given in Section III is inherently inefficient. The standard zones may comprise from 53 to 90% of the total depending upon the design of the test and the number of samples assayed in a test. A large plate design of the approximate accuracy of the petri dish assay uses about 25% of the zones for standards. This apparent greater efficiency of the large plate assay makes it worthy of more consideration than it has received in the U. S. The inherent difficulties of the test seem an insufficient cause of its rejection. The difficult step, the plating-out, probably can be made much easier to do by a slight change in design and in details of manipulation. See Chapter 6.3, Section III, for a description of a large plate method with simplified design.

V. Hydroxamic Acid Method

A. Introduction

Several chemical or physical methods for determining penicillins in purified preparations have been described. The most useful methods are: the iodometric,[14,15] the penicilloic acid,[16] the isotope dilution,[17] and the hydroxamic acid.[18,19] Infrared absorption by the β-lactam ring and titration of the carboxyl group set free by opening of the β-lactam ring are also used for special purposes. The hydroxamic acid method is the most generally applicable of these because it can be applied to rather impure samples without preliminary purification.

B. Principle of the Method

Penicillin reacts with hydroxylamine to form hydroxamic acid which gives a purple complex with ferric ion. Under carefully standardized conditions, the purple complex is proportional to concentration of the penicillin. The reaction with hydroxylamine occurs at the carbonyl group of the intact β-lactim ring to open the ring and to form the hydroxamic acid. Penicilloic acid does not give the reaction because it no longer contains

[14] J. F. Alicino, *Ind. Eng. Chem. Anal. Ed.* **18**, 619 (1946).

[15] J. F. Alicino, *Anal. Chem.* **33**, 648 (1961).

[16] S. C. Pan, *Anal. Chem.* **26**, 1438 (1954).

[17] M. Gordon, A. J. Virgona, and P. Numerof, *Anal. Chem.* **25**, 1208 (1954).

[18] J. H. Ford, *Ind. Eng. Chem. Anal. Ed.* **19**, 1004 (1947).

[19] G. E. Boxer and P. M. Everett, *Anal. Chem.* **21**, 670 (1949).

the β-lactam ring. The formation of hydroxamic acid is not specific to penicillins since many lactones, esters, anhydrides, and amides also react with hydroxylamine to form hydroxamic acids. Ketones and aldehydes react to give oximes that form colored complexes with ferric ion. Ford [18] eliminated the effect of the interfering substances and made the method specific for penicillin by the use of a blank which differed from the test solution in only one respect; the penicillin had been hydrolyzed to penicilloic acid by penicillinase. The specificity of the method is entirely dependent upon the specificity of the enzyme penicillinase which, so far as is known, does not hydrolyze any other than the amide linkage of the β-lactam ring of penicillins.

Although hundreds of thousands of determinations have been made by the method, very little has been published on the theory. Boxer and Everett[19] improved the method of Ford.[18] Mørch[20] studied the influence of concentration of reagents, pH, and time of reaction on the intensity of the color. Mørch found that: $(P) = k_1(E_{490})$ at constant (Fe^{3+}) and pH, $1/E_{490} = k_2 + k_3(H_3O+)$ at constant (P) and (Fe^{3+}), and $1/E_{490} = k_4 + k_5/(Fe^{3+})$ at constant (P) and pH, where (P) and (Fe^{3+}) are the concentrations of penicillin and ferric ions, respectively, and E_{490} is the extinction at 490 mμ. He also found that fading of the purple color was caused by reduction of the ferric ion by excess hydroxylamine. He wrote that the hydroxylamine concentration should be not less than 0.5 N if the formation of penicillin hydroxamic acid is to be completed in a reasonable time. These studies show the necessity for careful control of operational details. Each operator needs to standardize his technique and practice it until he can reproduce his results with an error of $\pm 1\%$ for a given lot of reagents. The method given here is essentially that of Boxer and Everett.

C. Reagents

Hydroxylamine hydrochloride, 5 M. Dissolve 87.0 gm. of the salt in distilled water and dilute to 250 ml. The solution is stable in the cold.

Sodium hydroxide solution. Dissolve 44 gm. sodium hydroxide in water, add 5.0 gm. sodium acetate (anhydrous) and dilute to 250 ml.

Iron reagent. Dissolve 50 gm. ferric ammonium sulfate dodecahydrate (reagent grade) and 23 ml. of concentrated sulfuric acid in water, dilute to 250 ml.

Penicillinase. Prepare a solution with a concentration of 10,000 u./ml. water. The solution is stable for more than 1 week if kept refrigerated.

Hydroxylamine buffer. Mix 1 volume of the 5 M hydroxylamine solution with 1 volume of the sodium hydroxide solution and add 4 volumes of water and 4 volumes of 95% ethyl alcohol. This reagent is stable for only about 4 hours and should not be used if it is any older.

[20] P. Mørch, *Dansk Tidsskr. Farm.* **28,** 157 (1954).

D. Procedure

Pipet 1.00 ml. of sample into each of two test tubes or colorimeter tubes, add 1 drop (about 0.05 ml.) of penicillinase solution to one tube, let the tubes stand at room temperature for 10 minutes. The penicillin in the tube containing the penicillinase will be converted to penicilloic acid and become the blank. At the end of 10 minutes, add 5.0 ml. of the hydroxylamine buffer solution to each tube, let stand for 15 minutes, and then add rapidly 1.0 ml. of the iron reagent to each tube, mix, and read in a colorimeter or spectrophotometer at a standard time (usually within 1 minute). The time of reaction between the penicillin and hydroxylamine buffer should be the same for sample and standard.

The blank and its corresponding sample are read in an Evelyn colorimeter (540 filter) or a Klett-Summerson colorimeter (50 filter) or in a spectrophotometer at 490 mμ.

A calibration curve is made for each penicillin and instrument. Beer's law is obeyed for a Beckman Model B spectrophotometer up to about 2 mg. of penicillin G in the 1-ml. sample (1-cm. cell depth). The slope of the calibration curve will be slightly different with different lots of reagents. The pH of the hydroxylamine buffer should be measured before the alcohol is added and adjusted, if necessary, to a value considered standard for the laboratory. If highest accuracy is needed, include several standards along with the sample. The best procedure requires that the amount of penicillinase solution added to the blank be added to the sample after the hydroxylamine buffer to compensate for the slight turbidity and color from the pencillinase.

E. Interfering Substances

Anything which will react with hydroxylamine, chelate ferric ion, reduce ferric to ferrous ion, or change the pH of either the hydroxylamine step or the iron step, will interfere in a way not completely compensated for by the blank. Several of these interferences can be nullified by the technique of the internal standard. Fortunately, most of the interferences are the kind compensated by the blank. Fluoride, phosphate, ketones, and aldehydes interfere. Ford showed that a saturated solution of amyl-acetate, acetamide at 100 mg./ml. and 2.5 mM. phosphate buffer did not interfere with the assay when the penicillinase blank was used.

Substances such as fluoride, citrate, and phosphate, which chelate ferric ion and thus decrease its concentration will decrease the intensity of the purple color. Interference[21] with penicillin G assay by pH 7.0 citrate and phosphate buffers was investigated. Citrate buffer at 0.02 M or lower

---

[21] From the analytical laboratory of the Antibiotic Manufacturing and Development Division, Eli Lilly and Company.

did not affect the assay. At 0.2 M concentration in the sample it caused a decrease of 3%. Influence of phosphate buffer at 0.002 M was not measurable. The buffer caused a reduction of 3% at 0.02 M and a reduction of 20% at 0.2 M. The calibration curves (at least to 2000 u./ml.) were straight in the presence of the buffers. Therefore, the interference caused by a phosphate or citrate buffer can be eliminated by preparing both the standards and samples in the same buffer. The standard curve so prepared will have a slope less than the usual curve prepared with standard sample dissolved in water.

F. Relative Color Intensities

Different penicillins give slightly different color intensities as was shown by Ford. Examples for several penicillins are given in Table III.

TABLE III

OPTICAL DENSITY AT 490 mμ OF 1 mg. OF THE PENICILLINS ASSAYED AS DESCRIBED WITH A BUFFER REACTION TIME OF 15 MINUTES AND pH 6.8 AND 7.5

		Optical density	
Penicillin	Mol. wt.	6.8	7.5
6-Aminopenicillanic acid	216	0.412	0.428
Benzylpenicillin, sodium	356	0.273	0.300
Phenoxymethylpenicillin, free acid	350	0.268	0.295

G. Discussion

Time of reaction with hydroxylamine is important and should fall within the flat portion of the curve (Fig. 1). Since fading begins as soon as the iron reagent is added, the purple color should be measured at a fixed time after the addition of the iron. The fading occurring in a few minutes is shown in Fig. 1.

The concentration of hydroxylamine buffer is designed for a sample volume of 1 ml. Should the sample be so dilute (0.5 mg./ml.) that small optical densities are obtained, the sample size should be increased to 3 ml., the 4 volumes of water omitted from the hydroxylamine buffer, and 3 ml. of hydroxylamine buffer added to the sample. In this way the larger sample is accommodated without changing concentration of reagents or total volume (7 ml.). The standard is treated the same as the sample.

The pH of the hydroxylamine reagent should be between 6.4 and 7.5 before the alcohol is added. The higher the pH, the greater is the amount of hydroxamic acid formed and the longer is the time required for the maximum to be attained. Each penicillin should be investigated and the reaction time selected to fall within the region of no change. Experiments

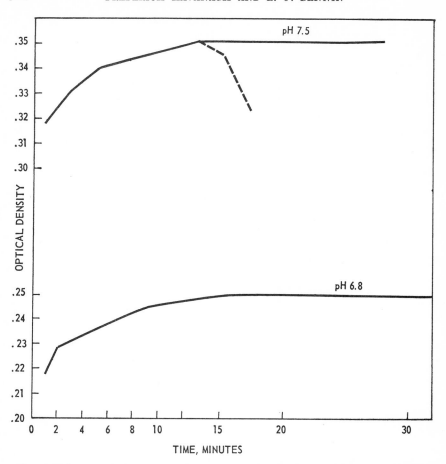

Fig. 1. Influence of time of reaction between hydroxylamine buffers and two concentrations of phenoxymethyl penicillin (penicillin V) upon the concentration of iron complex as measured at 490 mμ. The dashed line originating at the 13-minute point of the pH 7.5 line is the time course of the fading of the color of the iron complex.

in these laboratories[21] showed that the apparent reaction between hydroxylamine and penicillin was essentially complete for six penicillins once the reaction time had reached 13 minutes. The reaction time could be prolonged with certain penicillins to as long as 40 minutes with no decrease in optical density of the iron color.

Although a reaction time of 15 minutes between penicillin and hydroxylamine was required to give the maximum amount of hydroxamic acid, the same may not be true of the blank. A blank[21] typical of a penicillin fermentation broth was prepared by treatment with penicillinase. The sample of the blank was reacted with the hydroxylamine buffer for 1, 3, and 15 minutes before the iron solution was added. The optical densities were the same and no purple color was evident. The blanks slowly increased with time. The properties of the blank suggest that its optical absorption came from original colored substances and from turbidity caused by precipitation of unknown materials in the acid iron solution. These results indicate that a reaction time of 15 minutes between blank and hydroxylamine buffer may be unnecessary and could be omitted with a consequent shorting of the total time of the assay by 10 minutes because the penicillinase treatment of the blank and the hydroxylamine reaction with the sample could proceed simultaneously.

The pH of the solution should be about 1.0 after the iron reagent has been added. If it is not low enough, add more sulfuric acid to the iron reagent.

Some of the synthetic penicillins are more stable toward penicillinase than biosynthetic penicillins and require more penicillinase or a longer time to effect complete conversion into penicilloic acid. Test each new penicillin for completeness of conversion and adjust conditions to obtain it.

The penicilloic acid obtained from 1 mg. of pure penicillin gives an optical density indistinguishable (<0.01) from that of the reagent blank (water substituted for the penicillin solution).

H. Preparation of Samples

1. Fermentation Beers

Filter to remove mycelium, other solids, and oils. Dilute the clear or nearly clear filtrate with water to 0.5 to 1.0 mg./ml. The pH should be between 6.5 and 7.5.

2. Process Samples

Filter, if necessary, and dilute with water to 0.5 to 1.0 mg./ml., adjusting pH to 6.5 to 7.5.

3. Pharmaceutical Preparations

Under this heading are put those preparations not suitable for direct determination because of interference from impurities (compounds with acid-base character or with competitive complex formation character, or inhibitors of penicillinase) which cause alteration in the slope of the curve relating optical density to penicillin concentration, thus making the use of a standard curve deceptive.

Purification[20] and concentration of the penicillin through extraction by amyl acetate can be effected with a loss of about 3%. Recovery of added penicillin is a test of operating technique. Mix from 5 to 20 ml. of sample (2.5–5 mg. of penicillin) with 20 ml. of water and 40 ml. amyl acetate in a 125-ml. separatory funnel, cool to 0° to 5°C., and, added to 20 ml. ice cold glycine buffer (45 gm. glycine and 50 ml. conc. HCl per liter), shake 30 seconds. Draw off the aqueous layer and discard, add 5–10 gm. anhydrous sodium sulfate to the funnel to dry the amyl acetate, and filter. Shake 30.0 ml. of the filtered amyl acetate with 5.0 ml. of 2% sodium bicarbonate solution, remove, and filter the aqueous layer. Use two 1-ml. portions of the bicarbonate extract for the assay. The concentration of penicillin as read from the standard curve needs to be multiplied by $[1.03 \ (5)]/x$ to obtain the concentration of the sample where x is the volume of the sample taken for extraction.

Recovery of added penicillin is determined by adding 2.5–5 mg. of penicillin to x ml. of sample and proceeding as above to the bicarbonate extraction step. Extract 30 ml. of amyl acetate with 10 ml. of 1% sodium bicarbonate and determine the concentration of the penicillin in the extract. The multiplying factor now is $[1.03 \ (10)]/x$. Each operator should determine his recovery factor; it might not be the 1.03 found by Mørch.

According to Mørch the deviation between duplicate determination for broth is less than 2% and the added penicillin can be recovered with ±3%.

6.11 Penicillin V

FREDERICK KAVANAGH

Eli Lilly and Company, Indianapolis, Indiana

I. Photometric Method

The only difference between this assay and that for penicillin G is the standard. Penicillin V (phenoxymethylenepenicillinic acid, phenoxymethyl penicillin) is assayed in terms of the crystalline free acid (mol. wt. 350.38) although the potassium salt (mol. wt. 372.5) is the compound usually found in preparations. The range is from 0.07 to 0.4 μg./ml. with the 50% point at about 0.17 μg./ml. Standard solutions of 0, 0.10, 0.18, and 0.36 μg./ml. are the minimum number needed to define the log-probability standard curve.

Pure potassium salt of phenoxymethyl penicillin theoretically contains 940.6 μg. or 1594.3 units per milligram of dry salt.

II. Plate Method

Use penicillin V free acid (1695 units/mg.) as the standard in the penicillin G assay. The same concentrations (units/milliliter) of penicilin V free acid are used as of penicillin G in the assay.

6.12 Polymyxin

L. J. DENNIN

Eli Lilly and Company, Indianapolis, Indiana

I. Introduction

The test organism used for the cylinder plate assay of polymyxin is *Bordetella bronchiseptica*. The organism is relatively insensitive to the antibiotic under normal conditions, however, and it is generally necessary to increase the sensitivity of the method by the use of a high salt buffer. The alternate serum assay method also requires use of this buffer as well as a reduced "base" agar layer.

II. Test Organism

Maintain the stock culture of *B. bronchiseptica* (ATCC 4617) by weekly transfers to fresh sterile slants of G. & R. agar medium No. 9. Incubate the freshly prepared slants at 32° to 35°C. for 16 to 24 hours and store at 4° to 6°C. until ready for use in the preparation of the inoculum suspension.

III. Standard Solutions

Dry the polymyxin working standard for 3 hours at 60°C. and a residual pressure of 5 mm. Hg or less in a tared weighing bottle fitted with capillary tube. Accurately weigh a quantity of the dried standard to make a solution containing 10,000 units/ml. of polymyxin activity (base). Transfer this portion to a suitable container and bring to volume with 10% phosphate buffer pH 6. Store the stock standard solution at 4° to 6°C. and use for a period not to exceed 2 weeks.

IV. Sample Preparation

Polymyxin samples normally present little difficulty to the assayist from the standpoint of sample preparation. The antibiotic is readily soluble in 10% pH 6 phosphate buffer and is relatively insoluble in diethyl ether. This provides a system for removal of polymyxin from preparations such as ointments. The tetracycline antibiotics may cause interference in the assay of polymyxin but are separated from it by making use of selective solubilities. Oxytetracycline, oxytetracyline hydrochloride, and tetracycline are soluble in acetone; while tetracycline hydrochloride or chlortetracycline hydrochloride are soluble in acetone containing 0.3% piperidine. Polymyxin is relatively insoluble in both solvents.

V. Mechanics of the Assay

A. Design

Refer to Penicillin, Chapter 6.10, Section III.E.1.

B. Media

For conventional assays use a "base" layer consisting of 21 ml. of G. & R. No. 9, and a "seed" layer consisting of 4 ml. of G. & R. No. 10. For serum assays reduce the volume of the "base" layer to 10 ml.

C. Inoculum

Prepare an agar slant of the culture as described in Section II. Wash the growth on this slant from the agar surface using 2–3 ml. of sterile 0.85% sodium chloride solution. Transfer the suspension to a Roux bottle containing approximately 300 ml. of sterile G. & R. No. 9 and distribute it evenly over the surface of this agar with the aid of sterile glass beads. Incubate the inoculated Roux bottle at 32° to 35°C. for 24 hours. Harvest the resultant growth from the Roux bottle by washing the agar surface with approximately 50 ml. of sterile 0.85% sodium chloride solution. Adjust the inoculum suspension to contain approximately 1×10^8 viable cells per milliliter and store at 4° to 6°C. for a period not to exceed 2 weeks. Inoculate the "seed" agar by adding 0.2–0.3 ml. of this suspension per 100 ml. of liquefied and cooled (48°C.) agar and pour the plates immediately.

D. Standards

Dilute samples of the stock standard solution in sterile 10% pH 6 phosphate buffer or in appropriate biological diluents which, in turn, have been diluted (1:2) with 20% pH 6 phosphate buffer (see Table I). In the latter assay, samples must also be diluted (1:2) using the 20% buffer as the diluent. If dilutions greater than 1:2 are required, the diluent should be the mixed buffer—biological fluid diluent used for the standard

TABLE I

10% Phosphate buffer solution (G. & R. No. 6)	pH 6.0
Dibasic potassium phosphate	20 gm.
Monobasic potassium phosphate	80 gm.
Distilled water to make	1000 ml.
20% Phosphate buffer solution	pH 6.0
Dibasic potassium phosphate	40 gm.
Monobasic potassium phosphate	160 gm.
Distilled water to make	1000 ml.

response curve. The choice of standard levels will, of course, determine to what extent the stock standard solution is diluted. In my experience, the conventional assay has a linear response (semilog plot) in the range from 5–20 u./ml. and the more sensitive method normally yields a linear response in the range 5–10 u./ml. assuming a careful selection of the "control" diluent. Standard levels should be evenly spaced on the logarithmic scale.

E. Samples

See Penicillin, Section III.E.5.

F. Incubation

Incubate the assay plates for 16 to 18 hours at 38°C.

VI. Measuring the Response

See Penicillin, Section III.F.

VII. Computation of Answers

See Penicillin, Section III.G.

6.13 Ristocetin

ROLAND L. GIROLAMI

Abbott Laboratories, North Chicago, Illinois

I. Introduction

Ristocetin[1] is an antibiotic active against gram-positive bacteria and is produced by an actinomycete, *Nocardia lurida*.[2] The antibiotic is a

---

[1] The trade name of Abbott Laboratories for ristocetin is Spontin.

[2] W. E. Grundy, A. C. Sinclair, R. J. Theriault, A. M. Goldstein, C. J. Rickher, H. B. Warren, Jr., T. J. Oliver, and J. C. Sylvester, *Antibiotics Ann.* **1956/57, 687** (1957).

mixture of two closely related crystalline compounds designated ristocetin A and ristocetin B. The two components have a comparable antimicrobial spectrum but ristocetin B is 3–4 times more active than ristocetin A. The ristocetins can be distinguished by paper strip chromatography and paper strip electrophoresis.[3] Commercial preparations of the antibiotic contain >90% ristocetin A.

The complete structure of ristocetin has not been established but the antibiotic is amphoteric and contains amino and phenolic groups as well as four sugars: arabinose, glucose, mannose, and rhamnose. The molecular weight is approximately 4000. The antibiotic is readily soluble in acidic aqueous solutions, less soluble in neutral aqueous solutions and generally insoluble in common organic solvents. It has excellent stability at low pH but is rapidly inactivated above pH 7.5.

II. Turbidimetric Method

A reproducible and reliable turbidimetric assay method for ristocetin has been described by Eisenberg and Kirshbaum of the U. S. Food and Drug Administration.[4] Ristocetin B is approximately 2.5 times more active on a weight basis than is ristocetin A by this assay procedure. The plate assay method (Section IV) does not give a differential response. Samples which are known to contain substantial amounts of ristocetin B (established by paper chromatography) must be run at higher estimated potencies. For example: a sample which contains 20% ristocetin B may assay 600 μg./mg. by the plate assay method but would assay 780 μg./mg. of ristocetin A equivalent by the turbidimetric method.

The method has certain disadvantages. The region in which culture response is proportional to antibiotic concentration is somewhat narrow. It is necessary, therefore, that potency be known with a fair degree of accuracy. If the concentration is unknown the sample should be assayed at several estimated potencies or assayed first by the plate procedure.

A. Culture

The organism used in the turbidimetric assay of ristocetin is *Staphylococcus aureus* ATCC 6538P (FDA 209P). The choice of organism is mainly one of convenience: it has good growth characteristics and acceptable sensitivity but little or no selectivity of response for this antibiotic.

Transfer the culture as needed from lyophilized stocks to test tube slants of Antibiotic Medium No. 1 (Difco). Incubate for 24 hours at

---

[3] J. E. Philip, J. R. Schenck, and M. P. Hargie, *Antibiotics Ann.* **1956/57,** 699 (1957).
[4] W. Eisenberg and A. Kirshbaum, Personal communication.

37°C. Transfer from slants less than 1 week old to 100 ml. of Antibiotic Assay Broth [BBL (Baltimore Biological Laboratory, Inc.)] in a 300-ml. Erlenmeyer flask and incubate on a rotary shaker (150 r.p.m.) for 8 hours at 37°C. Refrigerate the culture overnight. Dilute the culture with sterile distilled water to give a concentration of 2.1×10^8 cells per milliliter, normally a dilution between 1:3.2 and 1:4. For preparation of assay inoculum add 4% by volume of the diluted culture suspension to sterilized test broth (Antibiotic Assay Broth, BBL).

B. Standards

Ristocetin A has been assigned a potency of 1000 μg./mg. The working standards are kept under refrigeration within a desiccator. The antibiotic has excellent stability under these conditions.

To prepare standard solutions, accurately weigh a sample of ristocetin A standard in a sterile weighing bottle and add a sufficient quantity of sterile, 1% phosphate buffer pH 6.0 to give a solution containing 500 μg./ml. The stock solution may be kept at 5°C and used for several weeks. Make further dilutions from the stock solution in buffer to give standard solutions containing 5, 6, 8, 10, 12, 14, 16, and 20 μg./ml.

C. Sample Preparation

The commercially available, pharmaceutical form of ristocetin is a sterile, lyophilized powder. The excellent solubility and acid stability of ristocetin makes sample preparation relatively simple. Dilute all unknowns in phosphate buffer pH 6.0 to a theoretical concentration of 10 μg./ml.

D. Mechanics of Assay

Add a 1-ml. aliquot of each standard solution and each unknown to triplicate, sterile test tubes (18 × 150 mm.). Prepare culture controls by adding 1 ml. of phosphate buffer alone to triplicate tubes and blank controls by adding 1 ml. of phosphate buffer plus 1 drop of 40% formaldehyde solution to triplicate tubes. Add 9 ml. of inoculated test broth (Section II.A) to all tubes. Incubate in a water bath at 37°C for 3 hours or until the concentration of culture controls reaches approximately 1.5×10^8 cells per milliliter. After sufficient growth has been obtained add 1 drop of 40% formaldehyde solution to each tube.

E. Reading Responses

Determine the turbidity (per cent transmission) of all tubes with an electrophotometer at 650 mμ using the blank controls to set the instrument to 100% transmittance. Care should be taken to ensure uniform culture suspension before determining turbidity.

F. Calculation

1. Standard Curve

Average the three readings for each concentration of the standard and construct a standard curve by plotting the standard concentrations logarithmically and the transmission percentages arithmetically on semilogarithmic paper. An "S" shaped curve is obtained.

2. Unknowns

Compare the average transmission readings of the unknowns to the standard curve to obtain concentrations in micrograms per milliliter. Multiply by appropriate dilution factors to obtain the potency of the sample. Good reproducibility is obtained when the unknown falls in the 8 to 12 μg./ml. region of the standard curve. Samples which fall outside of this region are reassayed after an appropriate adjustment is made in sample dilution.

III. Blood Assay Method

A. Culture

The organism used for the determination of ristocetin concentration in blood samples is a sensitive, saprophytic strain of *Corynebacterium* called Abbott GT 768.

Maintain the culture by biweekly serial transfer to slants of Antibiotic Medium No. 1 (Difco). Incubate at 37°C for 24 hours. Use slants as inoculum for 100 ml. Antibiotic Assay Broth (BBL) in a 300-ml. Erlenmeyer flask and incubate on a rotary shaker (150 r.p.m.) at 37°C for 18 hours. Dilute the culture 1:1,000,000 with sterile Brain Heart Infusion Broth (BBL).

B. Standard

Prepare an 80 μg./ml. stock standard in phosphate buffer pH 6.0 as described in Section II.B. Dilute further with an equal quantity of normal animal serum to give a 40 μg./ml. stock solution. Test the serum prior to use in the assay for the presence of substances which affect growth of the test organism.

Prepare a serial twofold dilution series of the stock solution in sterile Brain Heart Infusion Broth (BBL) using 0.5 ml. broth blanks (13 × 100 mm. tubes). Transfer 0.5 ml. from tube to tube and discard 0.5 ml. from the last tube. A 10-tube series is routinely used. Inoculate each of the tubes with 0.5 ml. of the diluted broth culture. The first tube of the series contains 10 μg./ml., the second 5 μg./ml., etc. Shake the tubes well and incubate at 37°C for 18 to 24 hours. Include controls for broth and serum sterility. A detailed discussion of the serial dilution method is given in Chapter 3.

C. Samples

Allow blood specimens to clot at room temperature, rim the clot with a sterile glass rod and refrigerate for 15 minutes. Centrifuge the samples and remove the serum aseptically. Run a serial twofold dilution series for each sample as described for the standard series.

D. Calculation

After incubation read the standards and samples. The greatest dilution in each dilution series which inhibits growth of the culture is the end point.

Table I illustrates the results obtained in a typical blood assay of ristocetin and the method used for calculation of antibiotic concentration.

TABLE I

RISTOCETIN BLOOD ASSAY AND CALCULATION[a]

Tube	Dilution	Concentration of standard (μg./ml.)	Culture growth	
			Standard	Unknown
1	1:4	10.00	—	—
2	1:8	5.00	—	—[c]
3	1:16	2.50	—	+
4	1:32	1.25	—	+
5	1:64	0.62	—	+
6	1:128	0.31	—[b]	+
7	1:256	0.16	+	+
8	1:512	0.08	+	+
9	1:1024	0.04	+	+
10	1:2048	0.02	+	+

[a] Sample potency = concentration of standard at end point multiplied by dilution of sample at end point, or $0.31 \times 8 = 2.48$ μg./ml.

[b] End point of standard.

[c] End point of unknown.

IV. Plate Method

A plate assay procedure is used for the determination of ristocetin in crude materials such as fermentation beers or recovery solutions where the potency cannot be accurately estimated. In this procedure ristocetins A and B give comparable responses. The results of this assay together with chromatographic information on the composition of an unknown sample, especially those with an appreciable quantity of ristocetin B, allows the estimation of potency for the accurate determination of the antibiotic by the turbidimetric method.

The sensitivity of the plate assay method for ristocetin is less than expected based on results in liquid medium. The diffusion of the antibiotic through an agar medium may be restricted by a number of factors including the large molecular size of the compound and agar-antibiotic complexing.

A. Culture

The test organism in the agar plate-paper disk method for the assay of ristocetin is the Illinois strain of *Bacillus subtilis* ATCC 10707. Stock cultures are maintained in lyophilized form.

For the production of a spore suspension of the culture, transfer from lyophilized stocks to test tube slants of B-1 agar (see Table II for composition). Incubate at 37°C for 24 hours. Transfer from slants to 100 ml. of B-1 broth (see Table II) in 500-ml. Erlenmeyer flasks and incubate on a rotary shaker (240 r.p.m.) at 32°C for 24 hours.

TABLE II

MEDIA LIST

Buffer
   Phosphate buffer pH 6.0 (1%)
      K_2HPO_4 2 gm.
      KH_2PO_4 8 gm.
      Distilled water to 1000 ml.
Media
   Antibiotic Medium No. 1 (Difco)
   Antibiotic Assay Broth (BBL)
   Brain Heart Infusion Broth (BBL)
   Antibiotic Medium No. 5 (Difco)

B-1 Agar		*B-1 Broth*
Tryptone	3 gm.	Same as B-1 Agar without
Beef extract	3 gm.	the agar.
Glucose	1 gm.	
Yeast extract	1 gm.	
Agar	15 gm.	
Tap water to 1000 ml.		

Sporulation medium

Peptone	1.0 gm.
Yeast extract	3.0 gm.
Glucose	1.5 gm.
$MnCl_2 \cdot 4H_2O$	0.1 gm.
Distilled water to 1000 ml.	

Use the 24-hour shaken culture at a level of 5% as inoculum for 100 ml. of sporulation medium (see Table II) in 500-ml. Erlenmeyer flasks. Incubate on a rotary shaker (240 r.p.m.) at 28°C for 6 to 8 days. Observe cultures microscopically and harvest when approximately 90% sporulation is evident. Recover the spores by centrifugation and resuspend in 1/50 the original volume. Pasteurize the spore suspension for 30 minutes at 65°C and then count for viable spores by a suitable plating method. The suspension may be refrigerated and used for at least 2 to 3 months.

B. Standards

Accurately weigh a sample of ristocetin A standard and dilute in phosphate buffer pH 6.0 to give standard solutions containing 5.0, 12.5, 25 (reference point), and 50 μg./ml.

C. Samples

Most samples may be diluted directly in phosphate buffer pH 6.0 to a theoretical potency of 25 μg./ml. Fermentation beers are first adjusted to pH 2.0 with H_2SO_4 and held for at least 1 hour before dilution in buffer to a theory of 25 μg./ml. The acidification releases that portion of the antibiotic held on or within the cells.

D. Mechanics of Assay

1. Preparation of Plates

Add 20 ml. of sterile Antibiotic Medium No. 5 (Difco) to flat-bottomed petri plates. Cover the plates and allow the base layer to harden. Inoculate an additional quantity of the same medium, cooled to 60°C, with sufficient spore suspension to give approximately 50,000 spores/ml. Add 5 ml. of the seeded medium to each plate and distribute evenly. Allow plates to cool, invert, and store in a refrigerator until used. Plates may be held up to 48 hours. Slight variations in the seeding level may be required as judged by assay performance.

2. Standard Curve

The plate assay is run using filter paper disks (Schleicher and Schuell Company No. 740-E, 12.7 mm.) to which is added 0.08 ml. of the sample to be assayed. Place one disk of each concentration of the standard solutions on each of 12 plates. Invert the plates and incubate for 16 to 18 hours at 30°C. Measure the size of inhibition zones with a suitable reader. Construct a standard curve with the average zone diameters plotted arithmetically against the standard concentrations plotted logarithmically on semilogarithmic paper. The slope of this curve, that is, the increase in zone diameter for each twofold increase in antibiotic concentration, should be 2.2–2.5 mm.

3. Unknowns

Use two or more plates for each unknown sample. Place two disks of the sample (theory, 25 μg./ml.) and two disks of the 25 μg./ml. reference standard alternately on each plate. Invert the plates and incubate for 16 to 18 hours at 30°C. Measure the zone size and average the zone diameters for the two standards and for the two samples. Correct for plate variation: if the plate standard gives a larger average zone size than that corresponding to the 25 μg./ml. reference point of the standard curve, subtract this difference from the average sample diameter; if the plate standard gives a smaller average zone size, add the difference to the average sample zone size. Use the corrected sample zone size and read the antibiotic concentration directly from the standard curve. Multiply by appropriate dilution factors to obtain the potency of the sample.

V. Salt Plate Method

A modification of the plate assay procedure described in Section IV is required in order to make the plate and turbidimetric assay procedures comparable. In the modified procedure 5% NaCl is added to both the base and the seed plating medium. The addition of this quantity of salt slows the diffusion of ristocetin A and has no effect on diffusion of the B component. The activity of ristocetin B by this method is 2.5 times that of A on a weight basis, approximately the same ratio of activity obtained in the turbidimetric method.

The standard curve in the salt plate method is run at 12.5, 25, 50, and 100 μg./ml. with the reference point at 50 μg./ml. Other assay details are the same as in the regular plate method.

6.14 Streptomycin

FREDERICK KAVANAGH

Eli Lilly and Company, Indianapolis, Indiana

I. Introduction

Streptomycin ($C_{21}H_{39}N_7O_{12}$, mol. wt. 581.6) is a basic antibiotic active against gram-positive and gram-negative bacteria. It is soluble in water and insoluble in organic solvents. The usual salts are the hydrochloride ($C_{21}H_{39}N_7O_{12} \cdot 3HCl$) and the sulfate [$C_{21}H_{39}N_7O_{12})_2 \cdot 3H_2SO_4$].

The assay is in terms of the free base with a theoretical assay of 842 μg./mg. for the chloride and 798 μg./mg. for the sulfate.

II. Turbidimetric Assay

The details of the assay except the range of standards is the same as for Dihydrostreptomycin, Chapter 6.5. The range is from 10 to 25 μg./ml. for *Klebsiella pneumoniae* and from 3 to 40 for *Staphylococcus aureus*. The answers are obtained in terms of free base.

III. Plate Assay

Proceed as for Dihydrostreptomycin, Chapter 6.5.

6.15 Tetracyclines

FREDERICK KAVANAGH

Eli Lilly and Company, Indianapolis, Indiana

The tetracycline antibiotics (chlorotetracycline, demethylchlorotetracycline, oxytetracycline, and tetracycline) may be assayed by chemical and antibacterial methods. The generally used antibacterial methods are the *Staphylococcus aureus* turbidimetric and the *Bacillus cereus* var. *mycoides* cylinder plate method given in Grove and Randall.[1] Colorimetric, spectrophotometric, and fluorometric methods suitable for relatively pure and concentrated solutions of tetracyclines are also given in Grove and Randall. Recently, Kohn[2] described sensitive fluorometric methods suitable for determinations in body fluids and soft tissues over the usual pharmacologic range of drug concentrations. Sensitivities of the methods ranges from 0.1 μg. for demethylchlorotetracycline to 0.5 μg. for oxytetracycline. Agreement between the fluorometric and the *B. cereus* plate method was satisfactory.

[1] D. C. Grove and W. A. Randall, "Assay Methods of Antibiotics," 238 pp. Medical Encyclopedia, New York, 1955.

[2] K. W. Kohn, *Analyt. Chem.* **33,** 862 (1961).

6.16 Thiostrepton

JOSEPH D. LEVIN AND JOSEPH F. PAGANO *

The Squibb Institute for Medical Research, New Brunswick, New Jersey

I. Introduction

Thiostrepton,[1] a polypeptide antibiotic produced by *Streptomyces azureus*,[2-4] is highly active against gram-positive bacteria. Thiostrepton is a useful antibiotic in the treatment of bovine mastitis, particularly if penicillin is contraindicated. About 75% of bovine mastitis currently is caused by staphylococci. Penicillin-resistant staphylococci are sensitive to thiostrepton.

Although thiostrepton is not water soluble, it is soluble in chloroform, dioxane, dimethylacetamide, formamide, dimethylformamide, and dimethylsulfoxide. Chloroform was used for extraction of ointment-type products

---

* *Present address:* Sterling-Winthrop Research Institute, Rensselaer, New York.

[1] The trade names of E. R. Squibb & Sons for thiostrepton-containing products are Gargon and Neothion.

[2] J. F. Pagano, M. J. Weinstein, H. A. Stout, and R. Donovick, *Antibiotics Ann.* **1955/1956,** 554 (1956).

[3] J. Vandeputte and J. D. Dutcher, *Antibiotics Ann.* **1955/1956,** 560 (1956).

[4] B. A. Steinberg, W. P. Jambor, and L. O. Suydam, *Antibiotics Ann.* **1955/1956,** 562 (1956).

but was not useful as an assay diluent. Dioxane, formamide dimethyl-formamide, and dimethylsulfoxide were tried as solvents and assay diluents; however, dioxane and dimethylformamide were not suitable because of apparent instability of the solvent lots. As an example, thiostrepton (1 mg./ml.) dissolved in one particular lot of dioxane lost 75% of its activity when held 8 days at 5°C. Other lots of dioxane did not produce this effect. Formamide and dimethylsulfoxide were useful solvents and assay diluents for thiostrepton. Dimethylsulfoxide was selected for routine use because of the increased sensitivity of the test organism to the antibiotic in the presence of this diluent.

The steeper the dose-response curve, the smaller the standard error for a given standard deviation of a response. The slope[5] of the dose-response curve was greater at pH 9 than at pH 6.6 or 8. Since the standard deviations of a response were the same at all values of pH, the standard error was smallest at pH 9. The sensitivity of the assay was also considerably greater at pH 9 than at lower pH. Sodium chloride[5] at 1 to 3% increased both slope of the dose-response line and sensitivity of the assay. A small inoculum size increased sensitivity and slope somewhat, but also increased difficulty in reading the plates because the resulting minimal growth made the edges of the zones indefinite. The combination of pH of medium, sodium chloride concentration, and inoculum level given are the optimum when both sensitivity and standard error are considered.

II. Test Organism

Maintain the test organism, *Staphylococcus aureus* (ATCC 6538P), on stock slants of G. & R. 5 medium. Grow the inoculum for the test by making a loop transfer from a stock slant to Antibiotic Assay Broth [BBL (Baltimore Biological Laboratory)] (G. & R. 3). Incubate for 20 hours at 37°C. without shaking, at which time there are approximately 10^9 viable cells. The agar medium used in the assay is G. & R. 5 medium with 1.0% sodium chloride added and then the pH adjusted to 9.0 ±0.1 before sterilizing. Pour the medium into 3-quart Pyrex baking dishes with inside dimensions of $8 \times 13 \times 1.7$ inches. Cover with stainless metal lids. Use two layers of medium; a 250-ml. base layer and a top layer of 150 ml., seeded with 0.1% of the inoculum.

III. Preparation of Standards and Samples for Assay

Prepare the crystalline antibiotic for assay by dissolving in dimethyl-sulfoxide and diluting in 80% aqueous dimethylsulfoxide to the concen-

[5] J. D. Levin, H. Stander, and J. F. Pagano, *Antibiotics & Chemotherapy* **10**, 422 (1960).

trations required for the test (5.0, 2.5, and 1.25 u./ml.). Dilute all solutions with 80% aqueous dimethylsulfoxide to the proper concentrations. Extract thiostrepton, formulated in a polyethylene preparation,[6] with chloroform prior to dilution with 80% aqueous dimethylsulfoxide.

IV. Assay Design

The assay for thiostrepton is modeled after a three-dose procedure described for nystatin.[7] The antibiotic-containing disks are distributed in a $6 \times 9 \times 9 \times 3$ pattern (six columns, nine rows, nine compounds, and three levels). The nine compounds include two independently prepared standard solutions and seven unknowns. The three levels are in a fixed ratio of 4:2:1 (high, medium, and low). These are duplicated within plate as well as between plates. The same technician who sets the disks on any one series of plates should also prepare and set the standard solutions as well as the unknowns for that series. Incubate the plates for 18 hours at 37°C. Use a six-point calculation as described for nystatin. If available, a digital computer may be utilized for the following computations: (1) calculations of potency; (2) analyses of variance; (3) determinations of standard errors; (4) comparisons of one standard with the other; and (5) calculations of F tests for significance of means, slopes, and assay validity (parallelism, curvature, and opposed curvature).

Apply diluted samples to the assay plates by means of $\frac{3}{8}$-inch filter paper disks (Schleicher and Schuell No. 740E). Incubate the assay plates for 18 hours at 37°C. Measure the resulting zones of inhibition in milliliters with the aid of a projection device modified to project the image directly in front of the technician. Magnification is approximately five diameters (see Nystatin, Chapter 6, Part II). An automatic zone reader also may be used (see Chapter 5).

V. Thiostrepton in Milk

The assay for thiostrepton was modified to obtain increased sensitivity.[5] It was carried out in petri dishes with cylinders as reservoirs for the samples. Prepare a single 15-ml. layer of inoculated assay medium containing 0.005 unit thiostrepton per milliliter. Hold the plates containing samples and standard at 5°C. overnight for prediffusion of the antibiotic prior to incubation at 37°C. As little as 0.2 unit thiostrepton per milliliter of whole raw milk was detectable by this method. Prepare all samples for test by adding an aliquot of the sample of milk to an equal quantity of dimeth-

---

[6] The trade name of E. R. Squibb & Sons for polyethylene preparation is Plastibase (99.5% mineral oil and 0.5% polyethylene).

[7] Nystatin, Large Plate Method, Chapter 6, Part II.

ylsulfoxide. If further dilution is required, dilute each sample in a freshly prepared mixture of 50% dimethylsulfoxide and 50% normal raw milk. Prepare the thiostrepton standard in dimethylsulfoxide and dilute with an equal volume of normal raw milk. Calculate sample potencies from a standard dose-response curve in the usual way.

6.17 Tylosin

FREDERICK KAVANAGH AND L. J. DENNIN

Eli Lilly and Company, Indianapolis, Indiana

I. Introduction

Tylosin is a macrolide antibiotic produced by an organism similar to *Streptomyces fradiae*. It is active against gram-positive bacteria, certain gram-negative organisms, mycobacteria and pleuropneumonia-like organisms. Desmycosin, an acid degration product, is similar in antibacterial spectrum and order of activity to tylosin.[1,2]

Tylosin ($C_{45}H_{77}NO_{17}$, mol. wt. 903) free base is only slightly soluble in water (5 mg./ml. at 25°C.) and soluble in the common organic solvents, including amyl acetate and chloroform.[3] Its salts are water soluble. It is stable in the pH range of 5.5 to 7.5 for 3 months at 25°C. It is unstable at pH less than 4 or greater than 9. In many respects its chemical properties are similar to those of erythromycin, but it differs in having much greater stability in acid solutions. Stability in acid soluton is intermediate between that of erythromycin and streptomycin.

The range of the photometric assay is from 0.7 to 3 μg./ml. of sample solution and of the plate assay from 0.1 to 2 μg./ml.

II. Photometric Assay

A. Test Organism

Staphylococcus aureus H.

B. Standard Solutions

Weigh the dry standard (Section III.C) (at least 10 mg.) and dissolve in pH 7 buffer to form a solution with a concentration of about 1 mg./ml. of free base. Store in the refrigerator. The solution may be used for several weeks. A stock solution of 1 mg./ml. did not lose measurable activity upon storage in the refrigerator for 5 months.

Dilute solutions (<10 μg./ml.) of tylosin are prepared and handled in Pyrex glassware because it seems to be adsorbed by some surfaces of soft glass. Dilute solutions (1 μg./ml.) can lose one-half of their activity in a few minutes if exposed to a large surface of soft glass.

C. Activity and pH

The activity of tylosin against *S. aureus* increases by a factor of about 4 for an increase in pH of 1 unit in the region of pH 6.5 to 7.5. The pH refers to the initial pH of the assay broth. The routine assays are performed with pH 7.0 broth.

[1] W. M. Stark, W. A. Daily, and J. M. McGuire, *Sci. Rept. Ist. super. sanità* 1, 340 (1961).

[2] J. M. McGuire, W. S. Boniece, C. E. Higgens, M. M. Hoehn, W. M. Stark, J. Westhead, and R. N. Wolfe, *Antibiotics & Chemotherapy* 11, 320 (1961).

[3] R. L. Hamill, M. E. Haney, Jr., M. Stamper, and P. Wiley, *Antibiotics & Chemotherapy* 11, 328 (1961).

D. Preparation of Sample for Assay

Treatment of Samples

Fermentation beers. Filter, or centrifuge, and dilute with sterile pH 7 buffer to the assay concentration of 1.4 μg./ml.

E. Inoculum

Prepare as for penicillin G assay. Add 15 ml. of the standardized inoculum to each liter of assay medium (1 liter of basal turb broth + 10 ml. 50% sterile glucose solution).

F. Range of the Assay

The best range of the test is from 0.8 to 3 μg./ml. of sample for tylosin, and from 0.8 to 4.0 μg./ml. for desmycosin. The dose-response line is straight (log-probability) from 0.8 to 3 μg./ml. of tylosin solution (0.5 ml. of sample solution per tube). Desmycosin is about three-fourths as active as tylosin and has a dose-response curve parallel to that of tylosin. Presence of desmycosin in tylosin does not distort the dose-response line in the log-probability form.

An incubation of time of 3 to 3.5 hours at 36° to 38°C. should be long enough to give a satisfactorily turbid zero tube.

G. Computation of Answers

The answers, obtained as μg./ml., are multiplied by the dilution to obtain the potency of the undiluted samples and are in terms of free base.

III. Plate Assay

A. Introduction

The plate method for tylosin is used principally in the assay of animal feeds or feed premixes where highly colored extracts are encountered, or where the slightly greater sensitivity (4 × that of the turbidimetric method) is advantageous.

B. Test Organism

Sarcina lutea (ATCC 9341) is maintained by the procedure described under Penicillin, Chapter 6.10, Section III.B.

C. Standard Solutions

Dry the working standard for 4 hours at 70°C. and at a residual pressure of 5 mm. Hg or less in a tared weighing bottle fitted with a capillary tube. Weigh a suitable quantity of the dried standard to make a solution containing 1000 μg. of tylosin activity (base) per milliliter. Transfer this quantity of standard to a suitable container, add 4–5 ml. of reagent grade absolute methyl alcohol to dissolve the tylosin, and adjust to volume with sterile pH 7 phosphate buffer. This stock standard solution is stored at 4° to 6°C. and may be used for a period not to exceed 1 month.

D. Sample Preparation

1. Fermentation Beers, Solutions, etc.

These sample types do not require purification by extraction and are diluted to approximately 0.5 μg./ml. of tylosin activity with methanolic buffer solution (use Pyrex vessels) (Table I).

TABLE I

BUFFER SOLUTIONS

Methanolic buffer solution	
pH 8 phosphate buffer	60 ml.
Methyl alcohol, reagent grade	40 ml.
pH 8 phosphate buffer	
K_2HPO_4	16.7 gm.
KH_2PO_4	0.5 gm.
Water, distilled to make	1000 ml.

2. Animal Feeds and Feed Premixes

Accurately weigh 10 gm. of premix or 20 gm. of feed and transfer to a 200- to 250-ml. homogenizer cup. Add 90 ml. of warm (70–80°C.) pH 8 phosphate buffer and place the sample on a steam bath for 10 minutes. Transfer the homogenizer cup to the homogenizer and blend for 5 minutes. At the end of this blending time, add 60 ml. of reagent grade methyl alcohol and blend an additional 5 minutes. Filter the sample through No. 1 Whatman filter paper. Dilute an aliquot of the filtrate, if necessary, with the methanolic buffer solution to a concentration of approximately 0.5 μg./ml. of tylosin activity. This dilution represents the test solution in which the methanol-buffer ratio has been made equal to that of the standards used in preparing the dose-response curve.

E. Mechanics of the Assay

1. Design

The design of this assay is similar to that described for penicillin G (see Penicillin, Section III.E.1).

2. *Medium*

The two-layer agar system is used in the assay. The "base" layer consists of 10 ml. of G. & R. agar medium No. 11, and the "seed" layer a 5-ml. aliquot of inoculated medium of this same composition.

3. *Inoculum*

Prepare an agar slant of the culture as described in Section B. Wash the growth from one slant into a flask containing 100 ml. of sterile G. & R. broth medium No. 3. Incubate the broth suspension at 22° to 25°C. on a mechanical shaker for a period of 40 to 48 hours. The resultant suspension constitutes the stock inoculum and is used for a period not to exceed 1 week when kept under continuous refrigeration (4–6°C.).

Inoculate the "seed" agar by adding 0.5 ml. of a 1:10 dilution of the stock inoculum suspension per 100 ml. of liquefied and cooled (48°C.) agar and pour the plates immediately. The dilution of the stock suspension may be made with either sterile distilled water or with sterile 0.85% sodium chloride solution.

4. *Standards*

Immediately prior to use, dilute samples of the stock standard solution in the methanolic buffer solution. The choice of standard levels will, of course, determine to what extent these aliquots are diluted. The assay yields a linear response in the range from 0.1 to 2.0 μg./ml. The dose levels should be spaced evenly on the logarithmic scale.

5. *Samples*

Refer to general considerations under the plate assay of Penicillin, Section III.D and to instructions in Sections III.D.1 and III.D.2 of this method.

6. *Incubation*

Incubate the tylosin plates for 16 to 18 hours at 30°C. The general considerations discussed under the assay of penicillin regarding incubation apply to this assay as well (see Penicillin, Section III.E.6).

F. Measuring the Response

Refer to Penicillin, Section III.F.

G. Computation of Answers

See Penicillin, Section III.G.

6.18 Vancomycin

FREDERICK KAVANAGH

Eli Lilly and Company, Indianapolis, Indiana

I. Introduction

Vancomycin (Vancocin, Eli Lilly & Company) is a large molecular weight (about 3000) antibiotic of unknown structure active against gram-positive bacteria. It is a basic substance with many free hydroxyl and amino groups. Its salts are quite soluble in water. The hydrochloride is soluble in methanol and insoluble in other organic solvents in general.

The antibacterial activity of vancomycin increases with increase in pH of the assay medium.

The plate method is the one of choice for routine assays of unknowns because its range is from 20 to 200 units/ml.

II. Turbidimetric Method

The turbidimetric method is of value only for control purposes because the range of the assay is so small. If the concentration of the antibiotic is known accurately enough to place it within the range of the assay, then it is known accurately enough for most purposes. The useful range is from 5 to 8 u./tube.

Reynolds[1] reported that the mode of action of vancomycin was similar to that of penicillin. Details of interference with cell wall synthesis were significantly different from those of penicillin since there was no cross resistance.

The very steep dose-response curve for accumulation of N-acetyl aminosugar esters (*Staphylococcus aureus* as test organism) was similar to that of the turbidimetric assay method described here.

A. Test Organism

Staphylococcus aureus Heatley strain used for assay of penicillins.

B. Standard Solutions

Vancomycin has not been prepared in a chemically pure form, consequently the standard preparation has an arbitrarily assigned potency.

Place a given amount of Vancomycin reference standard to represent 25,000 units of activity in a 25-ml. volumetric flask, dissolve in a little sterile water, and dilute to volume with sterile water. This stock solution of 1000 u./ml. may be used for one week to prepare working standards if kept in the refrigerator. Vancomycin is quite stable in the pH range 3–9 in the cold.

C. Preparation of Sample for Assay

Dissolve bulk samples in sterile water. Dilute the contents of ampoules with water. Dissolve ointments made with water soluble bases in water. Dissolve petroleum base ointments in petroleum ether, and remove the vancomycin which is insoluble in the solvent, by extracting with water.

D. Mechanics of Assays

1. Design

Prepare the standard response curve by pipetting 0.50 ml. of standard solutions of the following concentrations: 0, 8, 10, 11, 12, 13, 14, and

[1] P. E. Reynolds, *Biochim. Biophys. Acta* **52,** 403 (1961).

16 u./ml. into the culture tubes. Make 8 tubes of the 0 concentration and at least 4 of each of the other concentrations. Once the operator is confident that the log-probability response curve is straight over the concentration range 10–14 u./ml., he may wish to omit the 11- and 13- u./ml. tubes.

Dilute the samples to an estimated concentration of 11 to 12 u./ml. Accurate determination of potency is possible only between the concentrations of 10 and 14 u./ml. If the first assay falls outside these limits, put the sample on another assay at the newly estimated potency. If only a few samples are to be assayed and time is important, use several dilutions selected to be on each side of the estimated value. For example, if the estimate is 12 u./ml., assay at 6, 8, 10, 12, 14, 16, and 18 u./ml.; and one dilution should fall on the curve. If an estimate within ±50% of the potency cannot be made, then use the plate assay to establish the range of potency.

2. Medium and Inoculum

Add 15 ml. of standard inoculum $(300 \cdot 10^6$ cells/ml.) to each liter of erythromycin turb broth and proceed as for Penicillin assay (Chapter 6.10).

III. *Bacillus subtilis* Plate Assay

A. Introduction

The turbidimetric method of assay of vancomycin has such a short range that it is useless for assaying of solutions of unknown concentrations. A plate assay is used for these solutions. The range is from 5 to 40 u./ml. for a *B. subtilis* (ATTC 6633) assay and from 20 to 200 u./ml. for a *Staphylococcus aureus* (209P) assay. Details of the *Bacillus subtilis* assay will be given here with only an outline of the *Staphylococcus aureus* assay in Section IV.

B. Test Organism

Maintain stock cultures of *Bacillus subtilis* (ATCC 6633) by monthly transfers to fresh sterile slants of G. & R. agar medium No. 1. Incubate the freshly prepared slants at 37°C. for 16 to 18 hours and store at 4° to 6°C. until ready for use in the preparation of the inoculum suspension.

C. Standard Solutions

Dry the working standard for 3 hours at 60°C. and a residual pressure of 5 mm. Hg. or less in a tared weighing bottle fitted with an attached capillary tube. Weigh a quantity of the dried standard to make a solution containing 500 u./ml. of vancomycin activity (base). Transfer this quantity of standard to a suitable container and bring to volume with pH 8 phosphate buffer. Store the stock standard solution at 4° to 6°C. for a period not to exceed 2 weeks.

D. Sample Preparation

Little difficulty should be experienced in the preparation of vancomycin samples for assay. The antibiotic is readily soluble in pH 8 phosphate buffer and relatively insoluble in such solvents as benzene, toluene, and carbon tetrachloride. With most sample types, simple dilution to the test level (approx. 10 u./ml.) with pH 8 phosphate buffer is the only preparatory step necessary.

E. Mechanics of the Assay

1. Design

See Penicillin, Section III.E.1.

2. Medium

The "base" agar layer consists of 10 ml. of G. & R. agar medium No. 5, while the "seed" layer requires 4 ml. of this same agar.

3. Inoculum

Prepare according to plate assay for Hygromycin B, Chapter 6.8, Section V.C.

Determine the concentration of inoculum required for "seeding" of assay plates by running test plates each time a new spore suspension is prepared. In my experience, this is generally in the range 0.1–0.3 ml. of a 1:10 dilution of the stock suspension per 100 ml. of liquefied and cooled (48°C.) agar.

4. Standards

Dilute samples of the stock standard solution in sterile pH 8 phosphate buffer immediately prior to use on the day of the assay. The choice of standard levels will, of course, determine to what extent these samples are diluted. This assay gives a linear response in the range 5 to 40 u./ml. Standard levels should be evenly spaced on the logarithmic scale. The reference concentration used on each plate is 10 u./ml.

5. Samples

See Penicillin, Section III.E.5.

6. Incubation

Incubate the assay plates for 16 to 18 hours at 37°C. Also refer to Penicillin, Section III.E.6.

F. Measuring the Response

See Penicillin, Section III.F.

G. Computation of Answers

See Penicillin, Section III.G.

IV. *Staphlyococcus aureus* Plate Assay

Prepare the inoculum of *S. aureus* 209P as described in Penicillin, Section II.A.2. The concentration of cells will be approximately 380 million per milliliter of broth. Add about 650 million cells to each 100 ml. of agar medium (at 48°C.) and measure 8 ml. into 85-mm. plastic petri dishes as rapidly as possible. A single layer of agar as thin as this gives excellent inhibition zones. Refrigerate the plates for 2 hours before using. Plates kept in the refrigerator for as long as 2 days give satisfactory zones.

Prepare standard solutions in the range 20–200 u./ml. as described in Section III.C with a reference concentration of 50 u./ml. to be put on each plate along with the samples. Pipet 0.09–0.10 ml. of sample diluted to an estimated concentration of 50 u./ml. onto a $\frac{1}{2}$ inch sterile Schleicher & Schuell No. 704E disks, or dip the disk into the liquid and remove the excess by touching the edge of the disk to a dry spot on the wall of the sample container. The latter procedure is about as good as the former and is considerably quicker. Incubate, measure zones, and compute answers as described in Section III.

TABLE I
MEDIUM COMPOSITION

	gm./liter
Beef extract	1.5
Yeast extract	6
Wilson Pharmaceutical Peptone S.P.	6
Agar	15

Adjust pH to 7.9 with sodium hydroxide
  before sterilizing

The medium used in this plate assay has the composition given in Table I. Before substituting for the particular peptone, test the medium made with the different peptone to be certain that it is as satisfactory.

6.19 Media and Buffers

FREDERICK KAVANAGH

Eli Lilly and Company, Indianapolis, Indiana

Media used in more than one assay are listed here. Plate methods usually employ one or two of the media given in Chapter 21 of Grove and Randall. They are listed here by their G. & R. numbers. The broth medium used in the photometric methods usually are different from the four G. & R. broth media. The buffers are not all identical with those of G. & R. Special media are given in the procedures in which they are used.

The quantities of ingredients are the number of grams needed to prepare 1 liter of medium unless a different volume is indicated. Distilled water is used in making all media. The pH after sterilization is given. Sterilize at 121°C. for 15 minutes following the usual good practices. For example, do not autoclave at 121°C. a quantity of 500 ml. of solidified agar medium contained in a 1-liter flask or bottle for 15 minutes and expect it to be sterile; melt the agar first and then sterilize. Minimum sterilization time usually is the best.

TABLE I

AGAR MEDIA

Media	gm.
AC	
Glucose	5
Malt extract	3
Yeast extract	3
Proteose peptone No. 3 (Difco)	20
Beef extract	3
Agar	15
pH	6.3

TABLE I (Continued)

Media	gm.
G. & R. No. 1	
Peptone	6
Pancreatic digest of casein	4
Yeast extract	3
Beef extract	1.5
Glucose	1
Agar	15
pH	6.5–6.6
G. & R. No. 2	
Peptone	6
Yeast extract	3
Beef extract	1.5
Agar	15
pH	6.5–6.6

G. & R. No. 4

No. 2 agar plus 1 gm. glucose per liter.

G. & R. No. 5

Same as No. 2 except the pH is adjusted so that it is 7.8–8.0 after sterilization.

G. & R. No. 9	
Pancreatic digest of casein	17
Papaic digest of soybean	3
Sodium chloride	5
Dibasic potassium phosphate	2.5
Glucose	2.5
Agar	20
pH	7.2–7.3
G. & R. No. 10	
Pancreatic digest of casein	17
Papaic digest of soybean	3
Sodium chloride	5
Dibasic potassium phosphate	2.5
Glucose	2.5
Agar	12

Boil to dissolve the medium and then add 10 ml. of polysorbate 80 (Tween 80).

pH	7.2–7.3

G. & R. No. 11

Same as No. 1 except the pH is adjusted to be 7.9–8.0 after sterilization

TABLE II

BROTHS

Broth	gm.
Brain heart infusion	
Calf brain, infusion from	200
Beef heart, infusion from	250
Peptone	10
Glucose	2
Sodium chloride	5
Dibasic sodium phosphate	2.5
pH	7.4
Heart infusion	
Beef heart, infusion from	500
Bacto tryptose	10
Sodium chloride	5
G. & R. No. 3	
Peptone	5
Yeast extract	1.5
Beef extract	1.5
Glucose	1
Sodium chloride	3.5
KH_2PO_4	1.32
K_2HPO_4	3.68
pH	6.95–7.05
Basal turb broth	
Peptone, Bacto	5
Yeast extract	1.5
Beef extract	1.5
NaCl	3.5
KH_2PO_4	1.3
K_2HPO_4	3.7

The pH before autoclaving is 7.0. Filter, fill 1 liter into 5-pint bottles, stopper with a cotton plug, and sterilize. Filter, if necessary, just before using. The filtration need not be a sterile one. This medium is essentially G. & R. No. 3 without the sugar.

	ml.
Nutrient broth	
Basal turb broth (sterile)	1000
Glucose solution, 50% (sterile)	10

TABLE II (Continued)

Broth	ml.
Assay broth, penicillin	
Basal turb broth (sterile)	1000
Glucose solution, 50% (sterile)	25
Tris buffer	10
pH	7.1
Assay broth, erythromycin	
Basal turb broth (sterile)	1000
Glucose solution, 50% (sterile)	25
Glucose solution	
Glucose	500 gm.
Water, to make	1000 ml.

Dissolve the sugar in warm water and dilute to volume. Sterilize in the autoclave.

TABLE III
Buffer Solutions

Buffer	
pH 6 buffer	
KH_2PO_4	8.2 gm.
K_2HPO_4	1.8 gm.
Water, distilled	1000 ml.
pH	

Make with analytical reagent grade chemicals. Sterilize.

pH 7 buffer	
KH_2PO_4	4.0 gm.
K_2HPO_4	13.6 gm.
Water, distilled	1000 ml.

Sterilize by autoclaving.

pH 8 buffer	
KH_2PO_4	0.75 gm.
K_2HPO_4	16.4 gm.
Water, distilled	1000 ml.

Sterilize by autoclaving.

Tris buffer	
Tris (Hydroxymethyl) aminoethane	90 gm.
Phosphoric acid (85%)	42 gm.
Dilute with water to	187 ml. and filter

Store in a warm place, e.g., 37°C., incubator.

TABLE IV
MEDIA EQUIVALENTS

G. & R.	Difco[a]	BBL[b]
Pancreatic digest of casein	Casitone	Trypticase
Papaic digest of soybean meal	Soytone	Phytone
Peptone	Bacto peptone	Gelysate
	Proteose peptone No. 3	Polypeptone
	Tryptose	Biosate

G. & R. No.	Difco	BBL
1	Penassay seed agar	Penicillin assay seed agar
2	Penassay base agar	Base agar
3	Penassay broth	Antibiotic assay broth
9	Polymyxin base agar	Polymyxin base agar
10	Pclymyxin seed agar	Polymyxin assay agar

[a] Difco Laboratories, Detroit, Michigan.
[b] Baltimore Biological Laboratories, Inc., Baltimore, Maryland.

Antibiotic Substances
Part II. Antifungal Assays

JOHN R. GERKE, JOSEPH D. LEVIN, AND JOSEPH F. PAGANO *

The Squibb Institute for Medical Research, New Brunswick, New Jersey

I. Introduction

Antifungal compounds produced by microorganisms became a subject of active investigation in the mid-1930's. Weindling and Emerson[1] reported on gliotoxin in 1936. Oxford et al.[2] in 1939 described the biological and chemical properties of griseofulvin, an antifungal agent which is used in the treatment of fungal diseases of plants, animals, and man. Actino-

---

* Present address: Sterling-Winthrop Research Institute, Rensselaer, New York.
[1] R. Weindling and O. H. Emerson, Phytopathology, 26, 1068 (1936).
[2] A. E. Oxford, H. Raistrick, and P. Simonart, Biochem. J. 33, 240 (1939).

mycin was reported by Waksman and Woodruff [3] in 1941. Since this time other antifungal substances have been discovered and, some such as trichomycin,[4,5] nystatin,[6-8] and amphotericin[9,10] have received wide usage in the treatment of human and animal diseases.

Assay methods for antifungal compounds have followed the patterns established for the antibacterial antibiotics, measuring their effects on growth of test organisms by microscopy, turbidimetry, agar dilution, agar diffusion, and respirometry. To illustrate the ingenuity of various investigators in adapting the basic methods to their specific requirements, a number of such modifications will be described.

Brian et al.[11] originally described griseofulvin as a "curling factor" because concentrations below that required to inhibit growth tended to cause the growing hyphal tips of the test fungus to curl. Assays were run using microscopic examination of the hyphal tips to determine the end point. To utilize this effect on hyphal tips in an assay and to avoid the tedium of reading the end point microscopically, we grew either *Trichophyton mentagrophytes* or *Microsporum audouini* in submerged culture in test tubes which were continuously agitated throughout the assay incubation period. Under these conditions the fungus grew in the form of macroscopically visible spheres. In the uninhibited culture the surface of the spheres was covered with hyphal tips projecting from the main body to produce a furlike appearance. When grown in the presense of griseofulvin, the character of the tips was altered so that the spheres appeared to have a smooth surface.

Tarbet and Sternberg[12] developed a method for the assay of cyclohexamide, ascosin, candicidin, nystatin, and rimocidin based on the ability of the antibiotic to inhibit formation of blastospores of *Candida tropicalis*. They used a hemocytometer for counting the blastospores.

Another variation of the microscopic type of assay was the *Botrytis*

[3] S. A. Waksman and H. B. Woodruff, *J. Bacteriol.* **42**, 231 (1941).

[4] S. Hosoya, N. Komatsu, M. Solda, T. Yuwaguchi, and Y. Sonada, *J. Antibiotics (Japan)* Ser. B **5**, 564 (1952).

[5] S. Hosoya, N. Komatsu, M. Solda, and Y. Sonada, *Japan. J. Exptl. Med.* **22**, 505 (1952).

[6] E. L. Hazen and R. Brown, *Science* **112**, 423 (1950).

[7] E. L. Hazen and R. Brown, *Proc. Soc. Exptl. Biol. Med.* **76**, 93 (1951).

[8] R. Brown, E. L. Hazen, and A. Mason, *Science* **117**, 609 (1953).

[9] W. Gold, H. A. Stout, J. F. Pagano, and R. Donovick, *Antibiotics Ann.* **1955/1956**, 579 (1956).

[10] J. Vandeputte, J. L. Wachtel, and E. T. Stiller, *Antibiotics Ann.* **1955/1956**, 587 (1956).

[11] P. W. Brian, P. J. Curtis, and H. G. Hemming, *Trans. Brit. Mycol. Soc.* **29**, 176 (1946).

[12] J. E. Tarbert and T. H. Sternberg, *Mycologia* **46**, 263 (1954).

alli spore-germination inhibition test of Brian and Hemming[13] for such antifungal antibiotics as gliotoxin, gladiolic acid, griseofulvin, and alternaric acid. Essentially, they inoculated conidia of *B. alli* into various concentrations of the antifungal agent in nutrient media and incubated portions overnight at 25°C. on microscope slides in a moist chamber. The end point was determined by microscopic examination.

Although a variety of techniques exist for measuring the growth response of a microorganism to an antifungal antibiotic, the most commonly used is that of reading the turbidity caused by the growth of the microorganism in a liquid medium. Turbidimetric assay methods depend upon inhibition or retardation of growth of a test microorganism by the substance to be measured. Simple readings of turbidity made visually in a series of tubes may suffice if the specific situation does not require discrimination of differences in antibiotic concentration of less than twofold. In general, degree of turbidity can be determined with greater precision and objectivity with a photometer than by eye. To illustrate this form of assay a serial dilution assay and several photometric methods will be described in detail in this chapter.

In the agar dilution method the antibiotic is incorporated into the unseeded molten agar medium. The test microorganisms are then seeded on the surface of the solidified agar. An advantage of the method is that more than one type of organism may be seeded on a single plate, directly comparing the effect of one antifungal agent upon several microorganisms.

Most antifungal agents can be assayed by the agar diffusion procedure. For many it is the method of choice. A variety of physical means may bring the antifungal antibiotics and the microorganisms together. In the most common method, a layer of seeded agar is poured in the bottom of a petri dish, small cylinders are placed on the surface, and the antibiotic solutions to be assayed are placed in these cylinders. The cylinders may be made of stainless steel, glass, or plastic. In place of cylinders, fish spine beads, depressions, holes cut in the agar, or filter paper disks can be used as reservoirs for the antibiotic. In many applications, large dishes are more convenient than small petri dishes.

Some investigators prefer physical arrangement other than the plate procedure. Tsubura,[14] in assaying trichomycin, placed the seeded agar in test tubes and added a solution of the antifungal agent to the tube to form a layer on the top. The agent diffused from the liquid into the agar. The response was measured as the distance from the liquid-agar interface to the place in the tube where growth was first apparent. A variation of

[13] P. W. Brian and H. G. Hemming, *Ann. Appl. Biol.* **32**, 214 (1945).
[14] E. Tsubura, *J. Antibiotics (Japan) Ser. B* **7**, 201 (1954).

this method as reported by Davis et al.[15-17] was to fill capillary tubes with seeded agar and immerse one end in a solution of the antibiotic to be assayed.

The theory of diffusion techniques is described in Chapter 1. It is sufficient to say here that the sensitivity of diffusion test methods can be influenced by a variety of factors. The size of the zone is increased by: (1) prediffusing the antibiotic in the agar by holding the plates for a period of time below the optimal growth temperature for the microorganism; (2) decreasing the thickness of agar layer; (3) reducing the numbers of organisms in the seed; (4) varying the nutrient and salt composition of the agar; and (5) adding a subinhibitory level of the antifungal agent to the agar prior to pouring the plates.

The size of the zones of inhibition may be measured in a number of different ways, either manually or, as described in Chapter 5, automatically. The zones may be measured directly by means of caliper or, for ease of reading, their image may be projected onto a calibrated scale. Sharpness of the edge of the zone of inhibition may be increased by incorporating an indicator compound in the agar; for example, the colorless, 2,3,5-triphenyl tetrazolium chloride (oxidized form) is reduced to a red, insoluble compound (formazan) by the dehydrogenase of many species of microorganisms.

Inhibition of respiration of microorganisms by antibiotics has been the basis for many microbiological analytical methods. Respirometers of various types have been used. The complexity of many of these instruments, including the commonly used Warburg respirometric apparatus, does not suit them to routine use for analysis of large numbers of samples. Two devices, the AutoAnalyzer instrumental system and the hypodermic syringe used as a respirometer, are being used routinely for respirometric antifungal assays. Methods employing the AutoAnalyzer instrumental system are described in Chapter 5. Methods employing the hypodermic syringe respirometer are presented in this chapter.

Because of our experience with certain of the polyene antifungal antibiotics, the methods used to illustrate antifungal assays are primarily designed for these antibiotics. To prevent needless repetition, the sources for reference standards and the description of the solubility and stability characteristics of these antibiotics will be described in the following paragraphs and not with each method.

The reference standard for nystatin is obtained from the United States Pharmacopeia, amphotericin A from E. R. Squibb & Sons, and ampho-

[15] W. W. Davis, T. V. Parke, and W. A. Daily, Science 109, 545 (1949).

[16] W. W. Davis and T. V. Parke, J. Am. Pharm. Assoc. Sci. Ed. 39, 327 (1950).

[17] W. W. Davis, T. V. Parke, W. A. Daily, L. M. Rushton, and J. M. McGuire, J. Am. Pharm. Assoc. Sci. Ed. 39, 331 (1950).

tericin B from the Food and Drug Administration. The activity of nystatin is expressed in units while the potencies of amphotericin A and B are expressed in terms of weight of pure material. The solubility of these polyenes in certain polar solvents, such as: dimethyl sulfoxide, dimethyl formamide, and dimethyl acetamide exceeds 1 mg./ml.; otherwise they are quite insoluble, particularly amphotericin B. Water solubility is poor except when aided by a surfactant as exemplified by the sodium desoxycholate complex of amphotericin B. Water solubility of nystatin is about 200 u./ml., amphotericin A about 100 μg./ml., and amphotericin B about 1 μg./ml. Acids and bases enhance solubility but rapidly inactivate the polyenes. Nystatin and the amphotericins are quite stable if handled properly. They are more stable in the solid form than in solution. Protection from air, i.e., replacing the air with nitrogen, decreases the rate of decomposition by a factor of 2 or more. At 5°C. in air they may lose about 5% of their potency in 3 months, whereas, at 40°C. the rate of loss is doubled. Light degrades these polyenes, especially when they are in dilute solution. Solutions in neutral solvents at 5°C. protected from light degrade at the rate of 10% per day. Consequently, extracts and dilutions are handled in brown or red glass and are generally assayed within 2 hours of preparation. In consideration of the potency losses to be expected when solids are stored for a long time, reference standards are maintained under nitrogen in the dark at −20°C. and samples of the reference standard for daily use are kept in tightly stoppered vials at 5°C. for a period not to exceed 2 months.

II. Turbidimetric Methods

A. Amphotericin B: Serial Dilution Assay Method

1. Introduction

The method is used primarily to assay amphotericin B in body fluids, for testing susceptibility of fungi isolated from clinical material, and other applications for which the requirement for precision is secondary to an ability to carry out a large number of determinations with minimal effort. To serve the largest number of clinical applications, it was designed to require no specialized equipment. (See Chapter 3 for a general discussion of serial dilution methods.)

Activity of amphotericin B is influenced by the presence of serum, urine, tissue extracts, etc. When assaying body fluids large amounts of sample must be added to the test medium because usually its antibiotic content is low. Since each tube of the test will contain different concentrations of body fluid it is essential to equalize the body fluid content of each tube.

This principle applies to any turbidimetric method in which the vehicle affects activity of the antibiotic.

2. Test Organism

a. *Stock culture.* Grow the test organism, *Candida albicans* (Squibb 1539) for 1 to 2 days at 25° or 37°C. (preferably the latter) on Sabouraud's dextrose agar [B.B.L. (Baltimore Biological Laboratories)] slants. Nonpathogens such as *C. stellatoidea* or *Saccharomyces cerevisiae* may be substituted. Keep the slants at 5°C. and transfer at least once each month.

b. *Inoculum.* Inoculate tubes of Sabouraud's dextrose broth (B.B.L.) from a stock slant and incubate for 1 to 2 days at 25° or 37°C. Keep at 5°C. until used, but do not use when more than 1 week old.

c. *Preparation of inoculated medium.* Add 100,000 units of penicillin and 100 mg. streptomycin to 100 ml. of sterile Sabouraud's dextrose broth to prepare the test medium. Add 1 ml. of the inoculum to 100 ml. of medium and dispense the mixture into clear, sterile, 13 × 100 mm. tubes, 0.8 ml. per tube. Plug the tubes with cotton and place in groups of four in racks. Keep the racks of tubes at 5°C. until needed for the test. At that time, take from the refrigerator only those racks that can be used conveniently in half an hour.

3. Preparation of Standards and Sample for Assay

Protect both standard and samples from light. The use of brown glass vessels for holding both standard and samples is recommended.

a. *Standards.* Prepare a solution of amphotericin B standard containing 100 μg. of activity/ml. of dimethyl sulfoxide. Make further dilutions of the standard in the appropriate normal body fluid (equilibrating fluid), i.e., serum, plasma, or 25% urine. Make three final standard solutions: 0.10, 0.25, and 0.60 μg./ml. The concentration of any of these may be increased or decreased, as needed, to obtain an end point within the dilution series.

b. *Samples.* Depending on the expected potency, make dilutions with the appropriate normal body fluid. Undiluted urine causes inhibition of the test organism. Therefore, dilute all urine samples to 25% with water before testing and dilute the standard solution to the levels of the test with 25% normal urine. Likewise, use 25% normal urine as the equilibrating fluid for urine samples.

4. Assay Design

a. *Dosing, incubation, and reading of assay.* Approximately 2 ml. each of sample and normal body fluid are required for each test. To each of the four tubes add a quantity of sample and compensating fluid as indicated in Table I. Assay each of the three standard solutions as a separate sample using the same scheme.

TABLE I

	Tube no.			
	1	2	3	4
Test dilution	2-fold	4-fold	8-fold	16-fold
Sample (cr standard) (ml.)	0.8	0.4	0.2	0.1
Equilibrating fluid (ml.)	0	0.4	0.6	0.7
Inoculated medium (ml.)	0.8	0.8	0.8	0.8

Incubate the racks of tubes overnight.

Determine the end-point tube by visual inspection of the four tubes of each test. The tube containing the smallest quantity of sample capable of inhibiting the test organism is the end-point tube.

b. *Calculation.* Compute the minimal inhibitory concentration (MIC) of each of the three standards as follows:

$$\text{MIC} = \frac{\text{concentration of standard solution}}{\text{dilution of the end-point tube}}.$$

To calculate potency of sample, average the MIC of standards and multiply by the dilution of the end-point tube of the sample.

B. Amphotericin B: Photometric Assay Method

1. *Introduction*

The photometric method of assay for measuring amphotericin B[18] is recommended where high precision is required. This method may be used for assaying pharmaceutical dosage forms, process development and control samples, comparison of standards, and samples from drug metabolism studies. Although the following procedure is adequate for most purposes, modifications of the method and factors that influence the activity of amphotericin and other polyenes are presented elsewhere in greater detail by Gerke and Madigan.[19]

The selective susceptibility of *Candida tropicalis* to amphotericin B permits quantitation of this agent in the presence of other antibiotics. For example, amphotericin A which occurs together with B in fermentation, is only 2% as active as B against this test organism. Although their ac-

[18] The trade name of E. R. Squibb & Sons, Division of Olin Mathieson Chemical Corporation for amphotericin B is Fungizone.

[19] J. R. Gerke and M. E. Madigan, *Antibiotics & Chemotherapy* **11**, 225 (1961).

tivities are additive, the minor effect of A can be ignored. Antibacterial agents such as streptomycin, penicillin G, tetracycline, neomycin, or gramicidin do not interfere.

2. Test Organism

a. *Stock culture.* Subculture the test organism, *C. tropicalis* (Squibb 1647) each week using yeast beef agar slants (Difco). Incubate the slant overnight and hold at 5°C. until needed.

b. *Inoculum.* Transfer the yeast from the slant to inoculum broth (100-ml./500-ml. Erlenmeyer flask). Composition of inoculum broth is: Penassay broth (dehydrated, Difco), 1.75%; glucose, 1.0%; yeast extract, 0.5%; and tryptone (Difco), 1.0%. Sterilize by heating with steam in an autoclave for 20 minutes at 121°C. The pH after sterilization is 6.5–6.7. Prewarm the medium to 37°C. before using. This not only serves as a sterility check, but eliminates growth lag during the time required for the media to come to incubation temperature. Incubate this first broth transfer overnight at 37°C. on a shaker reciprocating at 120 (1½ inches) strokes per minute. Inoculate a second flask with 25% (v./v.) from the first broth transfer. Incubate on the shaker for 3 hours; then chill rapidly. As a check on growth, centrifuge a sample of the second culture; the yield of cells should be 2% (v./v.). The second transfer assures rapid growth of the cells in the assay tubes. Blend the second transfer for 2 minutes in a Waring Blendor to break up cell aggregates formed during incubation. Return to 5°C. for approximately ½ hour to allow foam to break. Store at 5°C. until needed but not longer than 1 week.

3. Preparation of Standards and Samples

Protect all solutions from strong light and high temperatures. Do not hold dimethyl sulfoxide solutions but dilute within ½ hour at least tenfold in DMW-1 (3 volumes dimethyl sulfoxide plus 4 volumes methanol plus 3 volumes water).

a. *Standards.* Keep the amphotericin B standard under nitrogen at −20° to −30°C. Samples of the standard, for daily use, may be kept in a screw cap vial in air at 5°C. for up to 3 months without loss of potency. Prepare, fresh daily, a solution of amphotericin B in dimethyl sulfoxide at a concentration of about 1 mg. of activity per milliliter. Dilute this solution in DMW-1 to obtain the following concentrations: 36.0, 0.9, 0.6, 0.4, and 0.267 μg./ml.

b. *Samples.*

Solutions: Estimate the potency of the sample and prepare solutions of 0.6 and 0.4 μg./ml. DMW-1.

Suspensions: Add sufficient dimethyl sulfoxide so that there are at least 9 volumes of dimethyl sulfoxide per volume of suspension and the resultant solution contains 100 to 1000 μg. of amphotericin B activity per milliliter. Shake about $\frac{1}{2}$ hour. Dilute to an estimated 0.6 and 0.4 μg./ml. in DMW-1.

Solids: Add from 1 to 10 ml. of dimethyl sulfoxide for each milligram of amphotericin B activity. Shake about $\frac{1}{2}$ hour or, if necessary, triturate in dimethyl sulfoxide with a mortar. In complex formulations, it may be necessary to prewash the sample with hexane or small volumes of water or ethanol to remove other ingredients that physically prevent solution of the amphotericin B. Dilute the dimethyl sulfoxide extract in DMW-1 to an estimated 0.6 and 0.4 μg./ml.

4. Assay Design

Prepare 150 × 25 mm. tubes, each containing 0.06 ml.[20] of one of the standard solutions of amphotericin B or of sample. The standard solutions are 36.0 μg./ml. for adjusting photometer; 0.90, 0.60, 0.40, and 0.267 μg./ml. for calculation of potency (standard curve); and 0 (diluent only) for checking growth rate. Dilute the samples to estimated concentrations of 0.6 and 0.4 μg./ml. Include replicate tubes of each solution in each rack. Distribute the tubes of standards and samples in a random design (Fig. 1) in a test tube rack to compensate for within-rack location effects.

a. Preparation of inoculated assay medium, incubation, and reading. Immediately prior to use, inoculate a flask of assay broth (prewarmed to 37°C.) with 2.5% (v./v.) from the second transfer flask. The composition of assay broth is: Casitone (Difco), 0.9%; glucose, 2.0%; yeast extract, 0.5%; sodium citrate, 1.0%; potassium dihydrogen phosphate, 0.1%; and potassium monohydrogen phosphate, 0.1%. Sterilize by heating with steam in an autoclave for 20 minutes at 121°C. The pH after sterilization is 6.5–6.7. To tubes already containing 0.06 ml. of amphotericin B solution, add 10 ml. of inoculated assay medium. Incubate for 3 hours at 37°C. on a shaker reciprocating at 120 (1$\frac{1}{2}$ inches) strokes per minute. After incubation, add 1.0 ml. of an aqueous 10% phenol solution to each tube to terminate growth. Then read the per cent transmission of each tube with a Spectronic-20 photometer (or equivalent) at a wavelength of 600 to 660 mμ, using a cuvette of 17 mm. diameter. The tubes dosed only with diluent should read about 35% T.

b. Calculation. In calculating results, treat each rack of tubes as an independent unit. Average the per cent transmission readings obtained for each concentration of the standard. Plot the average readings (linear

---

[20] The 0.06-ml. dose size (per 10 ml. of assay broth) was chosen to permit convenient delivery of triplicate doses from one filling of a 0.2-ml. pipet. Since the susceptibility to amphotericin B is decreased threefold by 0.1% dimethyl sulfoxide in the medium, a larger volume is not desirable.

U_{10} L	U_{11} L	U_{12} L	S_3 L	U_{13} L	U_{14} L	U_{15} L	U_{16} L	U_{17} L	U_{18} L	U_{19} L	S_2 O
U_1 L	U_2 L	U_3 L	U_4 L	U_5 L	U_6 L	S_2 L	U_7 L	U_8 L	U_9 L	S_1 ∞	S_1 L
U_{12} H	S_3 H	U_{13} H	U_{14} H	U_{15} H	U_{16} H	U_{17} H	U_{18} H	U_{19} H	S_2 HH	U_{10} H	U_{11} H
U_3 H	U_4 H	U_5 H	U_6 H	S_2 H	U_7 H	U_8 H	U_9 H	S_1 HH	S_1 H	U_1 H	U_2 H
U_{13} L	U_{14} L	U_{15} L	U_{16} L	U_{17} L	U_{18} L	U_{19} L	S_2 LL	U_{10} L	U_{11} L	U_{12} L	S_3 L
U_5 L	U_6 L	S_2 L	U_7 L	U_8 L	U_9 L	S_1 LL	S_1 L	U_1 L	U_2 L	U_3 L	U_4 L
U_{15} H	U_{16} H	U_{17} H	U_{18} H	U_{19} H	S_2 HH	U_{10} H	U_{11} H	U_{12} H	S_3 H	U_{13} H	U_{14} H
S_2 H	U_7 H	U_8 H	U_9 H	S_1 HH	S_1 H	U_1 H	U_2 H	U_3 H	U_4 H	U_5 H	U_6 H
U_{17} L	U_{18} L	U_{19} L	S_2 LL	U_{10} L	U_{11} L	U_{12} L	S_3 L	U_{13} L	U_{14} L	U_{15} L	U_{16} L
U_8 L	U_9 L	S_1 LL	S_1 L	U_1 L	U_2 L	U_3 L	U_4 L	U_5 L	U_6 L	S_2 L	U_7 L
U_{19} H	S_2 O	U_{10} H	U_{11} H	U_{12} H	S_3 H	U_{13} H	U_{14} H	U_{15} H	U_{16} H	U_{17} H	U_{18} H
S_1 ∞	S_1 H	U_1 H	U_2 H	U_3 H	U_4 H	U_5 H	U_6 H	S_2 H	U_7 H	U_8 H	U_9 H

FIG. 1. Design for tube placement in the rack. Design of 12×12, 144-tube rack containing 4-dose standard curve with three independent preparations of standards and 19 unknowns at 2 doses.

Compound	Concentration of antibiotic solution added to tubes (µg./ml.)					
	Diluent blank (0)	0.267 (LL)	0.4 (L)	0.6 (H)	0.9 (HH)	36 (∞)
Standard no. 1 (S_1)	—	**[a]	***	***	**	**
Standard no. 2 (S_2)	**	**	***	***	**	—
Standard no. 3 (S_3)	—	—	***	***	—	—
Sample no. 1 (U_1)	—	—	***	***	—	—
.	.	.	.	.	.	.
.	.	.	.	.	.	.
.	.	.	.	.	.	.
Sample no. 19 (U_{19})	—	—	***	***	—	—

[a] Number of asterisks indicate the number of tubes.

scale) against the respective concentrations (logarithmic scale). The resultant curve is "S" shaped. Average the percentage transmission readings obtained for each dilution of the unknown and interpolate the concentration of each from the dose-response curve. To obtain the concentration in the diluted unknown, multiply the interpolated concentration by the dilution factor.

C. Nystatin: Photometric Assay Method

1. Introduction

The applications and conditions of this method are essentially the same as those of the amphotericin B photometric method.

The differences are described in the following paragraphs.

2. Test Organism

No differences.

3. Preparation of Standards and Samples

a. *Standards.* Prepare daily a solution of nystatin[21] in dimethyl sulfoxide at a concentration of about 1000 u./ml. (the reference standard is obtained from the United States Pharmacopeia) and dilute in water to obtain the following concentrations, 150, 15, 10, 6.66 and 4.44 u./ml.

b. *Samples.* Dilute dimethyl sulfoxide extracts or solutions of samples to an estimated 10 and 6.66 u./ml. with water.

4. Assay Design

Prepare 150 × 25 mm. tubes, each containing 0.6 ml. of one of the following solutions of nystatin: the standards (u./ml.) : 150, for adjusting the photometer; 15, 10, 6.66, and 4.44 for calculation of potency; diluent only, for checking growth rate; and the dilutions of samples, 10 and 6.66 u./ml.

a. *Preparation of inoculated assay medium, incubation, and reading.* Decrease the incubation temperature (from 37°C.) to 30°C. to take advantage of the increased susceptibility of the organism to nystatin at this temperature. Increase the incubation time to 4 hours to compensate for the decreased growth rate at 30°C.

b. *Calculation.* No differences.

III. Agar Dilution Assay Method

Amphotericin B

1. Introduction

The agar dilution method of assay is primarily suited for testing the susceptibility of fungi to antifungal agents. One of the more important

---

[21] The trade name of E. R. Squibb & Sons, Division of Olin Mathieson Chemical Corporation for nystatin is Mycostatin.

applications is the testing of clinical isolates. The antifungal agent or agents to be tested are diluted in agar to several concentrations. The fungi to be tested are streaked across the surface of the plates containing different concentrations of antifungal agent. The lowest concentration of the antifungal antibiotic that inhibits growth of the fungus indicates its susceptibility to the antifungal agent. Since quantitation with this method does not depend on the characteristics of diffusion of the antifungal agent through an agar medium, it circumvents the inherent errors of the disk technique for determining susceptibility. The use of agar allows testing of several isolates on each plate.

2. Test Organism

a. *Inoculum.* Isolate a pure culture from the specimen by standard techniques. Prepare a uniform suspension of the culture in sterile saline. For a control, prepare a suspension of *Candida albicans* from a stock slant. Keep the control suspension at 5°C. and use for not longer than 1 month.

b. *Preparation of assay plates.* Prepare assay agar of the following composition: Tryptone (Difco), 0.5%; malt extract, 0.3%; yeast extract, 0.3%; glucose, 1.0%; and agar, 1.5%. Dispense 300-ml. aliquots in 500-ml. flasks and sterilize by heating with steam in an autoclave for 20 minutes at 121°C. Dilute the amphotericin B solution in the flasks of melted assay agar to the following concentrations: 100, 50, 25, 12, 6, 3, 1.6, 0.8, 0.4 and 0.2 μg./ml. Prepare one flask without antibiotic for a control. Fill petri dishes from each flask, about 15 ml. per dish.

3. Preparation of Standards

Weigh 25–30 mg. of amphotericin B reference standard and dissolve in dimethyl sulfoxide to a concentration of 4 mg. of activity per milliliter of the solvent.

4. Assay Design

Fill a 2- or a 5-ml. pipet with the suspension of the test organism and streak it across the surface of one plate of each concentration. Streak 3 to 10 different suspensions on each plate. Incubate the plates for 3 to 4 days at room temperature. Visually examine the plates for growth. The smallest concentration of amphotericin B that causes inhibition is the minimum inhibiting concentration.

IV. Diffusion in Agar

A. Nystatin: Diffusion Assay on Large Plates

1. *Introduction*

The agar diffusion method of assay is recommended for measuring nystatin potencies where a high degree of precision is required. This method may be used for assaying pharmaceutical dosage forms, process development and control samples, comparison of standards, and samples from drug metabolism studies.

The arrangement of standards and samples on the agar plate and the method of calculation, patterned after a procedure by Bliss,[22] minimize error due to nonrandom effects and permit detection of other aberrant situations.

2. *Test Organisms*

One of two test organisms is used depending on the nature of the sample to be assayed. Use *Candida albicans* (Squibb 1539) for assaying fermentation beers and impure concentrates as they may contain cyclohexamide (Actidione) which does not affect this organism.[23] Use *Saccharomyces cerevisiae* (Squibb 1600) for all samples which might contain tetracycline as this antibiotic depresses the activity of nystatin against *Candida albicans* in the agar diffusion method.

a. Stock culture. Subculture the test organisms each month using an agar medium composed of: agar, 1.5%; glucose, 1.0%; tryptone (Difco), 0.5%; yeast extract, 0.5%; and malt extract, 0.5%. The medium is sterilized, prior to inoculation, by heating with steam in an autoclave for 20 minutes at 121°C. Incubate 16 to 18 hours at 37°C. and store at 5°C.

b. Inoculum. Prepare inoculum weekly by transferring from the stock culture to a 125-ml. Erlenmeyer flask containing 80 ml. of inoculum broth. The composition of inoculum broth is the same as that of the stock culture agar except that agar is not included. Incubate the inoculum flask for 16 to 18 hours at 37°C. and store at 5°C.

c. Preparation of large plates. Use 3-quart Pyrex baking dishes with outside dimensions of 15⅜ × 9 × 2 inches and fitted with stainless steel lids. Pour a 200-ml. base layer of assay agar in each dish. The composition of assay agar is: anhydrous Yeast Beef Agar (Difco or B.B.L.), 3.2%; glucose, 2.0%; sodium chloride, 1.0%; malt extract, 0.5%; and sufficient hydrochloric acid, added before sterilization, to adjust the pH to 5.5.

[22] C. I. Bliss, "The Statistics of Bioassay." Academic Press, New York, 1952.

[23] R. Brown and E. L. Hazen, *in* "Therapy of Fungus Diseases" (T. H. Sternberg and V. D. Newcomer, eds.), pp. 164–167. Little, Brown, Boston, Massachusetts, 1955.

Sterilize by heating with steam in an autoclave for 20 minutes at 121°C. The assay organism is highly dependent on the sodium chloride content of the agar. The size of zones of inhibition can be increased, at the expense of decreased growth, by increasing the salt content to 3 or 4%. In the absence of salt there are no zones of inhibition even at the 200-u./ml. level.

After the base layer has solidified, add 20 ml. of the inoculum to a liter of liquid assay agar tempered at 46°C. and pour 120 ml. of this seed agar over the base agar.

3. Preparation of Standards and Samples for Assay

Use brown or red glassware containers for all nystatin solutions to protect them from photoinactivation. Minimize contamination of samples and standards by using sterile glassware.

a. *Standards.* Maintain the reference standard in the dark at −20°C. sealed under nitrogen. For daily weighings use an aliquot of the reference standard and keep it at 5°C. over a desiccant when not in use. This is called the working standard. When the working standard has been in use for 2 months, discard it and replace with a fresh one taken from the reference standard. Prepare duplicate weighings of the working standard each day. Prepare buffered 70% propanol by mixing 7 volumes of *n*-propanol and 3 volumes of sterile .05 M pH 7 phosphate buffer (5.62 gm. K_2HPO_4 and 2.132 gm. KH_2PO_4 per liter of distilled water). The buffered propanol may become cloudy due to precipitation of the salts but this does not interfere with its function. To each weighing add glass beads (2–4 mm. diameter) and one or more milliliters of buffered 70% propanol per milligram of the standard and shake on a reciprocating shaker (200 1-inch strokes per minute) for 1 hour in order to dissolve the nystatin. Dilute each of the two weighings in buffered 70% propanol to 200, 100, and 50 u./ml., respectively. These levels are referred to as high, medium, and low.

b. *Samples.* Extract all solid samples or dissolve with buffered 70% propanol using at least 0.5 ml. per 1000 u. of activity. Add glass beads to aid extraction or solution and shake as above. In case of tablets, crush with a pestle in a mortar before extracting. For ointments with a hexane soluble base, first slurry 1 gm. of ointment in 5 ml. of hexane and then add sufficient buffered propanol to make up to 100 ml. before extracting. Extract nystatin from fermentation beers that contain mycelium by adding sufficient *n*-propanol to 30 ml. of beer to make a total volume of 100 ml. While stirring, lower the pH of the solution to 3.0 with 18% HCl. Continue stirring at this pH for 10 minutes and then readjust to pH 7.0 with 20% sodium hydroxide. Continue stirring for an additional 5 minutes and readjust pH if necessary. If solvents other than buffered

70% propanol are used, either the subsequent dilution in buffered 70% propanol must be large or the standards must be treated in the same way. For example, an assay of a dimethyl sulfoxide extract will be biased if the extract is not diluted with buffered propanol at least 25-fold.

Dilute all solutions or extracts with buffered 70% propanol to the three assay levels: 200, 100, and 50 u./ml.

4. Large Plate Design

Align under the seeded plate the template (Fig. 2) showing the 6 × 9 × 9 × 3 order of distribution (6 columns, 9 rows, 9 compounds, and 3 levels of disks containing standards and unknowns). The 9 compounds

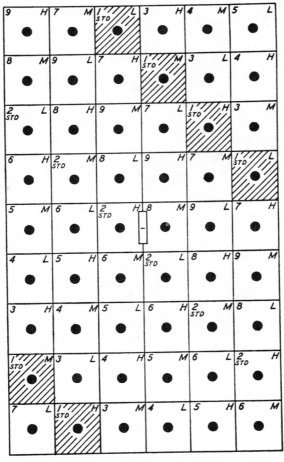

FIG. 2. Pattern for the arrangement of sample disks in the 6 × 9 × 9 × 3 assay design. S_1, standard 1; S_2, standard 2; 1 to 7, sample numbers; H, high dose; M, medium dose; L, low dose.

FIG. 3. A modified Baloptikon projector for reading large plates.

include two independent preparations of the standard and 7 unknowns. The three levels, 200, 100, and 50 u./ml., are in a fixed ratio of 4:2:1 (high, medium, and low). These are duplicated within each plate. Use ⅜-inch filter paper disks (Schleicher & Schuell 740-E) for applying the diluted nystatin solutions. Load each disk by holding it with a cover-glass forceps and immersing it in the appropriate solution. Remove excess liquid with a quick snap of the wrist immediately after removing the disk from the solution. Place each disk on the agar surface in successive order by rows or columns immediately after loading.

a. Incubation. Incubate the plates at 37°C. for 18 to 20 hours at a relative humidity of 60 to 80%.

b. Reading. Read the resulting zones of inhibition with the aid of a projector. The Baloptikon (Bausch & Lomb), modified as shown in Fig. 3, projects the zone of inhibition, magnified fivefold, on a calibrated ruler directly in front of the technician. The Zone Comparator (Technical Controls Incorporated) coupled with a digital system (Datex Corporation) automatically positions the plate for each zone and, by means of a paper tape printer, or card punch, records the zone diameter. The Zone Comparator is described by Haney *et al.* in Chapter 5.

d. Calculations. The IBM digital computer 650 is utilized for the following computations: (1) calculations of potencies, (2) analyses of variance, (3) determinations of standard errors, (4) comparison of one standard with the other, and (5) calculations of F tests for significance of means, slopes, and assay validity (parallelism, curvature, and opposed curvature).

The method for calculation of potencies is as follows:

$$\log \text{potency} = M + 2.301^* + \log D$$

where
$$M = \frac{4(0.301)(\sum U - \sum S)}{3(S_H + U_H) - (S_L + U_L)}$$

and $\sum U$ = sum of unknown responses

$\sum S$ = sum of standard responses

S_H = responses of high concentration of standard (sum)

U_H = responses of high concentration of unknown (sum)

S_L = responses of low concentration of standard (sum)

U_L = responses of low concentration of unknown (sum)

0.301 = log of 2, dosage interval

2.301* = log (high concentration of standard, 200 u./ml.)

D = dilution of high concentration of unknown.

B. Nystatin: Small Plate Assay Method

1. Introduction

The small plate assay method, a modification of the large plate method, combines the selectivity of a diffusion method with increased sensitivity. A thin layer of agar containing a subinhibitory concentration of nystatin enables subsequent detection of as little as 5 u. of nystatin per milliliter of sample on the disk used. Since the aqueous-buffer-propanol solvent diluent, employed for the large plate assay, produces zones of inhibition with this method, a less toxic solvent and diluent are employed. A simpler design and method of calculation, at the expense of some precision and control of accuracy, also make this method valuable when large numbers of assays are the prime objective. The differences of this method from the nystatin agar diffusion large plate method are described in the following paragraphs.

2. Test Organism

Candida albicans (Squibb 1539). The composition of the assay agar is: anhydrous Yeast Beef Agar (Difco or B.B.L.), 2%; sodium chloride, 4%; and dextrose, 1%. Adjust to pH 5.5 with hydrochloric acid before sterilization. Sterilize by heating with steam in an autoclave for 20 minutes at 121°C. After sterilization cool the liquefied agar to 46°C. and add 100 u. of nystatin per liter of agar. Add 10 ml. of inoculum per liter and pour a single 7-ml. layer in each 100-mm.-diameter flat-bottom glass petri dish. Cover with a lid containing an absorbent paper disk.

3. Preparation of Standards and Samples

Prepare a 4000 u./ml. solution by dissolving the reference standard with brief shaking in dimethyl sulfoxide (DMSO). Dilute this stock solution with 50% aqueous DMSO to concentrations of 80, 40, 20, 10, and 5 u./ml. These five solutions plus a diluent blank (50% DMSO) are used for establishing the standard curve.

Extract or dissolve samples with DMSO and, based on the expected potency, dilute to 20 u./ml. with 50% aqueous DMSO.

4. Small Plate Design

For each plate of the standard curve use six ⅜-inch paper disks, each one loaded with a different one of the six solutions used for the standard curve solutions. With a template under the plate as a guide, place the disks, as loaded, equidistant from each other on the surface of the agar. Make eight replicates. On each of a set of four more plates, place one disk loaded with the reference level, 20 u./ml., and one disk for each of five samples.

Calculate the potency of the samples from the standard-dose response curve in the usual way.

C. Amphotericin A: Large Plate Assay Method

1. Introduction

Amphotericin A and B may both occur in fermentations, and consequently, it is frequently necessary to measure the A in the presence of B. Amphotericin A and B are both active against the test organism but B diffuses poorly. Although it is necessary to use a mixed standard approximating the ratio of A to B expected for the sample, the selectivity of the diffusion process and of the test organism under the conditions of this method enables accurate assay of one part of A in a mixture containing up to 20 parts B. The design for the arrangement of standards and samples on the assay plate and the method of calculation offer protection from error due to nonrandom effects and permit detection of other aberrant situations.

The differences of this method from the nystatin agar diffusion large plate method are described in the following paragraphs.

2. Test Organism

a. Stock culture. Rhodotorula glutinis (Squibb 2358). Prepare stock slants weekly and incubate for 3 days at 24° ± 1°C.

b. Inoculum. Dispense 100-ml. quantities of inoculum broth in 500-ml. Erlenmeyer flasks. Prepare the inoculum twice weekly by suspending the cells on the stock slant in about 3 ml. of inoculum broth and transferring about 0.1 ml. of this to the flask of inoculum broth. Incubate the flask on a rotary shaker (180 r.p.m., 0.7 inch radius of gyration) for 3 days at 24° ± 1°C.

c. Preparation of large plates. The composition of assay agar is: Tryptone (Difco), 0.5%; malt extract, 0.3%; yeast extract, 0.3%; potassium chloride, 1%; and agar, 1.8%. Dispense 250-ml. quantities into flasks. Sterilize by heating with steam in an autoclave for 20 minutes at 121°C. Add 50 ml. of sterile 80% dextrose solution to each flask of sterile agar while still melted, cool to 46°C., and inoculate with 10 ml. of the working inoculum. Pour this into the large plates (described in Section IV.A); no second layer of agar is used.

3. Preparation of Standards and Samples

a. Standards. Use a crystalline amphotericin A reference standard. Prepare the standard solution in dimethyl sulfoxide (DMSO) at a concentration of 1 mg./ml. and shake briefly. Dilute in 80% DMSO (eight parts DMSO plus two parts water) to a 40, 20, and 10 μg./ml. (i.e., high, medium, and low working levels). If the standards are to be used to assay samples containing a mixture of amphotericin A and B and the amount of B is expected to be greater than 20% that of A, prepare a mixed standard by adding B to the A. The relative concentrations of A and B in the standard should approximate those in the sample within about ±15%. For example, if the relative content of B is expected to be between 60 and 90% use a standard mixture containing 75% B.

b. Samples. Dissolve or extract solid samples with 1 ml. of DMSO per milligram of expected activity in the sample. Extract fermentation broths by shaking one part of broth with four parts of DMSO for 15 minutes.

Dilute solutions and extracts to the levels of the assay, 40, 20 and 10 μg./ml. with 80% DMSO.

4. Assay Design

Incubate the plates for 2 to 3 days at 24° ± 1°C.

V. Respirometric Method of Assay

A. Nystatin

1. Introduction

The respirometric method of assay of nystatin or other polyene antifungal agents depends on the ability of these substances to inhibit respiration of susceptible yeasts. The amount of gas (CO_2) produced by the yeast is then a function of the antibiotic content of the sample.

The conventional equipment used for studying respiration of yeasts is exemplified by the Warburg respirometer. This equipment does not lend itself readily for use in routine analyses. With minor modifications, a hypodermic syringe can serve the purpose of a reaction vessel and a manometer, the two basic components of a Warburg respirometer. This can be done by closing the needle orifice with an undrilled needle nub. The sample containing inoculated nutrient medium and an antifungal agent is drawn into the syringe, the needle orifice closed, and the syringe and its contents incubated. Carbon dioxide produced by respiration of the yeast displaces the piston and its position on the calibrated barrel of the syringe is a measure of the quantity of carbon dioxide produced.

The respirometric method is used to assay turbid samples of antimicrobial agents which diffuse poorly in agar. It is useful for assaying such samples as tissue homogenates and citrated whole blood directly without extracting the antibiotic from the solids. Therefore the need for tedious extraction and clarification of samples, as might be required for a turbidimetric method, is eliminated.

2. Test Organism

Stock culture and inoculum. Prepare the *Candida tropicalis* inoculum by the method described in the amphotericin B photometric assay method (Section II.B). Just prior to use, centrifuge a 100-ml. aliquot of the inoculum in a calibrated tube to concentrate the cells. Discard the supernatant fluid and resuspend the cells with approximately 10 ml. of cold (5°C.) assay broth to produce a 20% (v./v.) cell suspension. The composition of assay broth is: Casitone (Difco), 0.9%; glucose, 7%; yeast extract, 0.5%; sodium citrate, 1%; potassium monohydrogen phosphate,

0.1%; and potassium dihydrogen phosphate, 0.1%. Sterilize by heating with steam in an autoclave for 20 minutes at 121°C. The pH after sterilization is 6.5 to 6.7.

3. Preparation of Standards and Samples

Incorporation of relatively large amounts of tissue in the assay medium causes significant changes in the rate of gas production by the yeast; consequently, standards are prepared in antibiotic-free tissue of the same type as the sample. The susceptibility also depends on the type of tissue. With various pork and chicken tissues,[24] the minimal concentrations detected varied from 3 to 12 units of nystatin per gram of tissue.

Grind all tissues in a meat grinder, rinsing the grinder with methanol between samples. Add 1 gm. each of penicillin and streptomycin per liter of assay broth to control bacterial contaminations. Prepare a 6000 u./ml. solution of nystatin in dimethyl sulfoxide. Add 0.5 u. of nystatin per milliliter of assay broth, i.e., 1.0 ml. of 500 u./ml. solution per liter of assay broth. With some samples, levels of nystatin less than 0.5 u./ml. enhance gas production. The addition of 0.5 u./ml. causes partial inhibition of respiration; therefore, any further addition of nystatin cannot stimulate gas production, but results in increased inhibition.

a. Standards for nonfatty tissue. To 3 gm. of nystatin-free tissue add 0.3 ml. of 16,000 u./ml. nystatin solution and homogenize (using a Omni-Mixer) with 30 ml. of assay medium to make a 160 u./gm. standard. Immerse the homogenizer in an ice bath when blending tissue with assay medium. Homogenizing for 1 minute is usually satisfactory. Add a drop or two of mineral oil to help dissipate foam after homogenizing. Keep homogenized tissues cold throughout the preparation procedure. To an additional 5 gm. of nystatin-free tissue add 0.5 ml. of dimethyl sulfoxide and homogenize with 50 ml. of assay medium to make a 0 u./ml. standard. The concentrations for the standard curve are 160, 40, 10, 2.5, and 0 u./gm. of tissue. Prepare the 40, 10, and 2.5 u./gm. concentrations by diluting serially the 160 u./gm. standard with the 0 u./gm. standard solution. Place 10 ml. of each solution in a 1 × 4 inch tube.

b. Samples for nonfatty tissue. Add 0.3 ml. of dimethyl sulfoxide and 30 ml. of assay medium to 3 gm. of tissue and homogenize. Pipet a 10-ml. aliquot into a 1 × 4 inch tube.

c. Standards and samples for fatty tissues. Prepare the same as for nonfatty tissues except replace the operation of homogenizing in assay medium with the following steps. After adding nystatin or dimethyl sulfoxide to the tissues, add about 6 gm. of sand and grind to a paste with a mortar and pestle. Slowly add 100 ml. of hexane, with continued grind-

---

[24] J. R. Gerke, J. Gentile, H. Stander, and J. F. Pagano, *Antibiotics Ann.* **1959/1960**, 563 (1960).

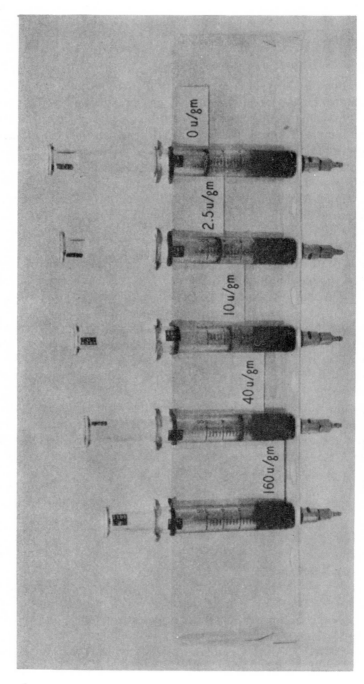

Fig. 4. Nystatin standard curve in syringe respirometers.

ing, to effect intimate contact between solvent and fat. Decant the hexane into a centrifuge bottle and precipitate suspended nystatin by centrifugal force. Discard the clear hexane supernate. Grind the residue in the mortar two more times with 25-ml. portions of hexane. Decant and precipitate the suspended nystatin, as above, and collect all the residues from each in one sample bottle. Vacuum dry the bottle, mortar, and their contents at room temperature by exposure to a pressure of 15 to 20 mm. Hg for 10 minutes. With 30 ml. of assay broth, rinse the contents of the mortar into the centrifuge bottle containing the residue. Shake ½ hour and subject to centrifugal force.

4. Assay Design

a. Preparation of syringes and incubation and reading of test. Lubricate the plungers of 5-ml. Luer-Lok Syringes (Becton, Dickinson & Company) with Blandol (L. Sonneborn Sons Inc., New York), or similar low viscosity mineral oil, to prevent gas leakage and to lessen friction. If properly lubricated, the plungers will slide down by gravity. Add 0.5 ml. of the 20% cell suspension to each 10-ml. aliquot of standard or sample, and mix. For each mixture draw 2 ml. into each of three syringes and cap each syringe with an undrilled needle nub (Becton, Dickinson & Company). Attach each syringe with a rubber band to a holder (Fig. 4), and place the holder in a 30°C. water bath. At 15-minute intervals during incubation, remove the holders from the bath and invert several times to aid release of dissolved carbon dioxide. Incubate about 2 to 4 hours, depending on the tissue assayed, until the syringes containing the 0 u./ml. standard produce about 3 ml. of gas. Read from the calibration on the syringe barrel the volume of gas produced.

b. Calculation. Construct a standard curve by plotting gas volumes versus concentration of nystatin, and from the curve, read the concentration of nystatin in each sample.

B. Amphotericin B

1. Introduction

The applications and conditions of this method are essentially the same as for the nystatin respirometric method. Detection of from 0.06 to 1 μg. of amphotericin B per gram of tissue sample is possible depending on the tissue used.

The differences of this method from the one for nystatin are described in the following paragraphs.

2. Test Organism

No differences.

3. Preparation of Standards and Samples

Instead of adding to each liter of cold assay broth 1.0 ml. of nystatin solution, add 0.5 ml. of a 100 μg./ml. solution of amphotericin B in dimethyl sulfoxide to give a final concentration of 0.05 μg. of amphotericin B per milliliter of assay medium. To the tissue samples add a sufficient volume of a 100 μg./ml. solution of amphotericin B in dimethyl sulfoxide to make standards containing, respectively, 4, 1, 0.25, and 0.0625 μg./gm. of tissue.

4. Assay Design

No difference.

CHAPTER 7

Vitamins

7.1 Introduction

FREDERICK KAVANAGH

Eli Lilly and Company, Indianapolis, Indiana

Microbiological vitamin assays were not created *de novo;* they arose in response to needs of plant and animal physiologists.

Animal physiologists studied the nutrition of animals, (rats, pigeons, chickens, cattle) fed purified or restricted diets. Response of the animals indicated the presence or absence of accessory factors in the diet. Such responses could be used as a form of bioassay to measure the quantity of growth factor in the diet. The assays were slow (an answer might take 6 weeks or more to obtain), expensive, and, by present standards, inaccurate. None the less, by 1924, three vitamins had been identified. They were the fat soluble vitamines A and D and water soluble vitamine B. Two components of vitamine B needed by the rat had been distinguished by 1933. They were vitamines B_1, the antineuritic factor, and B_2, the rat antipellagra factor. Other B vitamins were being suggested to an unbelieving world.

During the three decades of the animal investigations just described, growth requirements of strains of yeast were studied in several laboratories with the object of unraveling the bios-complex. Bios occurred in plant and animal tissues and was essential for the growth of domesticated yeasts. Bios fractions were water soluble and resembled vitamine B in this respect. The following members of the bios group, pyridoxine, pantothenic acid, nicotinic acid, biotin, and inositol were shown to be essential for growth of bacteria, yeasts, and other fungi before their vitamin activity for animals was known.

During the two decades prior to 1933, several investigators showed that vitamine B preparations, yeast extract among others, increased growth rate or were essential for growth of several fungi. Preparation of crystalline vitamin B_1 in 1932 and demonstration that it was essential for growth of the fungi *Phycomyces nitens* and *P. blakesleeanus* initiated the modern work on vitamins and growth of fungi. The availability of a pure vitamin known to occur in yeast extract triggered a burst of work and publication that lasted the remainder of the decade. Assay for vitamin B_1 and its intermediates, vitamin thiazole, and pyrimidine, with *Phycomyces* probably was the first modern microbiological assay where the results could be reported in terms of quantities of known substances with some certainty.

While all of this activity was going on in animal nutrition, physiology of fungi and vitamin isolations, the dairy bacteriologists were busy determining growth requirements of lactic acid bacteria. Taken as a group, the nutritional requirements of the bacteria were so complex that they needed 9 vitamins, 18 amino acids, and the purine and pyrimidine bases. The other groups of microorganisms mentioned here had simple growth requirements by comparison.

By 1940, thiamine (vitamin B_1), riboflavin (vitamin B_2), pyridoxine (vitamin B_6), pantothenic acid, biotin, inositol, nicotinic acid, and p-aminobenzoic acid had been isolated and purified. Their physiological activities had been determined in plants, animals, and bacteria and all except biotin synthesized. Folic acid was known, and extensive work with it had been initiated.

The remarkable achievements in the decade ending in 1941 would have been impossible without rapid microbiological assays. These assays permitted the chemists to make in 1 year progress that would have taken a decade had they been dependent upon animal bioassays.

An assay for riboflavin using lactic acid bacteria was published by Snell and Strong in 1939. Its rapidity and specificity stimulated application of lactic acid bacteria to the general problem of vitamin assays with the consequence that 8 vitamins, several other substances, and at least 18 amino acids can now be assayed with them. Assays with these bacteria played significant roles in the discovery and isolation of folic acid and vitamin B_{12}. At this time, an assay employing a strain of lactic acid bacteria is used, if at all possible, because of its speed and familiarity. In the pharmaceutical industry, the principal vitamin assays employing the lactic acid bacteria are those for folic acid, biotin, vitamin B_{12}, and pantothenic acid and derivatives. Other vitamins usually are measured by a chemical method. Yeast and fungus assays are rare.

Anyone starting vitamin assays with lactic acid bacteria should read the chapter on "Microbiological Methods in Vitamin Research" by Snell.[1]

[1] E. E. Snell, *in* "Vitamin Methods" (P. György, ed.), Vol. 1. Academic Press, New York, 1950.

Snell gives a summary of the history of each assay, its advantages and disadvantages and discusses problems inherent in microbiological assaying. It is a classic. The book by Barton-Wright[2] and the essay by Sokolski and Carpenter[3] contain information not in the article by Snell.

The following U.S.P. reference standards for vitamins are obtainable from U.S.P. Reference Standards, 46 Park Avenue, New York 16, New York: calcium pantothenate, cyanocobalamin (B_{12}), folic acid, nicotinamide, nicotinic acid, p-aminobenzoic acid, pyridoxine hydrochloride, riboflavin, and thiamine hydrochloride.

[2] E. C. Barton-Wright, "Microbiological Assay of the Vitamin-B Complex and Amino Acids," 179 pp. Pitman, New York, 1952.
[3] W. T. Sokolski and O. S. Carpenter, *Progr. in Ind. Microbiol.* 1, 93 (1959).

7.2 pH as Assay Response

FREDERICK KAVANAGH

Eli Lilly and Company, Indianapolis, Indiana

The growth response of species of *Lactobacillus* to vitamins may be measured in many ways. Common responses that have been used in assaying are: dry weight, turbidity, and titratable acidity. Silber and Mushett[1] suggested that pH at the end of the incubation period was a measure of quantity of growth substance. Elias *et al.*[2] compared U.S.P. XII methods with measurement of pH in the assay for nicotinic acid and riboflavin. They concluded that pH was as reliable an indication of quantity of vitamin as either titration or turbidity and much faster.

A requirement for the pH method is a stable pH meter which can be read to ±0.01 pH. A Beckman Model G pH meter with new batteries and desiccant was used for a number of years by the microbiological assay laboratories[3] of Eli Lilly & Company in assaying for nicotinic acid, pantothenic acid, folic acid, and vitamin B_{12}. During the last several years, the pH measurements were made with a modified Leeds and Northrup 7664 Stabilized pH Indicator fitted with the miniature glass electrode assembly dipping into a small cup which is emptied by opening a solenoid valve. The stability of the pH indicator is such that the scale can be expanded by a factor of 10. Scale expansion increases the precision of the readings but does not increase the accuracy. Precision is more important than accuracy because absolute values of pH are not needed. The pH range for the assays is from about 5.5 to 4.3.

[1] R. H. Silber and C. W. Mushett, *J. Biol. Chem.* **146**, 271 (1942).

[2] W. F. Elias, J. Merrion, A. Nagler, and M. Broome, *J. Lab. Clin. Med.* **30**, 622 (1945).

[3] J. T. Stephenson, personal communication.

A plot of pH against log concentration of growth substance gives a straight line over most of the range. Deviations occur at both the low and the high pH ends of the line. Replication of tube values is as good as with photometric measurements. The pH method has several advantages over the photometric procedure. It is free from the inherent difficulties of measuring turbidity of rod shaped bacteria and is not restricted to preparations free from color or suspended material. The samples, however, when diluted to assay level, must be free from substances contributing to the buffer capacity of the medium. Measurement of pH is much faster than titration of acid.

7.3 Specificity

FREDERICK KAVANAGH

Eli Lilly and Company, Indianapolis, Indiana

Vitamin assays, chemical and microbiological, are rarely as specific as the literature would lead one to believe. What is meant by specificity depends upon the complexity of the vitamin. If it is a single active substance, i.e., riboflavin, then the naturally occurring vitamin after appropriate preparation of the sample to release combined forms, can be assayed as a quantity of chemically pure standard. If the "vitamin" is a complex of related substances with somewhat different activities, then only activity in terms of an arbitrarily selected standard is possible. In this instance, specificity as meaning a particular chemical entity is impossible without more information than that usually furnished by an assay. Because biological assays measure responses and not quantities of substances, response to a standard substance is needed to permit conversion of responses to samples into quantity of substance. Uncertainty of interpretation increases rapidly as the number of chemically related structures with similar biological properties increases. Finally the situation of the folic acid group is reached; here, interpretation is possible only in general terms (see Chapter 7.5). Interpretation of assay responses to samples with a known history, i.e., some pharmaceutical preparations, may be specific because interfering substances are known to be absent.

Relative activities of different derivatives of a compound usually vary with the test organism (*Lactobacillus* sp., *Rattus rattus*, *Homo sapiens*). Ideally, in nutrition studies, the assay should be with the same species of organism in which action of the vitamin is of interest. This rarely is practical and recourse is had to a second best method, chemical or microbiological. These methods have the further advantages of speed and cost over animal assays.

Chemical assays are used for riboflavin and thiamine in both high and low potency preparations and for riboflavin, thiamine, folic acid, and nicotinic acid in very high potency preparations. Usually, fluorometric methods are used because of this high sensitivity. Introduction of two commercial fluorometers[1,2] at about the same time as the Snell and Strong[3] microbiological method for riboflavin permitted development of the two types of assays simultaneously. Fluorometric assays were applied to other vitamins where the degree of specificity was less than for riboflavin. Use of two basically different assay systems, neither necessarily specific, contributed to temporary confusion and ultimately to clarification of the problems presented by multiplicity of forms of a vitamin. However, at first, fluorometric and microbiological methods were used interchangeably without realizing their limitations. A group of investigators[4] followed adsorption of crystalline "folic acid" fluorometrically and adsorption of "folic acid" found in tomato juice and liver extract microbiologically. An adsorbent good for crystalline "folic acid" was a poor adsorbent for "folic acid" in the two natural products.

Adsorption of degradation products, the strongly fluorescent pteridines,[5] was measured in the fluorometric assay whereas "folic acid," not pteridines, was measured in the microbiological assay. The fluorometric assay did not measure the substance of interest because pure folic acid does not fluoresce appreciably under the conditions of Daniel and Kline.[4] This confusion of an impurity with the main substance results from the extreme sensitivity of the fluorometric method and emphasizes the necessity of first ascertaining that fluorescence is not caused by impurities.

Fluorometric methods, because of their sensitivity, are valuable in detecting impurities in nonfluorescent substances or even in fluorescing compounds if the impurity has a different excitation wavelength or emission spectrum or a different response to pH of the solution. If a decomposition product is fluorescent and devoid of microbiolgical activity as in the example of "folic acid" cited above, then fluorometric assays can be used in conjunction with the microbiological method to obtain material balance on a partially decomposed sample.

Fluorometric methods may require less work per sample and be less time consuming than microbiological assays. They have definite advantages for certain samples, i.e., pharmaceutical preparations, made with

[1] D. J. Hennessy and L. R. Cerecedo, *J. Am. Chem. Soc.* **61**, 179 (1939).

[2] F. Kavanagh, *Ind. Eng. Chem. Anal. Ed.* **13**, 108 (1941).

[3] E. E. Snell and F. M. Strong, *Ind. Eng. Chem. Anal. Ed.* **11**, 346 (1939).

[4] E. P. Daniel and O. L. Kline, *J. Biol. Chem.* **170**, 739 (1947).

[5] F. Kavanagh and R. H. Goodwin, *Arch. Biochem.* **20**, 315 (1949).

relatively pure vitamins. These samples are also the ones assayed most accurately by microbiological assays. The general tendency is to assay the easy samples by chemical methods and to assign the low potency and inherently difficult samples to the microbiological assay laboratory. Also there is a fundamental objection to chemical assays for biologically active compounds. The chemical methods may utilize properties unrelated to biological activity. For example, D- and L-forms of a substance may be indistinguishable in the chemical assay and yet only one of the enantiomorphic forms may be biologically active. Interpretation of a chemical assay can be certain for samples of known history, e.g., some pharmaceutical preparations, and uncertain for samples of natural origin. Usually microbiological methods of assay are better measures of general biological activity than are chemical assays.

7.4 Biotin

HELEN R. SKEGGS

Merck Sharp & Dohme Research Laboratories, West Point, Pennsylvania

The role of biotin in human nutrition is still unknown. It is widely distributed in nature, and is required by numerous microorganisms for growth.[1,2] Since no chemical or physical assay methods are available for its determination, microbiological assays are widely used. The *Lactobacillus arabinosus* assay described in 1944 [3] remains one of the most

[1] L. D. Wright, *Biol. Symposia* **12**, 290 (1947).
[2] P. György, R. S. Harris, and E. E. Snell, *in* "The Vitamins" (W. H. Sebrell, Jr. and R. S. Harris, eds.), Vol. 1, pp. 525–618. Academic Press, New York, 1954.
[3] L. D. Wright and H. R. Skeggs, *Proc. Soc. Exptl. Biol. Med.* **56**, 95 (1944).

specific means of assaying free biotin and is detailed here. Other organisms are capable of using naturally occurring derivatives of biotin[4-6] to a greater extent than is *L. arabinosus*, and have proved most useful in the elucidation of naturally occurring derivatives of biotin. A brief outline of the *L. casei* and *Neurospora crassa* procedures are also described.

I. Test Organism

Lactobacillus plantarum ATCC 8014 (*L. arabinosus* 17–5) is highly specific in its response to free biotin. Of the known derivatives of biotin, only oxybiotin,[7] *N*-biotinyl glycine,[8] and biotin-*d*-sulfoxide[9] have more than 5% of the activity of biotin, all being essentially equal in activity to biotin.

II. Standard Solution

A stock solution of 100 μg./ml. of *d*-biotin is prepared in 50% ethyl alcohol and stored in the refrigerator. Kept tightly stoppered to avoid evaporation, such solutions are stable for at least 1 year. At monthly intervals, or as needed, a working standard is prepared for daily use at 1 μg./ml. in 95% ethyl alcohol. For use in the assay, the working standard is diluted with distilled water in steps of 100 to a concentration of 0.4 mμg./ml., i.e., dilution of 1 to 100 followed by 4 to 100.

III. Stability of Compound

Biotin is indefinitely stable in alcoholic solutions if kept refrigerated. Water solutions are less stable. It can be autoclaved for as many as 2 hours in 6 N H_2SO_4 without loss of activity but longer autoclaving or more concentrated acid causes destruction of 20 to 40% of the activity. It is much less stable when autoclaved with 6 N NaOH.

[4] L. D. Wright, E. L. Cresson, H. R. Skeggs, T. R. Wood, R. L. Peck, D. E. Wolf, and K. Folkers, *J. Am. Chem. Soc.* **72**, 1048 (1950).

[5] L. D. Wright, E. L. Cresson, and C. A. Driscoll, *Proc. Soc. Exptl. Biol. Med.* **91**, 248 (1956).

[6] E. E. Snell, *in* "Vitamin Methods" (P. György, ed.), Vol. 1, p. 388 Academic Press, New York, 1950.

[7] F. J. Pilgrim, A. E. Axelrod, T. Winnick, and K. Hofmann, *Science* **102**, 35 (1945).

[8] L. D. Wright, H. R. Skeggs, and E. L. Cresson, *J. Am. Chem. Soc.* **73**, 4144 (1951).

[9] D. B. Melville, D. S. Genghof, and J. M. Lee, *J. Biol. Chem.* **208**, 503 (1954).

IV. Preparation of Samples for Assay

A. Extraction of Biotin

The free biotin content of a sample can be measured by diluting the sample with water to a concentration of between 0.1 and 0.4 mμg./ml. Most of the biotin in nature, however, occurs in a bound or conjugated form, and can be released by hydrolysis with sulfuric acid. Hydrochloric acid should not be used in place of sulfuric acid since under some conditions it destroys biotin. Assay of samples before and after acid hydrolysis can be used to distinguish between free and total biotin.

For example, a yeast extract sample contains 2.2 μg./gm. total biotin, of which 0.084 μg./gm. is free. In this case, the difference between the two values occurs as biocytin, ϵ-N-biotinyl-L-lysine. Samples are prepared for assay as follows:

(a) Free biotin: 200 mg. yeast extract diluted to 100 ml. with distilled water.

(b) Total biotin: 200 mg. yeast extract autoclaved 2 hours in 10 ml. 6 N H_2SO_4, cooled, neutralized with 10 N NaOH, diluted to 100 ml. and filtered if necessary. Dilute 5 ml. to 100 ml. with distilled water for assay.

In the analysis of tissue, 10 ml. of 6 N H_2SO_4 is sufficient to hydrolyze to 5 gm. of tissue.

B. Interfering Substances

The response of *L. arabinosus* to biotin can be markedly influenced by the presence of fatty acids and lipoidal material in the sample. Oleic acid, at relatively high concentrations (150 μg. or more) can replace biotin for growth in the presence of aspartic acid.[10] Fatty materials can be removed from most samples by simple filtration through paper at pH 4.5. When the concentration of such materials is high, however, samples should be extracted with equal volumes of ether prior to dilution and assay. Comparative assays before and after ether extraction give an indication of the degree of interference.

V. Preservation of the Test Organism

Lactobacillus plantarum (*L. arabinosus*) is a microaerophilic organism and is best maintained in agar stabs. Stock cultures are transferred in duplicate every month or 6 weeks, depending on the medium employed

---

[10] H. P. Broquist and E. E. Snell, *J. Biol. Chem.* **188**, 431 (1951).

with one culture reserved for stock, and the other used for preparation of inocula. Stabs are grown at 30° for 48 to 72 hours, then stored in the refrigerator. The medium given in the tabulation has proved very satisfactory for the maintenance of a vigorous culture, transferred every 6 weeks.

Bacto yeast extract	1%
Bacto tryptose	1%
Glucose	0.1%
K_2HPO_4	0.2%
$CaCO_3$	0.3%
Agar	1.5–2.0%

The culture also may be maintained on Micro Assay Culture Agar (BBL (Baltimore Biological Laboratory, Inc.) or Difco) or 1% glucose, 1% yeast extract, 1.5% agar stabs, but on such media should be transferred monthly since the higher glucose concentration leads to greater acid production that on prolonged storage reduces the vigor of the culture.

VI. Preparation of Inoculum

On the day prior to use in the assay, make a transfer from the stab culture into liquid medium, and incubate at 30°C. for 24 hours. A suitable medium is 1% glucose, 1% yeast extract, tubed and sterilized in 7 to 8 ml. quantities in $\frac{5}{8}$ × 5 inch culture tubes. Micro Inoculum Broth (Difco or BBL) also can be used, but is less preferred since it contains oleic acid which, if carried over into the assay tubes, could interfere with the performance of the assay. Centrifuge the 24-hour culture (cotton plug secured with a rubber band); discard the spent medium and resuspend the cells in approximately 10 ml. of sterile physiological saline. Dilute the cell suspension aseptically 1 to 100 with sterile saline and add 1 drop to each sterilized assay tube.

VII. Medium

The composition of the medium is given in Table I. The quantities of solutions given refer to solutions described under the vitamin B_{12} assay procedure. It is convenient when the number of samples to assay is large and continuous to prepare a large quantity of "basal" medium composed of all ingredients except the glucose and vitamin supplement which are added to the desired quantity of medium as needed. Shaken well with benzene and stored at room temperature, the basal medium can be used

TABLE I

COMPOSITION OF ASSAY MEDIUM FOR BIOTIN

	Per 100 ml. double strength medium	ml. of stock solutions[a]
Casein hydrolyzate	1.0 gm.	10 ml.
Na acetate (anhydrous)	1.2 gm.	1.2 gm.
Glucose	4 gm.	4 gm.
DL-Tryptophan	40 mg.	10 ml.
L-Cystine	20 mg.	10 ml.
Adenine sulfate	1 mg.	1 ml.
Guanine hydrochloride	1 mg.	1 ml.
Xanthine	1 mg.	1 ml.
Uracil	1 mg.	1 ml.
Salts A		1 ml.
KH_2PO_4	100 mg.	
K_2HPO_4	100 mg.	
Salts B		1 ml.
$MgSO_4 \cdot 7H_2O$	40 mg.	
$FeSO_4 \cdot 7H_2O$	2 mg.	
NaCl	2 mg.	
$MnSO_4 \cdot 4H_2O$	2 mg.	
Vitamin supplement		5 ml.
Pyridoxine hydrochloride	400 μg.	
Riboflavin	200 μg.	
Ca pantothenate	200 μg.	
Nicotinic acid	200 μg.	
Thiamine hydrochloride	400 μg.	
p-Aminobenzoic acid	100 μg.	1 ml.

Adjust pH to 6.6–6.8 and dilute to volume

[a] Made according to Table III in *Lactobacillus leichmannii* Assay for Vitamin B_{12} (Chapter 7.11).

for several months. The vitamin solutions should be stored in the refrigerator and renewed monthly. The composition given is double strength. Dispense 5 ml. into each assay tube.

VIII. Design of the Assay

Using tubes approximately ¾ by 6½ inches, standard and sample solutions are quantitatively measured and dispensed as outlined in Table II. Plug the tubes with cotton or use suitable glass or steel caps, autoclave for 15 minutes at 120°C (total time in autoclave should not exceed 30

TABLE II
Standard and Sample Solutions

Tube no.	ml. biotin std. 0.4 mμg./ml.	mμg. biotin per tube	Distilled H₂O − (ml.)
1	0	0	5
2	0	0	5
3	0.5	0.2	4.5
4	1.0	0.4	4
5	1.5	0.6	3.5
6	2.0	0.8	3
7	3.0	1.2	2
8	5.0	2.0	0
Yeast extract (Section IV. A. a)			
9	1 ml.		4
10	2		3
11	3		2
12	5		0
Yeast extract			
acid hydrolyzed (Section IV. A. b)			
13	1 ml.		4
14	2		3
15	3		2
16	5		0

minutes), cool the tubes to room temperature, inoculate each tube as described above with 1 drop of inoculum.

IX. Assay Conditions

Incubate the inoculated assay at 30°C. *Lactobacillus arabinosus* grows rapidly and tests may be read on a suitable turbidimeter after 24 hours incubation, or allowed to incubate for 65 to 72 hours and titrated. The longer the incubation the less chance there is of variation due to uneven temperatures in the incubator.

X. Reading the Assay

Assays that have been incubated for 40 hours or less are read on a turbidimeter or colorimeter. Growth is halted by cooling the tubes either in a refrigerator for 1 to 2 hours or by placing the tubes in an ice bath. Each tube is thoroughly shaken to ensure uniformity and turbidity or optical density determined with a suitable instrument set with a water blank or uninoculated material using a 540-mμ filter. Care must be taken to avoid fogging in the colorimeter cuvette.

Tests incubated for 65 to 72 hours are titrated with 0.1 N NaOH to determine the amount of acid produced. The contents of each tube are rinsed into a small (125 ml.) Erlenmeyer flask with distilled water, 1 or 2 drops of a 1% bromthymol blue solution in 50% alcohol solution added and titrated with 0.1 N NaOH just to a definite blue color. The same end point and same amount of bromthymol blue should be used throughout a single assay. Many people have difficulty with the blue-green range of bromthymol blue. The exact end point is less important than that a single individual titrate the complete test. Several laboratories have devised pH measurement systems and automatic rinsing devises that are very useful when large numbers of samples are measured.

TABLE III

EXAMPLE OF BIOTIN ASSAY RESULTS

Tube no.	Biotin standard (mμg./ml.)	Tur-bidity[a]	mμg./tube from std. curve	Calcu-lated (μg./gm.) original sample	Titra-tion[b]	mμg./tube from std. curve	Calcu-lated (μg./gm.) original sample
1	0	77			1.65		
2	0	83			1.70		
3	0.2	119			5.00		
4	0.4	141			6.30		
5	0.6	158			8.00		
6	0.8	175			8.90		
7	1.2	212			10.40		
8	2.0	269			12.35		
Yeast extract (Section IV. A. a)							
H₂O extract							
(mg./tube)							
9	2	116.5	0.20	0.100	4.70	0.18	0.090
10	4	131	0.32	0.080	6.05	0.32	0.080
11	6	150	0.50	0.083	7.60	0.53	0.088
12	10	172	0.72	0.072	8.85	0.77	0.077
Avg.				0.084			0.084
Yeast extract (Section IV. A. b)							
acid hydrolyzed							
(mg./tube)							
13	0.1	116.5	0.20	2.00	5.15	0.22	2.20
14	0.2	152	0.52	2.60	7.20	0.47	2.35
15	0.3	173	0.73	2.43	8.20	0.64	2.13
16	0.5	220	1.31	2.62	10.00	1.05	2.10
Avg.				2.41			2.20

[a] Klett readings after incubation of 24 hours.

[b] Milliliters of 0.1 N acid produced in an incubation of 72 hours.

XI. Computation of Answers

Typical standard curve values as obtained by titration or turbidity measurement on the Klett Summerson colorimeter are shown in Table III for the samples shown in Section VIII. From the biotin content found in the concentration of sample tested, the value per unit of original sample is calculated for each level of sample and averaged to give the final result. Grossly aberrant figures are excluded from the average and for greater accuracy several replications of each sample should be run. As a general rule, sample variation falls within ±10% of the average.

XII. Other Assay Methods for Biotin

A. *Lactobacillus casei* (ATCC 7649)

Lactobacillus casei frequently has been used in the analysis for biotin. It differs from *L. arabinosus* in its ability to use many of the conjugated forms of biotin[4,8] and in its inability to grow on biotin-*d*-sulfoxide, which is fully active for *L. arabinosus*.[9] In addition, desthiobiotin, which is inactive for *L. arabinosus*, is a competitive inhibitor of biotin for *L. casei*. The assay procedure is essentially the same as the *L. arabinosus* procedure with the following modifications:

(a) Inoculum and assay tubes are incubated at 37°C.

(b) The medium is supplemented with folic acid (1 μg./ml.) which is essential for *L. casei* and asparagine (0.5 mg./ml.).

(c) Growth of *L. casei* is slow in the absence of "strepoginin" concentrates and usually tests are routinely titrated after 72 hours incubation at 37°C.

B. *Neurospora crassa*

Neurospora crassa is even more versatile than *Lactobacillus casei* in its ability to utilize biotin derivatives for growth as Wright *et al.*[5] point out.

a. Test organism. All strains of *Neurospora crassa* require biotin for growth and may be used for its assay. The wild strain,[11] the cholineless strain and the *p*-aminobenzoic acid requiring strain have been used.

b. Standard solution. A stock solution prepared as for the *Lactobacillus arabinosus* assay is diluted to a concentration of 10 mμg./ml. with water. It is dispensed into 125-ml. Erlenmeyer flasks in quantities of 0.0, 0.05, 0.10, 0.15, 0.2, 0.3, and 0.5 ml.

[11] E. L. Tatum, M. G. Ritchey, E. V. Cowdry, and L. F. Wicks, *J. Biol. Chem.* **163**, 675 (1946).

c. Samples may be assayed following dilution with water to a concentration of approximately 5 to 10 mμg./ml. or after hydrolysis with acid as described for *L. arabinosus*.

d. Preservation of organism. *Neurospora crassa* is carried on agar slants of the composition shown in the tabulation, transferred monthly, grown at room temperature or 28° for 3 days, then refrigerated.

Sucrose	2.0 gm.
Tryptone	0.1 gm.
Yeast extract	0.1 gm.
KH$_2$PO$_4$	0.1 gm.
Agar	2.0 gm.
Distilled H$_2$O	100 ml.

e. Preparation of inoculum. Suspend a loopful of mycelia in approximately 15 ml. sterile physiological saline and aseptically inoculate each sterile assay flask with 1 drop.

f. Medium. The composition of the assay medium is given in Table IV.

TABLE IV

Neurospora BASAL MEDIUM

Sucrose	20 gm.
Ammonium tartrate	5 gm.
Ammonium nitrate	1 gm.
KH$_2$PO$_4$	1 gm.
MgSO$_4$·7H$_2$O	0.5 gm.
NaCl	0.1 gm.
CaCl$_2$	0.086 gm.
Trace elements solution	1 ml.
Water, distilled	1000 ml.
Trace elements solution	
Na$_2$B$_4$O$_7$·10H$_2$O	38 mg.
Na$_2$MoO$_4$·2H$_2$O	5 mg.
FeSO$_4$·7H$_2$O	100 mg.
CuSO$_4$·5H$_2$O	39 mg.
MnSO$_4$·4H$_2$O	8 mg.
ZnSO$_4$·7H$_2$O	880 mg.
Water	100 ml.

Add a few drops of HCl if cloudy

In the event that a nutritional mutant of *N. crassa* is used (i.e. cholineless, inositol-less, etc.) the medium must be supplemented with the required nutrient.

g. Design of assay. Sample and standard are pipetted into 125-ml flasks at the concentrations indicated above to a maximum of 0.5 ml. Add 25 ml. medium to each flask, sterilize by autoclaving for 5 minutes at 120°C., cool, and inoculate.

h. Assay conditions. Incubate the flasks at room temperature or 28°C. for 4 or 5 days.

i. Reading the assay. Using a stiff wire loop, scrape the mycelium from the surface of the liquid and sides of the flask, press out excess moisture on filter paper, and place mat in test tube or small beaker. Dry in an oven at 90° for at least 2 hours or overnight. Weigh dried mats directly on analytical balance.

j. Computation of answers. Plot a standard curve of mycelial weights against the concentration of biotin. From weights of sample flask mats, calculate results from standard curve.

7.5 The Folic Acid Group

E. EIGEN
Department of Microbiology
Colgate-Palmolive Company
New Brunswick, New Jersey

and

G. D. SHOCKMAN
Department of Microbiology
Temple University School of Medicine
Philadelphia, Pennsylvania

I. Introduction

A. Historical

Interest in the existence of hemopoietic factors in liver and yeast arose from the work of Wills *et al.* describing the treatment of tropical macrocytic anemia.[1-3] The blood picture in this disease is similar to that in Addisonian pernicious anemia but the clinical symptoms are not. Furthermore, the condition in women is complicated by pregnancy. Patients responded to oral ingestion of Marmite, an autolyzed yeast extract, and oral or parenteral administration of crude liver extracts. Refined liver extracts were lacking in the active factor(s). This anemia could be induced in monkeys by feeding a deficient diet[4] and could be treated in the same manner as for humans. The curative factor was distinct from the antipernicious anemia factor since several liver preparations active in preventing the latter disease had no effect on the monkeys with nutritional anemia. Wills and co-workers attempted to fractionate liver concentrates and found that the antipernicious anemia factor was precipitated in saturated ammonium sulfate leaving the antimonkey anemia factor behind.[5] Yeast extracts treated in the same manner yielded similar results.

Extensive research was conducted by many investigators on various species in the years following the initial work. The factor became known as factor U,[6] yeast Norite eluate factor,[7] vitamin B_c,[8] *Lactobacillus casei* factor, [9,9a] and folic acid.[10] Excellent reviews on the historical aspects of folic acid were prepared by Berry and Spies[11] and Jukes and Stokstad.[12]

Snell and Peterson[7] reviewed the nutritive requirements of lactic acid bacteria and showed that many required a growth factor found in liver and yeast. Since it could be adsorbed on Norite A, an activated charcoal, from which it could be eluted, they termed this material the Norite eluate factor. As noted above, this factor was found to be folic acid. This material

[1] L. Wills, *Brit. Med. J.* **I**, 1059 (1931).

[2] L. Wills and A. Stewart, *Brit. J. Exptl. Pathol.* **16**, 444 (1935).

[3] L. Wills, P. W. Clutterbuck, and P. D. F. Evans, *Biochem. J.* **31**, 2136 (1937).

[4] L. Wills and H. S. Billimoria, *Indian J. Med. Research* **20**, 391 (1932).

[5] L. Wills and P. D. F. Evans, *Lancet* **232**, 311 (1937).

[6] E. L. R. Stokstad and P. D. V. Manning, *J. Biol. Chem.* **125**, 687 (1938).

[7] E. E. Snell and W. H. Peterson, *J. Bacteriol.* **39**, 273 (1940).

[8] A. G. Hogan and E. M. Parrott, *J. Biol. Chem.* **132**, 507 (1940).

[9] E. L. R. Stokstad, *J. Biol. Chem.* **149**, 573 (1943).

[9a] E. L. R. Stokstad, B. L. Hutchings, and Y. SubbaRow, *J. Am. Chem. Soc.* **70**, 3 (1948).

[10] H. K. Mitchell, E. E. Snell, and R. J. Williams, *J. Am. Chem. Soc.* **63**, 2284 (1941).

[11] L. J. Berry and T. D. Spies, *Blood* **1**, 271 (1946).

[12] T. H. Jukes and E. L. R. Stokstad, *Physiol. Revs.* **28**, 51 (1948).

was also required for the growth of *Streptococcus faecalis* (ATCC **8043**) (known then as *Streptococcus lactis* R) and other lactic and propionic acid bacteria. The foundation for the microbiological assay of folic acid and related compounds was laid in these early works. Continued efforts in obtaining the active principle were successful finally when in 1943 two groups reported the isolation of pure pteroylglutamic acid (PGA) from liver.[8,13] The two groups used different methods for this isolation. One group[13,13a] started with autolyzed liver which was processed by adsorption on and elution from Amberlite IR-4, adsorption on and elution from activated charcoal, extraction of the free acid with butanol at pH 3 to 4, formation of a barium salt, extraction of the barium salt with hot water, formation of a zinc salt, and crystallization of the free acid from water.[12] The other group[9] started with an 80% alcohol-insoluble fraction of an aqueous extract of liver. The pure substance was obtained as follows[9a]: Adsorption on and elution from Norite, adsorption on and elution from Superfiltrol, formation of a barium salt with barium chloride and methanol, esterification of the barium salt with 0.2 N HCl methanol, extraction of the methyl ester from aqueous solution with butanol, chromatographic adsorption of the ester on Superfiltrol, and fractional precipitation of the ester from water and from methanol. The free acid was obtained by hydrolysis of the ester and was crystallized from hot aqueous solutions. The synthesis was accomplished by Angier *et al.*[14,15] in 1945. The structural formula of pteroylglutamic acid is shown in Table I, no. 5. The generic name is N-[4-{[(2-amino-4-hydroxy-6-pteridyl)methyl]amino}benzoyl] glutamic acid. PGA, therefore, is a pteridine compound containing a *p*-aminobenzoic and glutamic acid residue. The properties of this compound are described in Section I.C.

B. Terminology

There has been considerable confusion regarding the term "folic acid." It may refer to the compound mentioned above (PGA) or to an entire group of compounds that are related both structurally and biologically. For the purposes of this chapter we shall attempt to

[13] J. J. Pfiffner, S. B. Binkley, E. S. Bloom, R. A. Brown, O. D. Bird, A. D. Emmett, A. G. Hogan, and B. L. O'Dell, *Science* **97**, 404 (1943).

[13a] J. J. Pfiffner, E. S. Bloom, and B. L. O'Dell, *J. Am. Chem. Soc.* **69**, 1476 (1947).

[14] R. B. Angier, J. W. Boothe, B. L. Hutchings, J. H. Mowat, J. Semb, E. L. R. Stokstad, Y. SubbaRow, C. W. Waller, D. B. Cosulich, M. J. Fahrenbach, M. E. Hultquist, E. Kuh, E. H. Northey, D. R. Seeger, J. P. Sickels, and J. M. Smith, Jr., *Science* **102**, 227 (1945).

[15] R. B. Angier, J. W. Boothe, B. L. Hutchings, J. H. Mowat, J. Semb, E. L. R. Stokstad, Y. SubbaRow, C. W. Waller, D. B. Cosulich, M. J. Fahrenbach, M. E. Hultquist, E. Kuh, E. H. Northey, D. R. Seeger, J. P. Sickels, and J. M. Smith, Jr., *Science* **103**, 667 (1946).

TABLE I[a]

	Structural formula	Common names and synonyms	Abbreviations
1		p-Aminobenzoic acid	PAB PABA
2		2-Amino-4-hydroxy-6-dihydro-propylpteridine Crithidia factor Biopterin	
3		Pteroic acid	PA
4		N^{10}-Formylpteroic acid (pteroate) 10-Formylpteroic acid (pteroate) Rhizopterin Streptococcus lactis R factor SLR factor	N^{10}-CHO-PA or (–10) N^{10}- Formyl–PA or (10–) foPA

TABLE I (continued)

Structural formula	Common names and synonyms	Abbreviations
5	Pteroylglutamic acid Folic acid, folate, folacin (foliate) Vitamin B_c Liver *L. casei* factor	*PGA* FA
6	Tetrahydropteroylglutamic acid Tetrahydrofolic acid (-folacin) (-folate)	$PGA\text{-}H_4$, THFA Folate-H_4 FA-H_4, FH_4 Tetrahydro-PGA, THPGA
7	N^{10}-Formylpteroylglutamic acid (or 10-) N^{10}-Formylfolic acid N^{10}-Formylfolate (-folacin)	$N^{10}\text{-}CHO\text{-}PGA$, (folate, etc.) (or 10-) (-folic acid) (-folacin) (-folate) N^{10}-Formyl-PGA (or 10-) foPGA (fofolate, etc.)
8	N^5-Formyltetrahydropteroyl- glutamic acid 5-Formyltetrahydrofolic acid Folinic acid Folinic acid-SF Citrovorum factor Leucovorin	$N^5\text{-}CHO\text{-}PGA\text{-}H_4$ (or 5-) N^5-Formyl-PGA-H_4 N^5-Formylfolate-H_4 (etc.) (or 5-) 5-foFA-H_4, 5-CHO-FH_4 CF, LCF

Structural formulas (pteridine–p-aminobenzoyl–glutamic acid):

5:
```
CH2—COOH
|
CH2
|
CH—NH—OC—⟨C6H4⟩—NH—CH2(10)—⟨pteridine: 2-amino-4-OH; positions 1,2,3,4,5,6,7,8,9⟩—NH2
|
COOH
```

6: Tetrahydro form (H at 5,6,7,8; CH2(10)).

7: N^{10}-CHO (CHO on N-10).

8: N^5-CHO-tetrahydro form (CHO on N-5; asterisk at C-6).

TABLE I (continued)

Structural formula	Common names and synonyms	Abbreviations
9	N^{10}-Formyltetrahydropteroyl-glutamic acid 10-Formyltetrahydrofolic acid, (etc.)	N^{10}-CHO-PGA-H_4 N^{10}-Formyl-PGA-H$_4$ (etc.) (or 10-) 10-foFA-H$_4$, 10-CHO-FH$_4$
10	Pteroyltriglutamic acid Pteroyldiglutamylglutamic acid Pteroyl-γ-glutamyl-γ-glutamyl-glutamic acid Folyldiglutamate Teropterin Fermentation *L. casei* factor Vitamin M	*PTGA* PGAG$_2$, PGAGG, PG$_3$A Ter., Folate-G$_2$
11	Pteroylheptaglutamic acid Pteroylhexaglutamylglutamic acid Heptopterin Vitamin B$_c$ conjugate Yeast *L. casei* factor Norite eluate factor	*PHGA* PG$_7$A, PGAG$_6$

[a] Older and less widely used names are given in parentheses. We have attempted to restrict the abbreviations used in this chapter to those that are italicized.

refer to PGA or folacin for the substance described above (Table I, 5) and to use specific names where possible for the related compounds. The terms folic acid group, folic acid-like substances, or folic acid substances will be used to refer to either part of or the entire group of compounds.

The seemingly ever growing multiplicity of the folic acid group, it nomenclature and the abbreviations that have been used for these substances is illustrated in Table I (see also Table IV). For purposes of orientation and reference the formulae of some of the more prominent members of the group given in order of increasing chemical complexity are listed in Table I. A number of the names and abbreviations for these substances are also given. The reader may be assured that both the list of names and abbreviations are far from complete. Several relatively unsuccessful attempts at standardization of names and abbreviations have been made. In fact, some journals, such as the *Journal of Biological Chemistry,* are extremely reluctant to accept any abbreviations and will accept only the series of names that are etymologically derived from folic acid or folate and, therefore, include either of these as part of the name of the compound. An example of this would be the name 5-formyltetrahydrofolate for the substance that is often known as folinic acid, leucovorin, or citrovorum factor (no. 8 in Table I). We have attempted to restrict ourselves to a limited number of names and abbreviations. Exceptions will include names employed in various references where the exact substance referred to is not unequivocally clear and to certain conveniences in expressing information in tabular form.

Examples may be found in the literature where more than one chemical substance is included in each of the terms folic acid and citrovorum factor (or folinic acid). It can be extremely misleading to use a definite single name in reference to what is known to be a group of substances that is active in the promotion of growth of a test organism. One reason for this is the subsequent tendency to then equate activity on a weight for weight basis when, in reality, the molecular weights may, in some cases, be widely different. In addition to this is, of course, the knowledge that growth promoting activity may not even be equivalent on a molar basis (see Table III).

C. Folacin

1. General Properties of Folacin (PGA) [16,17] (see Table I, 5)

Empirical formula (free acid): $C_{19}H_{19}N_7O_6$; molecular weight: 441.42;

[16] E. L. R. Stokstad, *in* "The Vitamins" (W. H. Sebrell, Jr. and R. S. Harris, eds.), Vol. III. Academic Press, New York, 1954.
[17] "The Pharmacopeia of the United States of America," 16th revision, 1960.

color: yellow; crystal shape: spear-shaped leaflets; solubility in water free acid: 10 mg./liter at 0°C., 500 mg./liter at 100°C.; disodium salt: 15 gm./liter at 0°C.; optical rotation: $[\alpha]_D^{20} = +16°$ in 0.1 N NaOH at a concentration of 7.6 gm./liter; melting point: darkens and chars at about 250°C.; absorption maxima (pH 13): 256 mμ, $\epsilon = 30 \times 10^6$ cm.^{2}/mole; 282 mμ, $\epsilon = 26 \times 10^6$ cm.^{2}/mole; 356 mμ, $\epsilon = 9.8 \times 10^6$ cm.^{2}/mole; pK_a values: 5.0, 8.2.

2. Solubility

Folacin is insoluble in alcohol, acetone, benzene, chloroform, and ether. It dissolves readily in dilute solutions of alkali hydroxides and carbonates, and is soluble in hot, diluted HCl and in hot, diluted H_2SO_4. It is soluble in concentrated HCl and in concentrated H_2SO_4, yielding very pale yellow solutions.

3. Identification

The ratio A_{256}/A_{365} of a 1 in 6000 solution in 0.1 N NaOH = 2.80–3.00.

4. Stability

The stability of folic acid and folic acid-containing materials was studied extensively by Daniel and Kline.[18] Their concern was focused on the fate of this vitamin in natural products treated in various ways to release the free acid. They tried, in addition, to destroy folic acid in natural materials in order to prepare folic acid-free growth promoting supplements for use in basal media. The results obtained by these authors are presented in Table II. The gain in folic acid content of the antipernicious anemia preparation after autoclaving at alkaline pH was attributed by the authors as due, possibly, to the liberation of bound folic acid or other growth-stimulating compounds active for *Lactobacillus casei*. It should be noted that no folic acid was lost in tomato juice exposed to ultraviolet light. It has been reported, however, that irradiation of folic acid with such light results in its rapid inactivation.[19,20] Daniel and Kline do report that unbuffered solutions of the crystalline vitamin are less stable than solutions of natural sources of folic acid. Other reports indicate that folacin is much more stable in alkaline than in acid solutions.[21] It is advisable, therefore, to make stock solutions in 0.01 N NaOH.

[18] E. P. Daniel and O. L. Kline, *J. Biol. Chem.* **170,** 739 (1947).

[19] O. H. Lowry, O. A. Bessey, and E. J. Crawford, *J. Biol. Chem.* **180,** 389 (1949).

[20] E. L. R. Stokstad, D. Fordham, and A. de Grunigen, *J. Biol. Chem.* **167,** 877 (1947).

[21] M. I. B. Dick, I. T. Harrison, and K. T. H. Farrer, *Australian J. Exptl. Biol. Med. Sci.* **26,** 231 (1948).

TABLE II[a]

STABILITY OF FOLIC ACID UNDER DIFFERENT CONDITIONS

Substance	Concentration	Treatment[b]	Loss of folic acid (%)
	(μg./ml.)		
Crystalline folic acid	100	Standard, 0.01 N NaOH in 20% EtOH, stored in refrigerator at 4°, 3 mo.	0
	0.0002	Diluted standard in H_2O, filtered[c]	0
	10	Autoclaved, 121°, 30 min.	
		pH 1	100
		pH 3	40–45
		pH 4–12	10–30
	1	Autoclaved, 121°, 30 min., basal medium, pH 6.8	<0
		Sulfite, room temperature, 24 hr., pH 5	75–80
	(%)		
Liver extract preparation, high in folic acid	0.5	Autoclaved, 121°, 15 min., 0.05 N NaOH	0
		Autoclaved 121°, 30 min.	
		pH 1	70–80
		pH 3	35–40
		pH 4–12	10–20
	1	Autoclaved, 121°, 90 min., pH 1	90
		Autoclaved, 121°, 30 min., 0.2 N and 1.0 N HCl	d
		Heated in oven, 60°, 1 hr., pH 1	e
		Heated in oven, 60°, 1 hr., pH 3	f
		Heated on hot plate, 100°, 15 min., pH 3	g
		Sulfite, room temperature, 24 hr., pH 5	50–60
		Sulfite, autoclaved at 121°, 15 min., pH 5	20–25
		Irradiated, 500 watt Mazda lamp, 6 hr., pH 1, 100 ml. over 72 sq. in., 1 ft. distance	60
		Same; pH 5.5	<0
	2	Autoclaved, 121°, 15 min., 0.05 N NaOH	0
Liver extract, antipernicious anemia preparation	0.5	Autoclaved, 121°, 30 min.	
		pH 1	70
		pH 3	40
		pH 4	20
		pH 5–7	10
		pH 9	40 (Gain)
		pH 10–12	15 (Gain)
	1	Autoclaved, 121°, 90 min., pH 1	80–100
		Autoclaved, 121°, 90 min., pH 5.5	25–35
		Heated, 100°, 1 hr. in O_2, pH 10	50–60
		Sulfite, room temperature, 24 hr., pH 5	30–40
Tomato juice serum; filtered canned tomatoes	Undiluted	Irradiated, Hanovia lamp, 3 hr., pH 4.3, 100 ml. over 72 sq. in., 1 ft. distance	0
		Same; 7 hr.	0

[a] See ref. 18.

[b] Unless otherwise indicated, aqueous solutions adjusted to the desired pH by means of NaOH or HCl were used.

[c] Schleicher and Schüll no. 589 filter paper.

[d] Under these conditions the loss of folic acid was 5 times greater at 1 N HCl than at 0.2 N HCl.

[e] Less folic acid was lost in this case than in autoclaving at 121° for 30 minutes at pH 3.

[f] The loss of folic acid was less than under conditions of heating in an oven at 60° for 1 hour at pH 1.

[g] This treatment produced a loss equivalent to that resulting from heating in an oven at pH 3 for 1 hour at 60°.

5. Purification

When absolutely pure folic acid is not available it may become necessary to purify the commercial product. This type of material may contain impurities arising from photochemical decomposition. An excellent method of purification was presented by Sakami and Knowles.[22] The procedure combines chromatography on a cellulose column with filtration through charcoal. A suspension of 200 gm. of Whatman standard-grade cellulose powder in 0.1 M phosphate buffer (pH 7) saturated with isoamyl alcohol is poured into a tube, 7.5 by 55 cm., plugged with cotton. Nonabsorbent cotton contains traces of folic acid activity and thus a pure grade of cotton or glass wool should be used.[23] Pack the cellulose to a height of 40 cm. with the aid of suction, cover with a circle of heavy filter paper (Eaton-Dikeman No. 627-030) and compress with a plunger to a height of 37 cm. After washing the column with 500 ml. of buffer, add a solution of commercial folic acid (550 mg. suspended in 30 ml. of water and sufficient 1 N NaOH added to effect solution and yield pH 7). Carry out this and subsequent operations in the dark room in subdued light. Sodium folate is moved down the column with three 2-ml. portions of the isoamyl alcohol-saturated buffer. Continue the elution with the buffer at a rate of approximately 75 ml./hour. Collect the yellow folate band between 550 and 730 ml. of effluent. Collection of the folic acid fraction should not be started until the eluate is distinctly yellow to avoid contamination with p-aminobenzoylglutamic acid which just precedes the folate on the column. This procedure reduces the contamination from 1 to 2%, to 0.02% and removes most of the fluorescent material. The remainder of this latter contamination is removed as follows. Acidify the eluate from the cellulose column to pH 2 with HCl and centrifuge at 0°C. Suspend the residue in water and dissolve in NaOH as described earlier. The solution should be completely clear to insure fast movement through the charcoal column. Prepare the column by fusing a 25 by 200 mm. tube to the top of an 8 by 100 mm. chromatography tube. Plug the column with cotton, add a 1-cm. layer of cellulose powder to trap charcoal particles, pour in an aqueous slurry of a mixture of 0.5 gm. of a decolorizing charcoal (e.g., Darco G-60) and 1 gm. of cellulose powder, and put a cotton plug on top to hold the mixture in place. Wash the column with 25 ml. of 6 N HCl, then with 100 ml. of water. Pass the folic acid solution through the charcoal with the aid of suction and wash with water until the effluent is colorless. Acidify the effluent to pH 2 with 3 N HCl to precipitate folic acid and centrifuge at 0°C. Wash the residue 4 times with 50-ml. portions of 1% acetic acid with centrifugation at 0°C. be-

[22] W. Sakami and R. Knowles, *Science* **129**, 274 (1959).
[23] M. B. Sherwood and E. D. Singer, *J. Biol. Chem.* **155**, 361 (1944).

tween each washing. Suspend the residue in water and lyophilize. A pale yellow powder is obtained in 70 to 75% yield. This product is completely pure as determined by paper chromatography. The biological activity of folic acid will be found in Table III.

D. Folic Acid-Like Compounds

The study of folic acid has led to the discovery and synthesis of compounds related to the vitamin with activity of different magnitudes for various microorganisms and species. Some of these materials are found naturally while others are synthetic products. Some of the compounds of interest are the following.

1. Pteroyltriglutamic Acid [24] (Table I, 10)

This compound is a conjugated form of folic acid and has been known as Fermentation L. casei factor,[25,26] teropterin,[27] pteroyl-γ-glutamyl-γ-glutamyl glutamic acid,[28] and vitamin M.[29] As indicated in one of its synonyms, this product was found in a fermentation broth obtained from aerobic fermentation of a diphtheroid-type organism.

2. Pteroylheptaglutamic Acid [30] (Table I, 11)

This compound was isolated from yeast and was known as yeast L. casei factor[30] and vitamin B_c conjugate.[31] It is only very slightly active for L. casei, Streptococcus faecalis, and man, but does act as a growth factor for Tetrahymena geleii, the chick and rat (see Table III).

[24] C. W. Waller, B. L. Hutchings, J. H. Mowat, E. L. R. Stokstad, J. W. Boothe, R. B. Angier, J. Semb, Y. SubbaRow, D. B. Cosulich, M. J. Fahrenbach, M. E. Hultquist, E. Kuh, E. H. Northey, D. R. Seeger, J. P. Sickels, and J. M. Smith, Jr., *J. Am. Chem. Soc.* **70**, 19 (1948).

[25] B. L. Hutchings, E. L. R. Stokstad, N. Bohonos, and N. H. Slobodkin, *Science* **99**, 371 (1944).

[26] B. L. Hutchings, E. L. R. Stokstad, N. Bohonos, N. Sloane, and Y. SubbaRow, *Ann. N. Y. Acad. Sci.* **48**, 265 (1946).

[27] S. Farber, E. Cutler, J. W. Hankins, J. H. Harrison, E. C. Peirce, 2nd, and G. G. Lenz, *Science* **106**, 619 (1947).

[28] J. H. Boothe, J. H. Mowat, B. L. Hutchings, R. B. Angier, C. W. Waller, E. L. R. Stokstad, J. Semb, A. L. Gazzola, and Y. SubbaRow, *J. Am. Chem. Soc.* **70**, 1099 (1948).

[29] P. L. Day, W. C. Langston, and W. J. Darby, *Proc. Soc. Exptl. Biol. Med.* **38**, 860 (1938).

[30] J. J. Pfiffner, D. G. Calkins, E. S. Bloom, and B. L. O'Dell, *J. Am. Chem. Soc.* **68**, 1392 (1946).

[31] B. S. Binkley, O. D. Bird, E. S. Bloom, R. A. Brown, D. G. Calkins, C. J. Campbell, A. D. Emmett, and J. J. Pfiffner, *Science* **100**, 36 (1944).

3. 10-Formylpteroic Acid [32] (Table I, 4)

This acid was isolated from the fumaric acid fermentation of *Rhizopus nigricans* and was called Rhizopterin.[33] It is also known as the *Streptococcus lactis* R factor.[34] *Streptococcus lactis* R is known today as *Streptococcus faecalis* and responds to this compound whereas *Lactobacillus casei, Pediococcus cerevisiae,* chick, and rat do not. The nonformylated *pteroic* acid (Table I, 3) is a synthetic product[14] which is active for *Streptococcus faecalis,* but only slightly active for *Lactobacillus casei.* It is inactive for *Tetrahymena geleii, Pediococcus cerevisiae,* chick and rat.

The compounds discussed so far have been conjugated pteridines, i.e., they contain a pteridine nucleus with an aromatic substituent at carbon-6.

4. Crithidia Factor (Biopterin) (2-amino-4-hydroxy-6[1'2'-dihydroxypropyl-(L-erythro)]pterin) [35,36] (Table I, 2)

This compound is an unconjugated pteridine which is required in addition to folic acid for the growth of *Crithidia fasciculata,* a trypanosomid parasite of mosquitos.[37-39]

This factor has been found in a variety of sources including liver, culture supernatant of *Ochromonas malhamensis,*[37] human urine,[40] drosophila,[41] and royal jelly of the honeybee.[42] Crithidia factor may be involved in photosynthesis.[43] An assay procedure has been devised by Nathan *et al.*[44] Crithidia factor may be distinguished from folic acid by its stability to acid hydrolysis.

5. Leucovorin (Folinic Acid, Citrovorum Factor) (Table I, 8)

Another member of the folic acid family is leucovorin or folinic acid

[32] D. E. Wolf, R. C. Anderson, E. A. Kaczka, S. A. Harris, G. E. Arth, P. L. Southwick, R. Mozingo, and K. Folkers, *J. Am. Chem. Soc.* **69,** 2753 (1947).

[33] E. L. Rickes, L. Chaiet, and J. C. Keresztesy, *J. Am. Chem. Soc.* **69,** 2749 (1947).

[34] J. C. Keresztesy, E. L. Rickes, and J. L. Stokes, *Science* **97,** 465 (1943).

[35] E. L. Patterson, R. Milstrey, and E. L. R. Stokstad, *J. Am. Chem. Soc.* **78,** 5868 (1956).

[36] E. L. Patterson, R. Milstrey, and E. L. R. Stokstad, *J. Am. Chem. Soc.* **80,** 2018 (1958).

[37] H. A. Nathan and J. Cowperthwaite, *J. Protozool.* **2,** 37 (1955).

[38] H. A. Nathan, S. H. Hutner, and H. L. Levin, *Nature* **178,** 741 (1956).

[39] E. L. Patterson, H. P. Broquist, A. M. Albrecht, M. H. van Saltza, and E. L. R. Stokstad, *J. Am. Chem. Soc.* **77,** 3167 (1955).

[40] E. L. Patterson, M. H. van Saltza, and E. L. R. Stokstad, *J. Am. Chem. Soc.* **78,** 5871 (1956).

[41] H. S. Forrest and H. K. Mitchell, *J. Am. Chem. Soc.* **77,** 4865 (1955).

[42] A. Butenandt and H. Rembold, *Z. physiol. Chem.* **311,** 79 (1958).

[43] S. H. Hutner, A. Cury, and H. Baker, *Anal. Chem.* **30,** 849 (1958).

[44] H. A. Nathan, S. H. Hutner, and H. L. Levin, *J. Protozool.* **5,** 134 (1958).

(N^5-formyl-5,6,7,8-tetrahydrofolate). Sauberlich and Baumann[45], while surveying microorganisms for the microbiological assay of alanine, found an organism, believed to be *Leuconostoc citrovorum* (ATCC 8081), which would not grow on a synthetic medium suitable for the growth of *Leuconostoc mesenteroides* P-60 and other organisms. Addition of peptone, crude extracts of liver, or yeast extract to the medium permitted maximum growth. The factor in liver could not be replaced by large amounts of thiamine, pyridoxal, pyridoxamine, hydroxyproline, or the ash of yeast extract. Large amounts of folic acid stimulated the growth of the organism but only after some delay. The authors reasoned, therefore, that a similarity might exist in the structure of folic acid and the factor required by *Leuconostoc citrovorum*. Bioautographic studies by Winsten and Eigen[46] revealed the presence of at least 4 factors in liver extracts capable of supporting the growth of *L. citrovorum*. Sauberlich had shown that feeding folic acid to rats resulted in an increased excretion of the *L. citrovorum* factor (LCF).[47] With this in mind, Winsten and Eigen incubated folic acid with a rat stomach homogenate and demonstrated an increase in LCF activity. The rat stomach homogenate caused the disappearance of several of the factors revealed chromatographically in liver leaving only one active zone and increasing the activity of the preparation for *L. citrovorum*. These authors suggested the possibility that LCF existed in the form of conjugates similar to those of folic acid. Concentrates of LCF prepared from liver were found to reverse competitively the inhibitory effects of methyl folic acid for *Lactobacillus casei* under conditions where only the folic acid group was effective.[48] Bond *et al.*[48] named the factor folinic acid because of its apparent relation to folic acid. LCF was found to have a distribution coefficient similar to folic acid in the system *n*-butanol-water at pH 2.0. Under these conditions LCF activity was lost but folic acid activity increased indicating the possible release of folic acid from LCF.[49] Further evidence to the fact that LCF was a derivative of folic acid was supplied by Shive *et al.*[50] who synthesized a very active material from the latter compound. This was accomplished by preparing crude formylfolic acid and hydrogenating it in the presence of ascorbic acid and platinum oxide. The material was not very active but after autoclaving the reaction mixture for 1 hour at 120°C. half-maximum growth

[45] H. E. Sauberlich and C. A. Baumann, *J. Biol. Chem.* **176**, 165 (1948).
[46] W. A. Winsten and E. Eigen, *J. Biol. Chem.* **184**, 155 (1950).
[47] H. E. Sauberlich, *Federation Proc.* **8**, 247 (1949).
[48] T. J. Bond, T. J. Bardos, M. Sibley, and W. Shive, *J. Am. Chem. Soc.* **71**, 3852 (1949).
[49] H. P. Broquist, E. L. R. Stokstad, and T. H. Jukes, *J. Biol. Chem.* **185**, 399 (1950).
[50] W. Shive, T. J. Bardos, T. J. Bond, and L. L. Rogers, *J. Am. Chem. Soc.* **72**, 2817 (1950).

of *Leuconostoc citrovorum* was obtained with an amount of the reaction mixture equivalent to 10 to 40 $\mu\mu g$. of the original folic acid per milliliter of assay medium. Shive *et al.* did not isolate the product from the reaction mixture. This was achieved finally by Brockman *et al.*[51] Folic acid or its N^{10}-formyl derivative was reduced catalytically over platinum in formic acid at 0° to 30°C. Impurities were removed on Magnesol at pH 7, active material was adsorbed on Darco G-60 at pH 4. After elution, fractional crystallization of the barium salt, and chromatographic separation on Magnesol columns, a crystalline product was obtained. It was shown to be 5-formyl-5,6,7,8-tetrahydropteroylglutamic acid. It is also known as 5-formyltetrahydrofolic acid[51] and leucovorin.[52] Like pteroylglutamic acid it exists in conjugated forms,[46,53] these conjugates having been found in marine alga,[54] bacteria,[55,56] blood,[57] yeast extract, and liver.[58] Some properties of leucovorin are:[59] formula (free acid): $C_{20}H_{23}N_7O_7$; molecular weight: 464; absorption maxima (pH 13): 242 mμ; 282 mμ ($\epsilon = 32.6 \times 10^6$ cm.²/mole); pK values: 3.1, 4.18, 10.4. The calcium salt, $C_{20}H_{21}CaN_2O_7 \cdot 5H_2O$ is a yellowish white to yellow powder, very soluble in water and practically insoluble in alcohol.[17] Leucovorin is destroyed in dilute acid solution, e.g., pH 2.0, even at room temperature.[49]

Leucovorin was isolated from horse liver by Keresztesy and Silverman[60] using the following procedure: Autolysis of ground liver, adsorption on and elution from charcoal, precipitation and removal of water-acid insoluble materials, extraction into butanol at pH 3, precipitation of impurities in the aqueous ammoniacal extract of the butanol extract with methanol, adsorption on Dowex 1 column and subsequent elution, adsorption of active fraction on charcoal and subsequent elution, adsorption on alumina column from aqueous alcohol solution followed by elution, and fractional crystallization of the barium salt. This product from liver had twice the activity for *L. citrovorum* as the synthetic product. This discrepancy was caused by the synthetic compound existing as two dia-

[51] J. A. Brockman, Jr., B. Roth, H. P. Broquist, M. E. Hultquist, J. M. Smith, Jr., M. J. Fahrenbach, D. B. Cosulich, R. P. Parker, E. L. R. Stokstad, and T. H. Jukes, *J. Am. Chem. Soc.* **72**, 4325 (1950).

[52] H. P. Broquist, E. L. R. Stokstad, and T. H. Jukes, *J. Lab. Clin. Med.* **38**, 95 (1951).

[53] O. P. Wieland, B. L. Hutchings, and J. H. Williams, *Arch. Biochem. Biophys.* **40**, 205 (1952).

[54] L. E. Ericson, *Arkiv Kemi* **6**, 503 (1953).

[55] M. T. Hakala and A. D. Welch, *Federation Proc.* **14**, 222 (1955).

[56] B. E. Wright, *J. Am. Chem. Soc.* **77**, 3930 (1955); *J. Biol. Chem.* **219**, 873 (1956).

[57] G. Toennies, P. Phillips, and E. Usdin, *Federation Proc.* **17**, 322 (1958).

[58] V. M. Doctor and J. R. Couch, *J. Biol. Chem.* **200**, 223 (1953).

[59] F. M. Huennekens and M. J. Osborne, *Advances in Enzymol.* **21**, 369 (1959).

[60] J. C. Keresztesy and M. Silverman, *J. Am. Chem. Soc.* **73**, 5510 (1951).

stereoisomers[61] due to the asymmetric carbon atom at position 6 (see Table I, 8). Apparently only the LL-isomer is biologically active.

Leucovorin can replace the folic acid requirement for *Streptococcus faecalis, Lactobacillus casei*, chick, and man, but folic acid cannot replace the need for leucovorin by *Pediococcus cerevisiae*.

Leucovorin is 8 times as active as folic acid in reversing the inhibition of the growth of *Streptococcus faecalis* by 4-aminopteroylglutamic acid (aminopterin).[62] It is characterized by the United States Pharmacopeia (U.S.P. XVI) as an antidote for folic acid antagonists (e.g., aminopterin, amethopterin).

A summary of the activity of the compounds discussed above for various species can be found in Table III.

TABLE III

ACTIVITY OF FOLIC ACID AND RELATED COMPOUNDS

Organism	PGA[a]	PTGA	PHGA	10-Formyl pteroic acid	Pteroic acid	Leucovorin (synthetic)	References
T. geleii[b]	0.3 mμg.	Equal to PGA on a molar basis		—	Inactive	—	63, 64
L. casei	0.09 mμg.	0.06 mμg.	0.01 μg.	>0.02 μg.	c	0.17 mμg.[d]	62, 65–67
S. faecalis	0.18 mμg.	4.2 mμg.	0.10 μg.	0.035 mμg.	c	0.37 mμg.	16, 62, 66, 67
P. cerevisiae	30 mg.	Inactive	—	Inactive	Inactive	0.15 mμg.	16, 53, 62
Chick	Active	Active	Active	Inactive	Inactive	Active	12, 33, 62
Rat	Active	Active	Active	Inactive	Inactive	—	12, 33
Man	Active	Active	Slightly active	—	—	Active	62, 68

[a] PGA = pteroylglutamic acid; PTGA = pteroyltriglutamic acid; PHGA = pteroylheptaglutamic acid.
[b] Requirements of microorganisms are expressed as the amount of factor *per milliliter* stated to cause ½ maximum growth. Since this quantity (½ maximum growth) is totally dependent on the particular method of measurement, results are comparable only within one laboratory. The figures given should therefore be used only as an approximate guide.
[c] Pteroic acid is 50–100% as active as folic acid for *S. faecalis* (depending on the time of incubation) and 0.01% as active for *L. casei* [12].
[d] Naturally occurring leucovorin is as active as folic acid for *S. faecalis* and *L. casei* [53].

[61] D. B. Cosulich, J. M. Smith, Jr., and H. P. Broquist, *J. Am. Chem. Soc.* **74**, 4215 (1952).

[62] H. P. Broquist, J. A. Brockman, Jr., M. J. Fahrenbach, E. L. R. Stokstad, and T. H. Jukes, *J. Nutrition* **47**, 93 (1952).

[63] G. W. Kidder and V. C. Dewey, *Proc. Natl. Acad. Sci. U. S.* **33**, 95 (1947).

[64] G. W. Kidder and V. C. Dewey, *Arch. Biochem.* **21**, 66 (1949).

[65] E. L. R. Stokstad, B. L. Hutchings, and Y. SubbaRow, *J. Am. Chem. Soc.* **70**, 3 (1948).

[66] B. L. Hutchings, E. L. R. Stokstad, N. Bohonos, N. H. Sloane, and Y. SubbaRow, *J. Am. Chem. Soc.* **70**, 1 (1948).

[67] E. L. R. Stokstad and B. L. Hutchings, *Biol. Symposia* **12**, 339 (1947).

[68] E. Scholz, E. Williams, A. Ellis, and T. Spies, *Rev. intern. vitaminol.* **20**, 157 (1948).

6. Other Members of the Folic Acid Group

The substances described in the previous sections have been relatively well characterized. They have either been isolated from natural materials, synthesized, or both. In addition, there is evidence for the existence of a number of other forms of this vitamin in natural materials. In this respect, it must be remembered that unlike many of the other B vitamins, the coenzymatically active form (or forms) of the folic acid group has yet to be elucidated.

Table IV, which is based on one kindly provided by Dr. G. Toennies, not only indicates the microbiological activity of the compounds that have been well characterized but also indicates some of the possibilities for the future. It would not be surprising to find some of the blanks have been filled by newly found derivatives of folic acid by the time this volume is issued.

The relationship of pteroic acid, pteroylglutamic acid, and the glutamic acid conjugates to derivatives of folinic acid (5-formyl compounds) and to other members of the "family" can be seen both structurally and in relation to their growth promoting activity for 3 bacteria. For instance, it can be seen that compounds that are active in promoting the growth of *Pediococcus cerevisiae* are all tetrahydro derivatives and in fact tetrahydro PGA can support the growth of this organism efficiently.

We would, therefore, like to insert a few words of caution that are essential to the interpretation of microbiological assays for folic acid in natural materials. Conjugases can remove glutamic acid residues but one will still be left with an assortment of substances [i.e., those listed in Table IV—such as all the possible derivatives of pteroic acid (PA), pteroylglutamic acid (PGA), pteroyldiglutamic acid (PDGA), and pteroyltriglutamic acid (PTGA)] some of which are active for only one test organism but many of which are active for several or all of the test organisms listed. However, the quantitative responses of these test organisms to these substances will differ (see Table III), making quantitation difficult, or at times impossible.

II. Preparation of Samples for Assay

A. Conjugases

Folic acid usually occurs as a conjugate in natural products. Enzymes (called conjugases) have been prepared which are capable of releasing free folic acid from these conjugates.[69] Two main types of conjugases

[69] J. J. Pfiffner, D. G. Calkins, B. L. O'Dell, E. S. Bloom, R. A. Brown, C. J. Campbell, and O. D. Bird, *Science* **102**, 228 (1945).

TABLE IV

ACTIVITY OF VARIOUS FOLACIN DERIVATIVES FOR THREE MICROORGANISMS[a]

Parent compound	None (parent compound)			N^5—CHO			N^5—CHNH			N^{10}—CHO			N^5N^{10}=CH			N—CH$_2$OH		
	LC	SF	PC	LC	SF	PC	LC	SF	PC	LC	SF	PC	LC	SF	PC	LC	SF	PC
PA	−	+	−							−	+	−						−
H$_2$																		
H$_4$					+													
PGA	+	+	−							+	+	−				+	+	
H$_2$		+	+							+	+	−	+	+		+	+	
H$_4$	+	+	+	+	+	+	+	+	+	+	+	+	+	+	+	+	+	+
PDGA	+	+	−															
H$_2$																		
H$_4$				+	+	+	+	+										
PTGA	+	−	−															
H$_2$																		
H$_4$				+	−	+												
PHGA	−	−	−															
H$_2$																		
H$_4$																		

[a] PA = pteroic acid; PGA = pteroylglutamic acid; PDGA = pteroyldiglutamic acid; PTGA = pteroyltriglutamic acid; PHGA = pteroylheptaglutamic acid; H$_2$ = dihydro; H$_4$ = tetrahydro; —CHO = formyl; —CHNH = forminyl; =CH = methenyl (N^5N^{10}=CH = methylene (N^5N^{10}—CH); —CH$_2$OH = hydroxymethyl; PGAH$_4$ = anhydroleucovorin); —CH$_2$OH = hydroxymethyl; LC = L. casei; SF = S. faecalis; PC = P. cerevisiae; (+) = active; (−) = inactive.

have been described. The first, a carboxypeptidase[30] occurs widely in nature.[70] It has been found in rat liver,[71] hog kidney,[70] and chick pancreas.[72] Its pH optimum is 4.5. The second conjugase, a γ-glutamic acid carboxypeptidase, was isolated from chick pancreas.[73,74] Its pH optimum is 7–8. This enzyme preparation is used most widely for the assay of natural materials containing folic acid conjugates.[75] The material may be prepared as an aqueous suspension or as a dried material in the following manner.[76]

1. Wet Preparation

Grind chicken pancreas weighing approximately 4 gm. in a glass mortar or a high-speed blendor and suspend in 0.2 M phosphate buffer at pH 7.2 (2.723 gm. KH_2PO_4 and 0.560 gm. NaOH diluted to 100 ml. with water).

2. Dry Preparation

Grind the pancreas in 5 volumes of cold acetone. Place cheesecloth over a Buchner funnel containing filter paper. Squeeze the fine material through the cheesecloth and filter off the acetone with the aid of suction. Wash the fine material on the Buchner funnel several times with cold acetone and air-dry. Grind the final dry residue to a fine powder and store at or near 0°C. (The preparation may be obtained from Difco Laboratories, Detroit, Michigan.)

3. Purification of Conjugases

The preparations mentioned above contain measurable amounts of folic acid-active material. The usual procedure when using these substances to release bound folic acid is to conduct an assay on them with the microorganism being used for the test and subtract the appropriate blank from the treated samples. The assay on the conjugase preparations will only yield a value in terms of folic acid activity and will not be an indication of the type or types of compound responsible for this activity. In most cases it is only important to know the folic acid activity blank value of the enzyme preparation. Should it be necessary to prepare a conjugase relatively free from folic acid activity (e.g., for the assay of materials of relatively low folic acid group content), then the following procedures for

[70] O. D. Bird, S. B. Binkley, E. S. Bloom, A. D. Emmett, and J. J. Pfiffner, J. Biol. Chem. **157**, 413 (1945).

[71] V. Mims, J. R. Totter, and P. L. Day, J. Biol. Chem. **155**, 401 (1944).

[72] V. Mims and M. Laskowski, J. Biol. Chem. **160**, 493 (1945).

[73] A. Kazenko and M. Laskowski, J. Biol. Chem. **173**, 217 (1948).

[74] W. Dabrowska, A. Kazenko, and M. Laskowski, Science **110**, 95 (1949).

[75] H. R. Skeggs, Bacteriol. Revs. **21**, 257 (1957).

[76] "Official Methods of Analysis of the Association of Official Agricultural Chemists" (W. Horowitz, ed.), 8th ed. A.O.A.C., Washington D. C., 1955.

the preparation of either chick pancreas or hog kidney conjugase can be used.

Purified conjugase preparations were made by Laskowski *et al.*[77] from chicken pancreas by a procedure involving treatment with 0.1 M tricalcium phosphate suspension (tricalcium phosphate gel), precipitation with alcohol, and successive concentration and salting-out with sodium sulfate. Mims and Laskowski[72] effected a further purification by autolyzing ground chicken pancreas at pH 8, centrifuging, and treating the supernatant with ammonium sulfate, removing the salt by dialysis, precipitating the proteins with cold ethanol, followed by fractional precipitation with ammonium sulfate to yield a highly purified conjugase preparation. Calcium ion at a concentration of 0.01 M activated this preparation.[72]

Another method for the purification of the chick pancreas enzyme where the starting material is the chick pancreas acetone powder prepared as described above or as obtained from Difco Laboratories is as follows:[77,78] Suspend 10 gm. of acetone powder in 30 volumes of 0.1 M phosphate buffer of pH 7 and stir 1 hour at room temperature. Incubate at 37° to 38°C. for 17 hours overnight under toluene so that the enzyme can act on endogenous substrate and then centrifuge at 2500 r.p.m. for 20 minutes. Add to the supernatant after centrifugation an equal volume of 0.1 M tricalcium phosphate gel. Stir the suspension for 30 minutes in an ice bath and centrifuge for 20 minutes at 2500 r.p.m. (cold). Cool the supernatant to below 5°C, add an equal volume of absolute ethanol dropwise and store the mixture overnight below 5°C. Suspend the gray-white precipitate obtained by centrifugation in 100 ml. of 0.1 M, pH 7 phosphate buffer and stir 1 hour at 0°C. in an ice bath. Centrifuge for 20 minutes at 2500 r.p.m. to remove insoluble material. Mix the supernatant with Dowex 1 × 8 (chloride) and stir for 1 hour in the cold. Remove the Dowex 1 by centrifugation at 3000 r.p.m. for 30 minutes and pour the supernatant through gauze. The almost clear supernatant contains the conjugase which can then be stored in the frozen state.

A similar procedure can be used for the purification and removal of substrate from the hog kidney enzyme.[58,78] In this case the starting material is fresh defatted hog kidney.

Remove the fat layers from the hog kidney and homogenize it in 3 volumes of 0.32% (2 × 10^{-2} M) cysteine hydrochloride at pH 5.4. Incubate the suspension at 37° to 38°C. for 2 hours under toluene in a nearly full stoppered bottle, then mix with Hyflo Super Cel and filter twice through a coarse sintered glass filter. After centrifugation at 0°C. and

[77] M. Laskowski, V. Mims, and P. L. Day, *J. Biol. Chem.* **157**, 731 (1945).

[78] K. Iwai, S. Nakagawa, and O. Okinaka, *Mem. Research Inst. Food Sci. Kyoto Univ.* **19**, 17 (1959) (in English).

2500 r.p.m. for 20 minutes, adjust the almost clear filtrate to pH 4.5 with HCl and treat with 15 gm. Dowex 1 × 8 (chloride) for 1 hour in an ice bath, then centrifuge at 0°C. for 20 minutes at 2500 r.p.m. The active enzyme can be stored in the frozen state.

Conjugases prepared from sources other than chicken pancreas may present problems when used for liberation of folic acid. Bird et al.[79] found inhibitors of hog kidney conjugase activity in yeast extract. Subsequently, this inhibition was found to be due to nucleic acids. The inhibition could be reversed by cysteine and 2,3-dimercaptopropanol (BAL), suggesting the presence of an active sulfhydryl group.[80] Thymus and yeast nucleic acids inhibited conjugase preparations from hog intestine, rat liver, and human leukocytes but not the purified conjugase from chicken pancreas, suggesting that this latter enzyme is different from that in other tissues.

Iwai[81] investigated the folic acid conjugase in green leaves, e.g., leaves of the soybean plant, and found that it would release folic acid in the presence of cysteine, reduced glutathione, and thioglycolic acid but not ascorbic acid. The optimum pH for activity is 4.5 and conjugase activity is inhibited by p-chloromercuribenzoate which indicates the green leaf conjugase is a thiol enzyme similar to hog-kidney conjugase.[81] Cations activate this enzyme preparation (K^+, Na^+, NH_4^+, Mg^{++}, Ca^{++}, Sr^{++}, Cd^{++}, Ba^{++}, Pb^{++}, Mn^{++}, Co^{++}, Ni^{++}). Inhibition of activity is obtained with Ag^+, Cu^{++}, Zn^{++}, and Hg^{++}. Inhibition by zinc was the only one not reversed by the addition of cysteine. Ethylenediaminetetraacetate (EDTA·2Na) plus cysteine overcame the effect of zinc ion.[82]

It appears, therefore, that the use of conjugase preparations from various sources presents difficulties of which one must be aware in order to use them properly. Chicken pancreas conjugase preparations seem to present the least problems and, since it releases the most folic acid from natural materials, is the preparation of choice at the present time.

B. Specific Methods

1. Foods

The method for preparing samples of fruit, vegetables, grain products, meat, fish, etc., for assay will be found in Section III.A.11. This procedure does not measure all substances with folic acid activity, however,

---

[79] O. D. Bird, M. Robbins, J. M. Vandenbelt, and J. J. Pfiffner, *J. Biol. Chem.* **163,** 649 (1946).

[80] V. Mims, M. E. Swendseid, and O. D. Bird, *J. Biol. Chem.* **170,** 367 (1947).

[81] K. Iwai, *Mem. Research Inst. Food Sci. Kyoto Univ.* **13,** 1 (1957) (in English).

[82] K. Iwai and S. Nakagawa, *Mem. Research Inst. Food Sci. Kyoto Univ.* **13,** 10 (1957) (in English).

and those combined types of folic acid that cannot be measured by the test system must be released by a conjugase preparation. The following procedure using a crude conjugase preparation from chick pancreas[76,83] will liberate free folic acid.

Transfer a weighed sample containing 0.5–15 μg. of folic acid to a blendor containing 40 ml. of 0.2 M phosphate buffer, pH 7.2 (2.723 gm. KH_2PO_4 and 0.560 gm. NaOH diluted to 100 ml. with H_2O). Add approximately 50 ml. of H_2O and blend the mixture for 3 to 5 minutes. Add glycol distearate, capryl alcohol, or a similar compound to prevent foaming. Transfer the suspended material to an Erlenmeyer flask and heat the mixture in an autoclave for 15 minutes at 121°C. After cooling, add 20 mg. of dried chick pancreas enzyme (Section II.A.2) for each gram of sample taken for testing. Wet the enzyme preparation with a few drops of glycerine or methanol and suspend in 5 ml. of water before adding to the sample. For materials high in folic acid, e.g., liver and yeast, use 100 mg. of pancreas preparation for each gram of sample. Add a small amount of toluene or benzene and incubate the mixture at 37°C. for 24 hours, then heat in the autoclave for 5 minutes at 15 p.s.i. to drive off the solvent and inactivate the enzyme. Dilute the mixture to a convenient volume after cooling, filter or centrifuge, and further dilute to the desired concentration. The enzyme preparation contains folic acid activity and a suitable correction must be made. In samples heated to destroy the natural enzymes, 5 mg. of chick pancreas enzyme released the maximum amount of folic acid activity from 1 gm. of spinach.

The chick pancreas enzyme has been found to be superior to papain, takadiastase or hog-kidney enzyme.[75]

2. Whole Blood [84]

a. Reagents. Ascorbic acid–phosphate buffer (AP) solution (0.05 M phosphate buffer pH 6.1 with 0.05% ascorbic acid added). Dissolve 5.85 gm. KH_2PO_4 and 1.22 gm. K_2HPO_4 in distilled water and make up to 1 liter. The pH should be 6.10 ± 0.05. Because of its instability in solution the ascorbic acid is added to the phosphate buffer immediately before it is to be used.

b. Procedure. Dilute fresh whole blood 750-fold with the AP buffer, incubate at 37° to 38°C. for 90 minutes and precipitate the proteins by autoclaving at 15 p.s.i. for 1 to 2 minutes. Remove the precipitate by centrifugation and add the clear supernatant to the assay tubes in appropriate quantities.

The procedure can be carried out on a small sample of blood obtained

[83] E. W. Toepfer, E. G. Zook, M. L. Orr, and L. R. Richardson, *U. S. Dept. Agr. Handbook No.* **29** (1951).

[84] G. Toennies, E. Usdin, and P. M. Phillips, *J. Biol. Chem.* **221,** 855 (1956).

from a finger prick as follows:[85a] Take blood from a carefully cleaned and dried finger tip by means of a Thoma leucocyte diluting pipet. According to the measurements of Toennies et al.[85a] the sample would contain 0.0137 ml. with an average error of $\pm 2\%$. Dilute the blood sample immediately to the upper mark (0.130 ml.) with the AP buffer. If many samples are to be collected they can be kept on ice in this state until all samples are ready for assay. Add the diluted sample to 10 ml. AP buffer in an 18×150 mm. tube. At this stage the dilution is 1:750. Incubate the tubes containing the 1:750 dilution in AP buffer at 37° to 38°C. for 90 minutes. Then heat the tubes for 1 to 2 minutes at 15 p.s.i. in a preheated autoclave to coagulate proteins. Centrifuge for 3 to 5 minutes at 3000 r.p.m. to obtain clear supernatant which is then added to the assay tubes in appropriate quantities (see Section III.F). Treatment with a conjugase is apparently unnecessary because of the presence of a thiol enzyme in blood plasma which liberates folic acid activity from the erythrocytes.[85b] The plasma enzyme (plasma factor) has been partially purified [85b] from Fraction IV-4 of Cohn et al.[86] The resulting solution contains a mixture of folic acid-active compounds. Assay values with Lactobacillus casei are pronouncedly higher than with Streptococcus faecalis. A partial resolution of these compounds on TEAE (triethylaminoethyl) cellulose columns has been achieved by Usdin[87] (see also Section IV.C). Nine components active for S. faecalis, Lactobacillus casei, and/or Pediococcus cereviseae were isolated. Tentatively identified were: N^{10}-formyl pteroate, N^{10}-formyl pteroylglutamate, N^5-formyl-5,6,7,8-tetrahydropteroyl-γ-glutamylglutamate, and N^5-formyl-5,6,7,8-tetrahydropteroyl-di-γ-glutamylglutamate.

3. Serum

A similar procedure can be used to obtain folic acid activity from serum. Serum can be diluted 1:100 with the above buffer (Section II.B.2), incubated for 90 minutes at 37°C., and autoclaved to precipitate proteins.[88]

A further modification uses sodium phosphate instead of potassium phosphate and a different dilution of serum. It is claimed that the folic acid activity in human serum obtained by this extraction and the assay

[85a] G. Toennies, H. G. Frank, and D. L. Gallant, Cancer 9, 1053 (1956).

[85b] G. Toennies and P. M. Phillips, J. Biol. Chem. 234, 2369 (1959).

[86] E. J. Cohn, L. E. Strong, W. L. Hughes, Jr., D. J. Mulford, J. N. Ashworth, M. Melin, and H. L. Taylor, J. Am. Chem. Soc. 68, 459 (1946).

[87] E. Usdin, J. Biol. Chem. 234, 2373 (1959).

[88] J. M. Cooperman, A. L. Luhby, and C. M. Avery, Proc. Soc. Exptl. Biol. Med. 104, 536 (1960).

method used agrees with the clinical picture.[89] Serum is diluted 1:10 with a buffer prepared as follows:

Solution 1.   27.8 gm. NaH_2PO_4 dissolved in 1000 ml. distilled water.
Solution 2.   71.7 gm. $Na_2HPO_4 \cdot 12H_2O$ dissolved in 1000 ml. distilled water.

Take 212.5 ml. of solution 1 and add 37.5 ml. of solution 2, then dilute to 1000 ml. with distilled water. The pH is 6.1. Add 0.05% ascorbic acid to the buffer before use. Incubate the serum-buffer solution at 37°C. for 90 minutes and then autoclave for 10 minutes at 118°C. Remove the co-agulated proteins by centrifugation and use the clear supernatant for assay with *Lactobacillus casei.*

It should be remembered that, since blood plasma contains an enzyme that liberates "bound" folic acid activity from erythrocytes, folic activity values for serum will almost always be lower than those for whole blood.

III. Microbiological Assay Methods

A. *Streptococcus faecalis* Method (A.O.A.C.)

A method for determining folic acid with *S. faecalis* as assay organism has been established by the Association of Official Agricultural Chemists (A.O.A.C.).[90] The procedure was tested by numerous collaborative assays and modified several times. In its present form the assay is performed as follows on materials containing only free folic acid.

1. *Test Organism*

Streptococcus faecalis (ATCC 8043).

2. *Folic Acid Standard Solutions for Titrimetric Method*

a. *Stock solution I.* Weigh accurately, in a closed system, U.S.P. folic acid reference standard equivalent to 50 to 60 mg. of folic acid that has been dried to constant weight and stored in the dark over P_2O_5 in a desic-cator. Dissolve in approximately 30 ml. of 0.01 N NaOH, add approximately 300 ml. of water, adjust to pH 7–8 with HCl solution and dilute with additional water to make folic acid concentration exactly 100 μg./ml. Store under toluene in the dark at approximately 10°C.

b. *Stock solution II.* To 10 ml. of stock solution I, add approximately

---

[89] H. Baker, V. Herbert, O. Frank, I. Pasher, S. H. Hutner, L. R. Wasserman, and H. Sobotka, *Clin. Chem.* **5**, 275 (1959); V. Herbert, *J. Lab. Clin. Invest.* **40**, 81 (1961).
[90] "Official Methods of Analysis of the Association of Official Agricultural Chemists" (W. Horowitz ed.), 9th ed. A.O.A.C., Washington, D. C., 1960.

500 ml. of water, adjust to pH 7–8, and dilute with additional water to 1 liter. Store under toluene in the dark at approximately 10°C.; 1.0 ml. = 1.0 μg. of folic acid.

c. Stock solution III. To 100 ml. of stock solution II, add approximately 500 ml. of water, adjust to pH 7–8, and dilute with additional water to 1 liter. Store under toluene in the dark at approximately 10°C.; 1.0 ml. = 100 mμg. of folic acid.

d. Standard solution. Dilute 5.0 ml. of stock solution III with water to 500 ml.; 1.0 ml. = 1.0 mμg. of folic acid. Designate this as the titrimetric standard solution. Prepare fresh standard solution from stock solution III for each assay.

3. Folic Acid Standard Solution for the Turbidimetric Method

Dilute 5.0 ml. of folic acid standard stock solution III with water to 1 liter; 1.0 ml. = 0.5 mμg. of folic acid. Designate this as the turbidimetric standard solution. Prepare fresh standard solution for each assay.

4. Stock Solutions for Basal Medium

(Store all solutions in the dark at approximately 10°C. Proportionate quantities may be prepared.)

a. Acid-hydrolyzed casein solution. Mix 400 gm. of vitamin-free casein with 2 liters of constant-boiling HCl (approximately 20% HCl) and either reflux 8–12 hours or heat in an autoclave for 8–12 hours at 121° to 123°C. Remove HCl from the mixture by distillation under reduced pressure until a thick paste remains. Redissolve the paste in water, adjust the solution to pH 3.5 ± 0.1 with approximately 10% NaOH solution, and dilute with water to 4 liters. Add to the solution 80 gm. of activated charcoal, stir for 1 hour, and filter. Repeat the treatment with activated charcoal. Store under toluene. Filter the solution if a precipitate forms upon storage. (Some commercial sources of vitamin-free acid-hydrolyzed casein have been found satisfactory.) Commercial sources are not specified by the A.O.A.C. but we would suggest trying Hy Case SF (Sheffield Farms Company, Inc., New York, New York; see also Section III.F.3).

b. Adenine-guanine-uracil solution. Dissolve 0.7 gm. each of adenine sulfate, guanine hydrochloride, and uracil in 35 ml. of warm HCl (1 + 1), cool, and dilute with water to 700 ml. Store under toluene.

c. Asparagine solution. Dissolve 8 gm. of L-asparagine monohydrate in water and dilute to 800 ml. Store under toluene.

d. Manganese sulfate solution. Dissolve 2 gm. $MnSO_4 \cdot H_2O$ in water and dilute to 200 ml. Store under toluene.

e. Polysorbate 80 solution. Dissolve 25 gm. of polysorbate 80 (polyoxyethylene sorbitan monooleate, Tween 80) in alcohol to make 250 ml.

f. Salt solution B. Dissolve 20 gm. $MgSO_4 \cdot 7H_2O$, 1 gm. NaCl, 1 gm.

$FeSO_4 \cdot 7H_2O$, and 1 gm. $MnSO_4 \cdot H_2O$ in water and dilute to 1 liter. Add 10 drops HCl and store under toluene.

g. *Tryptophan solution.* Suspend 2.0 gm. of L-tryptophan (or 4.0 gm. DL-tryptophan) in 700 to 800 ml. of water, heat to 70° to 80°C., and add HCl (1 + 1) dropwise, with stirring, until the solid dissolves. Cool, and dilute with water to 1 liter. Store under toluene.

h. *Vitamin solution.* Dissolve 10 mg. of *p*-aminobenzoic acid, 40 mg. of pyridoxine hydrochloride, 4 mg. of thiamine hydrochloride, 8 mg. of calcium pantothenate, 8 mg. of nicotinic acid, and 0.2 mg. of biotin in approximately 300 ml. of water. Add 10 mg. of riboflavin dissolved in approximately 200 ml. of 0.02 N acetic acid. Then add a solution containing 1.9 gm. of anhydrous sodium acetate and 1.6 ml. of acetic acid in approximately 40 ml. of water, and dilute with water to 2 liters. Store under toluene.

i. *Xanthine solution.* Suspend 0.4 gm. xanthine in 60 to 80 ml. of water, heat to approximately 70°C., add 12 ml. of 6 N NH_4OH, and stir until the solid dissolves. Cool, and dilute with water to 400 ml. Store under toluene.

5. Liquid Culture Medium

Dissolve 15 gm. of peptonized milk, 5 gm. of water-soluble yeast extract, 10 gm. of anhydrous dextrose, and 2 gm. of anhydrous KH_2PO_4 in approximately 600 ml. of water. Add 100 ml. of filtered tomato juice, and adjust to pH 6.5–6.8 with NaOH solution. Add, with mixing, 10 ml. of polysorbate 80 solution (Tween 80, Atlas Chemical Company, Wilmington, Delaware), and dilute with water to 1 liter. Add 10-ml. portions of the solution to test tubes, plug with cotton, sterilize for 15 minutes in an autoclave at 121° to 123°C., and cool the tubes as rapidly as practicable to keep color formation at a minimum. Store in the dark at approximately 10°C. (Difco liquid culture medium for A.O.A.C. microbiological assays has been found satisfactory.)

6. Agar Culture Medium

To 500 ml. of liquid culture medium add 5.0–7.5 gm. of agar and heat on steam bath with stirring until the agar dissolves. Add approximately 10-ml. portions of the hot solution to test tubes, plug with cotton, sterilize for 15 minutes in an autoclave at 121° to 123°C., and cool the tubes in an upright position as rapidly as practicable to keep color formation to a minimum. Store in the dark at approximately 10°C. (Difco agar culture medium for A.O.A.C. microbiological assays has been found satisfactory.)

7. Suspension Medium

Dilute a measured volume of the basal medium stock solution (Section

III.A.8) with an equal volume of water. Add 10-ml. portions of the diluted medium to test tubes, plug with cotton, sterilize for 15 minutes in an autoclave at 121° to 123°C., and cool tubes as rapidly as practicable to keep color formation to a minimum. Store in the dark at approximately 10°C.

8. Basal Medium

The basal medium for 250 ml. (proportionate quantities may be prepared) is given in the accompanying tabulation.

Ingredients (stock solutions)	ml.
(a) Acid-hydrolyzed casein solution	25
(b) Adenine-guanine-uracil solution	2.5
(c) Asparagine solution	15
(d) Manganese sulfate solution	5
(e) Polysorbate 80 solution	0.25
(f) Salt solution B	5
(g) Tryptophan solution	25
(h) Vitamin solution	50
(i) Xanthine solution	5

Solids	gm.
Cysteine	0.13
Dextrose, anhydrous	10
Glutathione	0.0013
K_2HPO_4, anhydrous	1.6
Na citrate dihydrate	13

Using solutions prepared in Section III.A.4 and the quantities listed above, the medium is prepared by mixing in the following order: acid-hydrolyzed casein solution (a); tryptophan solution (g); adenine-guanine-uracil solution (b); xanthine solution (i); asparagine solution (c); vitamin solution (h); and salt solution B (f). Add approximately 50 ml. of water, and add, with mixing, the cysteine, anhydrous dextrose, Na citrate dihydrate, anhydrous K_2HPO_4, and glutathione. When solution is complete, adjust to pH 6.8 with NaOH solution, add, with mixing, the polysorbate 80 solution (e), and the $MnSO_4$ solution (d), and dilute with water to 250 ml. Some commercial sources of basal media have been found satisfactory (e.g., Difco-Bacto-folic acid A.O.A.C. medium).

9. Stock Culture

Stab cultures are prepared in the agar culture medium (Section III.A.6). Incubate 6–24 hours at any selected temperature between 30° and 40°C.

held constant to within ±0.5°C., and finally store in the dark at approximately 10°C. Before using a new culture in the assay make several successive transfers of the culture in a 1 to 2 week period. Prepare fresh stab cultures one or more times weekly and do not use for preparing the inoculum if it is more than 1 week old.

The activity of a slow-growing culture may be increased by daily or twice-daily transfer of the stab culture, and is considered satisfactory when definite turbidity in liquid inoculum broth can be observed 2–4 hours after inoculation. A slow-growing culture seldom gives a suitable response curve and may cause erratic results.

10. Inoculum

Make a transfer of cells from the stock culture of *Streptococcus faecalis* to a sterile tube containing 10 ml. of liquid culture medium (Section III.A.5) and incubate for 6 to 24 hours at any selected temperature between 30° and 40°C. held constant to within ±0.5°C. Centrifuge the culture under aseptic conditions and decant the supernatant liquid. Suspend the cells from the culture in 10 ml. of sterile suspension medium (Section III.A.7). The cell suspension so obtained is the inoculum.

11. Assay Solution

Place a measured quantity of the sample in a flask and a volume of water equal in milliliters to not less than 10 times the dry weight of the sample in grams; the resulting solution must contain not more than 1.0 μg. of folic acid per milliliter. Add the equivalent of 2 ml. of NH_4OH (6 N)/100 ml. of liquid. If the sample is not readily soluble, comminute it so that it may be evenly dispersed in the liquid; then agitate vigorously and wash down the sides of the flask with 0.1 N NH_4OH. Autoclave the mixture for 15 minutes at 121° to 123°C. and cool. If lumping occurs, agitate the mixture until the particles are evenly dispersed. Dilute the mixture to a measured volume with water, and let any undissolved particles settle, or filter or centrifuge if necessary. Take an aliquot of the clear solution, add water, adjust to pH 6.8, and dilute with additional water to a measured volume containing approximately 1.0 mμg. of folic acid per milliliter. Designate this as the assay solution.

12. Assay Solution for Turbidimetric Assay

Proceed as above except that where reference is made to folic acid concentration of approximately 1.0 mμg./ml., replace by concentration of approximately 0.5 mμg./ml. Designate the solution so obtained as the assay solution.

13. Cleaning of Glassware

By suitable means meticulously cleanse approximately 20 × 150 mm. hard-glass test tubes and other necessary glassware. Sodium lauryl sulfate U.S.P. has been found to be satisfactory as a detergent. Test organisms are highly sensitive to minute amounts of growth factors and to many cleansing agents. Therefore, it may be preferred to follow cleansing by heating 1–2 hours at approximately 250°C.

14. Assay Procedure

Using the appropriate standard solution, assay solution and basal medium stock solution and inoculum, proceed as follows:

Prepare tubes containing the appropriate standard solutions as follows: To test tubes add, in duplicate (or replicate), 0.0 (for uninoculated blanks), 0.0 (for inoculated blanks), 1.0, 2.0, 3.0, 4.0, and 5.0 ml., respectively, of the standard solution.

Prepare tubes containing appropriate assay solution as follows: To similar test tubes add, in duplicate (or replicate), 1.0, 2.0, 3.0, and 4.0 ml., respectively, of the assay solution.

To each tube of standard solution and assay solution add water to bring the total volume to 5.0 ml., then add 5.0 ml. of basal medium and mix. Cover the tubes suitably to prevent bacterial contamination and sterilize (10 minutes for titrimetric, or 5 minutes for turbidimetric method) in an autoclave at 121° to 123°C., reaching this temperature in not more than 10 minutes. Cool as rapidly as practicable to keep color formation at a minimum. Take precautions to maintain uniformity of sterilization and cooling conditions throughout the assay. Too close packing of tubes in the autoclave, or overloading of it, may cause variation in heating rate.

Aseptically inoculate each tube, except 1 set of duplicate (or replicate) tubes containing 0.0 ml. of standard solution (uninoculated blanks), with 1 drop of the appropriate inoculum. Incubate for a time period designated under titrimetric method, or turbidimetric method, at any selected temperature between 30° and 40°C. held constant to within ±0.5°C. Contamination of assay tubes with any foreign organism invalidates the assay.

15. Titrimetric Method

Incubate the tubes for 72 hours, and then titrate the contents of each tube with 0.1 N NaOH, using bromothymol blue indicator, or to pH 6.8 measured electrometrically. Disregard the results of the assay if the response at the inoculated blank level is equivalent to a titration of more than 1.5 ml. greater than that of uninoculated blank level. Response at the 5.0-ml. level of standard solution should be equivalent to a titration of approximately 8–12 ml.

Prepare a standard concentration-response curve by plotting the titration values, expressed in milliliters of 0.1 N of NaOH for each level of standard solution used, against the quantity of reference standard contained in the respective tubes.

Determine the quantity of vitamin for each level of the assay solution by interpolation from the standard curve. Discard any observed titration values equivalent to less than 0.5 ml. or more than 4.5 ml., respectively, of the folic acid standard solution. Proceed as under Calculation.

16. Turbidimetric Method

(Not applicable in the presence of extraneous turbidity or color in amount that interferes with turbidimetric measurements.)

a. Calibration of the photometer. Using the inoculum, the standard stock solution, and the suspension medium proceed as directed below.

Add aseptically 1 ml. of inoculum to approximately 300 ml. of sterile suspension medium containing 1.0 ml. of the standard stock solution and incubate the mixture for the same period and at the same temperature to be employed in the determination. After incubating, centrifuge and wash the cells 3 times with approximately 50-ml. portions of 0.9% NaCl solution; then resuspend the cells in the NaCl solution to make 25 ml.

Evaporate a 10-ml. aliquot of the cell suspension on a steam bath and dry to constant weight at 100°C. in a vacuum oven. Calculate the dry weight of the cells in milligrams per milliliter of suspension after correcting for the weight of NaCl.

Dilute a second measured aliquot of the cell suspension with 0.9% NaCl solution so that each milliliter is equivalent to 0.5 mg. of dry cells. To test tubes add in triplicate, 0.0 (for blanks), 0.5, 1.0, 1.5, 2.0, 2.5, 3.0, 4.0, and 5.0 ml., respectively, of this diluted cell suspension. To each tube add the NaCl solution to make 5.0 ml. Then add 5.0 ml. of the basal medium, mix (one drop of suitable antifoam agent may be added; 1–2% solution of Dow Corning Antifoam AF Emulsion or Antifoam B has been found satisfactory), and transfer to an optical cell. With the blanks set at 100% transmittance, measure the per cent transmittance of the contents of each tube under the same conditions to be used in the assay. Prepare a curve by plotting per cent transmittance readings for each level of diluted cell suspension used against the cell content (milligrams of dry weight) of the respective tubes.

Repeat the calibration steps at least 2 more times for the photometer to be used in the assay. Draw a composite curve best representing 3 or more individual curves, relating the per cent transmittance to milligrams of dried cell weight for the photometer under the conditions of the assay. Once an appropriate curve for the particular instrument is established, all subsequent relationships between per cent transmittance and cell weight are ascertained directly from this curve. Respective assay limits expressed as milligrams of dried cell weight per tube are so determined.

b. Determination. Incubate the tubes 16–24 hours until maximum turbidity is obtained, as demonstrated by a lack of significant change during a 2-hour additional incubation period in tubes containing the highest level of standard solution.

Determine the transmittance of tubes as follows: Mix thoroughly the contents of each tube (one drop of a suitable antifoam agent solution may be added; a 1–2% solution of Dow Corning Antifoam AF Emulsion or Antifoam B has been found satisfactory), and transfer to an optical cell. Agitate the contents, place the cell in a photometer set at any specific wavelength between 540 and 660 mμ, and read the per cent transmittance when the steady state is reached. The steady state is observed a few seconds after agitation when the galvanometer reading remains constant 30 seconds or more. Allow approximately the same time interval for the reading on each tube.

With transmittance set at 100% for the uninoculated blank level, read the per cent transmittance of the inoculated blank level. If this reading corresponds to a dried cell weight greater than 0.6 mg. per tube, discard the assay. Then with the transmittance reset at 100% for the inoculated blank level, read the per cent transmittance for each of the remaining tubes. Disregard the results of the assay if the per cent transmittance observed at the 5.0-ml. level of the standard solution is equivalent to that for a dried cell weight of less than 1.25 mg. per tube.

Prepare a standard concentration-response curve by plotting the per cent transmittance readings for each level of the standard solution used against the quantity of reference standard contained in the respective tubes. Typical turbidimetric and titrimetric response curves[91] are shown in Fig. 1.

Determine the quantity of vitamin for each level of the assay solution by interpolation from the standard curve. Discard any observed transmittance values equivalent to less than 0.5 ml. or more than 4.5 ml., respectively, of the standard solution. Proceed as under Calculation below.

17. Calculation for Both Titrimetric and Turbidimetric Methods

For each level of assay solutions used, calculate the vitamin content per milliliter of assay solution. Calculate the average of the values ob-

[91] H. W. Loy, Jr., W. P. Parrish, and S. Stephen, *J. Assoc. Offic. Agr. Chemists* **39,** 172 (1956).

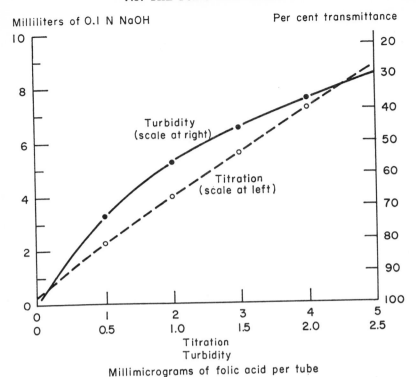

FIG. 1. Response of *S. faecalis* to folic acid as measured by titration and turbidity.[91]

tained from tubes that do not vary by more than ±10% from this average. If the number of acceptable values remaining is less than ⅔ of the original number of tubes used in the 4 levels of assay solution, data is insufficient for calculating the potency of the sample. If the number of acceptable values remaining is ⅔ or more of the original number of tubes, calculate the potency of the sample from the average of them.

B. *Lactobacillus casei* Method

Methods utilizing *L. casei*, although yielding results comparable to those obtained with *Streptococcus faecalis*, are not in general use because of the slower growth of this organism. Growth response of *Lactobacillus casei* is measured turbidimetrically after incubation for 40 hours compared with 16 hours for *Streptococcus faecalis. Lactobacillus casei*, however, is more sensitive to pteroylglutamic acid than *Streptococcus faecalis* and is the organism of choice when low potency materials are assayed. Thus, Baker *et al.*[89] utilized *Lactobacillus casei* to determine folic acid activity in serum. The procedure yielded results which correlated with clinical diagnosis of the folic acid status in humans.

Most methods utilizing *L. casei* are based on the original work of Teply and Elvehjem.[92] Baker *et al.*[89] used a slight modification of a method devised by Flynn and co-workers[93] which in turn was derived from the Teply and Elvehjem method. The following is the medium (Table V) and procedure of Flynn and co-workers.[93]

TABLE V

BASAL MEDIUM STOCK SOLUTION (DOUBLE STRENGTH) FOR *L. casei*[a]

Ingredients[b]	Quantity	Ingredients	Quantity
	(gm.)		(mg.)
Acid hydrolyzed casein	5.0	Guanine·HCl	5
Sodium acetate	20.0	Adenine·SO$_4$	5
Glucose	20.0	Uracil	5
K$_2$HPO$_4$	0.5	Xanthine	10
KH$_2$PO$_4$	0.5	Thiamine·HCl	0.2
MgSO$_4$·7H$_2$O	0.2	Riboflavin	0.5
NaCl	0.01	Calcium pantothenate	0.4
FeSO$_4$·7H$_2$O	0.01	Nicotinic acid	0.4
MnSO$_4$·H$_2$O	0.1	Pyridoxine·HCl	2
		p-Aminobenzoic acid	0.5
	(mg.)	Biotin	0.01
Asparagine	300	Adjust pH to 6.8 and dilute to 500 ml.	
L-Tryptophan	100	with distilled H$_2$O.	
L-Cysteine	250		
Glutathione	2.5		
Tween 80	50		

[a] Reprinted from Flynn *et al.*[93]

[b] Glucose, cysteine, and glutathione are added as solids. The rest of the ingredients are added as solutions similar to those found in the procedure using *S. faecalis*. The MnSO$_4$·H$_2$O is added to the medium after adjusting the pH.

1. Stock Culture

Maintain *L. casei* (ATCC 7469) as stabs in Difco Bacto yeast dextrose agar to which 0.6% sodium acetate is added. Incubate the stabs at 33°C. for 24 hours.

2. Preparation of Inoculum

Transplant the organism into the assay medium (single strength) to which has been added 0.5 μg. folic acid and 0.5 gm. Wilson's liver frac-

[92] L. J. Teply and C. A. Elvehjem, *J. Biol. Chem.* **157**, 303 (1945).

[93] L. M. Flynn, V. B. Williams, B. L. O'Dell, and A. G. Hogan, *Anal. Chem.* **23**, 180 (1951).

tion L per 1000 ml. Incubate the cultures for 16 to 18 hours at 33°C. and centrifuge. Wash the cells twice with sterile "vitamin-free medium" (the basal medium without the vitamins, tryptophan, cysteine, glutathione, and Tween), filter through a thin pad of cotton, and dilute to contain approximately 1.7×10^6 organisms per milliliter. The standardization of the inoculum may be obtained using the procedure outlined in the method utilizing *Streptococcus faecalis* (see Section III.A). The basal medium of Baker *et al.*, which is essentially identical in composition to that of Flynn *et al.*, is available in dehydrated form from the Baltimore Biological Laboratory, Inc.

3. Procedure

The assay may be conducted as outlined in the *S. faecalis* method. The standard curve ranges between 0 and 3.0 mμg. of folic acid per milliliter final concentration. Incubate the assays at 33° to 37°C. for 40 hours when assaying turbidimetrically and for 72 hours when assaying by the titrimetric method.

C. *Tetrahymena geleii* W Method

The organism *T. geleii* is a ciliated protozoan with unique and well-studied physiological properties. Its nutritional requirements are very similar to those of higher animals while its growth rate is greater than these same animals. Kidder and Dewey have prepared an excellent review on the biochemistry of this organism.[94] The organism contains a folic acid conjugase and is capable of utilizing pteroyltriglutamic acid, pteroylheptaglutamic acid and pteroylglutamic acid equally on a molar basis.[64,95] It is possible, therefore, to assay for folic acid in some natural materials without prior treatment of the samples with a conjugase to liberate free folic acid.

The following assay procedure utilizing *T. geleii* W was described by Jukes,[96] and is a modification of methods found in the literature.[94,97]

The test organism, *T. geleii* W, is maintained by weekly subculture in the following transfer medium: 2% proteose peptone, 0.05% Wilson's liver fraction L, 1% dextrose, pH 6.8. To prepare an inoculum of *T. geleii*, dilute a 72-hour culture with sterile 0.9% NaCl solution in order to add conveniently the equivalent of 0.0005 ml. of the original culture per assay tube.

[94] G. W. Kidder and V. C. Dewey, *in* "Biochemistry and Physiology of Protozoa" (A. Lwoff, ed.), Vol. 1. Academic Press, New York, 1951.

[95] G. W. Kidder, *Ciba Foundation Symposium, Chem. & Biol. Pteridines, 1954.*

[96] T. H. Jukes, *Methods of Biochem. Anal.* **2**, 121 (1955).

[97] V. C. Dewey, R. E. Parks, Jr., and G. W. Kidder, *Arch. Biochem.* **29**, 281 (1950).

Carry out the microbiological assay in 12×100 mm. Pyrex test tubes at a final volume of 2 ml. Add test substances and bring to a final volume of 1 ml. Pteroylglutamic acid is used as the assay standard. Add 1 ml. of standard solutions of the following concentrations: 0.01, 0.03, 0.06, 0.1, 0.2, 0.3, 0.6, 1.0, 3.0, and 10.0 mμg./ml. to a series of tubes. Then add 1 ml. of double-strength basal medium (Table VI). Plug the assay tubes or

TABLE VI

BASAL MEDIUM (DOUBLE STRENGTH) FOR ASSAY OF FOLIC ACID WITH *T. geleii* W[a]

Component	μg./ml.	Component	μg./ml.
DL-Alanine	110	DL-Thioctic acid	0.01
L-Arginine	206	Thiamine hydrochloride	1
L-Aspartic acid	122	Biotin	0.0005
Glycine	10	Choline chloride	1
L-Glutamic acid	233	$MgSO_4 \cdot 7H_2O$	100
L-Histidine	87	$Fe(NH_4)_2(SO_4)_2 \cdot 6H_2O$	25
DL-Isoleucine	276	$MnCl_2 \cdot 4H_2O$	0.5
L-Leucine	344	$ZnCl_2$	0.05
L-Lysine	272	$CaCl_2 \cdot 2H_2O$	50
DL-Methionine	248	$CuCl_2 \cdot 2H_2O$	5
L-Phenylalanine	169	$FeCl_3 \cdot 6H_2O$	1.25
L-Proline	250	Guanylic acid	30
DL-Serine	394	Adenylic acid	20
DL-Threonine	326	Cytidylic acid	25
L-Tryptophan	72	Uracil	10
DL-Valine	162		
Calcium pantothenate	0.10		mg./ml.
Nicotinamide	0.10		
Pyridoxine hydrochloride	1.00	Tween 80	10
Pyridoxal hydrochloride	0.10	Dextrose[b]	2.5
Pyridoxamine hydrochloride	0.10	K_2HPO_4	1.0
Riboflavin	0.10	KH_2PO_4	1.0
		Na acetate	1.0

[a] Reprinted from Jukes.[96]
[b] Autoclaved separately and added aseptically.

cap and sterilize for 10 minutes at 121°C. and cool and inoculate as previously described. Tip the assay racks to about a 75° angle by resting the side of the rack on a pipet and incubate the tubes in this position to provide greater contact with air. After incubation for 72 hours at 23° to 25°C., estimate the growth turbidimetrically and calculate the potency of the samples from the response of the organism to pteroylglutamic acid. Maximum growth is reached with about 1 mμg. of pteroylglutamic acid per tube.

D. *Bacillus coagulans* Method

Bacillus coagulans is a thermophilic organism which responds to folic acid and its conjugates. The use of this organism for the assay of folic acid was suggested by Baker *et al.*[98]

The assay may be conducted as suggested by Campbell and Sniff[99] and others.[98]

1. Organism

Bacillus coagulans (ATCC 12245). Maintain stock cultures by monthly transfer on nutrient agar slants.

2. Inoculum broth.

See the accompanying tabulation.

Basamin[a]	3.0 gm.
Glucose	2.0 gm.
Na_2HPO_4	2.5 gm.
KH_2PO_4	1.0 gm.
NaCl	1.0 gm.
NH_4Cl	1.0 gm.
$MgSO_4$	5.0 mg.
$FeSO_4$	5.0 mg.
$CaCl_2$	5.0 mg.
H_2O to make 1000 ml.	pH 7.2
Sterilized at 121°C. for 15 minutes	

[a] Basamin is a yeast autolysate manufactured by the Anheuser Busch Company.

3. Basal Medium

The composition of the basal medium is given in Table VII.

4. Preparation of Inoculum

Harvest cells from a 48-hour basamin broth culture incubated at 55°C. by centrifugation and wash 3 times with sterile distilled water. Adjust the washed cell suspension to an optical density of 0.40 at 525 mμ with distilled water using a Bausch & Lomb Spectronic 20 colorimeter. The diameter of the tube used is not specified but it is stated that this corre-

[98] H. Baker, S. H. Hutner, and H. Sobotka, *Proc. Soc. Exptl. Biol. Med.* **89**, 210 (1955).

[99] L. L. Campbell and E. E. Sniff, *J. Bacteriol.* **78**, 267 (1959).

TABLE VII
Basal Medium for Assay of Folic Acid with *B. coagulans*

Constituent	Amount	Constituent	Amount
	(mg.)		(μg.)
L-Arginine	10.5	Pantothenic acid	100
L-Glutamic acid	10	Pyridoxal	7.5
L-Histidine	4.5	Biotin	0.9
DL-Isoleucine	14		
L-Leucine	19.2		(mg.)
L-Lysine	19.5		
DL-Methionine	6.0	Glucose	400
L-Tryptophan	6.0	Sodium acetate	50
DL-Valine	14.4	Na_2HPO_4	250
		KH_2PO_4	100
	(μg.)	NH_4Cl	100
		NaCl	100
Thiamine HCl	15		
Riboflavin	15		(ml.)
Nicotinic acid	150		
		Mineral supplement[a]	0.1
		Distilled water	100
		Adjust pH to 7.2	—

[a] Mineral supplement: $MgCl_2$, 0.5 gm.; $FeCl_3$, 0.5 gm.; $CaCl_2$, 0.5 gm.; distilled water, 100 ml.

sponds to 0.17 to 0.20 mg. dry weight of cell material per milliliter. Prepare a fresh inoculum for each determination.

5. Assay

a. *Standard curve.* 0.5–50 mμg. folic acid per flask.

b. *Procedure.* Pipet test solutions into small flasks (10–25 ml. capacity) and adjust the volume to 2.5 ml. Add 2.5 ml. of basal medium (Table VII) and inoculate the flasks with 0.1 ml. of the inoculum. Baker *et al.*[98] incubated the assays at 55°C. for 18 hours. When this high temperature was used, sterilization of the medium was unnecessary. Measure growth response turbidimetrically at 525 mμ.

There are several advantages to this procedure. The high incubation temperature eliminates the need for aseptic technique. Since the organism responds to conjugates of folic acid (PGA, folinic, N^{10}-formyl PGA, PTGA), some samples can be assayed without prior treatment to release folic acid. It does not respond to pterioc acid or p-aminobenzoylglutamic acid. A response is obtained with p-aminobenzoic acid which may be overcome by the addition of 0.01% sulfanilamide to the basal medium.

p-Aminobenzoic acid is metabolized by higher animals and excreted in the urine as *p*-aminohippuric acid to which the organism does not respond. Therefore interference due to the presence of *p*-aminobenzoic acid conjugate in urine is negligible. The response to *p*-aminobenzoic acid is at least as great as to folic acid.[98] Disadvantages of the method include the possibility that nonspecific stimulation might occur when natural materials are assayed. The interpretation of data obtained with materials containing mixtures of such compounds would be impossible. In spite of these drawbacks the method seems to merit further study.

E. *Pediococcus cerevisiae* Method for the Determination of Leucovorin

The following procedure is based on the method recommended by the U.S.P. for the microbiological assay of calcium leucovorin[17] and can be used for the assay of natural materials. Free leucovorin can be liberated from its conjugates with chick pancreas conjugase in the same manner as described for bound folic acid.

1. Test Organism

Pediococcus cerevisiae (ATCC 8081) was thought at one time to be a strain of *Leuconostoc* and was known for many years as *Leuconostoc citrovorum*. It was determined later by Felton and Niven[100] to be a typical strain of *Pediococcus cerevisiae*.

2. Standard Solution

Dissolve dried U.S.P. calcium leucovorin reference standard (synthetic) in sufficient 0.1 *N* NaOH to give a concentration equivalent to 100 μg. of leucovorin (anhydrous free acid) per milliliter. The Reference Standard contains 5 moles of H_2O and thus contains 77.1% of the free acid. Dilute this solution further with water to a concentration of 0.02 μg./ml. and store in a freezer. On the day of the assay thaw this solution and dilute an aliquot with water to give a concentration of 0.2 mμg./ml.

3. Stock Solutions for Basal Medium

(Store all solutions in the dark in the refrigerator.)

a. Acid-hydrolyzed casein solution. See Section III.A.4.*a.* Hy Case SF might also be suitable for use here.

b. Cystine-tryptophan solution. Suspend 4.0 gm. of L-cystine and 1.0 gm. of L-tryptophan or 2.0 gm. of the DL form in 700 to 800 ml. of water, heat to 70° to 80°C., and add dilute hydrochloric acid (1 + 1) dropwise with stirring until solution occurs, cool, and dilute to 1000 ml. Store under toluene in a refrigerator at a temperature not below 10°C.

---

[100] E. A. Felton and C. F. Niven, Jr., *J. Bacteriol.* **65**, 482 (1953).

c. *Adenine-guanine-uracil solution*. See Section III.A.4.*b*.

d. *Xanthine solution*. See Section III.A.4.*i*.

e. *Riboflavin-thiamine hydrochloride-biotin solution*. Prepare a solution in 0.02 N acetic acid containing 50 μg. of riboflavin, 50 μg. of thiamine hydrochloride, and 0.1 μg. of biotin per milliliter. Store under toluene in a refrigerator.

f. *p-Aminobenzoic acid-nicotinic acid-pyridoxine hydrochloride solution*. Prepare a solution in neutral 25% alcohol to contain 10 μg. of p-aminobenzoic acid, 250 μg. of nicotinic acid, and 100 μg. of pyridoxine hydrochloride per milliliter. Store in a refrigerator.

g. *Calcium pantothenate solution*. Dissolve 50 mg. of D-calcium pantothenate in 500 ml. of water. Store under toluene in a refrigerator.

h. *Folic acid solution*. Dissolve 50 mg. of folic acid in 5 ml. of 0.1 N NaOH and dilute to 100 ml. Dilute 1 ml. of this solution to 100 ml.

i. *Salt solution A*. Dissolve 25 g. of KH_2PO_4 and 25 gm. of K_2HPO_4 in water and dilute to 500 ml., add 5 drops of concentrated HCl, and store under toluene.

j. *Salt solution B*. See Section III.A.4.*f*.

4. Basal Medium Stock Solution

See the accompanying tabulation.

Acid-hydrolyzed casein solution	50 ml.
Cystine-tryptophan solution	25 ml.
Dextrose, anhydrous	20 gm.
Sodium acetate, anhydrous	20 gm.
Adenine-guanine-uracil solution	10 ml.
Xanthine solution	10 ml.
Riboflavin-thiamine hydrochloride-biotin solution	10 ml.
p-Aminobenzoic acid–nicotinic acid–pyridoxine hydrochloride solution	10 ml.
Calcium pantothenate solution	5 ml.
Folic acid solution	2 ml.
Salt solution A	10 ml.
Salt solution B	10 ml.

Dissolve the anhydrous dextrose and sodium acetate in the solutions previously mixed and adjust to a pH of 6.8 with N sodium hydroxide. Finally add water to make 1000 ml.

5. Stock Culture of Pediococcus cerevisiae

Stab cultures may be prepared in Difco-Bacto micro assay culture agar (Difco Company, Detroit, Michigan). Incubate at 37°C. for 24 hours and store in a refrigerator. Prepare fresh stab cultures every 2 weeks.

6. Inoculum

Place 5 ml. of basal medium stock solution into a series of test tubes and add 5 ml. of water containing 10 mg. of yeast extract or other source of leucovorin (e.g., leucovorin 2 mμg./tube). Cover the tubes and sterilize in an autoclave at 121°C. Cool and inoculate with cells from the stock culture. After incubation at 37°C. for 24 hours, centrifuge the culture and discard the supernatant liquid. Suspend the cells in 10 ml. of sterile 0.9% NaCl solution and centrifuge again. Resuspend the cells in sterile saline and dilute them 1:10 with saline.

7. Assay Solution

Prepare samples of natural origin the same as for folic acid. Treat conjugated forms with chick pancreas enzyme as noted in the section "Preparation of Samples for Assay" (see Sections II.B.1 and III.A.11).

8. Assay Procedure

The procedure is similar to the methods for folic acid. Incubate the assays at 37°C. for 24 hours for the turbidimetric assay and 72 hours for the titrimetric procedure. The U. S. Pharmacopeia currently recommends the turbidimetric assay. Calculations are made in the usual manner. Prepare the standard curve by adding in duplicate 0, 1.0, 1.5, 2.0, 3.0, 4.0, and 5.0 ml., respectively, of the final standard solution to separate tubes. Dilute the sample to contain the equivalent of approximately 0.2 mμg. of synthetic leucovorin per milliliter. Since synthetic leucovorin has one-half the activity of the natural form, divide assay values obtained with natural products by 2 to obtain the result in terms of biologically active material.

F. Recommended Turbidimetric Assay of Folic Acid Activity with *Lactobacillus casei, Streptococcus faecalis,* or *Pediococcus cerevisiae* with a Single Medium

1. Introduction

This method has been used with some success to determine the folic acid activity of human blood.[85a,101] Using either of the media given below, with actively growing log phase inocula that are grown in growth-limiting concentrations of folacin (or folinic acid) plus the associated technics to be mentioned below, accurate and somewhat differential assays can be obtained. In addition, these assays seem to be somewhat more sensitive than some of the other methods that have been used.

[101] G. Toennies, H. G. Frank, and D. L. Gallant, *Growth* **16**, 287 (1952).

2. Test Organisms

The test organisms used are: *Lactobacillus casei*, ATCC 7469; *Streptococcus faecalis*, ATCC 8043; *Pediococcus cerevisiae*, ATCC 8081 (*Leuconostoc citrovorum*).

3. Assay Media

A medium prepared from stock solutions or one prepared from all dry ingredients can be used. The latter has the disadvantage of being somewhat less sensitive than the former but has two great advantages. The first is the ease and convenience of the use of a dry medium when large numbers of assays are being run frequently. The second is an added consistency of one variable (the medium) from assay to assay using the same batch of dry ingredients. This last point should not be overemphasized, however, since the dry medium will have a slightly decreasing ability to support good growth with age, particularly when stored improperly.

a. Basal medium I prepared from stock solutions.[101,102] The composition of this medium is given in Table VIII. Commercially available enzymatic casein hydrolysate of low folic acid activity such as casein hydrolysate enzymatic, 5%, vitamin free (Nutritional Biochemicals, Cleveland 28, Ohio) can be used. If preferred, or if this is not available, a standard method for the preparation of enzymatic hydrolysate can be used and the hydrolysate treated with activated charcoal to remove vitamin contamination. If this is done, a high grade of "vitamin free" casein should be used as starting material. A 5% solution of Hy Case SF plus 400 mg./liter of DL-tryptophan can be substituted for the enzymatic casein hydrolysate. This is the chief nitrogen source of the medium. To this is added the 9 stock solutions that are described in Table IX. These solutions should be renewed at least every 6 months and stored in the refrigerator.

Prepare the medium as follows: For each 100 ml. of double strength basal medium, weigh out 4.1 gm. (50 mM) sodium acetate (anhydrous), 4.0 gm. glucose, 696 mg. (4 mM) K_2HPO_4, 680 mg. (5 mM) KH_2PO_4, and 50 mg. ascorbic acid. Dissolve without heating in approximately 25 ml. of distilled water and add 2 ml. each of the 9 solutions listed in Table IX, and then 25 ml. of the 5% solution of enzymatically hydrolyzed casein. Checking with a pH meter add enough 1 N NaOH to bring the pH to 6.45. The amount of NaOH required can be obtained from a titration curve of the medium and then for subsequent batches of medium the

[102] E. Usdin, G. D. Shockman, and G. Toennies, *Appl. Microbiol.* **2,** 29 (1954).

TABLE VIII

COMPOSITION OF BASAL MEDIUM I (DOUBLE STRENGTH)[a]

	Per liter of basal medium
	(gm.)
Enzymatic casein hydrolysate[b]	12.5
Sodium acetate	41
Glucose	40
K_2HPO_4	6.96
KH_2PO_4	6.80
Ascorbic acid	0.50
	(mg.)
DL-Alanine	400
L-Asparagine	200
L-Cystine	200
$MgSO_4 \cdot 7H_2O$	400
NaCl	20
$FeSO_4 \cdot 7H_2O$	20
$MnSO_4 \cdot 4H_2O$	20
Adenine	20
Guanine	20
Uracil	20
Xanthine	20
Nicotinic acid	1
Calcium pantothenate	1
Thiamine HCl	1
Pyridoxamine HCl	0.4
p-Aminobenzoic acid	0.2
Riboflavin	1.
Biotin	0.02

[a] References 101 and 102.

[b] Casein hydrolysate enzymatic, 5%, vitamin free (Nutritional Biochemicals, Cleveland 28, Ohio). An equivalent amount of Hy Case SF (Sheffield Farms Company, Inc., New York, New York) plus 400 mg./liter of DL-tryptophan can be substituted for the enzymatic hydrolysate. Hy Case SF is a "salt free" acid hydrolysate of casein.

TABLE IX

STOCK SOLUTIONS FOR BASAL MEDIUM I[a]

Label	Substance	Amount	Added solvent	Final volume (ml.)
		(gm.)		
ALA	DL-Alanine	5.0		250
ASP	L-Asparagine	2.5		250
CYS	L-Cystine	2.5	100 ml. 2 N HCl	250
SAL	MgSO$_4$·7H$_2$O	5.0		
		(mg.)		
	NaCl	250		
	FeSO$_4$·7H$_2$O	250		
	MnSO$_4$·4H$_2$O	250	1 ml. 2 N HCl	250
AGU	Adenine sulphate	435		
	Guanine·HCl	310		
	Uracil	250	25 ml. 2 N HCl[b]	250
XAN	Xanthine	250	10 ml. conc. NH$_4$OH	250
VIT	Nicotinic acid	50		
	Ca pantothenate	50		
	Thiamine HCl	50		
	Pyridoxamine HCl	20		
	p-Aminobenzoic acid	10		1000[c]
RIB	Riboflavin	12.5	0.75 ml. acetic acid	250
BIO	Biotin	10	2 ml. ethanol, then 0.01 N HCl	10,000[d]

[a] Reference 101.

[b] Use heat to dissolve.

[c] Save 250 ml. and discard remainder or prepare 250 ml. by dilution of an aliquot of higher concentration.

[d] Since biotin concentration is frequently a problem its concentration can be increased fivefold.

required amount can be added and the pH checked. The pH of the basal medium should be between 6.40 and 6.50. In order to obtain a completely clear medium it can be left in the refrigerator overnight and then filtered through a medium sintered glass filter. The medium can be made when needed or larger amounts can be made and stored in the frozen state for at least 3 months without losing full activity.

b. Basal medium II prepared from dry ingredients.[84] The quantities of the dry constituents used and the final composition of basal medium II are given in Table X. The ingredients have been divided into two groups, the minor and the major constituents of the medium.

TABLE X

BASAL MEDIUM II PREPARED FROM DRY INGREDIENTS FOR THE ASSAY OF FOLIC ACID ACTIVITY WITH *L. casei, S. faecalis,* OR *P. cerevisiae*

Minor constituents	Amount (mg.)	μg./ml. basal medium
p-Aminobenzoic acid	5	0.2
Biotin	5	0.2
Pyridoxamine dihydrochloride	10	0.4
Thiamine hydrochloride	25	1.0
Calcium pantothenate	25	1.0
Riboflavin	25	1.0
Nicotinic acid	25	1.0
Xanthine	500	20
Uracil	500	20
$MnSO_4 \cdot 4H_2O$	500	20
$FeSO_4 \cdot 7H_2O$	500	20
NaCl	500	20
Guanine hydrochloride	620	20[a]
Adenine sulfate	870	20[a]

Major constituents	Amount (gm.)	mg./ml.
L-Cystine	5	0.2
L-Asparagine	5	0.2
DL-Tryptophan	10	0.4
$MgSO_4 \cdot 7H_2O$	10	0.4
DL-Alanine	10	0.4
Ascorbic acid	12.5	0.5
KH_2PO_4	23.7	0.95
K_2HPO_4	143.5	5.74
Hy Case SF[b]	312.5	12.5
Dextrose	1000	40
Sodium acetate (anhydrous)	1025	41

[a] Free base.
[b] Hy Case SF is a salt free acid-hydrolyzed casein sold by Sheffield Farms Company, Inc., New York, New York.

In order to obtain thorough mixing so that a sample of the dry medium taken at a later time will be representative, the minor constituents are added serially to a mortar and ground together. The minor constituents are then added to the major ones and mixed very thoroughly in a ball mill or dry blender of any of a variety of types. Again, thorough mixing is of great importance. Blending or mixing for 24 to 48 hours is therefore advisable. The acid hydrolyzed casein is hygroscopic as are some of the other constituents of the medium. Therefore, the mixture must be protected from moisture during blending and during subsequent storage (e.g., over $CaCl_2$). This can be avoided to some extent by not including the hydrolyzed casein in the mixture and adding it separately when the assay is being performed. Both the mixture and the Hy Case SF can then be kept satisfactorily in a room with low humidity. This also makes it easier to check the blank of each batch of Hy Case SF (see next paragraph). When prepared separately use 9.0 gm. of the mixture plus 1.25 gm. of Hy Case SF per 100 ml. of basal medium.

An additional word of caution. Hy Case SF has proven to be generally reliable and relatively free from folic acid. However, it is advisable to check each batch of Hy Case SF for folic acid blank and for growth response before incorporating it into a large batch of dry medium. Other hydrolyzed caseins including those hydrolyzed and dried in the laboratory might also prove to be usable in this medium. However, Hy Case SF has proven to be superior both in blank levels and growth response to a number of other commercial acid hydrolyzed products.

Basal medium is prepared from the dry mixture by dissolving 10.25 gm. in fresh distilled water to make 100 ml. This gives a solution of about pH 6.50 and 0.05 M in respect to phosphate.

4. Standard Solutions

Since folacin (PGA) has been shown to be more stable in alkaline than in acid solutions,[21] the following routine will give reliable results.

Standard I. Prepare a 7.20×10^{-4} M solution of pteroylglutamic acid in 0.01 N NaOH fresh every 3 months by dissolving 69.0 mg. of dry folacin (92% PGA) in 40 ml. of 0.05 N NaOH and distilled water to make 200 ml. (345 μg./ml. = 317.4 μg./ml. PGA).

Standard II. Prepare a 7.20×10^{-6} M solution in 10^{-4} M NaOH every month by diluting 2.0 ml. of standard I to 200 ml. with distilled water (= 3.17 μg./ml. PGA).

Standard III. Prepare a 7.20×10^{-8} M solution in 10^{-4} M NaOH every week by diluting 2.0 ml. of standard II to 200 ml. with distilled water (= 31.7 mμg./ml. PGA).

Standards IV and IVa. Prepare a 7.20×10^{-10} M solution in 10^{-8} M NaOH every 2 days by diluting 2.0 ml. of standard III to 200 ml. with distilled water (= 0.317 mμg./ml. PGA) (standard IV) or with 0.05 M phosphate buffer (standard IVa). These are used as standards for assays with *Streptococcus faecalis*. Standard IV is used with basal medium I and standard IVa with basal medium II.

Standard V. Prepare a 1.80×10^{-10} M solution from standard IV by dilution with 3 volumes of distilled water (= 0.079 mμg./ml. PGA). This standard is used for assays with *Lactobacillus casei* and basal medium I.

Standard Va (3.60 × 10⁻¹⁰ M solution). For use with *L. casei* and basal medium II, prepare standard Va by diluting standard IVa with an equal volume of 0.05 M phosphate buffer (pH 6.08 − 6.12) (= 0.158 mμg./ml. PGA).

For the analysis of blood samples containing ascorbic acid–phosphate buffer (AP buffer), prepare standards IV and V (or Va) by dilution with AP buffer instead of water starting with standard III. These are designated standards IVAP, VAP, and VaAP, respectively (see Section II.B.2).

Folinic acid standards can be prepared from synthetic calcium or barium folinic acid or from an ampule of leucovorin. When synthetic materials are used, twice the molar concentrations of the diluted solutions should be employed. Folinic acid standards should also be prepared in 0.01 N alkali. The final standard solution should contain 3.6 × 10⁻¹⁰ M pure folinic acid for use with either *L. casei* or *Pediococcus cerevisiae*.

Standards of other folic acid-active compounds, such as teropterin

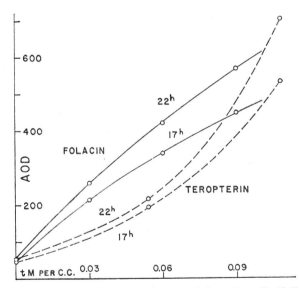

Fig. 2. Standard response of *L. casei* to pteroylglutamic acid (full lines) and to pteroyltriglutamic acid (teropterin) (dashed lines) for assays in basal medium I. Comparable response will be obtained in basal medium II to about twice the quantity of PGA. One trillimole per cubic centimeter (tM/cc.) represents 1 × 10⁻¹² moles/ml. (0.441 mμg. PGA) (abscissa). The ordinate is given in AOD (Adjusted Optical Density) which represents optical density adjusted to agree with Beer's law (see Chapter 8 on Amino Acid Assays for details) (1 AOD unit = 0.4 μg. bacterial dry weight/ml.).

(PTGA) can be prepared in the same way as the folic standards. *Lactobacillus casei* will respond to teropterin but will typically give a concave rather than a slightly convex standard curve (see Fig. 2).

5. Inocula

Superior results will be obtained if log phase inocula are used for all of these test organisms. This is particularly important for use with *L. casei* to obtain rapid and precise assays.

Log phase inocula can be grown in the assay medium and, in fact, the cultures can be carried in this medium routinely, and stored in an ice bath in the refrigerator. Since *Lactobacillus casei* is the least stable and most troublesome of the three test organisms, one should be certain to have reserve cultures as stab cultures in rich organic media or, preferably, in the lyophilized state. For further information on these technics see Chapter 8.

Inocula for all three test organisms can be grown in the presence of relatively low concentrations of PGA (or folinic acid). A tube consisting of 3 ml. of basal medium and 3 ml. of standard IV or IVa (final concentration 3.6×10^{-10} M PGA) will be found to be satisfactory for both *L. casei* and *Steptococcus faecalis*. Zero levels will be slightly reduced if the inocula are grown in 2.7×10^{-10} M PGA. A supply of such tubes can be sterilized and stored for several months in the frozen state. Such a tube can be inoculated from a log phase culture grown to a density of about 1.2×10^8 cells per milliliter (approximately 48 μg. bacterial dry weight per milliliter), and then placed in an ice bath until ready for use. If a log phase culture is not available, repetition of the procedure for 2 subsequent times will ensure a log phase inoculum. Inocula for assays should be grown on the day that they are to be used. Since this takes only 3–5 hours once the culture has been grown logarithmically, this is not at all inconvenient. Log phase cultures of *S. faecalis* can be stored for relatively long periods of time (weeks) and can be easily and rapidly restored to log growth. *Lactobacillus casei* log phase cultures should be transferred more often when not being used and will require more time to recover log phase growth at maximum growth rate. For instance, after storage for 3 days *L. casei* may require two such transfers as described above to regain maximal growth rate.

Inocula grown in the presence of a reduced amount of the substance being assayed need not be washed or centrifuged but can be added to the assay tubes (6 ml.) at the rate of one drop (approximately 0.05 ml.) per tube without introducing a significant blank level.

6. Cultural Conditions and Turbidimetric Readings

These are the same as those that have been described for amino acid assays (see Chapter 8). Calibrated tubes (18 × 150 mm.) are used to which 3 ml. of basal medium and 3 ml. of solution containing the standard or sample are added and autoclaved for 2½ minutes at 121°C. in a pre-heated autoclave. Determine the color blank in the turbidimeter before inoculation. Incubate the tubes aseptically and incubate in a constant temperature water bath at 37.7 ±0.05°C. For assays with *L. casei* and *Pediococcus cerevisiae* read turbidity (preferably o.d. scale, see Chapter 8) after 22 to 23 hours of incubation. For maximal response, *Streptococcus faecalis* assays should be read after 11 to 13 hours incubation since the culture tends to lyse upon longer incubation. Subtract the color blank readings to obtain the net value due to the growth of the test organism.

7. Responses

The response of *Lactobacillus casei* to PGA and to teropterin is shown in Fig. 2 and those for *Streptococcus faecalis* in Fig. 3. These are re-

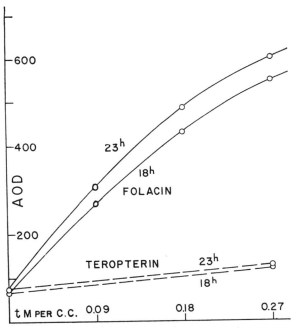

FIG. 3. Standard response of *S. faecalis* to pteroylglutamic acid (full lines) and to pteroyltriglutamic acid (dashed lines) in basal medium I. Ordinate and abscissa are the same as for Fig. 2.

sponses obtained in basal medium I. The responses in basal medium II are slightly reduced from those illustrated. The response of *Pediococcus cerevisiae* to folinic acid is nearly linear in either of these media and quantitatively is very similar to that of *Lactobacillus casei*.

8. Comments

For use with samples containing buffer or the ascorbic acid-phosphate buffer (AP buffer), the concentration of the buffer substances are adjusted to a constant value by making the final dilutions of the standard in buffer or in AP buffer and by adjusting the volume to 6 ml. with buffer or with AP buffer instead of distilled water.

Basal medium II differs from basal medium I in the amount of phosphate buffer that it contains. This is because the former medium was designed specifically for the analysis of the folic acid activity of blood samples that, after appropriate dilution, contained 0.05 M phosphate buffer.[84] Proper adjustment of the buffer concentration of the basal media is recommended for the use of either basal medium II with aqueous standards and samples, or basal medium I with standards and samples containing phosphate buffer.

All glassware must be washed and handled carefully. Washing with a good low-residue detergent such as Haemo-Sol (Meinecke & Company, Inc., 225 Varick Street, New York 14, New York), and rinsing several times in fresh distilled water has proven to be quite satisfactory. Difficulties that could be attributed to unclean glassware have been encountered only occasionally.

These methods, particularly with *L. casei* as test organism, have proven to be most satisfactory for the assay of folic acid activity in blood.[85a] Several modifications of the method have appeared. One involves the use of a total volume of 2 ml. in 13 × 100 mm. culture tubes.[88] In this way smaller samples can be assayed.

Using medium II (prepared from dry ingredients with Hy Case SF) net zero levels of about 28 μg. bacterial dry weight per milliliter (7×10^7 cells per milliliter) can be expected and a response with *L. casei* of about 280 to 320 μg. bacterial dry weight per milliliter (7×10^8 cells per milliliter) at a concentration of PGA of 1.8×10^{-10} M in the assay tubes (0.079 mμg./ml.). The response of *Pediococcus cerevisiae* to folinic acid will be only slightly less than that of *Lactobacillus casei* to PGA.

If the *L. casei* response falls significantly below that level (as it may on occasion), start with another lyophilized culture or obtain a fresh (or lyophilized) culture from the American Type Culture Collection.

IV. Agar Diffusion Methods

A. General

Because of the nature of lactic acid bacteria (see Chapter 8) agar diffusion methods are not ordinarily employed for the assay of vitamins.

However, they can be used in conjunction with paper chromatographic methods. This process has been termed Bioautography.[103]

Evaluation of the number and types of growth factors present is particularly important when one is dealing with related compounds such as the folic acid group. This is true particularly since the various known substances of this group do not demonstrate the same quantitative stimulation for one or more of the test organisms (see Sections I.B, C, and D and Tables III and IV). When assaying any crude natural material one can expect to find more than one member of the folic group present. Bioautography and information such as that contained in Tables III and IV can, therefore, assist in the evaluation and interpretation of the quantitative results of tube assays.

B. The Bioautographic Technique

This technique combines the use of paper chromatography with the response of microorganisms to growth or inhibitory factors. The procedure was utilized by Winsten and Eigen[103] to separate and identify members of the vitamin B_6 group, i.e., pyridoxine, pyrodoxal, and pyridoxamine. Later, studies were conducted on Vitamin B_{12},[104,105] leucovorin,[46] and deoxyribonucleosides.[105] The procedure depends on the separation on a paper chromatogram of a mixture of different forms of a growth factor. After separation has been achieved, the positions of the different forms on the paper strip are revealed by the use of a microbial indicator, that is, a microorganism capable of utilizing at least one, and, preferably several forms of the factor for growth. The positions of the various compounds on the chromatogram are compared with knowns, when available, and may, thereby, be identified.

The location of the growth factors is demonstrated by placing the dried chromatogram on an agar plate seeded with an organism which will respond to the growth factor(s). The nutrient agar contains all the factors necessary for the growth of the organism with the exception of the growth factor under investigation. In making the agar plate, pour a bottom layer of the nutrient agar, for example 300 ml. of the agar medium (unseeded), on a glass plate or dish 11 × 18 inches and allow to harden. Cool a 200 ml. portion of the nutrient agar to 48° to 50°C. and seed with a suspension of the test organism in 10 ml. of sterile physiological saline. Pour the seeded agar on the hardened underlayer and allow to cool. Place the paper chromatogram on the surface of the moist agar, cover the plate and place in an incubator until adequate growth is obtained. After the incubation period remove the paper strip and zones of growth will be observed along

[103] W. A. Winsten and E. Eigen, *Proc. Soc. Exptl. Biol. Med.* **67**, 513 (1948).
[104] W. A. Winsten and E. Eigen, *J. Biol. Chem.* **177**, 989 (1949).
[105] W. A. Winsten and E. Eigen, *J. Biol. Chem.* **181**, 109 (1949).

the locus of the chromatogram indicating the presence of the growth factors.

This technique has been applied to the study of the folic acid group, its analogs and antagonists by various workers.[102,106,107,108] These workers found, in addition, that the inclusion of a tetrazolium compound in the agar medium increases the sensitivity of the method due to the formation of highly colored formazans by the bacteria. Usdin et al.[102] studied this system and refined it to a point where they could detect 10^{-6} µg. of folic acid. Since this method is particularly applicable to the folic acid group, it will be given here in detail.

C. Bioautography with Tetrazolium Salts

1. Medium

The medium employed is suitable for the growth response of *Streptococcus faecalis* (ATCC 8043), *Lactobacillus casei* (ATCC 7469), and *Pediococcus cerevisiae* (ATCC 8081). The compositions of suitable double strength basal media are given in Table VIII or X. Dilute the basal medium with an equal volume of water, add agar to a concentration of 1.75%, cover, and sterilize at 121°C. for 2.5 to 3 minutes. The agar should not contain significant amounts of folic acid compounds (Difco-Bacto agar is suitable). Separately place a 1.75% aqueous suspension of agar in a suitable size Erlenmeyer flask, cover, and autoclave for 2.5 to 3 minutes at 121°C. The flask containing the aqueous agar suspension and the flask containing the agar growth medium should contain a magnetic stirring bar and a thermometer capable of measuring temperatures above 121.5°C. Remove the flasks from the autoclave and stir the suspensions magnetically to mix and dissolve the agar and then place the flask in a 47°C. water bath. Maintenance of this temperature is important because *Lactobacillus casei* will be killed at a higher temperature and the agar will solidify at a lower temperature. In addition, at significantly higher temperatures the tetrazolium salts may be reduced. When temperature equilibrium has been reached, as indicated by the thermometer in the agar, add 6 ml. of a 2% aqueous solution of triphenyltetrazolium chloride (TTC) per 100 ml. to the aqueous agar suspension only. Thus the concentration of TTC in the top layer will be 0.12%. This is then mixed with magnetic stirring. It should be noted that the TTC is added only to the aqueous agar suspension and not to the agar medium.

---

[106] J. E. Ford and E. S. Holdsworth, *Biochem. J.* **53**, xxii (1953).

[107] C. A. Nichol, S. F. Zakrzewski, and A. D. Welch, *Proc. Soc. Exptl. Biol. Med.* **83**, 272 (1953).

[108] E. Usdin, P. M. Phillips, and G. Toennies, *J. Biol. Chem.* **221**, 865 (1956).

2. Inoculum

An inoculum with a high concentration of actively growing (log phase) bacteria favors more rapid and sharper responses than does an inoculum of the customarily used overnight grown bacteria. Methods that can be employed to grow such an inoculum have been described in Section IX.G of Chapter 8.

The inoculum cultures may be grown in centrifuge tubes or bottles, so that washing and resuspension can be performed conveniently. Grow the inoculum for each 100 ml. of agar medium in 80 ml. of assay medium to a density representing 80–100 μg. of bacterial dry weight per milliliter. This is equivalent to about 2 to 2.5×10^6 bacteria per milliliter. An initial inoculum of 10^8 cells per milliliter will yield the above required bacterial population in 3.5 to 6 hours depending on the test organism. Chill the cells and wash aseptically twice in the cold with sterile distilled water, suspend in 5 ml. of sterile distilled water, and add with magnetic stirring to the agar medium (47°C. $\pm 1°$). For each 100 ml. of agar medium, this gives an inoculum concentration of 64 to 80 μg./ml. (approximately 1.6 to 2×10^8 bacteria per milliliter). Background growth and color can be minimized further by growing the inoculum in the presence of relatively low amounts of folacin or derivatives (e.g., 0.12 mμg./ml.) (see also Section III.F).

3. Preparation of the Bioautographs

Pour a layer approximately 3 mm. thick of the seeded agar medium into an appropriate dish. This may consist of a large petri dish, or a shallow glass baking dish of a suitable size. For larger chromatograms, a dish can easily be constructed of plate glass surrounded by ordinary 1-inch aluminum or copper molding fastened to the plate glass with plastic, plastic coated cloth, or paper masking tape. When plastic masking tape is used it need be replaced only after several usages. Since a large inoculum is used the plates do not have to be sterilized, ordinary cleanliness will be sufficient to prevent gross contamination.

After the layer of seeded agar medium solidifies, place the paper or paper chromatograms containing folacin and derivatives on it. When the paper is completely wetted by contact with the agar medium, enough of the aqueous agar TTC solution is poured on top to form a layer 6–7 mm. thick. When the top layer has solidified, place the plates in a 37°C. incubator. Zones of growth colored red by the reduced TTC formazan are detectable after about 9 hours incubation and are fully developed after 15 hours.

4. Discussion

As mentioned above the method is extremely sensitive for the detection of the folic acid group of compounds. Traces of contamination of folic acid substances in pure materials and in supposedly folic acid-free materials can be detected.[102] The method is applicable to the study of folic acid antagonists and their detection in natural products, but to our knowledge has not as yet been applied. A few notes of caution. TTC will be reduced by a number of reducing agents such as the cysteine in the growth medium. Oxygen from the air will inhibit or prevent the formation of the red formazan. These effects are minimized by the use of the TTC in the top layer. Relatively long exposure to light or sun will cause the entire plate to turn red. Other tetrazolium salts can be used. However, TTC appears to be superior to "blue tetrazolium" and "neotetrazolium" in regard to intensity of color, degree of toxicity, and suitability for the photographic recording of results.

The use of all three of the test organisms mentioned above is recommended if differentiation between the various members of this group is desired.

5. Application

Usdin[87] applied this procedure to the study of folic acid-active materials in blood. The chromatograms were run on Schleicher and Schuell No. 589 green ribbon paper. Five solvent systems were used as follows:

Solvent A: 5% citric acid + NH_4OH to pH 9.0, saturated with isoamyl alcohol.

Solvent B: n-Butanol-water-ethanol-acetic acid (52:28:20:0.3, v./v.).

Solvent C: 5% Na_2HPO_4 saturated with benzyl alcohol.

Solvent D: Water-saturated sec-butanol + 3% acetic acid.

Solvent E: 15% $Na_2HPO_4 \cdot 12H_2O$.

Table XI shows the behavior of 13 compounds in the 5 solvent systems and the organisms used for bioautography.

This method can be used for the demonstration of antifolic compounds by including a small amount of folic acid in the medium. Zones of inhibition will indicate the position of these compounds. Under the proper conditions both growth stimulating and growth inhibiting compounds can be detected.

D. Pad-Plate Method

A technique related to bioautography may be used for the quantitative determination of folic acid. This is the pad-plate method adapted from assays for antibiotics.[96] Plates containing folic acid-free agar-containing medium seeded with the test organism are prepared as described pre-

TABLE XI

R_F VALUES FOR FOLIC ACID AND RELATED COMPOUNDS[a]

Compound	Microbiological activity			R_F in solvent system[b]				
	L. casei	S. faecalis	P. cerevisiae	A	B	C	D	E
N[10]-formylpteroate	−	+	−	0.48	0.54		0.44	0.61
N[10]-formylfolate	+	+	−	0.66	0.26	0.83	0.28	0.73
N[5]-formyl-H$_4$-folate	+	+	+	0.65	0.34	0.78	0.36	0.71
Folate	+	+	−	0.17	0.17	0.52	0.13	0.18
Pteroate	−	+	−	0.05			0.08	0.01
Folyl-γ-glutamate	+	+	−	0.31			0.12	
Folyl-di-γ-glutamate	+	−	−	0.41	0.03	0.80	0.10	0.48
N[5]-formyl-H$_4$-pteroate	−	+	−	(0.39)				
5,10-CH$_2$-H$_4$-folate	+	+	+	0.65	0.25	0.84		
N[5]-formyl-H$_4$-folylglutamate	+	−	+	(0.82)				
N[5]-formyl-H$_4$-folyldiglutamate	+	+	−	(0.93)				
Thymine	+	+	−	0.70	0.58	0.75		
Thymidine	+	+	−	0.79	0.59	0.83		

[a] See ref. 87.
[b] Numbers in parentheses are predicted R_F values.

viously. Filter paper disks (obtained from Schleicher and Schuell, Keene, New Hampshire, catalog no. 740-E, 6.5 mm.) are dipped in standard solutions having concentrations of 0.6, 3, and 15 μg. of folic acid per milliliter. Three disks are used for each concentration and are placed on the agar plate approximately 30 mm. apart. Similarly, nine disks are prepared from three dilutions of each sample being tested. Folic acid diffuses from the disks into the surrounding medium resulting in the formation of circular zones of growth after incubation for 16 to 24 hours at 37°C. The diameters of these zones are proportional to the logarithm of folic acid concentration of the samples. The diameters of the standards are measured and plotted on semilogarithmic paper against the concentrations of folic acid. In the paper cited above,[96] the average diameters for the three levels of standard were 19.5, 22.6, and 25.88 mm., respectively, which gave a straight dose-response line. The folic acid content of the unknown solutions is determined by interpolation from the standard curve. Although this method is not in general use, it may be of value where many samples are to be screened rapidly for their folic acid content.

V. Chemical Methods for Folic Acid

The quantitative analysis of folic acid may be achieved by use of a modification of the method of Bratton and Marshall.[109,110] Folacin is reduced at an acid pH with zinc dust or zinc amalgam in the presence of gelatin to release p-aminobenzoylglutamic acid. Cleavage may also be accomplished with potassium permanganate.[17] The aromatic amine is diazotized with sodium nitrite and coupled with N-(1-naphthyl)-ethylenediamine after destroying excess nitrite with ammonium sulfamate. The color formed is measured at 550 mμ in a suitable spectrophotometer or photoelectric colorimeter. The color intensity, incidentally, does not follow Beer's law exactly. The method is not sensitive, requiring 10–40 μg. of folacin for construction of the standard curve. It is nonspecific since any aromatic amine will develop a color in this procedure. Thus, the method is useless for the determination of the folic acid family in natural materials.

Another chemical method makes use of permanganate oxidation of pteroylglutamic acid (PGA) to 2-amino-4-hydroxypteridine-6-carboxylic acid, a strongly fluorescent compound which can be measured in a suitable fluorometer.[111] The fluorescence peak is at 470 mμ and is directly proportional to the PGA content over a range of at least 0.01 to 10 μg./ml. This

[109] A. C. Bratton and E. K. Marshall, *J. Biol. Chem.* **128,** 537 (1939).

[110] B. L. Hutchings, E. L. R. Stokstad, J. H. Booth, J. H. Mowat, C. W. Waller, R. B. Angier, J. Semb, and Y. SubbaRow, *J. Biol. Chem.* **168,** 705 (1947).

[111] V. Allfrey, L. J. Tepley, C. Geffen, and C. G. King, *J. Biol. Chem.* **178,** 465 (1949).

is still not sensitive enough for determinations on most natural products. In addition, interfering fluorescent related compounds (e.g., xanthopterin, isoxanthopterin, and pteroic acid) must be removed prior to assay.

The oxidation product may be purified and isolated by chromatography on Florisil. High concentrations of tyrosine or tryptophan give erroneously high results. Assays conducted on lettuce, carrots, potato, and beef muscle gave results approximately 30% higher than microbiological assay with *Lactobacillus casei*. The method, therefore, is of little use for assay of most natural products for folacin content.

VI. Critical Evaluation of the Microbiological Assay of Folic Acid and Related Compounds

The assays described above are subject to the same pitfalls and precautions as the other microbiological assays described in other sections of this volume. There are complications, however, which are peculiar only to the assay of the folic acid family of compounds. At best each assay is only accurate when the factor occurs free. Unfortunately, folic acid-active compounds occur in both free and conjugated forms in natural materials. These various forms have different degrees of activity for the various test organisms ranging from zero to 100%. Table III illustrates this point by showing, for example, that pteroylheptaglutamic acid is active for *Lactobacillus casei*, *Streptococcus faecalis*, the chick and rat and only slightly active for man. On the other hand, pteroylglutamic acid and pteroyltriglutamic acid are active for all the organisms listed except *Pediococcus cerevisiae*. The chick, rat, and *P. cerevisiae* do not respond to 10-formylpteroic acid which is slightly active for *Lactobacillus casei* but very active for *Streptococcus faecalis*. As research continues in this field more varieties of the folic acid group may be found to complicate the assay situation still further.

For "folic acid activity" today we have little knowledge of what is actually being assayed. For instance how can data be judged or compared when the possibility exists that only a fraction of the potentially active material is being detected? Correlations with other types of information might be useful but are certainly far from conclusive. A vast number of infectious agents have been "discovered" by correlation with their occurrence during a clinical syndrome only later to find an equal or greater correlation with "normalcy." We are certain that the reader is aware of similar analogies in areas with which he is most familiar.

In this connection we would like to quote two sentences from a recent review by Hutner *et al.*[111a] "The bacterium used, *Lactobacillus casei*, had

---

[111a] S. H. Hutner, L. Provasoli, and H. Baker, *Microchem. J. Symposia Ser.* **1**, 95 (1961).

long been used in folic research but gave erratic clinical correlations when used to assay folic acid in whole blood. This inaccuracy disappeared when serum was used rather than whole blood—evidently the "formed elements" (red and white cells) of blood are rich in compounds (probably purines and pyrimidines) bypassing the folic requirement for the assay organism and so masking folic deficiencies."

First, the cellular elements of blood, particularly the erythrocytes, have been known to contain derivatives of the folic acid group for a number of years.[84,108] Some of the supporting data are given in Section IV.C. Second, it has also been known that human blood plasma contains a conjugase type of enzyme that has been called "plasma factor." [84,109] Third, Baker et al.[89] report values of 7.5 to 24 mμg./ml. of "PGA" (our quotes) in the serum of normal human subjects while Toennies et al.[85a] report 25 to 330 mμg./ml. (approximately 12 to 150 mμg./ml. calculated as PGA) of folic acid activity in a group of 100 normal human subjects. Without commenting on the number of subjects examined, it is clear that "the disappearance of an inaccuracy" as far as a clinical correlation is concerned which is based on what may be less than one sixth of the total folic acid activity that is present in the circulatory fluid of the body may be merely fortuitous. In fact, the total folic acid activity of blood may well exceed the "totals" that have thus far been obtained with L. casei assays.

The example mentioned also serves to emphasize the recommendations made in Section I.B that it may be wiser to relate activity to the whole group of substances and again preferably on a molar basis.

How should one then go about selecting and/or discarding one or more of the methods that are given in this chapter either in detail or in outline form or, for that matter, one or more of the many other variations that are available? To a great extent this is left up to the individual and to the purpose of the determinaions. Therefore, the following is given to merely serve as a guide.

"Total" folic acid. Total is, of course, an abstract figure. The methods of choice are those that employ L. casei as test organism. However, it must be remembered that L. casei "misses" certain members of the folic acid group even after treatment with a conjugase (see Table IV). The method using L. casei that is given in Section III.F is recommended as a first approach. This recommendation is based not only on our familiarity with the method but also because the same medium can be used with equal or greater facility for assays with *Streptococcus faecalis* and *Pediococcus cerevisiae*. We would like to qualify our recommendation with the following suggestions. Bioautographic methods (Sections IV.B and C) or assays with a test organism other than *Lactobacillus casei* should also be used at least in a preliminary fashion to obtain some

information as to what kind of folic acid group compounds may be present in the samples that are being assayed.

Pteroylglutamic acid only. For materials that are known to contain only free (unconjugated) PGA (e.g., purified fractions, vitamin preparations, known mixtures), the well-proven A.O.A.C. method using *Streptococcus faecalis* is recommended (Section III.A). Again, conservatism would indicate that some certainty should exist that only PGA of the folic acid group is present in the sample. Bioautography or the response of one of the other test organisms would serve as a check. While the A.O.A.C. folic acid method is somewhat cumbersome to follow and use, it should be almost foolproof for its stated purpose. An alternate to the A.O.A.C. procedure is that given in Section III.F, using *S. faecalis* as test organism. Suitability for use with the other two test organisms is again an advantage of this method.

Folinic acid only. The reasons given in the previous paragraph are equally valid for the assay of folinic acid by the U.S.P. method (Section III.E). A reminder that more than one member of the folic acid group (see Table IV) will support the growth of *Pediococcus cerevisiae* is certainly in order. Second choice falls to the method in Section III.F, using *P. cerevisiae* as test organism. This method would be particularly recommended when the content of more than one member of the folic acid group is of interest.

The two other assay methods that are described in Section III (III.C, using *Tetrahymena geleii* and III.D, using *Bacillus coagulans*) are not recommended for routine use at the present time. Both of these methods may have potential advantages over the more widely used ones, but further investigations are required in order to be able to judge the comparative adequacies and deficiencies of either of these two methods.

The use of conjugases, as thus far described, is very far from completely satisfactory. For the assay of some types of materials they must certainly be employed—but their limitations must also be kept in mind.

The assay of natural products is inadequate today because of the occurrence of a multiplicity of derivatives of PGA and their presence as conjugated forms. The variable response of the organisms to the derivatives and conjugates known thus far makes the development of an adequate method of liberation of the free compounds of utmost importance. The conjugase preparations known today are inadequate for this task. In a collaborative study conducted by the Association of Official Agricultural Chemists (A.O.A.C.) the folic acid content of soy flour determined with *Lactobacillus casei* and *Streptococcus faecalis* was close to the value determined by chick assay. The assay of turnip greens and mustard greens, on the other hand, did not agree with the chick assay even after treatment with chick pancreas conjugase.[112] On the basis of this and other studies the A.O.A.C. recommended that their method, using *S. faecalis,* be used only to determine *free* folic acid and that a study be made to find a suitable method for the assay of the bound forms.[113] There is no doubt that there is room for improvement in this area. One suggestion is the possibility of the use of more than one purified conjugase preparation, either together or in series. The enzymes are apparently quite different and combined use may lead us a step closer to the determination of "total" folic acid activity.

An additional need is the availability of pure substances of some of the lesser known members of the group for use as standards.

New approaches to the problem are being investigated. As an example, Hutner *et al.* feel that an assay for *p*-aminobenzoic acid might satisfy the need for an assay of total folic acid activity of samples containing the various conjugated and unconjugated forms of the vitamin.[43,114]

[112] L. M. Flynn, *J. Assoc. Offic. Agr. Chemists* **33**, 633 (1950).

[113] H. W. Loy, Jr. *J. Assoc. Offic. Agr. Chemists* **41**, 591 (1958).

[114] S. H. Hutner, H. A. Nathan, and H. Baker, *Vitamins and Hormones* **17**, 1 (1959).

7.6 Agar Plate Assays for Pantothenic Acid, Inositol, and Pyridoxine

L. J. DENNIN

Eli Lilly and Company, Indianapolis, Indiana

I. Introduction

To our knowledge, the use of agar plate methods for vitamin assay was first reported in 1947 by Bacharach and Cuthbertson[1] and, shortly there-

[1] A. L. Bacharach and W. F. J. Cuthbertson, *Analyst* **73,** 334 (1948).

after by Genghof *et al.*[2] The former workers investigated the use of *Lactobacillus fermentum*, *L. casei*, and *L. arabinosus* in procedures for thiamine, riboflavin, nicotinic acid, and biotin. The latter workers reported on the use of *Saccharomyces cerevisiae* and *Lactobacillus arabinosus* in plate methods for biotin. This work was followed by that of Jones and Morris[3] in which assay with *L. fermentum* was reportedly improved by alteration of the basal medium and by preliminary incubation of the test plates at 37°C. for 1.5 hours. It was reported that the latter technique accomplished significant improvement in zone definition, which was attributed to drying of the agar. In 1949, Jones and Morris[4] applied the agar plate method to the assay of the vitamin B₆ complex. They utilized the yeast, *Saccharomyces carlsbergensis* (Fleischmann strain 4228), in a medium which was a slight modification of that described by Atkin *et al.*[5] for the yeast turbidimetric assay of vitamin B₆. The principal changes made in the assay medium were the addition of nicotinic acid, tryptophan, and agar. The addition of tryptophan reportedly resulted in uniform response of the yeast to pyridoxine, pyridoxal, and pyridoxamine, and also improved the definition of the enhancement zones. These investigators reported that in the absence of tryptophan the organism response was comparable for pyridoxine and pyridoxal, but was of a different magnitude for pyridoxamine. Jones,[6] and Morris and Jones,[7] followed their early work with additional plate applications and subsequently described a yeast-plate method for inositol, and bacterial-plate methods for biotin, nicotinic acid, and pantothenic acid. Furthermore, they recognized the possibility of utilizing the yeast-plate method for pantothenic acid, but apparently felt that its relative insensitivity precluded its practical application.

The following methods represent adaptations by Baker *et al.*[8] of the yeast-plate techniques for the assay of pyridoxine, pantothenic acid, and inositol. Modifications of the original methods were made based upon our experiences and applications. The primary purpose of the assay has been the control of pharmaceutical preparations, and we have made no attempt to apply these methods to a wide variety of sample types. Acceptance of the methods in our laboratory was made on the basis of results comparable with those obtained using generally accepted techniques circa 1954. It is

[2] D. S. Genghof, C. W. H. Partridge, and F. H. Carpenter, *Arch. Biochem.* **17,** 413 (1948).

[3] A. Jones and S. Morris, *Analyst* **74,** 333 (1949).

[4] A. Jones and S. Morris, *Analyst* **75,** 613 (1950).

[5] L. Atkin, A. S. Schultz, W. L. Williams, and C. N. Frey, *Ind. Eng. Chem. Anal. Ed.* **15,** 141 (1943).

[6] A. Jones, *Analyst* **76,** 588 (1951).

[7] S. Morris and A. Jones, *Analyst* **78,** 15 (1953).

[8] B. W. Baker, H. B. Shafer, and J. T. Stephenson, personal communication.

recommended that others contemplating use of these methods make similar comparisons across their sample spectrum before adopting them.

The three assay systems are so nearly identical that they will be presented as a single method. Variations required for the different procedures will be indicated under the appropriate topics within the body of the general method.

II. Test Organism

Maintain stock cultures of *S. carlsbergensis* (ATCC 9080) by weekly transfers to fresh sterile slants of the following composition: dextrose 40 gm., peptone 10 gm., agar 20 gm., distilled water to make 1000 ml. Incubate the freshly prepared slants at 30°C. for 18 to 24 hours and store at 4° to 6°C. until ready for use in the preparation of the inoculum suspension.

III. Standard Solutions

Dry the appropriate working standard (*d*-calcium pantothenate, inositol, or pyridoxine hydrochloride) for 3 hours at 60°C. and a residual pressure of 5 mm. Hg or less in a tared weighing bottle fitted with a capillary tube. Weigh a quantity of the dried standard sufficient to make a stock solution containing 100 μg. (activity) per milliliter (if *d*-calcium pantothenate or pyridoxine hydrochloride) or 1000 μg. (activity) per milliliter (if inositol). Transfer the weighed portion to a suitable container and bring to volume with sterile distilled water. Store the solutions at 4° to 6°C. and use for a period not to exceed 1 month.

IV. Sample Preparation

The exploration of all extraction procedures required for release of these vitamins from the numerous sample types which may confront the assayist is far beyond the scope of both this discussion and my experience. The methods outlined below have been found satisfactory for a reasonably wide range of pharmaceutical preparations; no other claims are made for them. The rather extensive literature in the field of vitamin assays should be consulted for further guides in this area.

A. Calcium Pantothenate, Pantothenic Acid

Pharmaceutical preparations containing this vitamin have been successfully assayed using the following three basic extraction procedures:

(a) simple aqueous dilution, (b) aqueous extraction at 121°C. for 30 minutes, and (c) enzymatic digestion of the sample using Mylase P (Wallerstein) for 3 hours at 50°C. in a sodium acetate buffer at pH 4.5. If the latter method is used, readjust the pH of the treated sample to that of the stock standard solution prior to assay. Make subsequent dilutions with distilled water so that the final dilution contains approximately 1.0 μg. of d-calcium pantothenate activity per milliliter. If the vitamin is present as pantothenic acid, or if it is desirable to express the results in terms of the acid rather than the calcium salt, multiply the results in terms of the calcium salt by 0.916.

B. Pyridoxine, Pyridoximers

Pharmaceutical preparations containing this vitamin have been successfully assayed using the following three basic extraction procedures: (a) simple aqueous dilution, (b) aqueous extraction at 121°C. for 60 minutes, and (c) acid extraction in $1 N$ hydrochloric acid at 121°C. for 60 minutes. If the latter method is used, readjust the pH of the treated sample to that of the stock standard solution prior to assay. Make subsequent dilutions with distilled water so that the final dilution will contain approximately 0.2 μg. of pyridoxine activity per milliliter.

C. Inositol

Pharmaceutical preparations containing this vitamin may be prepared for assay by one of the following three basic extraction procedures: (a) simple aqueous dilution, (b) aqueous extraction at 121°C. for 60 minutes, and (c) acid extraction in 3% sulfuric acid at 121°C. for 120 minutes. If the latter method is used, readjust the pH of the treated sample to that of the stock standard solution prior to assay. Make subsequent dilutions with distilled water so that the final dilution will contain approximately 50 μg. of inositol activity per milliliter.

V. Mechanics of the Assay

A. Design

Refer to Penicillin, Chapter 6.10, Section III.E.1.

B. Media

The composition of the various basal media for conducting these assays is shown in Tables I and II. For the sake of convenience, stock solution of the various salts and vitamins may be made and stored under refrigeration until needed. Our experience has shown that these solutions are stable for at least 1 month. Experience has also indicated that the

TABLE I

BASAL MEDIUM

Dextrose	100 gm.
Potassium phosphate, monobasic	1.1 gm.
Potassium chloride	0.85 gm.
Calcium chloride	0.25 gm.
Magnesium sulfate, heptahydrate	0.25 gm.
Ferric chloride	0.50 mg.
Manganese sulfate, monohydrate	0.50 mg.
Potassium citrate	10.0 gm.
Citric acid	2.0 gm.
Acid hydrolyzed casein	10.0 gm.
Thiamine hydrochloride	0.50 mg.
Biotin	0.016 mg.
Nicotinic acid	5.0 mg.
Ammonium phosphate, dibasic	2.0 gm.
L-Asparagine	1.5 gm.
DL-Tryptophan	0.1 gm.
Agar	20.0 gm.
Distilled water to make	1000 ml.
Final pH	5.2 ±0.1

TABLE II

MEDIA FOR SPECIFIC ASSAYS

	Pantothenic acid assay	Pyridoxine assay	Inositol assay
Inositol	50 mg.	50 mg.	—
Pyridoxine hydrochloride	5 mg.	—	5 mg.
d-Calcium pantothenate	—	5 mg.	5 mg.
Basal medium (Table I)	1000 ml.	1000 ml.	1000 ml.

medium itself gives best results if not more than 3 days old, and then only if it is stored under constant refrigeration.

Dissolve the ingredients in the distilled water with the aid of heat. Dispense into appropriate containers (500 ml. in a 1500-ml. flask), plug with gauze-covered nonabsorbent cotton, and immediately sterilize by steaming (Arnold sterilizer) for 30 to 40 minutes. Refrigerate until just prior to use.

All three of the above assays use a single 10-ml. layer of seeded agar. Apply the seeded agar to the plates and allow to harden. *Cover the plates and place in a 30°C. incubator for 2 to 3 hours before applying the test solutions!* In this investigator's experience this drying or preincubation period significantly improves the character and definition of the growth zones. This experience appears to be consistent with the observations of

Jones and Morris[3] in the *Lactobacillus fermentum* thiamine assays. It may well be that the mechanism here is more complex than mere drying of the agar.

C. Inoculum

On the day prior to assay, make a transfer from one of the stock culture slants to a flask containing approximately 100 ml. of sterile medium of the following composition: Dextrose 20 gm., yeast extract 1.25 gm., potassium phosphate, monobasic 0.5 gm., and distilled water to make 1000 ml. Incubate the culture at 28°C. on a mechanical shaker for 18 to 20 hours. Centrifuge, decant, and wash the cells twice with 100 ml. of sterile distilled water. Resuspend the washed cells in 100 ml. of sterile distilled water.

Inoculate each 100 ml. of previously melted and cooled (48°C.) assay medium with 3.0 ml. of the inoculum suspension prepared as directed above.

D. Standards

Dilute aliquots of the stock standard solutions in distilled water immediately prior to use on the day of the assay. The choice of standard levels will, of course, determine to what extent these aliquots are diluted. The pantothenic acid assay gives a linear semilog response in the range 0.5–5.0 μg./ml.; the pyridoxine assay in the range 0.10–1.0 μg./ml.; and the inositol assay in the range 25–100 μg./ml.

E. Samples

As previously indicated (Section IV), dilute all samples to the assay range with distilled water. Exercise care to make certain that the pH of all final sample dilutions is the same as the pH of the standard solutions.

F. Incubation

Incubate the plates at 30°C. for 16 to 18 hours. Refer also to Penicillin, Section III.E.6.

VI. Measuring the Response

Refer to Penicillin, Section III.F.

The essential difference between these assays and those for antibiotics is that the resultant zones are areas of growth rather than areas of inhibition of growth.

VII. Computation of Answers

A. Standard Curve

Refer to Penicillin, Section III.G.1.

B. Potency of Sample

Refer to Penicillin, Section III.G.2.

C. Validity of Assay

Refer to Penicillin, Section III.G.3.

7.7 Pantothenic Acid and Related Compounds

ORSON D. BIRD

Parke-Davis Research Laboratories, Ann Arbor, Michigan

I. Introduction

A. Scope of Procedures to Be Discussed

A variety of microbiological methods have been developed for assaying pantothenic acid and pantethine, the vitamin moieties derived from coenzyme A (Co A) present in natural products, and panthenol, a related compound with pantothenate activity used widely in pharmaceutical products. These methods will be discussed broadly and enzymatic procedures outlined for releasing Co A components from the complexes in which they occur in natural products. In addition, preferred procedures, both for releasing free forms from complexes and for assaying these forms, which have been found highly satisfactory in the writer's laboratory, will be described in some detail.

B. Nature of Pantothenate Complexes in Natural Products

The first definitive microbiological assay for pantothenic acid, an assay which made possible its concentration and establishment as a vitamin,[1] employed the Gebrüder Mayer strain of *Saccharomyces cerevisiae* growing in a synthetic medium. The assay lacked specificity, as witnessed by the fact that after long incubation all assay tubes, including the blanks, grew out. This was because this species of yeast actually produced its own supply of pantothenate, but at a rate so slow that during a short incubation period a growth stimulus was produced by the vitamin. Later, a highly specific assay for pantothenic acid was developed using the yeast *S. carlsbergensis*.[2] Under the conditions imposed by the medium used in the assay this yeast species has an absolute requirement for pantothenate, and is not simply stimulated by it. Meanwhile, the emphasis had been placed on lactobacilli as assay organisms for pantothenic acid, largely because they were more specific in their response and less care was required in conducting assays with them. Snell has reviewed the development of the lactobacilli assays.[3] *Lactobacillus casei* was the first among this group to be used as an assay organism for pantothenic acid, but due to its complex nutritional requirements came to be replaced by *Lactobacillus plantarum* (old name, *L. arabinosus*) as the organism of choice.

Experience in assaying mammalian tissues before and after autolysis convinced early workers in the field that pantothenic acid occurred in these tissues in bound forms unavailable to the microorganisms used for assay.[4] Treating natural products with commercial enzyme preparations possessing amylolytic and phosphorolytic activity (takadiastase, Mylase P, etc.) prior to assay also increased the potency obtained[5]; but it re-

[1] R. J. Williams, C. M. Lyman, G. H. Goodyear, J. H. Truesdail, and D. Holaday, *J. Am. Chem. Soc.* **55**, 2912 (1933).

[2] L. Atkin, W. L. Williams, A. S. Schultz, and C. N. Frey, *Ind. Eng. Chem. Anal. Ed.* **16**, 67 (1944).

[3] E. E. Snell, in "Vitamin Methods" (P. György, ed.), Vol. 1, p. 327. Academic Press, New York, 1950.

[4] E. Rohrman, G. E. Burget, and R. J. Williams, *Proc. Soc. Exptl. Biol. Med.* **32**, 473 (1934).

[5] H. H. Buskirk and R. A. Delor, *J. Biol. Chem.* **145**, 707 (1942).

mained for the development of a chick assay for pantothenic acid [6] to demonstrate the true discrepancy between microbiological assay values and the total content of this vitamin in tissues and other natural products. As an example of the spread in potencies indicated by these two methods, a sample of dry Brewer's yeast was assayed microbiologically in the writer's laboratory, without enzyme treatment, and gave a value of 25 μg. pantothenic acid per gram. The chick assay for this sample indicated 204 μg. pantothenic acid per gram.

The reason for the discrepancy between microbiological and chick assays for pantothenic acid was largely explained when Lipmann and co-workers[7] demonstrated that Co A, which had been recognized for some time and was known to function in acetylation reactions, contained bound pantothenate which could be released effectively by the combined actions of a liver enzyme and alkaline phosphatase.[8] This observation was used by Novelli and Schmetz[9] as a basis for a quantitative procedure for releasing the total pantothenate combined in the tissues as Co A. They concluded that the pantothenate found by microbiological assay without enzyme treatment plus that occurring as Co A, and hydrolyzed by these two enzymes, represented the total pantothenate present in natural products. This method of releasing bound pantothenate has recently been scaled down so that the pantothenate content of microgram quantities of tissues, actually microtome slices, can be determined.[10] However, the concept that all pantothenate occurs either as the uncombined vitamin or combined as Co A is not strictly true since other complexes derived from Co A, as well as phosphorylated forms of both pantothenate and pantethine, have been demonstrated to occur widely.[11,12] In the writer's laboratory as high as 80% of the pantothenate activity of *Escherichia coli* cells grown under certain conditions was shown to be present as pantothenyl phosphate. Brown[12] has devised sequential treatments with liver enzyme, alkaline phosphatase, and acid phosphatase followed by differential assays with *Lactobacillus helveticus* and *Saccharomyces carlsbergensis*, which account quite well for all the known forms of pantothenate found in natural products.

[6] T. H. Jukes, *Biol. Symposia* **12**, 253 (1947).

[7] F. Lipmann, N. O. Kaplan, G. D. Novelli, L. C. Tuttle, and B. M. Guirard, *J. Biol. Chem.* **167**, 869 (1947).

[8] G. D. Novelli, N. O. Kaplan, and F. Lipmann, *J. Biol. Chem.* **177**, 97 (1949).

[9] G. D. Novelli and F. J. Schmetz, *J. Biol. Chem.* **192**, 181 (1951).

[10] R. M. Twedt, H. C. Lichstein, and D. Glick, *Arch. Biochem. Biophys.* **85**, 374 (1959).

[11] E. E. Snell, G. M. Brown, V. J. Peters, J. A. Craig, E. L. Wittle, J. A. Moore, V. M. McGlohon, and O. D. Bird, *J. Am. Chem. Soc.* **72**, 5349 (1950).

[12] G. M. Brown, *J. Biol. Chem.* **234**, 379 (1959).

C. Compounds with Pantothenate Activity Used in Pharmaceutical Products

Chemically made calcium or sodium salts of D-pantothenic acid are widely used in pharmaceutical products, the calcium salt perhaps being used most widely in solid products due to its ease of handling, and the sodium salt being widely used in aqueous dispersions of combined oil and water soluble vitamins in which calcium ions need to be avoided in order to prevent precipitation of certain components. These products present no problem in the preparation of assay solutions, except to ensure that the respective pantothenate salt present is completely extracted into the water solution which is to be assayed.

Pantothenyl alcohol, referred to commercially as panthenol, has come to be widely used as a source of pantothenate activity for pharmaceutical vitamin products because it is more stable than the pantothenate salts, especially in liquid multivitamin products which must be slightly acid to preserve their thiamine content. Panthenol itself has no pantothenate activity; in fact, it is an inhibitor of many lactic acid bacteria. This inhibitory property of panthenol was reported early in the microbiological work on pantothenic acid [13] and has been made the basis of the microbiological assay for this compound [14] which has been recommended as the preferred method described below. Panthenol has been demonstrated to be quantitatively converted to pantothenic acid in the animal body,[15] and to be equivalent to pantothenic acid in humans.[16,17]

II. Methods of Treating Natural Products to Release Total Pantothenate Activity

A. Hot Water Extraction

The method for treating powdered dry yeast to determine its content of readily available pantothenic acid or pantethine will be described. This procedure would be applicable to the treatment of any dry natural product, after it had been ground to a fine powder.

Suspend 1 gm. finely powdered dry yeast in 90 ml. water in a 100-ml. centrifuge tube and autoclave 5 minutes after reaching 121°. Cool, centrifuge, and decant supernatant into a 100-ml. volumetric flask. Suspend residue in 10 ml. water, recentrifuge, and combine supernatant with extract. Dilute to mark and mix.

[13] E. E. Snell and W. J. Shive, *J. Biol. Chem.* **158**, 551 (1945).

[14] O. D. Bird and L. McCready, *Anal. Chem.* **30**, 2045 (1958).

[15] H. Pfaltz, *Z. Vitaminforsch.* **13**, 236 (1943).

[16] E. Burlet, *Z. Vitaminforsch.* **14**, 318 (1944).

[17] S. H. Rubin, J. M. Cooperman, M. E. Moore, J. Scheiner, *J. Nutrition* **35**, 499 (1948).

Assay for pantothenic acid either as in Section III.A.1, using *S. carlsbergensis,* or Section IV.A, using *Lactobacillus plantarum.* Assay for pantethine as in Section III.B.1, using *L. helveticus.*

B. Intestinal Phosphatase Treatment

Alkaline phosphatase treatment[8] removes the phosphate-containing moiety (triphosphoadenosine) from Co A, leaving pantethine, the fragment of Co A containing pantothenic acid and β mercaptoethylamine joined through a peptide linkage. Pantetheine, the reduced sulfhydryl form, and pantethine, the oxidized disulfide form, can both be assayed by the *L. helveticus* method described in Section III.B.1.

C. Liver Enzyme Treatment

An enzyme found in avian liver and first described by Novelli *et al.*[8] specifically hydrolyzes the peptide linkage in pantethine which joins the pantothenic acid and β mercaptoethylamine moieties. When pantethine, or pantetheine, is treated with this enzyme, pantothenic acid is liberated. However, if Co A is treated directly with this enzyme a highly phosphorylated conjugate of pantothenic acid described by Cheldelin and co-workers[18] results. For practical assay purposes combined treatment with alkaline phosphatase and liver enzyme is usually carried out simultaneously as described in Section II.D.

D. Combined Enzyme Treatment to Release Total Pantothenate Activity

1. Preferred Enzyme Treatment

The following combined enzyme treatment is based on the procedure of Novelli and Schmetz.[9] Each sample is treated in a test tube or flask containing the following:

1 ml. of intestinal phosphatase preparation containing 20 mg. alkaline phosphatase, purified (Pentex, Inc., Kankakee, Illinois), in water suspension.

2 ml. of Dowex-treated liver enzyme.

1 ml. of 0.2 M tris (hydroxymethyl)aminomethane buffer adjusted to pH 8.3 with NaOH, or 1 ml. of commercial standard buffer mixture containing boric acid-potassium chloride-sodium hydroxide (Hartman-Leddon Company, Philadelphia, Pennsylvania).

Dilute to 10 ml. with water an amount of sample containing 5–15 μg. bound pantothenic acid, add 2 drops toluene and incubate either 3 hours in a water bath at 37°, or overnight in an air incubator at 37°. Bring to

[18] T. E. King, L. M. Loche, and V. H. Cheldelin, *Arch. Biochem.* **17,** 483 (1948).

a boil, cool, and centrifuge. Run assay on supernatant according to Section III.A.1 or IV.A.

Dowex-treated liver enzyme is prepared as follows:

Weigh and mince chilled fresh chicken liver and transfer to a chilled Waring blendor along with 20 ml. chilled acetone for each gram of the liver. Homogenize 2 minutes and filter in a Büchner funnel with coarse paper. Wash residue in the funnel with chilled acetone and then with peroxide-free ethyl ether. Dry the powder in a vacuum desiccator over fresh phosphorus pentoxide. Remove connective tissue by passing the powder (should be pink) through a 40-mesh sieve. Rub the sieved powder into 10 volumes of ice-cold 0.02 M potassium bicarbonate solution to give a smooth suspension. Centrifuge at 1500g for 30 minutes in the cold and store the supernatant until needed at $-20°$. Wash Dowex-1 (X 10 cross linkage, 200–400 mesh) twice with 10-volume amounts of N HCl and decant after centrifugation. Wash the acid-treated resin 8–10 times with 10-volume amounts of water until the pH is about 5.0 and leave the resin in a slurry which just can be pipetted with a 10-ml. serological pipet. To 1 volume of acid-washed Dowex-1 add sufficient M tris buffer at pH 8.3 to bring pH to 8.0. Add 1 volume of ice-cold liver enzyme and stir the suspension for 5 minutes in an ice bath. Centrifuge the suspension at 4200g in the cold. Decant the supernatant and repeat the Dowex treatment. Centrifuge, decant the supernatant, and store at $-20°$.

2. Alternate Enzyme Treatment

This procedure was described by Schweigert and Guthneck[19] and was designed especially to deal with Co A concentrates and samples of organ meats. In place of the liver enzyme preparation described in Section II.D.1 is substituted the more simply made aqueous extract of hog kidney described by Bird et al.,[20] which is prepared as follows: Trim a fresh hog kidney of all visible fat, slice, and grind in a Waring blendor with the addition of 3 volumes of water. Centrifuge the resulting suspension at 4200g and filter the supernatant through Super-Cel. Treat the filtrate with Dowex-1 as described in Section II.D.1 and store frozen in small amounts in test tubes until used.

Moist tissue samples are homogenized with water in a Potter-Elvehjem glass homogenizer, followed by heating 3–5 minutes in a boiling water bath and rehomogenizing. Dry samples are finely ground and suspended in about 10 parts water.

Carry out enzyme treatment as follows: To a test tube add 0.5 ml. sample containing 1–5 μg. pantothenic acid equivalent, 0.2 ml. intestinal phosphatase (prepared as in Section II.D.1), 1.25 ml. Dowex-treated hog

[19] B. S. Schweigert and B. T. Guthneck, J. Nutrition 51, 283 (1953).
[20] O. D. Bird, M. Robbins, J. M. Vandenbelt, and J. J. Pfiffner, J. Biol. Chem. 163, 649 (1946).

kidney preparation, and 0.1 ml. 0.1 M NaHCO$_3$. Add water to 2.0 ml. (pH should be 7.1–7.3), then add a few drops of toluene and incubate at 37° in a water bath for 4 hours, or overnight in an air incubator. Dilute to 10 ml., bring to a boil, cool, and centrifuge. Run assay on supernatant.

3. Alternate Enzyme Treatment for Clinical Samples

Baker et al.[21] have shown that the pantothenic acid present in blood, urine, and spinal fluid requires different treatment to ensure full microbiological activity than do assay samples from other sources. Treatment with Clarase was the most effective way to release the pantothenic acid activity of blood and spinal fluid, being better than autolysis, acid hydrolysis, treatment with Mylase P, combinations of Clarase and papain, or liver and alkaline phosphatase. No activity was found without prior enzyme treatment. However, in urine no enzyme treatment tried gave a higher assay value than assaying the urine directly without any enzyme treatment.

a. *Pantothenate in blood.* Dilute citrated blood 1:10 with the following enzyme-buffer solution: dissolve 0.2 gm. Clarase (Fisher Scientific Company, New York) in 100 ml. citrate buffer (5 gm. potassium citrate monohydrate and 1 gm. citric acid in 1000 ml. distilled water, pH 5.6), and add preservative (1 part of chlorobenzene, 1 part of 1,2-dichloroethane and 2 parts of *n*-butyl chloride). Incubate for 3 days at 37°. Heat in autoclave at 121° for 15 minutes to stop enzymatic action and to coagulate the proteins. Filter and add 0.25, 0.5, and 1.0 ml. of the supernatant to individual tubes and assay. Include control tubes to estimate pantothenate contamination of the enzyme.

b. *Pantothenate in urine.* Dilute urine 1:20 with citrate buffer without enzyme and heat in autoclave at 121° for 10 minutes to coagulate any protein present. Filter and add 0.5, 1.5, and 2.5 ml. amounts to tubes for assay.

c. *Pantothenate in cerebrospinal fluid.* Treat cerebrospinal fluid the same as whole blood except that it is diluted 1:5 with enzyme-buffer solution. Add 0.5, 1.5, and 2.5 ml. amounts of the supernatant to tubes for assay.

III. Assay Procedures for Components of Co A in Natural Products

A. Pantothenic Acid

1. Saccharomyces carlsbergensis Method

This assay procedure[2] is essentially the same as the assay for pyri-

---

[21] H. Baker, O. Frank, I. Pasher, A. Dinnerstein, and H. Sobotka, *Clin. Chem.* **6**, 36 (1960).

TABLE I

ASSAY MEDIUM

Combine indicated amounts of the following in the order given to make double strength medium:

1. Sugar and salts solution	500	ml.
2. Potassium citrate buffer	100	ml.
3. Vitamin solution	10	ml.
4. Asparagine solution	125	ml.
5. Ammonium sulfate	7.5	gm.
Add water to make	1000	ml.

No pH adjustment necessary.

Stock Solutions for Medium:
1. Sugar and salts solution

	gm./liter
Glucose, anhydrous	200
KH_2PO_4	2.2
KCl	1.7
$MgSO_4 \cdot 7H_2O$	0.5
$CaCl_2 \cdot H_2O$	0.5
$MnSO_4$	0.01
$FeCl_3$	0.01

2. Potassium citrate buffer

	gm./liter
Potassium citrate	100
Citric acid	20

3. Vitamin solution

	mg./liter
Thiamine·HCl	50
Pyridoxine·HCl	50
Biotin	4
Inositol	5000

4. Asparagine solution

	gm./liter
L.-Asparagine	30

doxine employing the same organism[22] except that asparagine and pyridoxine have been included in the medium, while pantothenic acid is omitted. The presence of asparagine causes *S. carlsbergensis* to have a growth requirement for pantothenate which pantethine will not satisfy.

[22] L. Atkin, A. S. Schultz, W. L. Williams, and C. N. Frey, *Ind. Eng. Chem. Anal. Ed.* **15,** 141 (1943).

Hence an assay procedure employing this organism is the one of choice for determining the pantothenic acid which occurs in natural products together with other pantothenate forms (see Table I for assay medium).

Plan of assay. (1) Add to 22-mm. test tubes in duplicate the following amounts of pantothenic acid (computed in the case of the standard and estimated in the case of the samples to be assayed): 50, 100, 150, 200, 300, 400 mμg. per tube.

(2) Add 5 ml. medium to each tube and make up to 10 ml. with water.

(3) Heat in flowing steam in sterilizer for 10 minutes and cool.

Preparation of inoculum. Inoculate Difco malt agar slants with *S. carlsbergensis* ATCC-9080 and incubate 24 hours at 30°. These slants may be stored in refrigerator 1 month. Prepare fresh inoculum by scraping yeast from slants and suspending 0.3 mg. moist yeast (as determined from a previously prepared standard turbidity curve) in each milliliter of 0.85% saline. Add 1 drop to each assay tube.

Incubation and evaluation. Incubate tubes at 30° for 16 to 18 hours on a reciprocating shaker, with approximately 150 excursions of 4 inches each per minute, in as nearly a horizontal position as possible without spilling culture. Read turbidity in Evelyn photoelectric, or equivalent, colorimeter, using a 640-mμ filter, and calculate unknowns from a standard curve.

2. Lactobacillus plantarum Method

This assay procedure, described in detail in Section IV.A, is suitable for determining uncombined pantothenate when it is known, from the nature of the sample, that pantethine is absent. Pantethine has about 40% as much activity as pantothenate for *L. plantarum*. Phosphorylated forms of pantothenate and pantethine also would tend to interfere with this assay as a true index of the amount of uncombined pantothenic acid present.

3. Other Methods

Assay methods employing *L. casei* were the first developed for pantothenic acid, and were used widely for several years until assays utilizing *L. plantarum* were developed. *Lactobacillus casei* assay procedures are typified by that of Strong et al.[23] They suffer from the fact that *L. casei*, in addition to responding to pantothenate-related forms, also requires more growth factors and growth stimulators than *L. plantarum* and therefore is more responsive to extraneous materials in natural sources of pantothenic acid.

[23] F. M. Strong, R. E. Feeney, and A. Earle, *Ind. Eng. Chem. Anal. Ed.* **13**. 566 (1941).

Another assay for pantothenic acid, employing the protozoan *Tetrahymena pyriformis*, has recently been applied to clinical material by Baker *et al.*[21] This method is too new for its advantages and disadvantages to have become known. One obvious disadvantage is a long incubation period.

B. Pantethine

1. *Lactobacillus helveticus Method*

This assay employs a modification of the medium developed by Craig and Snell [24] in their study of the relative needs of various microorganisms for the then recently isolated LBF factor,[25] which later became known as pantethine (reduced form: pantetheine), the chemical preparation of which has been described.[26] The particular Lactobacillus species used in this assay is not the one employed in the original demonstration of the existence of the LBF factor,[25] but one which was demonstrated in the later study[24] to have optimum characteristics for assaying pantethine. The preferred assay procedure described is essentially that described by Snell and Wittle,[27] although some modifications have been adopted.

Setting up assay. Any one of several compounds can be used as a standard for this assay: either highly purified pantethine (an oil), or the crystalline mercury mercaptide of pantetheine,[26] or crystalline *S*-benzoylpantetheine.[28] Dilute standard and samples so that 1 ml. of each final dilution will contain 0.020 μg. pantethine equivalent. Pipet in duplicate into 18- or 22-mm. test tubes (depending on type of colorimeter used for evaluation of turbidity) the following:

Standard—0, 0.5, 1.0, 1.5, 2.0, 2.5, 3.0, 4.0, 5.0 ml. per tube

Samples—1.0, 1.5, 2.0, 3.0, 4.0 ml. per tube

Dilute contents of all tubes to 5 ml. with water

Assay medium (see Table II)

Comment. It is convenient to make up this medium in 4× strength (twice as concentrated as indicated above) and freeze in bottles of a size which hold enough medium for one assay, thus making it necessary to thaw only enough medium for one assay at a time. Glass bottles may be used and the concentrated and frozen medium will keep for many months.

[24] J. A. Craig and E. E. Snell, *J. Bacteriol.* **61**, 283 (1951).

[25] W. L. Williams, E. H. Jorgensen, and E. E. Snell, *J. Biol. Chem.* **177**, 933 (1949).

[26] E. L. Wittle, J. A. Moore, R. W. Stipek, F. E. Peterson, V. M. McGlohon, O. D. Bird, G. M. Brown, and E. E. Snell, *J. Am. Chem. Soc.* **75**, 1694 (1953).

[27] E. E. Snell and E. Wittle, *in* "Methods in Enzymology" (S. P. Colowick and N. O. Kaplan, eds.), Vol. III, p. 918. Academic Press, New York, 1957.

[28] R. Schwyzer, *Helv. Chim. Acta* **35**, 1903 (1952).

TABLE II
ASSAY MEDIUM

Add to each tube 5 ml. of the following double-strength medium:

Glucose	2	gm.
Sodium acetate·3H$_2$O	2	gm.
Dibasic sodium phosphate	0.25	gm.
Difco Casamino acids	4	gm.
Potassium chloride	0.25	gm.
Salts B	2	ml.
Adenine	2	mg.
Guanine	2	mg.
Uracil	12	mg.
DL-Tryptophan	20	mg.
L-Cystine	20	mg.
L-Cysteine	20	mg.
Asparagine	20	mg.
Tween-oleic acid mixture[a]	2	ml.
Pyridoxal·HCl	80	μg.
Thiamine·HCl	80	μg.
Riboflavin	160	μg.
Nicotinic acid	160	μg.
p-Aminobenzoic acid	80	μg.
Folic acid	6	μg.
Biotin	0.8	μg.

Bring to volume of 80 ml. with water, adjust to pH 6.0, and dilute to 100 ml.

Salts B		
MgSO$_4$·7H$_2$O	10	gm.
NaCl	0.5	gm.
FeSO$_4$·7H$_2$O	0.5	gm.
MnSO$_4$·4H$_2$O	0.5	gm.

Dissolve in water, dilute to 250 ml., add 2 drops concentrated HCl.

[a] Tween–oleic acid mixture: Mix 2 gm. Tween 40 and 20 mg. oleic acid and make up to 20 ml. with water.

Sterilization of medium. Several methods are available for protecting the assay tubes against contamination after sterilization. If only a few tubes are involved, the simplest procedure is to plug them individually with cotton. With large numbers of tubes it is advisable to have stainless steel rectangular covers made to fit over all the tubes in a rack. A single piece of absorbent cotton is wedged into the cover so as to form a cushion covering the tubes. The whole cover can be removed briefly from the rack of sterilized tubes while they are inoculated. Another alternative

is to cover each rack of tubes with a folded linen towel before sterilization. This can be lifted carefully during inoculation and replaced during
incubation. Place the covered tubes in a preheated autoclave and quickly
raise the temperature to 118°. Keep the temperature at this level for 3
minutes, reduce to atmospheric pressure as rapidly as possible, and cool
the tubes by immersing in cold water.

TABLE III

EXAMPLE OF DATA FOR AN ASSAY SAMPLE

	Standard			
ml. per tube	μg. pantethine per tube	Turbidity readings		
0	0	100	100	
0.5	0.01	89.5	89	
1.0	0.02	81.5	81	
1.5	0.03	74.5	73.5	
2.0	0.04	71	69	
2.5	0.05	64	66	
3.0	0.06	57.5	63	
4.0	0.08	56.5	56	
5.0	0.10	52	50.5	

Sample A (estimated at 150 mg. pantethine per milliliter)				
ml. per tube	μg. pantethine per tube (est.)	Turbidity readings	Read from curve	Calculated (mg./ml.)[a]
1.0	0.02	84	0.016	120
1.0	0.02	84	0.016	120
1.5	0.03	78.5	0.023	115
1.5	0.03	79	0.0225	112
2.0	0.04	73.5	0.0305	114
2.0	0.04	74	0.0298	111
3.0	0.06	66	0.0463	115
3.0	0.06	68	0.0415	104
4.0	0.08	61	0.0613	115
4.0	0.08	59	0.0683	128[b]
			Determined mean potency	114

[a] 0.016/0.02 × 150 = 120, etc.

[b] Omitted from calculations because greater than 10% from mean.

Assay organism. The assay organism is *Lactobacillus helveticus* ATCC
12046. This was originally referred to as strain H-80.[24] Stock cultures
are maintained as stabs in solid medium. (To 500 ml. single strength

assay medium add 125 μg. pantethine and 7.5 gm. agar. This makes 50 tubes). Make transfers weekly with a loop into broth medium (same as above omitting agar), incubate 18 hours, and from this broth culture make a series of transfers, using 0.5 ml. of inoculum for each and incubate 18 hours to give a week's supply of inoculum. Store these in the refrigerator during the week and for each day's inoculum centrifuge one of these second transfers, decant the medium, and suspend cells in sterile saline. Add enough saline to give a transmission reading of 80 in an Evelyn colorimeter using a 620-mμ filter. Add 1 drop of this cell suspension to each assay tube.

Incubation. Incubate 40 hours at a constant temperature as near to 38° as possible. This organism has been found to function best in this assay using an incubation temperature slightly above 37°.

Evaluation of results. Read the culture turbidity in an Evelyn, or equivalent, photoelectric colorimeter using a 620-mμ filter. Construct a response curve from the readings obtained with the standard tubes and calculate assay results by an appropriate procedure such as that referred to as the alternate method in Section IV.A.

For an example of data for an assay sample, see Table III.

2. Limitations of Lactobacillus helveticus Method

In developing their assay for pantethine, Craig and Snell [24] recognized that the lactobacilli investigated for their response to this compound also responded in a minor degree to pantothenic acid. The particular species (*L. helveticus* H-80) actually recommended for the assay required about 100 times as much pantothenic acid as pantethine to give the same response. Brown [29] studied this characteristic of this strain of *L. helveticus* and found that gradually reducing the pantethine and increasing the pantothenic acid in the medium in which the organism was grown, produced a culture which would respond to equal amounts of either pantothenic acid or pantethine. But he could never produce a culture which would respond to either compound but not the other. Thus, he concluded that the culture normally is a mixture of organisms, some of which require pantothenic acid and some pantethine. This is the reason for maintaining the inoculum culture in a medium containing pantethine but no pantothenic acid, so as to favor the cells requiring the former at the expense of those requiring pantothenic acid. So far, no bacterium has been observed which requires pantethine exclusively. This makes it impossible to assay the pantethine content of a product containing too high a proportion of pantothenic acid in relation to its pantethine content.

[29] G. M. Brown, *J. Biol. Chem.* **226**, 651 (1957).

IV. Assay Procedures for Pantothenic Acid and Panthenol in Pharmaceutical Products

A. Recommended Method for Pantothenic Acid

This method is based on that described in U. S. Pharmacopeia XVI [30] and employs *L. plantarum*.

Preparation of standard pantothenic acid solution. Make a stock solution by weighing out, in a closed system to prevent absorption of water, 54.4 mg. of U.S.P. calcium pantothenate reference standard (equivalent to 50 mg. pantothenic acid), previously dried for 2 hours at 105°, and dissolve in about 500 ml. water in a 1000-ml. volumetric flask. Add 10 ml. of 0.2 N acetic acid and 100 ml. of 0.2 N sodium acetate. Add water to the mark so that 1 ml. will contain 50 μg. pantothenic acid. Store under toluene in the refrigerator. On the day of the assay, dilute 0.5 ml. of above solution to 1000 ml. with water. Each milliliter represents 0.025 μg. pantothenic acid.

Setting up assay. Dilute samples so that each milliliter contains an estimated 0.025 μg. pantothenic acid. Add in duplicate to standardized test tubes the following amounts of standard and sample dilutions:

Standard—0, 1.0, 1.5, 2.0, 3.0, 4.0, 5.0 ml.

Samples—1.5, 2.0, 3.0, 4.0 ml.

Add water to all tubes to make 5 ml.

Assay medium—The medium recommended is the dehydrated Difco, or similar product, rehydrated according to directions on the package. Or the medium can be prepared according to U.S.P. XVI [30] in 4× concentration and stored frozen (see Table IV).

Sterilization of medium. Plug tubes individually with cotton, or cover all tubes in a rack as a unit as described in Section III.B.1. Place the covered tubes in a preheated autoclave and raise the temperature as quickly as possible to 121°. Hold at this temperature for 5 minutes, then reduce to atmospheric pressure as soon as possible. Total time in sterilizer should not be over 12 to 14 minutes. Cool tubes immediately by placing racks in cold water.

Assay organism. The assay organism is *L. plantarum* ATCC 8014. Stock cultures are maintained as stabs made monthly in Difco Micro Assay Culture Agar. Inoculate with a needle from these stabs into single strength assay medium to which has been added 0.01 μg. pantothenic acid per milliliter. Incubate 16 hours at 37° and dilute culture with sterile

---

[30] Calcium Pantothenate Assay, *in* "U. S. Pharmacopeia," Vol. XVI, p. 871 (17th rev.), 1960.

TABLE IV
ASSAY MEDIUM

Each 250-ml. double-strength U.S.P. calcium pantothenate assay medium contains:

Acid-hydrolyzed casein	2.5 gm.
L-Cystine	0.1 gm.
L-Tryptophan	25 mg.
Tween 80	25 mg.
Dextrose, anhydrous	10 gm.
Sodium acetate, anhydrous	5 gm.
Adenine sulfate	5 mg.
Guanine·HCl	5 mg.
Uracil	5 mg.
Riboflavin	100 μg.
Thiamine·HCl	50 μg.
Biotin	0.2 μg.
Nicotinic acid	50 μg.
Pyridoxine·HCl	200 μg.
p-Aminobenzoic acid	50 μg.
Salt Solution A	5 ml.
Salt Solution B	5 ml.

Salt solution A

25 gm KH$_2$PO$_4$
25 gm. K$_2$HPO$_4$
Water to 500 ml.
5 drops HCl

Salt solution B

10 gm MgSO$_4$·7H$_2$O
0.5 gm. NaCl
0.5 gm. MnSO$_4$·4H$_2$O
0.5 gm. FeSO$_4$·7H$_2$O
Water to 500 ml.
5 drops HCl

Add 5 ml. double-strength medium to all tubes.

saline to read 85% transmission. Add 1 drop of this inoculum per tube. (It is not necessary to wash the inoculum before dilution.)

Incubation. Incubate in a carefully regulated air incubator, or a water bath, at 37° for 16 to 18 hours.

Evaluation of results. Read the culture turbidity in an Evelyn, or equivalent, photoelectric colorimeter using a 620-mμ filter. Then carry out the following procedure:

1. For each level of the standard add the duplicate transmittances and subtract from 200.

2. Plot these figures on the ordinate of cross-section paper against the logarithm of the milliliters of standard preparation per tube on the abscissa, using for the ordinate either an arithmetric or a logarithmic scale, whichever gives the better approximation to a straight

line. Draw the straight line or smooth curve which best fits the plotted points.

3. For each level of assay sample add the duplicate transmittance readings and subtract from 200.

4. Read from the standard curve the logarithm of the volume of the standard corresponding to each of these values of the assay sample determined in step 3.

5. Subtract the value determined in step 4 from the logarithm of milliliters of sample added to this sample tube. Average these values to obtain $\overline{X}$, which is the log-relative potency of the sample.

6. Subtract $\overline{X}$ from the log of the estimated potency of the sample, and find the antilog of the resulting figure. This is the potency of the sample.

For an example of summarized data for an assay sample, see Table V.

TABLE V

EXAMPLE OF SUMMARIZED DATA FOR AN ASSAY SAMPLE

	Sample A; estimated potency: 7.5 mg./ml.		
Sample readings	Log of reading from curve	Log of ml. sample added	Difference
47.5	0.1620	0.1760	−0.0140
60	0.3050	0.3010	+0.0040
73	0.4320	0.4770	−0.0450
87	0.5620	0.6020	−0.0400
			$X = -0.0950$
	Log of 7.5 = 0.8751		$\overline{X} = -0.0238$
	−0.0238		
	0.8513		
	Potency = antilog of 0.8513 = 7.10 mg./ml.		

An alternate, usually sufficiently precise, method consists in plotting the standard responses against concentration on ordinary graph paper and connecting the points with the smoothest possible curve. Read the unknowns (individually, not averaging duplicate tube results) from this curve and calculate the mean of all results obtained. Eliminate those individual tube results which deviate from this mean by more than ±10% and calculate a new mean with the remaining results. If two-thirds of the original individual tube results remain the assay is considered valid.

B. Methods Available for Panthenol

Several microbiological, and also chemical, assays for panthenol have been developed. One of the early assay procedures[31] involved both biological and microbiological steps. Since lactic acid bacteria do not respond directly to pantothenyl alcohol, the sample was fed to rats who converted the compound to pantothenic acid which was collected in the urine over a 24-hour period and assayed in the conventional manner with *L. plantarum*. This assay required several days but was quite specific, not being affected by pantoic acid or pantoyl lactone (split products of panthenol).

A direct assay of panthenol, based on its growth-stimulating effect on *Acetobacter suboxydans*, had been suggested earlier.[32] However, this method required exhaustive removal of extraneous pantothenic acid and pantoic acid from the sample before assay since both these substances have far greater growth stimulating activity for this organism than does panthenol.

A microbiological procedure permitting the assay of panthenol in the presence of pantoyl lactone as well as pantoic acid was developed by DeRitter and Rubin.[33] Any lactone present originally is removed by continuous ether extraction of the sample which is in aqueous solution, following which the panthenol in the aqueous residue is hydrolyzed to pantoic acid and assayed with *A. suboxydans*. By this procedure a preliminary assay prior to hydrolysis indicates preformed pantoic acid present and a correction can be made for pantothenic acid originally present if it does not exceed 10%. This method was later improved[34] to eliminate the ether extraction for removal of pantoyl lactone or the separate assay required to correct for preformed pantoic acid. Instead, these interfering substances are separated by passing through an ion exchange resin column.

A recently developed assay procedure[35] utilizes the growth stimulating effect of pantoic acid on an *Escherichia coli* mutant. The growth of this organism should be more easily controlled than that of *Acetobacter suboxydans*. However, the method still suffers from the necessity to hydrolyze all panthenol present in the assay sample to pantoic acid, which is the actual growth promotant measured.

C. Recommended Method for Panthenol

The recommended assay procedure[14] is based on the inhibition by

[31] L. Drekter, R. Drucker, R. Pankopf, J. Scheiner, and S. H. Rubin, *J. Am. Pharm. Assoc. Sci. Ed.* **37**, 498 (1958).

[32] M. Walter, *in* "Jubilee Volume, Emil Barell," p. 98. Reinhardt, Basel, 1946.

[33] E. DeRitter and S. H. Rubin, *Anal. Chem.* **21**, 823 (1949).

[34] M. S. Weiss, I. Sonnenfeld, E. DeRitter, and S. H. Rubin, *Anal. Chem.* **23**, 1687 (1951).

[35] C. G. Rogers and J. A. Campbell, *Anal. Chem.* **32**, 1662 (1960).

panthenol of the growth of the bacterium *Leuconostoc mesenteroides* under standard conditions, and is applicable directly to most types of pharmaceutical vitamin products without prior treatment of samples. Reasonable amounts of interfering pantothenic acid can be removed from assay samples by a simple resin treatment.

Setting up assay. To prepare standard tubes dilute D-panthenol (or an equivalent amount of DL-panthenol) just prior to each assay so that 1 ml. will contain 1 μg. Further dilute 0, 6, 10, 14, 18, 22, 26, 30, and 34 ml. amounts to 50 ml. so that 5 ml. additions of these dilutions to assay tubes (18 or 22 mm.) in triplicate will give 0, 0.6, 1.0, 1.4, 1.8, 2.2, 2.6, 3.0, and 3.4 μg. D-panthenol per tube. Dilute samples so that 1 ml. will contain an estimated 1 μg. D-panthenol and add in triplicate the following amounts to assay tubes: 1.4, 1.8, 2.2, 2.6 ml. Add water to all tubes to make 5 ml.

Assay medium. To each tube add 5 ml. of medium prepared by either of the following procedures:

(A) To prepare double strength assay medium from original ingredients make up the following solutions and store in the refrigerator (do not freeze) until needed, but not longer than 1 month:

(1) Dissolve 10 gm. of Difco Casamino acids in 80 ml. of water, adjust to pH 3.5 with concentrated HCl, add water to 100 ml. Add 1 gm. of Darco G-60, stir 15 minutes and filter. Repeat Darco treatment.

(2) DL-Tryptophan, L-cystine, L-asparagine: 2 gm. of each; adenine sulfate, guanine·HCl: 200 mg. of each; $MgSO_4 \cdot 7H_2O$: 8 gm., $MnSO_4 \cdot 4H_2O$, $FeSO_4 \cdot 7H_2O$, NaCl: 400 mg. of each; dissolve in a minimum of concentrated HCl and make to 100 ml. with water.

(3) Uracil: 1.2 gm.; folic acid: 600 μg.; dissolve in weak NaOH and make to 100 ml. with water.

(4) Biotin: 80 μg.; calcium pantothenate (calculated as the acid): 100 μg.; pyridoxal·HCl, thiamine·HCl, p-aminobenizoic acid: 8 mg. of each; nicotinic acid: 16 mg.; riboflavin phosphate: 23 mg.; dissolve in 20% ethanol and make to 100 ml. with same solvent.

(5) Tween 40: 20 gm.; oleic acid: 200 mg.; dissolve in 20% ethanol and make to 100 ml. with same solvent.

(6) KCl: 25 gm.; Na_2HPO_4: 25 gm.; dissolve in water and make to 200 ml.

These solutions are combined with additional components in the following order:

1. Water   500 ml.
2. Cysteine·HCl   200 mg.

3. Dextrose   20 gm.
4. Sodium acetate   20 gm.
5. Solution 1   50 ml.
6. Solution 2   10 ml.
7. Solution 3   10 ml.
8. Solution 4   10 ml.
9. Solution 5   10 ml.
10. Solution 6   20 ml.

Adjust to pH 6.0 and make to 1000 ml. with water.

(B) Prepare the desired amount of double-strength Difco Pantothenate Assay Medium according to directions and allow to cool to room temperature. (Difco dehydrated panthenol assay medium has not proved satisfactory in the author's laboratory.) For each 100 ml. of medium add 1 ml. of each of the following supplementary solutions:

(1) Pyridoxal·HCl: 8 mg.; calcium pantothenate (calculated as the acid): 75 μg.; dissolve in 10% ethanol and make to 100 ml. with same solvent.
(2) Tween 40: 10 gm.; oleic acid: 100 mg.; dissolve in 20% ethanol and make to 50 ml. with same solvent.

Store these solutions in the refrigerator and renew each month.

Sterilization. Plug tubes loosely with cotton or cover all tubes in each rack with a folded towel and autoclave 5 minutes at 121°. Release steam from sterilizer by fast exhaust (total heating time 12–14 minutes) and submerge tubes in cold water.

Assay organism. Maintain stock cultures of *Leuconostoc mesenteroides* P-60, ATCC No. 8042 (now referred to as *Streptococcus* sp.) in Difco Micro Assay Agar stabs. Make two stabs each month and incubate 24 hours at 37°. Refrigerate one and transfer the other daily in the same medium to provide inoculum for assays.

Inoculum. Transfer the growth from a daily stab culture to 10 ml. of single strength assay medium to which has been added 0.01 μg. calcium pantothenate per milliliter. Incubate 16–24 hours at 37°, centrifuge, and wash cells twice with 0.9% saline and finally suspend in 10 ml. saline. Dilute 1 ml. of cell suspension to 20 ml. with saline. Add exactly 1 drop of this inoculum to each tube (more or less may interfere with the inhibition response).

Incubation. Incubate tubes 16–18 hours at 37°, or longer if necessary to obtain readings of 45 to 55% transmission for negative control tubes (no panthenol) and 90–100% transmission for the highest level of panthenol. Place tubes in cold water to slow growth rate and read turbidity at once in an Evelyn, or equivalent, photoelectric colorimeter using a

620 mμ filter. If above readings are not obtained, adjust dilution of inoculum in future tests to give approximately these final readings.

Evaluation of results. Using the turbidity readings construct a response curve relating the degree of growth inhibition in each tube of standard to the amount of panthenol contained therein. Calculate the readings for the sample tubes from this standard curve for each sample and average them. Discard those varying by more than ±10% from the mean and determine a new mean from those remaining, provided two-thirds still remain.

See Table VI for an example of assay sample data.

TABLE VI
EXAMPLE OF DATA FOR AN ASSAY SAMPLE

μg. panthenol per tube	*Standard*		
	Turbidity readings		
0	53	54	54
0.6	57.5	58	58
1.0	62	63	63
1.4	68	68	68.5
1.8	75	75.5	76
2.2	84	85.5	88
2.6	91	91	92
3.0	94.5	94	95.5
3.4	97	97	96.5

Sample A (estimated at 8.0 mg. panthenol per gram)

μg. panthenol per tube (est.)	Turbidity readings	Read from curve	Calculated (mg./gm.)[a]
1.4	68.5	1.42	8.12
1.4	68	1.40	8.00
1.4	67	1.32	7.55
1.8	77.5	1.89	7.62
1.8	76	1.80	8.00
1.8	76	1.80	8.00
2.2	88	2.42	8.75
2.2	84.5	2.22	8.07
2.2	84	2.20	8.00
2.6	91.5	2.65	8.15
2.6	92	2.68	8.25
2.6	92.5	2.72	8.35
		Determined mean potency	8.07 mg./gm.

[a] $(1.42/1.4) \times 8.0 = 8.12$, etc.

Comment. In the case of samples which contain natural sources of pantothenic acid in addition to panthenol, or mixtures of panthenol and salts of D-pantothenic acid, add 2.5 gm. of Amberlite MB-1 (Rohm and Haas Company) to each 100 ml. of dilution containing 1 µg./ml. of D-panthenol and shake on a mechanical shaker for 30 minutes, decant the supernatant, and add to the assay tubes as usual. Experience will show how much pantothenate activity can be removed by this simple procedure.

7.8 Riboflavin

FREDERICK KAVANAGH

Eli Lilly and Company, Indianapolis, Indiana

Riboflavin may be determined through its influence upon growth of *Lactobacillus casei* as in the example in Chapter 2 or by chemical methods. The range of concentrations of the assays are: 0.5–2 μg./ml. for the large plate method and 0.005–0.3 μg./tube for a titrametric or a turbidimetric procedure. Barton-Wright[1] in a series of articles discussed the microbiological methods employed in his laboratory. His method for riboflavin, using *Streptococcus zymogenes* (ATCC 10100) as the test organism, had a range of 0.005–0.04 μg./tube in both the nephelometric and the acidimetric type of assays. His method employing *Lactobacillus helveticus* had a range from 0.05 to 0.25 μg. riboflavin per tube.

For many preparations, there is little reason to choose between chemical and microbiological methods except convenience. Both can be used in the millimicrogram range. Lowry and Bessey[2] modified the medium, incubated in a carbon dioxide atmosphere and scaled down to an incubation volume of 0.2 ml. Under these conditions the *L. casei* titrametric method had a range from 0.5 to 20 mμg. per tube with an over-all precision of $\pm 3\%$. The micromethod was more precise than the macromethod. Good results were obtained with 0.1 to 0.5 mμg. in an incubation volume of 0.05 ml. If the method employing *Streptococcus zymogenes* could be scaled down in a similar manner, the riboflavin content of one liver cell (approximately 0.2 $\mu\mu$g. riboflavin) could be measured.

Chemical methods include direct photometric measurement of large con-

---

[1] E. C. Barton-Wright, *Lab. Practice* **10**, 543, 633, 715 (1961).
[2] O. H. Lowry and O. A. Bessey, *J. Biol. Chem.* **155**, 71 (1944).

centrations of relatively pure material and fluorometric methods for small concentrations. The fluorometric method may be as sensitive as a microbiological method. Burch *et al.*[3] used a sensitive fluorometer to measure riboflavin and its natural derivatives in whole blood, red cells, serum, and white cells and platelets. They could measure 0.2 mμg. of riboflavin in a cuvette volume of 0.5 ml. The proteins of blood seemed to interfere less with the chemical method than with the microbiological method. Chemical methods, which seem to be more popular than microbiological methods, are given in books on biochemical analysis and vitamin assays.

[3] H. B. Burch, O. A. Bessey, and O. H. Lowry, *J. Biol. Chem.* **175**, 457 (1948).

7.9 Turbidimetric Assay for Thiamine

HELEN R. SKEGGS

Merck Sharp & Dohme Research Laboratories, West Point, Pennsylvania

The use of microbiological assay procedures for thiamine has not been extensive for two reasons, as Hoff-Jorgensen[1] points out. Chemical methods are well established, and, although there is a place for a more sensitive and specific technique for the determination of thiamine, it has been difficult to obtain consistent results with *Lactobacillus fermenti* in the assay procedure described by Sarett and Cheldelin.[2] In part, this is due to the ability of *L. fermenti* to dispense with thiamine when grown in its absence. Hoff-Jorgensen has described an assay procedure employing the yeast, *Kloeckera brevis*, but a more recent method proposed by Deibel and associates[3] appears to be less complicated and is presented

[1] E. Hoff-Jorgensen and B. Hansen, *Acta Chem. Scand.* **9,** 562 (1955).
[2] H. P. Sarett and V. H. Cheldelin. *J. Biol. Chem.* **155,** 153 (1944).
[3] R. H. Deibel, J. B. Evans, and C. F. Niven, Jr., *J. Bacteriol.* **74,** 818 (1957).

here as the microbiological assay of the future for thiamine. Scholes[4] has adapted both the *Lactobacillus fermenti* and the *L. viridescens* procedures for use with small amounts of test materials and finds *L. viridescens* less susceptible than *L. fermenti* to inhibitory or stimulatory substances other than thiamine.

I. Test Organism

The assay organism is *Lactobacillus viridescens* strain S38A, ATCC 12706. It is a heterofermentative organism typical of the organisms that are isolated from discolored cured meats. In general, these organisms are unique[5] in that they require manganese and citrate for growth even in complex laboratory media. The organism does not utilize the thiazole and pyrimidine moieties of thiamine either singly or together. The activity of cocarboxylase (thiamine pyrophosphoric acid ester) is almost equal to its thiamine content. The active acetaldehyde thiamine intermediate described by Krampitz *et al.*[6] has about 80% of the activity of thiamine for both *L. viridescens* and *L. fermenti*. (Presumably this is a DL compound and only one isomer is active. Thus, the active isomer would be 1.6 times as active as thiamine.)

II. Standard Solution

Thiamine hydrochloride is used as the standard, freshly diluted with water to a concentration of 10 mμg./ml. for each assay.

III. Stability of Compound

In the dry form thiamine hydrochloride is stable and heating for 24 hours at 100°C does not destroy its potency. In aqueous solutions it can be sterilized at 110°, but if the pH of the solution is above 5.5 it is rapidly destroyed. In the assay medium, at pH 6.0, there is no significant loss of activity during autoclaving recommended, 15 lb. (120°) for 5 minutes. Thiamine is destroyed by alkali and precipitated by tannins and by reagents which precipitate alkaloids.

IV. Preparation of Sample for Assay

Weigh a representative sample and suspend in 25 volumes of 0.1 N H_2SO_4. Steam the mixture or reflux at 100° from 30 to 45 minutes. Cool,

[4] P. M. Scholes, *Analyst* **85**, 883 (1960).

[5] J. B. Evans and C. F. Niven, Jr., *J. Bacteriol.* **62**, 599 (1951).

[6] L. O. Krampitz, G. Gruell, C. S. Miller, J. B. Bicking, H. R. Skeggs, and J. M. Sprague, *J. Am. Chem. Soc.* **80**, 5893 (1958).

and adjust the pH to 4.5 with 2.5 M sodium acetate. Add 100 mg. Mylase P (Wallerstein Laboratories) or other similar enzyme preparation per gram of sample; cover with toluene or benzene and incubate at 37° for 18 to 24 hours. Steam the digest for 10 minutes at 100°C., then dilute with water to a suitable volume. Filter through filter paper. Dilute the filtrate further with water to give a concentration of about 5 mμg. thiamine per milliliter. Deibel *et al.* suggest that less elaborate procedures may prove effective. In these laboratories, preliminary results on the assay of chicken plasma by direct aseptic addition to the assay gave results that agree well with the thiochrome assay results. In preparing to study the effectiveness of samples added to the test without heating, it was found that whereas a solution of 100 μg./ml. could be filtered through glass without detectable loss of activity, a thiamine solution passed through an ultrafine sintered glass filter at a concentration of 0.1 μg./ml. was adsorbed on the glass.

V. Preservation of Test Organism

The organism is maintained in stab cultures in a medium with the composition of APT broth (Table I) but with only 0.1% glucose and

TABLE I

Composition of APT Broth for *Lactobacillus viridescens*[a]

Compound	Amount (gm./liter) single strength
Tryptone	10
Yeast extract[b]	5
Sodium chloride	5
Sodium citrate	5
K_2HPO_4	5
Glucose[c]	10
$MgSO_4 \cdot 7H_2O$	0.8
$FeSO_4 \cdot 7H_2O$	0.04
$MnCl_2$	0.14
Tween 80	1 ml.

[a] pH adjusted to 6.7–7.0 for culture media; pH adjusted to 6.0 for assay medium.

[b] Yeast extract containing thiamine is used for culture medium or 1 mg. of thiamine is added to each liter. Thiamine deficient yeast extract is used in preparing the assay medium which is available from Difco (0808) in dehydrated form.

[c] Glucose concentration is decreased to 1 gm./liter for maintenance medium and 2% agar added.

1.5–2% agar. Stock cultures on agar can be held under refrigeration for 1 month.

VI. Preparation of Inoculum

The inoculum is grown at 30° for 16 to 20 hours in pH 6.7–7.0 APT broth prepared with yeast extract rich in thiamine or supplemented with thiamine. Wash the cells with 1 volume of saline, centrifuge, resuspend in 10 ml. saline, and dilute tenfold with saline for inoculum. Preliminary experiments indicate that washed cells may be resuspended and diluted in assay medium, frozen and stored for as long as 30 days, thawed, and used directly to inoculate the assay tubes.

VII. Media

Double strength APT broth is prepared as described in Table I with thiamine deficient yeast and adjusted to pH 6.0. Since the assay procedure was developed as a result of the prevalent deficiency of thiamine in commercial yeast extracts, Deibel and his colleagues were not concerned with the preparation of thiamine-free yeast extract. Since thiamine is so readily destroyed by heat at neutral pH, the enterprising assayist probably could readily prepare a suitable deficient yeast extract. Very satisfactory assay media is available from Difco in the dehydrated form (No. 0808) that can easily be fortified with thiamine for transfer of the organism.

VIII. Design of Assay

The standard 10-ml. volume assay design may be used. The standard curve is run at 0, 0, 5, 10, 15, 20, 30, and 50 mμg. of thiamine per tube. Water is added to a volume of 5 ml., followed by 5 ml. assay medium. Samples are diluted to approximately 5 mμg./ml. and added in increments up to 5 ml. In modifying the medium for greater sensitivity, Scholes uses a final volume of 2 ml., and an assay standard range up to 10 mμg. with standard and sample dispensed directly into colorimeter tubes.

Overheating the assay tubes is to be avoided. Sterilization for 5 minutes at 118° is recommended for the standard 10-ml. procedure, provided the autoclave reaches that temperature in 3 or 4 minutes, otherwise the time should be reduced. Scholes, dealing with smaller volumes finds that running a small autoclave up to 10 lb./sq. inch and shutting the steam off gives adequate sterilization for the assay period.

IX. Assay Conditions

Incubate the test for 20 hours at 30°. Incubation for longer than 24 hours is not advisable because the sterilization procedure is not sufficient to kill all bacterial spores and possible interference from contamination may occur. Furthermore, in the absence of contamination the cells may lyse with prolonged incubation.

X. Reading the Assay

Measure growth photometrically in any available colorimeter with a 660-mμ filter. Cool the tubes and shake them prior to reading. Titration values do not correlate well with turbidity readings or thiamine levels.

XI. Calculation of Results

A standard curve is drawn and sample values read from the curve, corrected for dilution factor, averaged, and recorded. A typical standard response is given in Table II.

TABLE II

RESPONSE OF *Lactobacillus viridescens* TO THIAMINE

mμg. thiamine per 10 ml. culture	Turbidity Klett Summerson[a]
0	118
5	156.5
10	182
15	213
20	240
30	272
50	332

[a] Instrument set at zero against distilled water.

7.10 Vitamin B₁₂ and Congeners

HELENE NATHAN GUTTMAN *

The Haskins Laboratories, New York, New York; Department of Biological Sciences, Goucher College, Towson, Maryland; and Department of Medicine, Medical College of Virginia, Richmond, Virginia

I. Introduction

This chapter is intended to supply methods for the microbiological assay of vitamin B₁₂ and its congeners using four assay organisms, which, in the judgment of the author, will fulfill the assay needs of most workers. No attempt has been made to give a comprehensive treatment of the

* Aided by a grant from the U.S. Public Health Service.

numerous aspects of the biology of the vitamin B_{12} group. Such information may be found in one or more recent reviews.[1-7]

Cyanocobalamin (vitamin B_{12}) is the largest (M. W. 1355) and most complicated water soluble vitamin known. This molecular complexity offers nature, and the organic chemist, many opportunities for modification of the molecule with potential concomitant modification of biological activity. The most frequent modification is substitution of other nucleotides for 5,6-dimethylbenzimidazole ribotide. The activity of nucleotide-substituted cobalamins as well as of cobinamide (Factor B), which is devoid of any nucleotide, for the four assay organisms discussed here is summarized in Table I.

Procedures Applicable to All Assays

1. Preparation of Glassware

All glassware should be scrupulously cleaned of traces of organic material because of the sensitivity of the assays. Failure to remove assayable material yields high blanks or uniformly high growth throughout the assay. It is also essential to free the glassware of traces of cleaning preparations because they frequently inhibit the growth of the test organisms. Such inhibition can appear at random if some, but not all, of the glassware contains the inhibitory material.

All glassware should be soaked in a good quality soap or detergent, washed in very hot water and rinsed with tap, and then distilled water. In some laboratories this treatment is sufficient, but in others it has been found necessary to add an acid wash, in which washed glassware is soaked for 1 to 2 minutes in 75% H_2SO_4, rinsed with tap water until the effluent is neutral, and then finally rinsed with distilled water. Special care should be taken to handle glassware only by the parts which will not later come in contact with assay components.

All clean glassware should be stored in covered cabinets or bins.

2. Distilled Water

Freshly distilled water, preferably prepared in an all-glass still, should be used. Stored distilled water often contains assayable quantities of

---

[1] H. G. Heinrich, ed., "Vitamin B_{12} and Intrinsic Factor. 1. Europäisches Symposion, Hamburg, 1956." 576 pp. Ferdinand Enke, Stuttgart, 1957, and H. G. Heinrich, ed., "Vitamin B_{12} and Intrinsic Factor. 2. Europäisches Symposion, Hamburg, 1961." 798 pp. Ferdinand Enke, Stuttgart, 1962.

[2] J. E. Ford and S. H. Hutner, Vitamins and Hormones **13**, 101 (1955).

[3] C. C. Ungley, Vitamins and Hormones **13**, 137 (1955).

[4] A. W. Johnson and A. Todd, Vitamins and Hormones **15**, 1 (1957).

[5] R. T. Williams, ed., "The Biochemistry of Vitamin B_{12}." 123 pp. Cambridge Univ. Press, London and New York, 1957.

[6] E. L. Smith, "Vitamin B_{12}," 196 pp. Wiley, New York, 1960.

[7] R. Gräsbeck, Advances in Clin. Chem. **3**, 299 (1960).

TABLE I

ACTIVITY OF COBALAMINS ACCORDING TO THE BASE OF THE NUCLEOTIDE

Assay organism	Base of the nucleotide											Coenzyme forms			B₁₂ bypass compounds	Range of assay of cyanocobalamin (μμg./ml.)
	5,6-Dimethyl benzimidazole[a]	Benzimidazole	5-Hydroxy benzimidazole	Adenine	2-Methyl adenine	Hypoxanthine	2-Methyl hypoxanthine	5,6-Diethyl benzimidazole	5-Carboxamide benzimidazole	2,3-Naphtha amidazole	None (Factor B)	5,6-Dimethyl benzimidazole	Benzimidazole	Adenine		
Ochromonas malhamensis	100	36	—	0	0	0	0	50	45	60–100	0	100	50	0	None	0.5–50.0
Euglena gracilis tube	100	+	—	100	60	—	—	—	—	—	0	—	—	—	None	0.3–50.0
plate	100		—								0					
Escherichia coli tube	100	200	50	10	50	160	40	100	88	70	20	100	100	100	Methionine	40.0–250.0
plate	100		100–150	100	100		280	+	+	+	100–250					500–50,000
Lactobacillus leichmannii	100	200	35	50	40	20–100	15–40	130	94	—	0	—	—	—	Deoxyribosides	1.0–20.0
Clinical activity	+	+	+	—	—	—	—	+ but low	+ but low	+	—	—	—	—	—	

[a] All other compounds are in relation to "vitamin B₁₂" arbitrarily expressed at 100.

vitamin B_{12}.[8] Vessels for holding distilled water should be used exclusively for that purpose.

3. Stock Solutions

Stock solutions of vitamin B_{12}, at 10.0 or 100.0 μg./ml., may be stored in the refrigerator for about 1 month without loss of activity. Solutions are most stable at pH 4–5. The vitamin is available commercially in the stable, hydrated crystalline form or as a triturate with various sugars (mannitol or sucrose most commonly).

If desired, an aqueous solution of B_{12} can be checked spectrophotometrically for the following specifications: In water the maxima are 550 mμ ($A_{1\,cm.}^{1\%}$ 64), 361 mμ ($A_{1\,cm.}^{1\%}$ 204), 278 mμ ($A_{1\,cm.}^{1\%}$ 115).

4. Preservation of Solutions

All stock solutions and media which are stored for any appreciable time should be frozen or refrigerated in stoppered bottles. To deter contamination of refrigerated solutions, add approximately 1% of the following volatile preservative[9] which is completely driven off on autoclaving: chlorobenzene: 1,2-dichloroethane: n-butyl chloride (1:1:2 v/v/v).

5. Standard Curves

A standard curve should be run with each assay (see Section II.A).

II. Turbidimetric Methods

A. Methods Applicable to All Turbidimetric Assays

Assay procedures should be followed in the sequence given below. Specific procedures are found in sections devoted to each of the assay organisms.

Prepare double strength assay medium.

Prepare dilutions of vitamin B_{12} from the stock solution. This preparation is to be done daily according to the following procedure:

1.0 ml. (10.0 μg./ml. stock solution) + 99.0 ml. water → 100.0 mμg./ml.
1.0 ml. (100.0 mμg./ml. solution) + 99.0 ml. water → 1.0 mμg./ml.
1.0 ml. (1.0 mμg./ml. solution) + 9.0 ml. water → 100.0 μμg./ml.
1.0 ml. (100.0 μμg./ml. solution) + 9.0 ml. water → 10.0 μμg./ml.

Use a sufficient number of vessels to provide at least duplicate tubes for each level of vitamin B_{12} and of unknown.

[8] W. J. Robbins, A. Hervey, and M. E. Stebbins, Ann. N. Y. Acad. Sci. **56,** 818 (1953).

[9] S. H. Hutner, L. Provasoli, and J. Filfus, Ann. N. Y. Acad. Sci. **56,** 852 (1953).

Add double strength assay medium to each assay vessel.

Add vitamin plus distilled water, to final volume, to the vessels for the standard curve. Details for one standard curve are in Table II.

TABLE II

PREPARATION OF VITAMIN B_{12} STANDARD CURVE

Tube no.	ml. double strength assay medium	ml. distilled water	ml. of B_{12} (dilution)	Final concn. B_{12} ($\mu\mu$g./ml.)
0^a	2.5	2.5	None	None
1, 2	2.5	2.5	None	None
3, 4	2.5	2.0	0.5 (of 10.0 $\mu\mu$g./ml.)	1.0
5, 6	2.5	1.0	1.5 (of 10.0 $\mu\mu$g./ml.)	3.0
7, 8	2.5	2.2	0.3 (of 100.0 $\mu\mu$g./ml.)	6.0
9, 10	2.5	2 0	0.5 (of 100.0 $\mu\mu$g./ml.)	10.0
11, 12	2.5	1.0	1.5 (of 100.0 $\mu\mu$g./ml.)	30.0
13, 14	2.5	2.2	0.3 (of 1.0 mμg./ml.)	60.0
15, 16	2.5	2.0	0.5 (of 1.0 mμg./ml.)	100.0
17, 18	2.5	1.0	1.5 (of 1.0 mμg./ml.)	300.0

[a] This tube is sterilized, but not inoculated, and the contents used to set the densitometer.

Add samples of unknown to be assayed to the remaining vessels. Be sure to choose at least three widely spaced values for each sample so that at least one level will fall within the linear range of the standard curve.

Prepare one extra blank vessel to be left uninoculated. The contents of this vessel will be used to set the densitometer to 100% light transmission or zero optical density.

Close the vessels with either screw caps, glass beakers, aluminum caps, stainless steel caps or Morton closures. Do not use cotton plugs. Autoclave at 121°C. for 15 minutes. Cool.

Prepare inoculum and inoculate each vessel except the one to be used to set the densitometer.

Incubate all assay vessels under the same conditions for the required time.

Set the densitometer with the medium prepared for that purpose.

Read the densities of the standards and unknowns.

Plot values for the standard curve on semilog paper with vitamin B_{12} concentration along the abscissa, log scale, and optical density along the ordinate, arithmetic scale.

Interpolate the optical density readings of each unknown on the standard curve to determine its B_{12} concentration.

Test Organisms

The four organisms given in Table I may be used.

B. *Ochromonas malhamensis* (ATCC 11532)

This alga is the most specific in its requirement for cobalamins. Only the forms of B_{12} which are active clinically satisfy the growth factor requirement of this organism. Thus growth factor activity for this organism has been referred to as "true-B_{12}" activity. This term should be used with caution in view of the reports of natural products which are active for *O. malhamensis* but not for some of the less exacting assay organisms[10,11] and the failure of some of the coenzyme forms of the vitamin to satisfy the growth factor requirement[12] (Table I).

The specificity and assay range for *O. malhamensis* are given in Table I. This is the assay of choice when high biological specificity or correlation with human clinical studies is desired.

Although this assay takes 4–7 days of incubation before evaluation is made, its specificity compensates for the time required.

1. Culture Vessels

Any of several types of vessels may be used as long as care is taken to provide a large surface/volume ratio during algal growth. For example, any of the vessels listed in Table III may be used.

TABLE III

CULTURE VESSELS FOR USE IN TURBIDIMETRIC ASSAYS

Size of vessel	Position during incubation	Final volume medium
16 × 125 mm. tube	Slanted	5 ml.
16 × 150 mm. tube	Slanted	5 ml.
20 × 125 mm. tube	Slanted	5 or 10 ml.
10-ml. microfernbach flask	Upright	5 ml.
25-ml. microfernbach flask	Upright	10 ml.
35-ml. microfernbach flask	Upright	10 ml.

2. Preparation of Samples for Assay

a. Blood. An accurate estimation of the amount of cobalamins in blood may be obtained by assaying the serum. Blood is obtained by venipuncture, allowed to clot and the serum aseptically drawn off. To measure

[10] H. A. Nathan and H. B. Funk, *Am. J. Clin. Nutrition* **7**, 375 (1959).

[11] H. A. Nathan, H. Baker, and O. Frank, *Nature* **188**, 35 (1960).

[12] H. A. Barker, R. D. Smyth, H. Weissbach, A. Mundt-Petersen, J. I. Toohey, J. N. Ladd, B. E. Volcani, and R. M. Wilson, *J. Biol. Chem.* **235**, 181 (1960).

free cobalamins, the serum is aseptically diluted with water or buffer and added to the assay tubes. To free bound cobalamins, the serum is diluted 1:10 (aseptic measures are not necessary) with a modification of the buffer of Baker et al.[13]: potassium metabisulfite, 0.01 % and *trans*-aconitic acid, 0.5% adjusted to pH 4.5 with triethanolamine. The metabisulfite acts as a stabilizing agent.

The diluted serum is heated in a 100°C. water bath for 30 minutes. The precipitated proteins, now free of cobalamins, are separated by centrifugation. The supernatant is diluted so that the vitamin content falls within assay range. Normal serum contains 200–800 $\mu\mu$g. B_{12}/ml.

b. Other body fluids and liquid samples. The method outlined for blood serum can be used. Some average values to use as a guide for dilutions are: urine 0–200 $\mu\mu$g./ml.; cerebrospinal fluid 0–30 $\mu\mu$g./ml.

c. Solid samples. Material obtained from biopsy, autopsy, or harvest of microbial crop is frozen until prepared for assay. Such materials to be assayed should be touched only with B_{12}-free utensils. The tissue is diluted 1:10 with pH 4.5 buffer (Section II.B.2.*a*) and homogenized. Hydrolysis and subsequent treatment is the same as for blood serum.

3. Media

a. Maintenance. The medium given in Table IV is suitable. Either

TABLE IV

MAINTENANCE MEDIUM

Component	gm./liter final medium
KH_2PO_4	0.5
$MgSO_4 \cdot 7H_2O$	0.2
NaCl	0.1
Yeast extract (or autolysate)	5.0
Trypticase	5.0
Sucrose	12.0
Liver extract[a]	0.05
$FeSO_4 \cdot 7H_2O$	0.001
Agar (only for semisolid medium)	5.0
Adjust to pH 5.0–5.5	

[a] Any commercial preparation may be used.

liquid or semisolid versions may be used. This medium may also be used to maintain the other three assay organisms with only a change in the pH of the medium.

[13] H. Baker, H. Sobotka, I. Pasher, and S. H. Hutner, *Proc. Soc. Exptl. Biol. Med.* **91,** 636 (1956).

The algae kept in liquid media should be transferred at 2 to 3 week intervals, depending upon the incubation temperature. A vigorous culture is one in which more than 50% of the organisms are growing near the surface. Limiting aeration, by keeping the maintenance tubes in an upright position, can be used to extend the time between transfers. After the appearance of growth, cultures in semisolid medium may be stored at 10°C. for approximately 3 months without loss of viability.

b. Assay. The medium in Table V, a modification of the one used by Hutner *et al.*[9] is adequate for assays in both the light and dark.

TABLE V

B$_{12}$ ASSAY MEDIUM FOR *Ochromonas malhamensis*

Component	Weight/liter double strength medium	
(NH$_4$)$_2$H citrate	2.4	gm.
L-Glutamic acid	6.0	gm.
DL-Methionine	1.2	gm.
L-Histidine · HCl	1.0	gm.
L-Arginine · HCl	1.0	gm.
Hycase—SF.	2.0	gm.
Sucrose (fine grain, grocery store grade)	30.0	gm.
CaCO$_3$	0.3	gm.
MgCO$_3$ (basic)	1.0	gm.
KH$_2$PO$_4$	0.6	gm.
NH$_4$HCO$_3$	0.8	gm.
FeCl$_3$	60.0	mg.
ZnSO$_4$ · 7H$_2$O	40.0	mg.
MnSO$_4$ · H$_2$O	6.0	mg.
CuSO$_4$ · 5H$_2$O	0.62	mg.
CoSO$_4$ · 7H$_2$O	5.0	mg.
H$_3$BO$_3$	1.14	mg.
(NH$_4$)$_6$Mo$_7$O$_{24}$ · 4H$_2$O	1.34	mg.
Thiamine · HCl	10.0	mg.
Biotin	0.02	mg.

If necessary, adjust to pH 4.8–5.2 with H$_2$SO$_4$ or KOH

The assay medium can be conveniently prepared as a dry mix in lots sufficient for 20 to 40 liters. The dry mix should be placed in tightly closed glass, or preferably plastic, bottles and stored in a cool dry cabinet. Storage under these conditions for as long as a year has no adverse effect on subsequent assays.

c. Inoculation. Cells grown in the liquid maintenance medium are ordinarily used. If B_{12} carry-over from such a preparation is too large, cells may be grown in assay medium (Table V) to which vitamin B_{12} (10.0 $m\mu g.$ %) has been added.

4. Assay Procedure

a. Preparation of inoculum. One or two drops of heavy top growth in liquid medium (Section II.3.*c*) are diluted with sterile distilled water so that when a sample of this suspension is viewed at 100× magnification, each field contains about 100 organisms. One drop of this dilute suspension is used to inoculate each assay vessel. If the growth in the tubes of liquid medium is too evenly spread to ensure the removal of sufficient organisms without carrying over significant amounts of vitamin B_{12}, the cells should be concentrated by centrifugation, the supernatant decanted, and the cell concentrate diluted as outlined above.

b. The standard curve. For an assay in which a total of 5.0 ml. of final medium is contained in each tube or flask, the standard curve, in duplicate, is prepared according to the outline in Table II. A general outline for executing an assay is given in Section II.A.

5. Incubation Conditions

Assays are incubated for 4 to 7 days at 24° to 28°C. in the dark or illuminated with warm white fluorescent lamps. Although the medium given here is sufficient for algal growth in both light and dark, workers who routinely incubate *Euglena gracilis* assays (Section II.C) in the light may prefer to do the *Ochromonas malhamensis* assay under the same conditions.

6. Evaluation of Assay Results

Algal growth may be measured with almost any densitometer equipped with a red filter. If a spectrophotometer is used, it should be set at 525 $m\mu$ so that the results indicate amount of cell mass (turbidity) and not green pigment production (which may vary with changes in assay method independent of amount of cobalamin present). The validity of results obtained with the particular measuring instrument may be tested by making accurately measured dilutions of a dense algal suspension and determining the range over which Beer's law is followed. Protocols of *Ochromonas* and *Euglena* assays done on the same samples will be found in Section C.6.

C. *Euglena gracilis* Z Strain (ATCC 12716)

This organism is not as specific as *Ochromonas malhamensis*, but it shares with *O. malhamensis* the property of responding only to cobalamins. The wide range of cobalamins which satisfy the growth factor requirement of *Euglena gracilis* (Table I) coupled with the failure, thus far, to bypass this requirement[13a] have made this the organism of choice for measuring "total B_{12}." There have been reports of natural materials which have more activity for *Ochromonas malhamensis* than for *Euglena gracilis*, but no explanation is available yet.[10,11]

Although this assay requires 4–7 days of incubation, its usefulness for measuring "total" cobalamins compensates for the time required.

1. Culture Vessels

The distribution of medium/tube or flask is given in Table III.

2. Preparation of Samples for Assay

The same methods are used as for *Ochromonas malhamensis* assay (Section II.B.2.). Many workers run both *O. malhamensis* and *Euglena gracilis* assays for all samples to measure both "true" and "total" cobalamin content. Prior to assay with *E. gracilis*, samples should be adjusted to pH 3.6 unless the sample to be added to the assay vessel will not alter the pH of the assay medium.

3. Media

a. Maintenance. The medium given in Table IV is suitable. If desired, the medium may be adjusted to pH 3.6 for use with *E. gracilis* but this pH adjustment is not essential.

b. Assay. The assay medium given in Table VI is a modification of the one used by Hutner *et al.*[14] The assay medium may be conveniently prepared as a dry mix sufficient for 20 to 40 liters of final assay medium. The dried ingredients should be placed in tightly closed glass, or preferably plastic, bottles and stored in a cool dry cabinet. Storage under these conditions for as long as a year has no adverse effect on subsequent assays.

c. Inoculation. Cells are grown in assay medium (Table VI) to which vitamin B_{12} (10.0 mμg./100 ml.) has been added.

4. Assay Procedure

a. Preparation of inoculum. Cells grown for inoculum are concentrated by centrifugation and resuspended in sterile distilled water so that a sample of this suspension viewed at 100× magnification contains about 100 organisms/field.

[13a] *Note added in proof:* Under the conditions of these assays, neither β-methyl-aspartate nor methyl-malonate bypass the B_{12} requirement of *E. gracilis* or *O. malhamensis* [H. A. Nathan and H. B. Funk, *Proc. Soc. Exptl. Biol. Med.* **109,** 213 (1962)].

[14] S. H. Hutner, M. K. Bach, and G. I. M. Ross, *J. Protozool.* **3,** 101 (1956).

TABLE VI

ASSAY MEDIUM[a] FOR *Euglena gracilis* Z STRAIN

Component	Weight/liter double strength medium
KH₂PO₄	0.6 gm.
MgSO₄·7H₂O	0.8 gm.
L-Glutamic acid	6.0 gm.
CaCO₃	0.16 gm.
NH₄HCO₃	0.72 gm.
Sucrose	30.0 gm.
DL-Aspartic acid	4.0 gm.
DL-Malic acid	2.0 gm.
Glycine	5.0 gm.
Succinic acid	1.04 gm.
Thiamine·HCl	12.0 mg.
FeCl₃	60.0 mg.
ZnSO₄·7H₂O	40.0 mg.
MnSO₄·H₂O	6.0 mg.
CuSO₄·5H₂O	0.62 mg.
CoSO₄·7H₂O	5.0 mg.
H₃BO₃	1.14 mg.
(NH₄)₆Mo₇O₂₄·4H₂O	1.34 mg.

Adjust to pH 3.6 with KOH and H₂SO₄

[a] This medium is available from Difco Laboratories, Detroit 1, Michigan, under the name 0532 Bacto-Euglena B-12 Medium.

b. The standard curve. The following values of vitamin B_{12} (in $\mu\mu$g./ml.) will give a standard curve with points evenly distributed (when the B_{12} concentrations are plotted on the log scale of semilog paper): none; 0.3; 0.6; 1.0; 3.0; 6.0; 10.0; 20.0; 30.0; 60.0. See Table II for details of preparation of a standard curve. A general outline for executing an assay is given in Section II.A.

5. Incubation Conditions

Assays are incubated for 4 to 7 days at 24° to 28°C. illuminated with warm white fluorescent lamps. Between 100 and 300 foot candles is satisfactory.

6. Evaluation of Assay Results

See Section II.B.6.

Typical protocols of standard curves and serum samples for the two

Assay ___B_12___

Organism ___O. malhamensis___

Date _____

Harvested _____

Distribution in ml./vessel ___5.0___   pH ___4.8-5.2___

Incubation at ___25°-27° C.___

Instrument used for reading ___Spectronic 20 set at 525 mμ___

General assay method: ___2.5___ ml. 2× strength dilution + ___2.5___ ml. 2× strength base/tube

Standard Curve				Operational information μμg./ 10 ml. H$_2$O (this is 2 × strength)
Tube no.	Contents	Reading		
1	B$_{12}$ none	0.04	0.04	0
2	0.3 μμg./ml.	0.07	0.07	6.0
3	0.6	0.10	0.12	12.0
4	1.0	0.13	0.15	20.0
5	3.0	0.23	0.25	60.0
6	6.0	0.39	0.37	120.0
7	10.0	0.51	0.49	200.0
8	20.0	0.74	0.70	400.0
9	30.0	0.86	0.85	600.0
10	60.0	1.00	1.02	1200.0

Comments:

Fig. 1. Protocol of standard curve for B$_{12}$ assay of *O. malhamensis*.

algal assays are given in Figs. 1–4. The samples, normal, d'Guglielmo's disease, and pernicious anemia, are representative of the range likely to be encountered in clinical practice.

D. *Escherichia coli* 113-3 (ATCC 11105)

This organism responds to Factor B, a wide variety of cobalamins and to methionine (Table I), which bypasses the cobalamin requirement. The advisability of using this organism to assay crude samples is questionable. If samples have been purified chromatographically or are known not to contain methionine, this organism provides a rapid (16–24 hours) assay for total cobalamin and Factor B.

Assay B$_{12}$ Date   Organism   *O. malhamensis*

Standard curve is on P

Patient's name and age   N. B.
Provisional diagnosis (before assay)   d' Guglielmo's disease
Body fluid being assayed and stock dilution   Serum 1:10

Tube no.	ml. fluid / ml. base	Readings		Final dilution factor	Calculated μμg. B$_{12}$/ml.	Operational ml./ 10 ml. H$_2$O (for 2 ×)
	0.1	0.74	0.75	100	2150	2.0
	0.4	0.74*	0.73*	25+	2100	8.0

Comments: *Reading of 1:4 dilution
+Estimated B$_{12}$ multiplied by 4 to compensate for 1:4 dilution

Patient's name and age   L. P.
Provisional diagnosis (before assay)   Pernicious Anemia
Body fluid being assayed and stock dilution   Serum 1:10

Tube no.	ml. fluid / ml. base	Readings		Final dilution factor	Calculated μμg. B$_{12}$/ml.	Operational ml./ 10 ml. H$_2$O (for 2 ×)
	0.2	0.10	0.09	50	25	4.0
	0.4	0.12	0.12	25	17	8.0

Comments:

Patient's name and age   S. D. F.
Provisional diagnosis (before assay)   Normal
Body fluid being assayed and stock dilution   Serum 1:10

Tube no.	ml. fluid / ml. base	Readings		Final dilution factor	Calculated μμg. B$_{12}$/ml.	Operational ml./ 10 ml. H$_2$O (for 2 ×)
	0.2	0.50	0.50	50	500	4.0
	0.4	0.73	0.71	25	550	8.0

Comments:

FIG. 2. Protocol of serum samples for B$_{12}$ assay of *O. malhamensis*.

1. Culture Vessels

Any of the vessels listed in Table III may be used. It is most convenient to use tubes containing 10 ml. of medium. Tubes need not be slanted.

2. Preparation of Samples for Assay

It is not recommended that this organism be used for the assay of

Assay ___B$_{12}$_____

Organism ___*E. gracilis*___   Date _____

                                                                Harvested _____

Distribution in ml./vessel ___5.0___   pH ____3.6___

Incubation at ____25°-27° C. under fluorescent lights_____

Instrument used for reading ____Spectronic 20 set at 525 mμ_____

General assay method: __2.5__ ml. 2× strength dilution + __2.5__ ml. 2× strength base/tube

Standard Curve			
Tube no. Contents	Reading		Operational information μμg./ 10 ml. H$_2$O (this. is 2 × strength)
1 B$_{12}$ none	0.10	0.09	0
2 0.3 μμg./ml.	0.16	0.14	6.0
3 0.6	0.21	0.22	12.0
4 1.0	0.24	0.26	20.0
5 3.0	0.48	0.48	60.0
6 6.0	0.75	0.68	120.0
7 10.0	0.95*	0.90*	200.0
8 20.0	1.70*	1.30*	400.0
9 30.0	1.80*	1.60*	600.0
10 60.0	2.00*	2.65*	1200.0

Comments: *Diluted 1:4 before reading in the Spectronic 20 and the reading
            multiplied by 4 to obtain the numbers recorded.

FIG. 3. Protocol of standard curve for B$_{12}$ assay of *E. gracilis*.

natural materials without prior knowledge of the methionine content. Amounts of methionine insufficient to bypass the vitamin B$_{12}$ requirement of *E. coli* may spare B$_{12}$ enough to invalidate the assay.

This organism provides an excellent rapid method for the assay of chromatogram eluates (see Sections III.A and B). Samples prepared according to the method outlined in Section II.B.2 should be adjusted to pH 6.5–7.0 and diluted for assay.

3. Media

a. Maintenance. The medium outlined in Table IV, but adjusted to pH 7 is used. For long term maintenance, use agar (1.5%) slants of this medium.

Assay B_{12} Date   Organism   *E. gracilis*

Standard curve is on P

Patient's name and age _____N. B._____
Provisional diagnosis (before assay) ___d' Guglielmo's disease___
Body fluid being assayed and stock dilution____Serum 1:40____

Tube no.	$\frac{ml.\ fluid}{ml.\ base}$	Readings		Final dilution factor	Calculated $\mu\mu g.\ B_{12}/ml.$	Operational ml./ 10 ml. H_2O (for 2 ×)
	0.05	0.38	0.40	800	2000	1.0
	0.1	0.60	0.59	400	2000	2.0

Comments:

Patient's name and age ___L. P.___
Provisional diagnosis (before assay) ___Pernicious Anemia___
Body fluid being assayed and stock dilution___Serum 1:10___

Tube no.	$\frac{ml.\ fluid}{ml.\ base}$	Readings		Final dilution factor	Calculated $\mu\mu g.\ B_{12}/ml.$	Operational ml./ 10 ml. H_2O (for 2 ×)
	0.1	0.20	0.19	100	52	2.0
	0.2	0.24	0.25	50	45	4.0

Comments:

Patient's name and age. ___S. D. F.___
Provisional diagnosis (before assay) ___Normal___
Body fluid being assayed and stock dilution ___Serum 1:10___

Tube no.	$\frac{ml.\ fluid}{ml.\ base}$	Readings		Final dilution factor	Calculated $\mu\mu g.\ B_{12}/ml.$	Operational ml./ 10 ml. H_2O (for 2 ×)
	0.1	0.68	0.65	100	560	2.0
	0.2	0.52*	0.51*	50[+]	580	4.0

Comments: *Diluted 1:2 before reading
[+]Estimated B_{12} multiplied by 2 to compensate for
1:2 dilution

FIG. 4. Protocol of serum samples for B_{12} assay of *E. gracilis*.

b. *Assay.* The medium in Table VII, a modification of the medium of
Burkholder[15] is used. The assay medium can be conveniently prepared as
a dry mix in lots sufficient for 20 to 40 liters of final medium. The dry mix
(note that neither glucose nor $(NH_4)_2SO_4$ is included in the dry mix)
should be placed in tightly closed glass, or preferably plastic, bottles and

[15] P. R. Burkholder, *Science* **114,** 459 (1951).

TABLE VII

ASSAY MEDIUM FOR *Escherichia coli* 113-3

Component	gm./liter final medium
KH$_2$PO$_4$	6.0
K$_2$HPO$_4$	14.0
Na$_2$citrate·2H$_2$O	1.0
MgSO$_4$·7H$_2$O	0.2
DL-Asparagine·H$_2$O	8.0
L-Arginine·HCl	0.2
L-Glutamic acid	0.2
Glycine	0.2
L-Histidine	0.2
L-Tryptophan	0.2
L-Proline	0.2
(NH$_4$)$_2$SO$_4$ (do not include in dry mix)	2.0
Glucose (added aseptically)	10.0

Adjust to pH 6.8–7.2

stored in a cool, dry cabinet. Glucose is prepared separately as a sterile 50% (w/v) solution. When assay vessels are ready for inoculation, 0.2 ml. of the sterile glucose solution is added aseptically to each vessel.

c. Inoculation. Cells are grown in assay medium (Table VII) to which vitamin B$_{12}$ (30.0 mμg./100 ml.) has been added.

4. Assay Procedure

a. Preparation of inoculum. Cells for inoculum are harvested by centrifugation, washed, and resuspended in sterile distilled water so that the resulting suspension is barely cloudy to the eye (optical density about 0.05). One drop of this dilute suspension is used to inoculate each assay tube.

b. The standard curve. The following values of vitamin B$_{12}$ (in $\mu\mu$g./ml.) give a standard curve with points evenly distributed (when the B$_{12}$ concentrations are plotted on the log scale of semilog paper): none; 1.0; 3.0; 6.0; 10.0; 30.0; 60.0; 100.0; 300.0. The response curve to methionine ranges from 0.3 to 100.0 μg./ml. See Table II for an example of the preparation of a standard curve.

5. Incubation Conditions

Assays and tubes for inocula are incubated for 16 to 24 hours at 37°C.

6. Evaluation of Assay Results

The methods outlined in Section II.B.6 are used.

E. *Lactobacillus leichmannii* (ATCC 4797 or 7830)

This organism responds to a wide variety of cobalamins (Table I) and to deoxyribosides, which bypass the cobalamin requirement. This assay should always be done in two parts: (a) for cobalamins plus deoxyribosides and (b) for deoxyribosides alone. When used in this fashion, the estimated cobalamins assayed agree with the estimates obtained by *Euglena gracilis* assay (Section II.C). This organism provides a rapid (24–48 hours) assay for total cobalamins and deoxyribosides.

1. Culture Vessels

See Section II.D.1.

2. Preparation of Samples for Assay

It is not recommended that this organism be used for the assay of natural materials unless assay for deoxyribosides is done in the same assay. Presence of amounts of deoxyribosides insufficient to bypass the vitamin B_{12} requirement of *Lactobacillus leichmannii* may spare B_{12} enough to invalidate the assay. Thus it is essential that even small amounts of deoxyribosides be accounted for.

To measure both cobalamins and deoxyribosides, samples are prepared according to the method outlined in Section II.D.2. To measure deoxyribosides alone, the samples are adjusted to pH 11 with KOH and autoclaved at 121°C. for 30 minutes, after which they are cooled, adjusted to pH 6.5–7.5 with H_2SO_4, and centrifuged. The supernatant is saved for assay. The difference between the results of the assay for cobalamin plus deoxyribosides and the assay for deoxyribosides is attributable to cobalamins.

3. Media

a. *Maintenance.* See Section II.D.3.a. For long-term maintenance, use stabs of this medium solidified with agar (1.5%).

b. *Assay.* The medium is commercially available from the Difco Company, 920 Henry Street, Detroit 1, Michigan, as a dry mix.

c. *Inoculation.* The assay medium with vitamin B_{12} (10.0 $\mu\mu$g./ml.) added is used.

4. Assay Procedure

a. *Preparation of inoculum.* The last of a series of at least three serial transfers should be used. Cells for inoculum are centrifuged, washed, and resuspended in sterile distilled water so that the resulting suspension is barely cloudy to the eye (optical density about 0.05). One drop of this dilute suspension is used to inoculate each assay tube.

b. The standard curve. The following values of vitamin B_{12} ($\mu\mu$g./ml.) give a standard curve: none; 1.0; 3.0; 6.0; 8.0; 10.0; 20.0. See Table II for an example of the preparation of a standard curve.

5. Incubation Conditions

Assays and tubes for inocula are incubated for 24 to 48 hours at 37°C.

6. Evaluation of Assay Results

The methods outlined in Section II.B.6 are used.

III. Plate Methods

Plate methods are most useful when, combined with chromatography, they are used to locate B_{12}-activity on paper chromatograms. Such procedures permit the identification of active compounds which are available in very small amounts. It is important to note that plate assays are cumbersome and impractical for routine quantitative estimations of cobalamins. The sensitive quantitative methods are outlined in Section II.

A. Methods Applicable to All Plate Assays

1. Bioautographs

After paper chromatograms which contain a mixture of substances are developed with solvents, the strips containing separated unknowns are laid on the seeded plates and allowed to remain there for the duration of the incubation period. The R_f of the active material is determined at the conclusion of the assay by the location of the areas of microbial growth. For quantitative estimation of the concentration of a compound in an active area, one of the turbidimetric methods (Section II) should be used. Unknowns can be identified by matching the R_f of the active area of the chromatogram with the R_f of a known which was run on a parallel chromatogram under the same conditions. It is not necessary to sterilize the chromatograms to be used with assay organisms whose assay growth period is 24 hours or less: the results will be evident before contaminating organisms appear.

2. Disk Assays

This method is included for historical purposes only. It is less sensitive, and requires more effort to execute than do the turbidimetric assays given in Section II.

The compounds to be tested, as well as the standards, are impregnated into filter paper disks of the type used for antibiotic assays. Disks 12.7 mm. diameter, which hold 0.1 ml. fluid, can be purchased from Schleicher and Schuell, Keene, New Hampshire. The impregnated disks are sterilized and placed on the surfaces of seeded plates.

3. Assay Procedure

Prepare and sterilize assay medium.

Sterilize petri or other dishes to be used as containers.

Prepare dilutions of vitamin B$_{12}$ to be used for standards (see Section II.A) and dilutions of unknowns.

Prepare and run chromatograms to be used for bioautographs.

Prepare inoculum and seed sterile assay medium.

Pour seeded assay medium into the sterile plates and allow the agar to solidify.

Place chromatograms or impregnated disks on the surface of the seeded agar.

Incubate all plates under the same conditions for the required time.

Mark the R_f of the active spots on the chromatogram or measure the diameter of the growth zones around the disks, using a Fisher-Lilly zone reader or a millimeter ruler.

Prepare a standard curve for the disk assays: B$_{12}$ concentration plotted against diameter of growth zones. Concentration of B$_{12}$ in each disk carrying a sample can be determined by interpolation of diameter of growth zone on the standard curve.

Auxiliary method for preparing plates. The sensitivity of the assay can be increased by preparing two-layered agar plates. The bottom or base layer contains unseeded assay agar and is made just thick enough to cover the surface of the plate. The top or seed layer contains the assay organism and at times indicators to intensify visualization of the growth zones.

B. *Escherichia coli* 113-3

This organism is the most useful for bioautography because of its low specificity (Table I) and speed of assay development. This low specificity and low sensitivity (Table I) makes disk assays useless for all but purified samples, and the sensitive turbidimetric assays (Section II) are recommended for purified materials.

1. Culture Vessels

Petri dishes or glass baking trays are used. The baking trays are available in several sizes ranging from 6-inch squares to rather heroic proportions.

2. Preparation of Samples for Assay

See Sections II.B.2 and III.A.

3. Media

a. *Maintenance.* See Section II.D.3.
b. *Assay.* The medium given in Table VII is solidified with 1.5% agar.
c. *Inoculation.* See Section II.D.3.

4. Assay Procedure

The general procedure is outlined in Sections III.A.3 and 3.a. The two-layered plates are used with either Andrade's indicator at 1.0 ml./100 ml. of medium or triphenyl tetrazolium chloride at 20 mg./ml. of medium added to the seed layer when the plates are poured. In addition to defining clearly the growth zone, indicators reduce the length of incubation time necessary before the assay can be read.

Inoculum is prepared by harvesting the overnight growth contained in a 10-ml. tube of inoculation medium (Section III.B.3) and washing the cells once with sterile distilled water. This procedure provides enough cells to inoculate 100 ml. of assay agar.

The range of the assay is given in Table I.

5. Incubation Conditions

See Section II.D.5.

6. Evaluation of Assay Results

See Section III.A.

C. *Lactobacillus leichmannii*

This organism is not as useful as is *Escherichia coli* for either bioautography or disk assays. *Lactobacillus leichmannii* is less sensitive, has a narrower assay range, and requires a longer incubation period. The lengthening of the incubation time makes it essential that all samples to be assayed be added aseptically to prevent contamination of the plates before the assay can be read.

1. Culture Vessels

See Section III.B.1.

2. Preparation of Samples for Assay

See Sections II.E.2 and III.A.

3. Media

a. *Maintenance.* See Section II.E.3.

b. *Assay.* An agar-containing assay medium is commercially available from the Difco Company, Detroit, Michigan.

c. *Inoculation.* See Section II.E.3.

4. Assay Procedure

The same methods outlined in Section III.B.4 are used.

5. Incubation Conditions

See Section II.E.5. It may be necessary to extend the incubation time longer than 24 hours if the plates are lightly seeded.

6. Evaluation of Assay Results

See Section III.A.

D. Euglena gracilis Z Strain

Euglena gracilis is theoretically the assay organism of choice because of the wide range of cobalamins which satisfy the growth requirement and the high sensitivity of the assay. However, such technical difficulties as the length of time required for the assay (usually 7 days), so that all samples must be sterilized before assay, and the special procedure which must be followed for preparation of the assay medium limit its usefulness. Labile substances should not be assayed with this organism. A special use for *E. gracilis* plate methods is given in Section III.D.7.

1. Culture Vessels

See Section III.B.1.

2. Preparation of Samples for Assay

See Sections II.B.2 and III.A. All samples should be sterilized before they are placed on the plates.

3. Media

a. *Maintenance.* See Section II.C.3.

b. *Assay.* Because of the low pH of the medium, the agar and liquid phases must be sterilized separately to prevent hydrolysis of the agar. Thus double strength liquid assay medium (Section II.C.3) and double strength agar (3.0%) are sterilized separately, allowed to cool to 50°C., and then mixed aseptically. The assay medium is now single strength.

The single strength agar assay medium is kept in a 50°C. water bath while it is being dispensed. Do not reheat this medium above 55°C.

c. *Inoculation.* See Section II.C.3.

4. Assay Procedure

a. *Preparation of inoculum.* Inoculum is grown as outlined in Section II.C. After concentration of cells by centrifugation, they are diluted so that there are 10^6 organisms/ml. One milliliter of the dilute suspension is sufficient to seed about 100 ml. of assay agar.

b. *Special procedures.* The general outline of the assay method is given in Section III.A. The inoculum should be added to the molten agar *just before* the plates are to be poured to minimize the length of time the inoculum must remain at the elevated temperature. These organisms are very heat sensitive and prolonged incubation at elevated temperatures will affect the quality of the assay. All samples should be sterilized and then added aseptically to the plates.

5. Incubation Conditions

See Section II.C.5.

6. Evaluation of Assay Results

See Section III.A.

7. Special Uses

To identify possible cobalamin-producing microorganisms in a mixed culture: The mixed culture is grown on appropriate agar medium and then layered with seeded *E. gracilis* assay agar. *Euglena gracilis* growth zones will appear over and around the cobalamin-producing colonies. The active colony may be isolated for further study.

E. Ochromonas malhamensis

Bioautography with *O. malhamensis* permits location of animal active material in one step.

1. Culture Vessels

See Section III.B.1.

2. Preparation of Samples for Assay

See Sections II.B.2 and III.A. All samples should be sterilized before they are placed on the plates.

3. Media

a. *Maintenance.* See Section II.B.3.

b. *Assay.* Because of the low pH of the medium, the agar and liquid phases must be sterilized separately to prevent hydrolysis of the agar. Thus double strength liquid assay medium (Section II.B.3) and double strength agar are sterilized separately, allowed to cool to 50°C., and then mixed aseptically. The assay medium is now single strength. The single strength agar assay medium is kept in a 50°C. water bath while it is being dispensed. Do not reheat this medium above 55°C.

c. Inoculations. See Section II.B.3.

4. Assay Procedure

a. Preparation of inoculum. Inoculum is grown as outlined in Section II.B. After concentration by centrifugation, the cells are diluted so that there are between 10^6 and 10^7 organisms/ml. One milliliter of the dilute suspension is sufficient to seed about 100 ml. of assay agar.

b. Special procedures. See Section III.D.4.*b.*

5. Incubation Conditions

See Section II.C.5.

6. Evaluation of Assay Results

See Section III.A.

7. Special Uses

To identify microorganisms, in mixed culture, which produce animal-active cobalamins: See Section III.D.7.

IV. Nonmicrobiological Methods

A. Spectrophotometry

This method is suitable for checking the purity of crystalline vitamin B$_{12}$ (Section I.A.3) and for the estimation of vitamin B$_{12}$ in relatively pure samples. Spectrophotometric methods are 10^5 to 10^6 times less sensitive than microbiological methods.

Spectrophotometry does not clearly distinguish between several of the cobalamins which can be differentially measured microbiologically. Other cobalamins which are active clinically may give false low values when estimated spectrophotometrically because their absorption maxima are at different wavelengths from those of the standard, vitamin B$_{12}$.

Methods for separation and subsequent spectrophotometric assay are discussed elsewhere.[16–19]

[16] M. M. Marsh and N. R. Kuzel, *Anal. Chem.* **23**, 1773 (1951).

[17] G. O. Rudkin, Jr. and R. J. Taylor, *Anal. Chem.* **24**, 1155 (1952).

[18] W. J. Mader and R. G. Johl, *J. Am. Pharm. Assoc. Sci. Ed.* **44**, 577 (1955).

[19] P. J. Van Melle, *J. Am. Pharm. Assoc. Sci. Ed.* **45**, 26 (1956).

B. Isotope Dilution

This method is specific, but only when coupled with rigorous purification.[20,21] A known amount of radioactive vitamin B_{12}, usually labeled with Co^{60}, is added to a sample, then vitamin B_{12} is isolated and purified. The purification is monitored by periodically measuring the radioactivity.

This method is also useful clinically for kinetic measurements of the excretion of vitamin B_{12} administered either orally or by injection.

[20] F. A. Bacher, A. E. Boley, and C. E. Shonk, *Anal. Chem.* **26**, 1146 (1954).
[21] E. L. Smith, *Analyst* **81**, 435 (1956).

7.11 Lactobacillus leichmannii Assay for Vitamin B₁₂

HELEN R. SKEGGS

Merck Sharp & Dohme Research Laboratories, West Point, Pennsylvania

The use of microbiological assays for the determination of vitamin B₁₂ spread much more rapidly than knowledge of the idiosyncrasies of the assay procedure or the complex nature of the cobalamins. The several assay methods employing lactobacilli were reviewed previously.[1] We are presenting the details of the assay procedure in use in these laboratories, along with some of the difficulties we have encountered in the hope that our experiences will be of value in solving problems encountered by others, regardless of the assay method employed. All microbiological assays require meticulous attention to details; but the extreme sensitivity

[1] L. D. Wright and H. R. Skeggs, *in* "Vitamin Methods" (P. György, ed.), Vol. II, p. 683 Academic Press, New York, 1951.

of the vitamin B_{12} assay makes it more susceptible than most to errors of omission on the part of the assayist.

Two reviews have appeared recently which cover admirably the complexities involved in the assay of vitamin B_{12}.[2,3]

I. Test Organism

Two strains of *Lactobacillus leichmannii*, ATCC 7830 [4,5] and ATCC 4797 [6,7] are used in the assay of vitamin B_{12}. Strain 7830 usually grows more rapidly and total growth is heavier than that obtained with strain 4797. Either strain may be used in the recommended medium.

The advantages of the *L. leichmannii* assays are their speed and sensitivity. The assay range approaches the sensitivity of *Euglena gracilis*[8] and exceeds that of *Ochromonas malhemensis*.[9] In many instances, failure of agreement between assays with various organisms relates more to discrepancies in sample preparation than to differences in organism responses.

The disadvantages of the *Lactobacillus leichmannii* assays lie in the ability of the organism to utilize the derivatives of deoxyribonucleic acid, the deoxyribotides and deoxyribosides (e.g., thymidine) and pseudovitamins B_{12}, Factor A, etc., for growth in the absence of vitamin B_{12}. Such interfering substances may be removed from the sample prior to assay,[10] distinguished by bioautography[11] or differential assay.[9,12,13] Frequently, interference by deoxynucleic acids may be ruled out by simple dilution since they can substitute for vitamin B_{12} only at levels above 1 μg./ml. of assay solution. The pseudovitamins B_{12} and Factor A, on the other hand, have been reported to have 20–100% of the activity of vitamin B_{12}.[3,9,12] These factors are of particular importance in the assay of samples that

[2] R. Gräsbeck, *Advances in Clin. Chem.* **3**, 299 (1960).

[3] W. H. Shaw and C. J. Bessell, *Analyst* **85**, 389 (1960).

[4] C. E. Hoffman, E. L. R. Stokstad, B. L. Hutchings, A. C. Dornbush, and T. H. Jukes, *J. Biol. Chem.* **181**, 635 (1949).

[5] "U. S. Pharmacopeia," Vol. XV, p. 885. Mack Publ., Easton, Pennsylvania, 1955.

[6] H. T. Thompson, L. S. Dietrich, and C. A. Elvehjem, *J. Biol. Chem.* **184**, 175 (1950).

[7] H. R. Skeggs, H. M. Nepple, K. A. Valentik, J. W. Huff, and L. D. Wright, *J. Biol. Chem.* **184**, 211 (1950).

[8] S. H. Hutner, M. K. Bach, and G. I. M. Ross, *J. Protozool.* **3**, 101 (1956).

[9] J. E. Ford, *Brit. J. Nutrition* **7**, 299 (1953).

[10] J. M. McLaughlin, C. G. Rogers, E. J. Middleton, and J. A. Campbell, *Can. J. Biochem. and Physiol.* **36**, 195 (1958).

[11] W. A. Winsten and E. Eigen, *Proc. Soc. Exptl. Biol. Med.* **67**, 513 (1948).

[12] E. Hoff-Jorgensen, *J. Biol. Chem.* **178**, 525 (1949).

[13] J. E. Ford and J. W. G. Porter, *Brit. J. Nutrition* **7**, 326 (1953).

have undergone microbial fermentation, notably feces, but are of little consequence in the assay of animal tissues or pharmaceutical preparations.

II. Standard Solutions

Cyanocobalamin is the reference standard. It may be either a solution of crystalline vitamin B_{12} in water or a triturate as supplied by the U.S.P. which is weighed according to directions and diluted with water. A convenient stock solution is prepared (20 μg./ml.) and stored in the refrigerator. It should not be exposed to excessive light. The assay standard is prepared by diluting from the stock to a concentration of 0.02 mμg./ml. For example, a 20 μg./ml. stock solution may be diluted in steps of 100 to give a solution of 0.00002 μg. (0.02 mμg.) per milliliter. The standard solution is dispensed into tubes according to the design of the assay being employed.

III. Stability of Cobalamins

Crystalline cyanocobalamin[14] itself is stable in air and is not affected by moisture. The anhydrous compound, however, is very hygroscopic and when exposed to moist air may absorb about 12% water. Although vitamin B_{12} is slowly decomposed by ultraviolet or visible light, the crystalline solid and its aqueous solutions show no significant decomposition during exposure at room temperature to normal indoor illumination or indirect sunlight during processing, but prolonged exposure to strong light should be avoided. Aqueous solutions at pH 4.6–7.0 show no decomposition during extended storage at 25°C. For optimum stability at elevated temperatures the solutions should be adjusted to pH 4.0–4.5. A rapid method for determining chemical compatibility with various substances in aqueous solution consists of heating a buffered pH 4.0 M Na acetate-acetic acid solution of crystalline vitamin B_{12} and the material to be tested for 4 hours at 100°C. If assays of the heated mixture differ by less than 10% from the control solution containing only vitamin B_{12}, the substance under test can be assumed to be compatible. Cyanocobalamin is decomposed in aqueous solutions containing thiamine and ascorbic acid unless a stabilizer such as a trace amount of iron is present, or a suitable vehicle such as equal parts of propylene glycol and glycerol is employed.

Most of the cobalamins present in animal tissue are noncyano and

[14] Vitamin B₁₂, Merck Service Bulletin, p. 7, 1958.

are considerably less stable than cyanocobalamin.[15,16] Such cobalamins readily convert to cyanocobalamin when exposed to cyanide although conditions under which conversion occurs may vary with the exact nature of the original complex. Free noncyanocobalamins convert to cyanocobalamin at room temperature, but protein bound cobalamins require prior denaturation of the protein by heat. The instability of such noncyanocobalamins is responsible for the use of a reducing substance in the *Lactobacillus leichmannii* assay methods. During early work with the assays, discrepancies in assay results were found between *L. lactis* and *L. leichmannii* assays and between laboratories. Hydroxocobalamin, for example, was reported to be from 20 to 100% as active as cyanocobalamin microbiologically, while equal in activity biologically. Many natural products such as liver extracts gave higher vitamin B_{12} assay results when added to sterile assay medium than when they were subjected to sterilization in the medium. The response to crystalline B_{12}, however, was improved when it was incorporated into the assay medium and autoclaved in the usual manner. The incorporation of reducing agents such as thioglycollic acid, thiomalic acid, or ascorbic acid prevented the loss of heat-labile activity in natural products during sterilization of the test, equalized results, and improved the growth response of the organisms. Protection of labile cobalamins could also be achieved by incorporation of cyanide or ethylenediaminetetra-acetic acid into the medium, although such agents depressed the total growth response. From this discussion it is apparent that the assay of vitamin B_{12} can be complicated by instability within the assay depending upon the ability of the reducing substance employed to cope with the extraneous material present in the sample being assayed.

IV. Activity and pH

As with most lactobacilli, *L. leichmannii* grows faster and more luxuriantly under slightly acid conditions. The optimum pH of the medium appears to be between 5.5 and 6.0, and most media are adjusted to be within that range.

V. Preparation of Sample for Assay

A. Extraction Procedure

The type of sample to be assayed determines the extraction procedure to be used.

[15] H. L. A. Tarr, *Can. J. Technol.* **30**, 265 (1952).

[16] G. Cooley, B. Ellis, V. Petro, G. H. Beavan, E. R. Holiday and E. A. Johnson, *J. Pharm. and Pharmacol.* **3**, 271 (1951).

(a) Pharmaceutical preparations containing vitamin B_{12} usually can be diluted with water and assayed directly. However, should values obtained by direct assay prove low, treatment with metabisulfite or cyanide may be indicated, particularly if the sample contains material to which the vitamin B_{12} may bind.

(b) Metabisulfite extraction[5,17] is useful for the extraction of crude materials such as fermentation broths or purified materials in which vitamin B_{12} may occur either loosely bound or in noncyano form.

Autoclave 1 gm. or 1 ml. of sample for 15 minutes at 121° in 250 ml. of $0.1\,M$ phosphate-citrate buffer at pH 4.5 (4.54 ml. $0.2\,M$ disodium phosphate + 5.46 ml. $0.1\,M$ citric acid) containing 0.1% sodium metabisulfite. Centrifuge, if necessary, and dilute the supernatant with water to the assay range.

(c) Treatment of the sample with cyanide is an alternative to procedure (b), and sometimes more successful. The exact conditions for optimum conversion of cobalamin activity with cyanide vary with the sample and the pH of the buffer; time and temperature of extraction may have to be varied for optimum results.[18] We have found the following to give reproducible answers with serum samples:

Mix 1 ml. serum and 10 ml. of $0.1\,M$ phosphate buffer (pH 6.0) containing 0.1 mg./ml. NaCN, autoclave at 120° for 15 minutes, dilute to 50 ml., and filter. (Serum is unique in that it can be assayed by simple dilution, provided the assay is titrated after 72 hours growth and the organism is maintained in a state of ability to utilize serum bound vitamin B_{12}. See comments under inoculum maintenance.) When high potency pharmaceutical samples are assayed, i.e., 1 mg./ml., the cyanide concentration of the buffer is increased to 1 mg./ml. and the buffer volume employed per milliliter of sample is 50–100 ml.

(d) Tissues pose a problem in that the cobalamin contained in them is largely in the noncyano form and bound to protein[16] requiring both release and stabilization. Shenoy and Ramasarma[19] devised a combined digestion and protection system which is useful for samples such as liver. Proceed as follows:

Suspend 1 gm. of homogenized liver in 50 ml. of water, add 0.5 ml. of 5% aqueous suspension of crude papain, and digest the mixture for 1 hour at 60°. Add 1 ml. of freshly prepared 5% aqueous sodium metabisulfite to the digest, steam for 5 minutes, cool, and dilute to 100 ml. Filter through paper and dilute the filtrate with water to the assay range.

[17] H. W. Loy, Jr., J. F. Haggerty, and O. L. Kline, *J. Assoc. Offic. Agr. Chemists* **35**, 169 (1952).

[18] H. R. Skeggs, C. A. Driscoll, J. Charney, and L. D. Wright, *Abstr. 119th Meeting Am. Chem. Soc.*, Cleveland, Ohio, p. 19A (1951).

[19] K. G. Shenoy and G. B. Ramasarma, *Arch. Biochem. Biophys.* **51**, 371 (1954).

In their hands, papain preparations do not usually contribute B_{12} or deoxyriboside activity whereas pancreatic preparations, although effective in releasing B_{12} activity, contribute enzyme blank problems. Gregory[20] recommended papain treatment for the liberation of bound vitamin B_{12} in milk samples.

The *L. leichmannii* assay is not suitable for the direct assay of fecal material that may contain pseudovitamin B_{12}.

B. Interference in the Assay

1. Extraction of Interfering Substances

The assay of materials that have been subjected to microbial fermentation frequently presents a problem in that they may contain appreciable amounts of deoxyribosides and pseudovitamins B_{12} in conjunction with vitamin B_{12} itself. McLaughlin et al.[10] have devised an extraction procedure, based on that of Bacher et al.[21,22] for the radioactive tracer assay, which effectively removes all of the interfering substances except vitamin B_{12} III. The recovery of added vitamin B_{12} in the procedure is slightly less than 100% (93.7% in McLaughlin's hands), so that for its routine use a correction factor should be established within a given laboratory. The procedure is as follows:

Prepare the following reagents:

(1) Sodium metabisulfite-phosphate solution (prepare just prior to use)
   1.29% Na_2HPO_4, 1.1% anhydrous citric acid, and 1% Na metabisulfite in distilled water.
(2) Cresol-carbon tetrachloride solution
   Mix equal volumes of carbon tetrachloride and freshly distilled cresol.
(3) Sulfuric acid, 5 N
(4) Phosphate-cyanide
   Dissolve 100 mg. of KCN in 1000 ml. of a saturated solution of Na_2HPO_4 and mix well.
(5) Butanol-benzalkonium chloride solution
   Mix 1 volume of a 12.8% solution of benzalkonium chloride with 9 volumes of normal butyl alcohol.

Procedure:
1. Autoclave approximately 1 gm. or 1–3 ml. sample for 15 minutes

[20] M. E. Gregory, *Brit. J. Nutrition* **8**, 350 (1954).

[21] F. A. Bacher, A. E. Boley, and C. E. Shonk, *Anal. Chem.* **26**, 1146 (1954).

[22] "U. S. Pharmacopeia," 15th Revision, 1st Suppl., Mack Publ., Easton, Pennsylvania, 1956.

(121°–123°C.) with 20 ml. of the bisulfite-phosphate solution in a 40-ml. Maizel-Gerson reaction vessel, cool, and centrifuge for 1 minute. Quantitatively transfer the supernatant liquid to a second 40-ml. Maizel-Gerson flask, washing the residue with 5 ml. water.

2. Add 5 ml. of the cresol-carbon tetrachloride solution to the bisulfite extract and shake gently for 1 minute. Avoid strong agitation to prevent gel formation. Centrifuge for 2 minutes at not less than 3000 r.p.m. With a hypodermic syringe, remove and *discard* the upper aqueous phase taking care not to remove any of the solid layer which may form at the interface of the two solvents.

3. Wash the organic solvent phase with 5 ml. of $5\,N$ sulfuric acid. Shake the flask gently for 1 minute and centrifuge 2 minutes. Remove and *discard* the upper aqueous phase. Wash once more with sulfuric acid.

4. Add 5 ml. of the phosphate-cyanide solution to the organic solvent

TABLE I

BASIC PROCEDURE FOR PREPARING MEDIUM FOR DEOXYRIBOSIDE ASSAY USING
Lactobacillus acidophilus (ATCC 11506) (*Thermobacter acidophilus* R 26)

Double strength medium[a]	
Acid hydrolyzed casein	10 ml.
Tryptophan solution	10 ml.
Cystine solution	10 ml.
Adenine, guanine, xanthine, uracil, orotic acid, thymine solutions	1 ml. each
Salts A and Salts B solutions	1 ml. each
DL-α-Alanine	100 mg.
Na acetate·3H$_2$O	2 gm.
Folic acid, biotin, PABA, Pyridoxal, vitamin B$_{12}$ solutions	1 ml. each
Tween 80	0.2 ml.
Glucose	4 gm.
Distilled water to make	100 ml.
(pH adjusted to 6.5 before sterilization)	
Vitamin supplement[b]	5 ml.

Standard thymidine solution: 10 mg. in 100 ml. water (100 μg./ml.); dilute to 4 μg./ml.

Standard curve						
Thymidine (μg./tube)	0	2	4	6	8	12	20
4 μg./ml. standard (ml.)	0	0.5	1.0	1.5	2.0	3.0	5.0
Water added (ml.)	5	4.5	4	3.5	3	2	0
Medium (ml.)	5	5	5	5	5	5	5

Autoclave tubes 15 minutes at 121°C. (total time in autoclave about 22 minutes)
Seed with a 1:1000 saline suspension of a 24-hour skim milk culture
Incubate 24 hours at 37°C; read on any suitable turbidimeter

[a] Solutions as prepared in Table III.
[b] For preparation, see item 12 in Table III.

phase. Shake the mixture gently for 1 minute, centrifuge for 2 minutes, and remove and *discard* the upper aqueous phase.

5. To the washed organic phase add 15 ml. of 2:1 butanol-benzalkonium chloride-carbon tetrachloride mixture and 5 ml. distilled water. Shake the mixture gently for 1 minute, centrifuge for 2 minutes, remove and *retain* the upper aqueous phase. Wash the organic layer a second time with distilled water, remove, and combine the upper aqueous layer with the previous aqueous extract.

6. Dilute the combined aqueous extract with distilled water to give a solution containing between 0.01 and 0.02 mμg. vitamin B_{12} per milliliter and assay. For samples known to be very low in activity and requiring little dilution, wash the combined aqueous extract with 5 ml. ethyl ether to remove traces of organic solvents that can interfere with the growth of the test organism.

2. Differential Deoxyriboside Assay

For the most part, vegetables and fruits are presumed to be lacking in vitamin B_{12} activity, but such substances may promote growth of *Lactobacillus leichmannii* due to deoxyribose nucleic acid (DNA) derivatives. *Lactobacillus acidophilus*[12] does not grow in response to vitamin B_{12}, but does respond to the DNA derivatives over the same range (2–20 μg.) as does *L. leichmannii*. Therefore, a sample active at equal concentrations for both organisms contains DNA derivatives. A condensed version of the *L. acidophilus* procedure is given in Table I.

Ochromonas malhemensis[9] grows only in response to the biologically active cobalamins and vitamin B_{12} III. The procedure is detailed elsewhere in this book.

VI. Preservation of Test Organisms

Microinoculum agar (Difco, BBL) has been recommended for maintenance of *Lactobacillus leichmannii*. We prefer, because the procedures have been satisfactory for 10 years, to maintain this particular type of lactic acid organism (i.e., strains requiring fatty acids and stimulated by tomato juice) in Bacto skim milk medium supplemented with 1% Bactotryptose. Stock cultures are transferred every 6 weeks into freshly prepared skim milk medium, incubated 24 hours at 37°, subcultured in duplicate into tryptose milk, and after a second incubation, one culture is returned to the refrigerator as the stock, the other kept available for inoculum preparation. This rather involved procedure is carried out to rid the stock culture of any toxic materials carried from the old culture and has proved very satisfactory. Previous experience had shown us that a single transfer after a month's storage occasionally resulted in the loss of a culture (usually *L. bulgaricus*), and this was presumed to be due to transfer of toxic material. Apparently the buffering capacity of milk is adequate to protect the viability of the organism for periods of 6 to 8 weeks under conditions of refrigeration.

VII. Preparation of Inoculum

The inoculum can be prepared directly from a 24-hour transfer in skim milk by preparing a 1:100 dilution in sterile physiological saline. Growth in the assay tubes is better, however, if the inoculum is grown in broth culture. We find it a distinct advantage to use the assay medium in single strength supplemented with 0.1 mμg. B$_{12}$ per milliliter and to transfer it daily for several days prior to use. With a rapidly growing organism, well conditioned to the assay medium, the assay response is more rapid and sensitive than with a milk culture inoculum. Such an inoculum, however, must be washed thoroughly to rid it of cell adsorbed vitamin B$_{12}$. Therefore, the culture is centrifuged, supernatant discarded, cells resuspended in sterile physiological saline, recentrifuged, and saline resuspension repeated twice more. The final saline suspension is diluted to 1:100 for use, one drop added aseptically to each sterile assay tube, as the inoculum.

The organism should not be maintained in continuous subculture in assay media for more than 2 weeks when crude samples are being analyzed. It was recently observed that continued subculture in a synthetic medium can diminish the capacity of the organism to utilize bound vitamin B$_{12}$ in serum.[23]

VIII. Media

The composition of our medium is given in Table II and the preparation of stock solutions described in Table III. It differs slightly from any of the published media as a result of additions made over the course of the years to improve the response. It should be pointed out that the original U.S.P. medium contained a tomato juice supplement which caused considerable variation. Under controlled growth conditions, guanylic acid gives a response equal to that provided by tomato juice, and we are of the opinion that supplementation of the medium with guanosine and guanylic acid obviates the need for tomato juice supplements while avoiding possible B$_{12}$ contamination from a crude supplement. Furthermore, if the ribose nucleic acid derivatives are responsible for the growth promoting effects

[23] H. R. Skeggs, "Developments in Industrial Microbiology," Vol. 2. Plenum Press, New York, 1961.

TABLE II

DOUBLE-STRENGTH ASSAY MEDIUM FOR VITAMIN B$_{12}$—*Lactobacillus leichmannii*

	Weight/100 ml.		ml. sol./100 ml.[a]	
Casein hydrolyzate	1	gm.	10	ml.
DL-Tryptophan	40	mg.	10	ml.
L-Cystine	20	mg.	10	ml.
Adenine sulfate	1	mg.	1	ml.
Guanine HCl	1	mg.	1	ml.
Xanthine	1	mg.	1	ml.
Uracil	1	mg.	1	ml.
Salts A			1	ml.
KH$_2$PO$_4$	100	mg.		
K$_2$HPO$_4$	100	mg.		
Salts B			1	ml
MgSO$_4$·7H$_2$O	40	mg.		
FeSO$_4$·7H$_2$O	2	mg.		
NaCl	2	mg.		
MnSO$_4$·4H$_2$O	2	mg.		
Na acetate (anhydrous)	1.2	gm.	1.2	gm.
Guanosine (dissolve with aid of NaOH and heat)	20	mg.	20	mg.
Guanylic acid	20	mg.	20	mg.
DL-α-Alanine	100	mg.	100	mg.
CaCl$_2$	20	mg.	20	mg.
Glucose	4	gm.	4	gm.
Vitamin supplement			5	ml.
Pyridoxine·HCl	400	µg.		
Riboflavin	200	µg.		
Ca pantothenate	200	µg.		
Thiamine·HCl	400	µg.		
Nicotinic acid	200	µg.		
p-Aminobenzoic acid	100	µg.	1	ml.
Biotin	1	µg.	1	ml.
Folic acid	100	µg.	1	ml.
Pyridoxal	100	µg.	1	ml.
Tween 80	0.2	ml.	0.2	ml.
Thiomalic acid	100	mg.	100	mg.

Adjust pH to 5.8 and dilute to 100 ml. with distilled water.

[a] From Table III.

of tomato juice, much of the variability encountered with Norite treated tomato juice eluates is explained, since the nucleic acid derivatives are fairly effectively adsorbed by Norite.

The inclusion of calcium chloride in the medium is dictated by past

TABLE III

PREPARATION OF STOCK SOLUTIONS

1. Casein hydrolyzate (100 mg./ml.): Reflux under a water condenser for 8 to 10 hours: 100 gm. Labco or NBI vitamin-free casein in a mixture of 500 ml. water and 500 ml. concentrated HCl, adding a few glass beads to prevent "bumping." Distill off the HCl under vacuum until a thick paste remains. Dissolve paste in about 800 ml. water and repeat distillation. Dissolve the paste in about 400 ml. water, adjust to pH 3.0 with 40% NaOH, and filter. Stir filtrate ½ hour with 10 gm. Darco G60; filter, dilute with water to 1000 ml., and store under benzene at room temperature. Vitamin-free casein hydrolyzates are available from Sheffield, Nutritional Biochemicals, etc.

2. Tryptophan (4 mg./ml.): Suspend 4 gm. DL-tryptophan in 500 ml. of water and heat to dissolve, add 5 ml. concentrated HCl, and dilute to 1000 ml. with water. Store at room temperature.

3. Cystine (2 mg./ml.): Suspend 2 gm. cystine in 500 ml. water containing 10 ml. conc. HCl and heat to dissolve, dilute to 1000 ml., and store at room temperature.

4. Adenine sulfate (1 mg./ml.): Dissolve 100 mg. adenine sulfate with heat in 100 ml. water containing 5 ml. conc. HCl.

5. Guanine·HCl (1 mg./ml.): Dissolve 100 mg. guanine·HCl with heat in 100 ml. water containing 5 ml. conc. HCl.

6. Uracil (1 mg./ml.): Dissolve 100 mg. uracil with heat in 100 ml. water containing 5 ml. 28% NH$_4$OH.

7. Orotic acid (4 mg./ml.): Dissolve 400 mg. in a minimum of 0.05 N sodium hydroxide with heating and dilute to 100 ml.

8. Thymine (4 mg./ml.): Dissolve 400 mg. in 100 ml. 1% hydrochloric acid.

9. Xanthine (1 mg./ml.): Dissolve 100 mg. xanthine with heat in 100 ml. water, containing 3 ml. 28% NH$_4$OH.

10. Salts A: Dissolve 10 gm. K$_2$HPO$_4$ and 10 gm. KH$_2$PO$_4$ in 100 ml. water, store under toluene in refrigerator.

11. Salts B: Dissolve 4 gm. MgSO$_4$·7H$_2$O, 200 mg. NaCl, 200 mg. FeSO$_4$·7H$_2$O, and 200 mg. MnSO$_4$·4H$_2$O in 100 ml. H$_2$O. Store in refrigerator. Solution becomes cloudy.

12. Vitamin mixture: Dissolve 4 mg. riboflavin, 4 mg. Ca pantothenate, 8 mg. pyridoxine, 8 mg. thiamine·HCl, and 4 mg. nicotinic acid in 100 ml. water. Store in refrigerator. Renew monthly.

13. *p*-Aminobenzoic acid (100 µg./ml.): Dissolve 10 mg. in 100 ml. 20% ethyl alcohol, store in refrigerator.

14. Folic acid (100 µg./ml.): Dissolve 10 mg. folic acid in 5 ml. 0.1 N NaOH; add 20 ml. ethyl alcohol, and dilute to 100 ml. Store in refrigerator. Renew monthly.

15. D-biotin (1 µg./ml.): Dissolve 10 mg. D-biotin in 100 ml. 50% ethyl alcohol. Dilute 1 ml. to 100 ml. with ethyl alcohol. Store in refrigerator.

16. Pyridoxal hydrochloride (100 µg./ml.): Dissolve 10 mg. pyridoxal·HCl in 100 ml. 20% ethyl alcohol. Store in refrigerator.

17. B$_{12}$ solution (10 µg./ml.): Store in refrigerator.

experience. A culture of *L. leichmannii* strain 4797 that was transferred daily in skim milk-tryptose medium acquired a need for calcium. The stock culture, infrequently transferred, exhibited no dependence on calcium. It was concluded, however, that in the presence of a latent requirement for calcium it would be advisable to include it in the assay medium. The precedent for this was set by the fact that the organism, when first studied, was stimulated only mildly by guanylic acid but subsequently became dependent on the purine nucleotide for optimal growth.

One further comment with respect to the medium. The reducing agent plays a dual role in the vitamin B_{12} assay. It maintains the proper O-R potential and protects noncyanocobalamins from destruction during autoclaving. The proper E_h of the medium is critical to assure uniform growth as was shown by Lees and Tootill.[24] Indeed, even in the presence of a reducing agent, variations in tube diameter may exert an effect on assay results. This was particularly noticeable in the direct assay of serum where the precipitated serum proteins interfere with the ability of the reducing agent to maintain the standard E_h in wide tubes. With the reducing agent present in the medium, responses to crystalline vitamin B_{12} are identical regardless of the diameter of the tube, but in the serum samples growth may be considerably less as the diameter of the tube increased. This is a particularly important point with respect to the use of dehydrated culture media which contain the readily oxidizable ascorbic acid as the reducing agent in a hygroscopic mixture. Dr. Kavanagh supplied the data in Fig. 1 that show the effects of deterioration of the reducing sub-

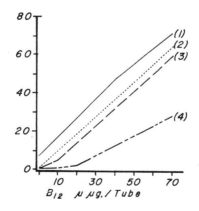

Fig. 1. Dose-response curves of four lots of commercially prepared (dehydrated) media. The ordinate is concentration of bacteria (arbitrary scale) after an incubation period of 48 hours. Lot 1 was contaminated by a small amount of vitamin B_{12}. Lot 2 gave an ideal curve. Lots 3 and 4 were deficient in reducing agent.

stance in dehydrated media. The response could be restored to normal with the addition of ascorbic acid.

[24] K. A. Lees and J. P. R. Tootill, *Biochem. J.* **50**, 455 (1952).

IX. Design of the Assay

The design of the assay is dependent on the type of sample being assayed and the goal of the procedure. For precision analytical control a 2 to 4 point standard curve at levels of 20, 40, 60, and 80 $\mu\mu$g. of sample and standard with 4 replications randomized in the racks provides a suitable design and sufficient internal compensation for good statistical significance.

When comparative assay values on large numbers of samples are required, it is much less time consuming and more expedient to employ the standard curve 0, 0, 10, 20, 40, 60, and 100 $\mu\mu$g. per tube, with samples diluted to approximately 15 $\mu\mu$g./ml. and run sequentially at 1, 2, 3, and 5 ml.

Attempts to demonstrate a significant difference in results due to the design of the experiment failed in these laboratories. (Such experiments in a small group are biased from the start because of the personnel reaction to a study of technique.) Variations within an assay amount to an intrinsic error of $\pm 7\%$ on a given day. The adoption of a statistically randomized design for precise analytical work, however, provides insurance against assay variation within an assay and has, in laboratories other than our own, yielded significantly better results.

Regardless of the design of the assay, standard and samples are diluted to 5 ml. with water, 5 ml. medium are added, the tubes plugged, and autoclaved at 120°C. for 10 minutes.

X. Assay Conditions

Assays are incubated at temperatures between 35° and 40°C. Growth at 40°C. is more rapid and consequently more suitable for short term assays. Ideally, water bath incubation is preferred, since air currents, especially in large incubators, tend to induce variations in temperature. Increasing the temperature or lengthening the time of incubation significantly decreases the within-rack variation (by 50%) and thus should be carefully adapted to the laboratory conditions. For example, in our laboratory we have a large circulated air incubator that is maintained at 37°C. for maximum versatility. The preferable incubation time under these conditions is 40 hours for turbidimetric assays and 72 hours for assays to be titrated.

XI. Special Precautions

Almost every laboratory has equipment that will lend itself to the assay of vitamin B$_{12}$, but special care must be taken of that equipment bearing in mind that the assay is designed to measure the difference between 1 and 2 $\mu\mu$g./ml. of a large molecule that binds readily to a variety of chemicals. Under the heading of special precautions, we place extra care in handling of all equipment and materials involved in the assay, and present a few of our experiences in dealing with it in the hope that knowledge of our errors will aid others in avoiding common problems.

The first lesson we had to learn was that there were no degrees of freedom in the cleanliness of equipment employed. Various ruses were tried, but in all cases the assay detected them. When we attempted to use detergent-washed glassware, spotty, erratic growth occurred, and finally high blanks. When a lazy glassware washer decided that 15 rinses after acid cleaning was foolishness, the assay refused to function in the presence of acid residue and we got no growth.

Pipets, test tubes, and volumetric flasks are all acid washed and thoroughly rinsed (13 times with tap water and 2 times with distilled). Erlenmeyer flasks are washed in detergent and subjected to the same rinsing, plus additional rinsing with distilled water just prior to use. The simplest criterion is used to detect clean glassware—droplets of water do not collect along the sides—it "drains clean."

Many laboratories have experienced difficulty with high blanks traced to the water supply. Deionized water cannot be substituted for distilled water—the demineralizing resins frequently become hosts of microorganisms capable of producing vitamin B_{12}. Even distilled water stored in containers that have not been thoroughly cleaned or that have been exposed to room dust can be a source of trouble. We are fortunate in having available a plentiful supply of freshly distilled water from a glass-lined apparatus which is ideal but not always available. Treatment of a suspect supply of water with activated charcoal to remove vitamin B_{12} is one way to trace the source of high blanks.

One of our colleagues consistently had beautiful results except for one or two racks of tubes out of five or six. The trouble? The Cornwall syringe used to deliver water, and used for nothing else but distilled water, was apparently harboring B_{12} producing germs in the rubber valves, and it took extra rinsing, through the first two racks, to wash out the B_{12}. It took us several years to outwit the organisms that contaminated the Brewer machine used for dispensing medium but finally the simple routine of rinsing the hoses immediately before and after use with distilled water and sterilization of the syringe and hoses between uses keeps us out of trouble. We reorganized our whole mode of operation when one of our girls turned out to be a "B_{12} carrier." For years we had set up samples and standard tubes in the morning and prepared medium according to what was required in the afternoon. Our "carrier" had become so contaminated by afternoon that blanks ran high, so the order had to be reversed. In the course of ferreting this problem out, the hand washing routine was developed, that is, wash hands between each step, which in turn increased the requirement of the personnel for hand lotion. Now, I am not reporting these things to be facetious. Each incident, in its turn, has cost hours of work to figure out and serves as an example of how seemingly inconsequential variants can interfere with assay performance and emphasizes the extreme sensitivity of the vitamin B_{12} assay.

XII. Reading the Assay

The inherent error in the assay is reduced by 50% when adequate time is allowed for incubation. Assays can be read on a suitable colorimeter after 40 hours incubation with less variation obtained than when they are read after 18 to 24 hours (see Chapter 4). When it is more convenient to allow longer incubation, or in the assay of samples that are turbid, assays are titrated with $0.1 N$ NaOH using brom thymol blue as an indicator (see biotin procedure). A standard curve is plotted, and sample activity read from the curve, multiplied by the dilution factor, averaged, and reported. The potencies of the undiluted sample obtained from the several dilutions should be within ±15% of the average for the set of dilutions. A typical standard curve is given in Table IV.

TABLE IV

RESPONSE OF *Lactobacillus leichmannii* TO VITAMIN B₁₂

mμg. B₁₂/10 ml.	Turbidity[a]	Acid produced[b]
0	26	1.85
0.01	46	4.22
0.02	70	5.81
0.03	85	7.11
0.04	103	8.19
0.06	135	9.78
0.10	180	11.43

[a] Turbidity as Klett-Summerson scale readings after an incubation of 40 hours. Instrument set at zero against distilled water.

[b] Acid production on incubation for 72 hours. The acid is measured as the milliliters of $0.1 N$ sodium hydroxide required to titrate the contents of a tube to pH 7.0.

CHAPTER 8

Amino Acids

GERALD D. SHOCKMAN

Department of Microbiology, Temple University School of Medicine, Philadelphia 40, Pennsylvania

I. Introduction

The use of microorganisms for the assay of amino acids occurred as a logical outgrowth of their earlier use for the assay of vitamins. Wood et al.[1] were the first to suggest that microorganisms could be used for the quantitative determination of amino acids. However, at that time (1940), adequate knowledge of the growth factor requirements of organisms suitable for the assay of amino acids was lacking. Using 17 amino acids in place of acid hydrolyzed casein and a tomato juice eluate as a source of growth factors, Kuiken et al.[2,3] reported that 9 amino acids were essential for growth of *Lactobacillus arabinosus* 17-5 and that 3 amino acids, leucine, isoleucine, and valine, could be quantitatively determined by the microbiological techniques similar to those previously used for vitamin assays.[3] From this has grown the ability to conveniently determine, with a good degree of precision, accuracy, and specificity, 18 amino acids. A variety of organisms and culture media

[1] H. G. Wood, C. Geiger, and C. H. Werkman, *Iowa State Coll. J. Sci.* **14,** 367 (1939–1940).

[2] K. A. Kuiken, W. H. Norman, C. M. Lyman, and F. Hale, *Science* **98,** 266 (1943).

[3] K. A. Kuiken, W. H. Norman, C. M. Lyman, F. Hale, and L. Blotter, *J. Biol. Chem.* **151,** 615 (1943).

can and have been used over the past 20 years,[4-10] and undoubtedly others that are superior in one respect or another will be found in the future. To some extent, uniformity in organism and media for a wide range of amino acids has been sought after with only limited success. Some organisms, such as *Leuconostoc mesenteroides* P-60, require a large proportion of the 18 amino acids for growth.[11,12] However, in many instances it is more suitable to use other organisms or more than one medium for at least some of the amino acid assays.

II. Comparison of Microbiological Assays to Other Available Methods

When the necessity arises to determine either the amino acid composition or the content of a particular amino acid in an unknown substance, a choice of chemical, physicochemical, chromatographic (paper or ion exchange column), biological, enzymatic, or microbiological methods must be made. The choice of method, of course, depends on the goal of the particular investigation. Table I lists some of the available methods with some of their particular advantages and limitations. In the opinion of the author, no single method should be relied on entirely. However, where a complete amino acid analysis is desired on a limited amount of material on a more or less continuous basis, ion exchange chromatography is now the primary method of choice where the equipment and personnel are, or can be made, available. For less continuous demands, where (1) the content of one or several amino acids is desired rather than a complete analysis, (2) the configuration of the amino acid is of interest, or (3) facilities for column chromatography are not available, properly selected and executed microbiological methods often prove to be far more than adequate. The complementary use of both column chromatography and microbiological assays has served to overcome some of the

[4] M. S. Dunn, *Physiol. Revs.* **29**, 219 (1949).

[5] E. E. Snell, *Advances in Protein Chem.* **11**, 85 (1945).

[6] B. S. Schweigert and E. E. Snell, *Nutrition Abstr. & Revs.* **16**, 497 (1947).

[7] E. C. Barton-Wright, "The Microbiological Assay of the Vitamin B-Complex and Amino Acids," p. 179. Pitman, New York, 1952.

[8] S. Saperstein, in "Amino Acid Handbook" (R. J. Block and K. W. Weiss, eds.), p. 50. C. C Thomas, Springfield, Illinois, 1956.

[9] J. P. Greenstein and M. Winitz, "Chemistry of the Amino Acids." Wiley, New York, 1961.

[10] H. H. Williams, *Cornell Univ. Agr. Expt. Mem.* **337**, 1955.

[11] M. S. Dunn, S. Shankman, M. N. Camien, W. Frankl, and L. B. Rockland, *J. Biol. Chem.* **156**, 703 (1944).

[12] B. F. Steele, H. E. Sauberlich, M. S. Reynolds, and C. A. Baumann, *J. Biol. Chem.* **177**, 533 (1949).

TABLE I
Comparison of Methods Used to Determine Amino Acids

	Advantages	Limitations
I. Chemical		
A. Isolation	Recovery	Large amounts of starting material required, quantitation often poor or unreliable
B. Colorimetric	Can sometimes be used on intact proteins (i.e., cysteine, tryptophan, tyrosine)	Very often lack sensitivity, accuracy, and/or specificity
II. Physicochemical		
Mass spectography	Highly accurate	Complex and expensive equipment; isotope (N^{15}) containing amino acids required
III. Chromatographic		
A. Paper	Rapid, very sensitive, comparison of the amino acid spectra of similar substances obtained rapidly	Limited accuracy and specificity; separation sometimes poor particularly with samples of crude materials
B. Ion exchange	Very accurate, specific, rapid, reproducible; only small amounts of material required, excellent for a complete amino acid analysis of an unknown	Elaborate equipment and skill in its operation and interpretation of results required; specificity must be checked by another method (i.e., paper chromatogram or microbiological assay); overly elaborate facilities required unless a complete analysis is required; knowledge of configuration not obtained
IV. Biological		
A. Animal		Costly, not usually sufficiently reliable or precise, slow
B. Enzymatic	Specificity for optical isomer	Availability of sufficiently specific enzyme preparations, sometimes necessary to grow organisms and isolate enzymes; enzymatic methods not available for some amino acids; interferences between amino acids present in a hydrolysate often a problem
V. Microbiological	Relatively as accurate as chemical methods; simple, can be specific, inexpensive, convenient; only small amounts of material required; sometimes can be made specific for one enantiomorphic form; instrumentation not elaborate; excellent for determination of the content of one or more particular amino acids in an unknown	Somewhat larger amounts of material needed than for chromatography when a number of amino acids are to be determined; continuous checks necessary for consistency, specificity, and interferences

limitations of either method alone. For instance, microbiological assays can be used to further identify a chromatographic peak or to ascertain the configuration of the amino acid in that peak. An alternative is the microbiological determination of some amino acids on the same hydrolysate used for ion exchange column chromatography. The results of the two methods can thus be compared and the configuration determined. An example of such a use is given in Table II.[13] These are analyses of *Streptococcus faecalis* (9790) cells and cell fractions. The chromatographic method used was the semiautomatic ion exchange column method of Moore *et al.*[14] Microbiological assays for the L-amino acids and D-alanine were all done by the methods recommended in Section X. It is clear that in most instances (valine, threonine, leucine, isoleucine, arginine, histidine, and lysine) there is excellent agreement between the two methods. For other amino acids, such as alanine, glutamic, and aspartic acids, lack of agreement for the whole cells and cell wall fraction has been attributed to the occurrence of the D-isomers of these 3 amino acids mainly in the bacterial cell wall, but not in the soluble portion of the cells.[13] The microbiological assays for these amino acids were examined severely for specificity for the L-isomer (see Sections IV.C.2, X.A.5, X.C.4, and X.E). In one case, alanine, the D-isomer could also be determined microbiologically. Here the total alanine determined chromatographically, and the sum of D- and L-alanine determined separately by microbiological methods show excellent agreement, particularly in the cell wall fraction which contains relatively large amounts of D-alanine. With the rare exceptions of a few chemical and microbiological methods using *Neurospora* and protozoa (see Sections III.B.4 and 5), which have not as yet been sufficiently explored, all methods for quantitatively determining amino acids in proteins have the same deficiency. They all require hydrolysis of one sort or another. Destruction during hydrolysis, particularly of some amino acids, as well as incomplete liberation, are important factors that cannot be overemphasized (see Section VII).

III. The Assay Organisms

A. General Considerations

A successful assay for amino acids requires some knowledge of the nutrition and physiology of the assay organism. This is necessary not only to correct occasional difficulties with the assay but also to provide a firmer basis for determination of validity of the assay. Quantitative determination of a particular amino acid (or growth factor) by micro-

---

[13] G. Toennies, B. Bakay, and G. D. Shockman, *J. Biol. Chem.* **234**, 3269 (1959).
[14] S. Moore, D. H. Spackman, and W. H. Stein, *Anal. Chem.* **30**, 1185 (1958).

TABLE II

Analyses of Exponential Phase *Streptococcus faecalis* (9790)[a] Cells[b]

Component	Whole cells			Soluble fraction			Wall fraction		
	Chromatography %	Bacterial[c] assay L- %	By difference[d] D- %	Chromatography %	Bacterial[c] assay L- %	By difference[d] D- %	Chromatography %	Bacterial[c] assay L- %	By difference[d] D- %
1. Glutamic acid	8.3 ± 0.3	6.6	1.7 ± 0.3	8.4 ± 0.2	6.0 ± 0.6	2.4 ± 0.6	5.4 ± 0.1	<0.5	>4.9 ± 0.1
2. Aspartic acid	6.3 ± 0.6	4.9	1.5 ± 0.6	6.5 *	5.1	1.4 ± 0.1	4.4	0.4	4.0 ± 0.1
3. Lysine	5.9 ± 0.2	5.9 ± 0.2	*	5.6 ± 0.1	5.8	*	5.0[e]	5.4 ± 0.2	*
4. Alanine	4.7 ± 0.3	4.1	0.6 ± 0.3	4.1 ± 0.1	3.9 ± 0.2	*	6.2	3.2	3.0 ± 0.1
5. D-Alanine	—	—	1.1 ± 0.1	—	*	*	*		3.0 ± 0.2
6. Ammonia	2.4 ± 0.3	—	—	2.3 ± 0.2	—	0.6 ± 0.1	2.3 ± 0.2	—	—
7. Leucine	3.9 ± 0.2	4.1	*	4.7 ± 0.2	—	—	0.4[e]	—	—
8. Glycine	3.8 *	—	—	4.5 ± 0.1	—	—	0.3 ± 0.1	—	—
9. Valine	3.7 ± 0.2	3.9	—	4.6 ± 0.3	4.4	*	0.1 ± 0.1	—	—
10. Isoleucine	3.4 ± 0.2	3.5	*	3.9 ± 0.1	4.0	*	*	—	—
11. Arginine	2.8 ± 0.2	2.7	*	3.1 ± 0.2	—	—		—	—
12. Threonine	2.7 ± 0.4	2.8	*	2.9 ± 0.2	2.9	*	0.4 ± 0.1	—	—
13. Phenylalanine	2.3 *	—	—	2.6 ± 0.2	—	—		—	—
14. Proline	2.1	1.6	0.5 (?)	2.6 ± 0.3	—	—	0.7 *	—	—
15. Tyrosine	1.8 ± 0.1	—	—	2.0 ± 0.1	—	—	*	—	—
16. Serine	1.6 ± 0.2	—	—	1.8 ± 0.1	—	—	0.2 *	—	—
17. Methionine	1.6 ± 0.1	—	—	1.4 ± 0.1	—	—	0.2	—	—
18. Histidine	1.1 ± 0.1	1.2	*	1.3 ± 0.1	—	—	0.2 ± 0.2 *	—	—
19. Cystine	0.3 ± 0.1	—	—	0.2 *	—	—	0.3 ± 0.2	—	—
20. Peptide substance[f]	51 ± 2	—	—	54 ± 1	—	—	27 ± 3	—	—
21. Peptide corrected[d]	—	—	—	50	—	—	—	—	—
22. N recovered[g]	73 ± 1	—	—	68 ± 1	—	—	93 ± 6	—	—
23. N content[h]	13.2			14.5			5.7		

[a] See reference 13.
[b] Data show amounts of substance listed (± average deviation), as percentage of dry weight (except for line 22). * indicates that zero values were obtained.
[c] Average deviation is ±0.1% unless shown otherwise.
[d] In the case of alanine also by direct bacterial assay (line 5).
[e] Only one determination.
[f] Sum of the values of lines 1–19, reduced by 14% (approximate correction for H₂O loss in peptide bond formation).
[g] Corrected for estimates of NH₃ (excess over 5% of protein N) and glycine (⅓ of amount found) produced by nucleic acid degradation.
[h] As percentage of N in sample chromatographed.

biological assay requires an organism that, in a specific medium, will not grow without that amino acid and whose growth (or acid production) will be proportional *only* to the quantity of the amino acid provided. There are exceptions to this. For instance, quantitation has been achieved on the basis of slow growth that is accelerated by the presence of the particular substance. The outstanding example of this is the stimulation of growth rates by peptides such as strepogenin.[15]

Several other characteristics of an assay organism are not necessarily required but are certainly desirable. The organism should be relatively easy to cultivate and stable in its growth requirements. Rapid growth is often an additional aid not only in the speed of obtaining the results of the assay but also may increase the precision of the assay. Where possible, organisms pathogenic to humans should be avoided since added effort will be needed to prevent possible infection. For convenience, an organism should be suitable for the assay of several amino acids, although the most exacting organisms are often either less stable or more difficult to cultivate. For the examination of the amino acid content of unknowns, particularly unpurified materials, it is advantageous to introduce into the assay tubes as little extraneous material as possible. Therefore, sensitivity of the organism for the particular amino acid may be an important factor.

Obviously, the selection of the test organism and the choice of method for measuring growth will be interdependent. Some organisms are better acid producers and therefore are sometimes considered to be more suitable for acidimetric methods. Organisms vary in their ease of suspension and in the uniformity of the turbidities obtained. Those that are more easily and uniformly suspended have a real advantage for the use of turbidimetric methods.

B. Specific Test Organisms

Amino acid assays have primarily been done with various organisms from the family Lactobacteriaceae (lactic acid bacteria). These include species of *Lactobacilli, Streptococci,* and *Leuconostoc.* Many other organisms, including species of *Clostridia, Eberthella typhosa,* auxotrophic mutants of *Escherichia coli, Neurospora crassa* mutants, protozoa of the *Tetrahymena* group and others, have been proposed, and in isolated instances used, for amino acid assays. Many of the reasons given at the time they were proposed or used are no longer valid, or today the disadvantages outweigh the advantages.

Actually only a limited number of lactic acid bacteria have been used for amino acid assays. *Lactobacillus arabinosus* 17-5 (8014)[2,3,16] and *L.*

---

[15] H. Sprince and D. W. Woolley, *J. Exptl. Med.* **80**, 213 (1944).
[16] S. Shankman, M. S. Dunn, and L. B. Rubin, *J. Biol. Chem.* **150**, 477 (1943).

casei (7469)[17,18] were the first to be proposed and used. This was quickly followed by the use of *Leuconostoc mesenteroides* P-60 (8042),[12,19] *Lactobacillus delbrueckii* 5 (9595),[20] and *Streptococcus faecalis* (9790).[21] The use of others, primarily *Leuconostoc citrovorum* (8081),[12,22] *Lactobacillus fermenti* (9338),[23-25] *L. delbrueckii* 3 (4913),[26,27] *L. brevis* (8257),[28] *L. leichmannii* 313 (7830) and 327 (4797),[29,30] and *L. delbrueckii* 780 (9649),[31] followed.

Since the lactic acid bacteria are of primary importance and usefulness for the assay of amino acids, their characteristics and use will be detailed in subsequent sections. A brief description of the characteristics and use of some other organisms will be given here since they are of historic importance and, while of limited usefulness, they may be important in some special situations.

Examination of the nutritional requirements of microorganisms would reveal a wide variety of organisms (bacteria, yeast, and fungi) that could be used for the assay of amino acids either as such or as auxotrophic mutants. The extensive use of lactic acid bacteria was due, at least in part, to the fact that the knowledge of their nutritional requirements was further advanced than that of other groups of microorganisms. Over the years they have remained the organisms of choice because of their many advantageous characteristics, some of which are mentioned above and many of which are detailed in Section IV.

1. Clostridia

Clostridium perfringens (*welchii*) BP6K (10543) has been used by

[17] J. R. McMahan and E. E. Snell, *J. Biol. Chem.* **152**, 83 (1944).

[18] S. Shankman, M. S. Dunn, and L. B. Rubin, *J. Biol. Chem.* **151**, 511 (1943).

[19] M. S. Dunn, M. N. Camien, S. Shankman, W. Frankl, and L. B. Rockland, *J. Biol. Chem.* **156**, 715 (1944).

[20] J. L. Stokes and M. Gunness, *J. Biol. Chem.* **157**, 651 (1945).

[21] J. L. Stokes, M. Gunness, I. M. Dwyer, and M. C. Caswell, *J. Biol. Chem.* **160**, 35 (1945).

[22] H. E. Sauberlich and C. A. Baumann, *J. Biol. Chem.* **177**, 545 (1949).

[23] M. S. Dunn, M. N. Camien, and S. Shankman, *J. Biol. Chem.* **161**, 657 (1945).

[24] M. S. Dunn, S. Shankman, and M. N. Camien, *J. Biol. Chem.* **161**, 669 (1945).

[25] M. S. Dunn, M. N. Camien, S. Shankman, and H. Block, *J. Biol. Chem.* **163**, 577 (1948).

[26] A. M. Violante, R. J. Sirny, and C. A. Elvehjem, *J. Nutrition* **47**, 307 (1952).

[27] J. C. Alexander, C. W. Beckner, and C. A. Elvehjem, *J. Nutrition* **51**, 319 (1953).

[28] M. S. Dunn, L. E. McClure, and R. B. Merrifield, *J. Biol. Chem.* **179**, 11 (1949).

[29] B. S. Schweigert, B. T. Guthneck, and H. E. Scheid, *J. Biol. Chem.* **186**, 229 (1950).

[30] B. S. Schweigert, B. A. Bennett, and B. T. Guthneck, *J. Biol. Chem.* **190**, 697 (1951).

[31] E. E. Snell, N. S. Radin, and M. Ikawa, *J. Biol. Chem.* **217**, 803 (1955)

Boyd et al.[32,33] for amino acid assays. The organism has been grown in a chemically defined medium consisting of amino acids, vitamins, salts, glucose, phosphate buffer, and ascorbic acid. For assay purposes, the organism can be grown at 45°C and sodium azide can be included in the medium. These two conditions essentially eliminate bacterial contamination and give the assays the advantage of allowing the use of nonaseptic conditions. The turbidity of the assays can be read at 16 to 20 hours. The organism is considered to be strictly anaerobic but the assays have been successfully done without anaerobic precautions in 10-ml. amounts in 15 × 150 mm. tubes providing ascorbic acid or cysteine is present and the glucose is not autoclaved with the medium. Standard curves for 13 amino acids, L-methionine, L-tryptophan, DL-threonine, DL-phenylalanine, L-histidine, L-tyrosine, L-leucine, DL-valine, L-arginine, DL-isoleucine, L-glutamic acid, DL-serine, and L-cystine have been obtained.[32] The organism has been used for the assay of 9 of these:[33] histidine, arginine, leucine, isoleucine, valine, methionine, threonine, tryptophan, and phenylalanine in hydrolysates of purified proteins. Except for the lack of response to D-tryptophan the stereospecificity of the assay has not been reported. The results were in good agreement with those obtained with Streptococcus faecalis 9790 by Stokes et al.[21] The advantages of the assay (speed, nonaseptic conditions) are outweighed by the potential danger of the organism as a pathogen and the relative lack of sensitivity of some of the assays.

2. Eberthella typhosa (Salmonella typhosa)

This organism was proposed and used for the assay of tryptophan by Wooley and Sebrell.[34] Today its use is only of historical interest. At the time it was proposed it had the advantage over the use of Lactobacillus arabinosus 17-5 in speed (turbidity read after 16 hours rather than titration after 3 days) and simplicity of medium. Eberthella typhosa has less complex nutritional requirements, and a completely synthetic medium containing pure amino acids could be conveniently used. Assays for L-tryptophan with Lactobacillus arabinosus and Eberthella typhosa gave identical results on hydrolyzed or enzyme digested proteins,[34] and both assays were in close agreement with the chemical determinations of Horn and Jones.[35] Therefore, there seems to be no analytical advantage to E. typhosa as an assay organism. The potential danger of the organism as a pathogen has limited the further exploration of its usefulness, partic-

[32] M. J. Boyd, M. A. Logan, and A. A. Tytell, J. Biol. Chem. 174, 1013 (1948).
[33] M. J. Boyd, M. A. Logan, and A. A. Tytell, J. Biol. Chem. 174, 1027 (1948).
[34] J. G. Wooley and W. H. Sebrell, J. Biol. Chem. 157, 141 (1945).
[35] M. J. Horn and D. B. Jones, J. Biol. Chem. 157, 153 (1945).

ularly since the use of lactic acid bacteria has improved as far as speed and simplicity are concerned.

3. *Escherichia coli*

Both auxotrophic mutants[36] and cultures inhibited by antimetabolites[37] have been used. The appeal of *E. coli* mutants is due to the specificity of mutation and simplicity of the medium that can be employed. The organism will grow on a medium consisting of inorganic salts, glucose or other even simpler energy sources such as acetate and ammonium salts. An auxotrophic mutant would, in addition, require the specific nutrient. However, quite often the addition of a hydrolysate or other type of unknown to such a simple medium causes difficulties. This was overcome by Lampen et al.[38] for the assay of methionine with an *E. coli* mutant by the addition to the assay medium of peroxide treated yeast extract. Reversion of the mutant to the prototropic strain which no longer requires the amino acid is always a problem with *E. coli* mutants.[36] The reversal of sulphonamide inhibition by methionine derivatives led to the discovery of the natural occurrence of the dimethylsulfonium derivative of methionine in cabbage and other plant juices.[37,39] The sulfonium compound was 3 times as active as methionine in reversing sulfonamide inhibition. It is also of interest that the methionine dimethylsulfonium salts are not utilized in place of methionine by some lactic acid bacteria such as *Streptococcus faecalis* (9790),[40] *Leuconostoc mesenteroides* P-60 (8042), or by some auxotrophic mutants of *Escherichia coli* and *Neurospora crassa*.[41] However, for some *Escherichia coli* mutants, *Lactobacillus arabinosus* 17-5 (8014) and *L. casei* (7469), these substances are more effective than methionine in promoting growth.[41] A number of amino acids requiring mutants have been isolated from a number of strains of *Escherichia coli*.[36,38,42] Those that are of particular interest include one that utilizes peptides of proline better than it does proline,[43] and one that has a requirement for methionine or *p*-aminobenzoic acid when incubated at 37°C but requires neither substance when incubated at 21° to 24°C.[44] The requirement for optical isomers of amino acids is sometimes quite

[36] R. R. Roepke, R. L. Libby, and M. H. Small, *J. Bacteriol.* **48**, 401 (1944).

[37] R. A. McRorie, G. L. Sutherland, M. S. Lewis, A. D. Barton, M. Glazener, and W. Shive, *J. Am. Chem. Soc.* **76**, 115 (1954).

[38] J. O. Lampen, M. J. Jones, and A. B. Perkins, *Arch. Biochem.* **13**, 33 (1947).

[39] F. Challenger and B. J. Hayward, *Biochem. J.* **58**, IV (1954).

[40] G. D. Shockman and G. Toennies, *Arch. Biochem. Biophys.* **50**, 1 (1954).

[41] R. A. McRorie, M. R. Glazener, C. G. Skinner, and W. Shive, *J. Biol. Chem.* **211**, 489 (1954).

[42] E. L. Tatum, *Proc. Natl. Acad. Sci. U. S.* **31**, 215 (1945).

[43] S. Simmonds and J. S. Fruton, *J. Biol. Chem.* **174**, 705 (1948).

[44] T. J. Bird and J. S. Gots, *Proc. Soc. Exptl. Biol. Med.* **98**, 721 (1958).

complicated. For instance, one methionineless mutant will grow on either D- or L-methionine. However, the growth response to the two isomers differs in both character and rate.[36]

The most interesting *E. coli* mutants, from the standpoint of amino acid assays, are those that require α,ϵ-diaminopimelic acid (DAP).[45] This amino acid has been found in a number of bacteria[46-50] but apparently has not been found in algae, fungi, higher plants, or animals. DAP was found to be a constituent of the bacterial cell wall [51-58] and to be a precursor of lysine in certain bacteria.[45,59-62] The compound has two isomeric centers so that 4 isomers are possible. The meso and LL-[47,61,63,64] forms have been reported to occur naturally. The DD-isomer has been found only in a *Micromonospora*.[65] One mutant that has an absolute requirement for DAP is stable enough to be used in an assay but the quantitation is complicated by lysis on a limiting amount of DAP and a stimulation of growth rate by small amounts of lysine but inhibition by excess lysine.[45,66,67] It has, however, been used successfully for the qualitative identification of DAP. It has been necessary to use ion exchange columns or the specific DAP decarboxylase enzyme or both for the quantitative determination of this amino acid.[61,68]

[45] B. D. Davis, *Nature* **169**, 534 (1952).
[46] E. Work, *Nature* **165**, 74 (1950).
[47] E. Work, *Biochem. J.* **49**, 17 (1951).
[48] J. Asselineau, N. Choucroun, and E. Lederer, *Biochim. et Biophys. Acta* **5**, 197 (1950).
[49] T. Gendre and E. Lederer, *Biochim. et Biophys. Acta* **8**, 49 (1952).
[50] E. Work and D. L. Dewey, *J. Gen. Microbiol.* **9**, 394 (1953).
[51] E. S. Holdsworth, *Biochim. et Biophys. Acta* **8**, 110 (1952).
[52] E. S. Holdsworth, *Biochim. et Biophys. Acta* **9**, 19 (1952).
[53] M. R. J. Salton, *Biochim. et Biophys. Acta* **10**, 512 (1953).
[54] M. R. J. Salton, *Nature* **180**, 338 (1957).
[55] C. S. Cummins and H. Harris, *Biochem. J.* **57**, XXXII (1954).
[56] C. S. Cummins and H. Harris, *J. Gen. Microbiol.* **13**, iii (1955).
[57] C. S. Cummins and H. Harris, *J. Gen. Microbiol.* **14**, 583 (1956).
[58] E. Work, *Nature* **179**, 841 (1957).
[59] D. S. Hoare and E. Work, *Proc. Intern. Congr. Biochem. 3rd Congr. Brussels Abst.* p. 37 (1955).
[60] D. L. Dewey and E. Work, *Nature* **169**, 533 (1952).
[61] D. L. Dewey, D. S. Hoare, and E. Work, *Biochem. J.* **58**, 523 (1954).
[62] R. F. Denman, D. S. Hoare, and E. Work, *Biochim. et Biophys. Acta* **16**, 442 (1955).
[63] D. S. Hoare and E. Work, *Biochem. J.* **60**, ii (1955).
[64] M. Ikawa and J. S. O'Barr, *J. Biol. Chem.* **213**, 877 (1955).
[65] D. S. Hoare and E. Work, *Biochem. J.* **65**, 441 (1957).
[66] L. E. Rhuland, *J. Bacteriol.* **73**, 778 (1957).
[67] P. Meadow, D. S. Hoare, and E. Work, *Biochem. J.* **66**, 270 (1957).
[68] E. F. Gale, *Methods of Biochem. Anal.* **4**, 285 (1957).

4. Neurospora

A variety of auxotrophic mutants of *Neurospora crassa* have been either used or suggested for use for the assay of amino acids.[69-73] This organism will grow on a relatively simple medium consisting of an energy and carbon source (i.e., sucrose), an inorganic source of nitrogen (i.e., ammonium salts), several inorganic salts, and one vitamin (biotin).[71,74,75] The organisms are aerobic and are cultivated in a high surface to volume ratio to permit access to air. Ordinarily, the initial pH of the growth medium used is about 5. The mold mycelium grows on the surface of liquid media and the extent of growth can be measured either from the diameter of mycelial growth or by the dry weight of mycelial growth. The former method is less time consuming, the incubation period being about 20 hours as against about 5 days for the latter method. Another method that has been proposed by Ryan[76] is the measurement of the germination of *Neurospora* spores. The number of spores germinating was shown to be proportional to the amount of proline, leucine, or lysine provided for the appropriate auxotroph. This method is apparently more sensitive and rapid than the other *Neurospora* assays. The *N. crassa* method has been used by Brand *et al.*[72] for the determination of proline and leucine in hydrolysates of purified proteins. At that time bacterial methods for proline were not available and an independent check on a bacterial method for leucine was desired and obtained.

The discovery of a *Neurospora* mutant that requires asparagine[77,78] might become of some interest as an analytical aid. The mutant does not respond to aspartic acid or to many other closely related compounds. D-Asparagine is equivalent to the L-isomer up to about half maximal growth. Higher concentrations are inhibitory. α-Ketosuccinamate, a proposed immediate precursor of asparagine, gives response equivalent to that of asparagine. The organism also responds to asparaginylglycine.

The observation that *Neurospora* mutants can utilize amino acids in unhydrolyzed proteins[70] has been used for the direct estimation of leucine

[69] F. J. Ryan, *Federation Proc.* **5**, 366 (1946).

[70] D. C. Regnery, *J. Biol. Chem.* **154**, 151 (1944).

[71] F. J. Ryan and E. Brand, *J. Biol. Chem.* **154**, 161 (1944).

[72] E. Brand, L. J. Saidel, W. H. Goldwater, B. Kassell, and F. J. Ryan, *J. Am. Chem. Soc.* **67**, 1524 (1945).

[73] E. Brand, *Ann. N. Y. Acad. Sci.* **47**, 187 (1946).

[74] E. T. Butler, W. J. Robbins, and B. O. Dodge, *Science* **94**, 262 (1941).

[75] G. W. Beadle and E. L. Tatum, *Proc. Natl. Acad. Sci. U. S.* **27**, 499 (1941).

[76] F. J. Ryan, *Am. J. Botany* **35**, 497 (1948).

[77] S. W. Tannenbaum, L. Garnjobst, and E. L. Tatum, *Am. J. Botany* **41**, 484 (1954).

[78] C. Monder and A. Meister, *Biochim. et Biophys. Acta* **28**, 202 (1958).

in certain foodstuffs.[79] Since complete hydrolysis of proteins without amino acid destruction is still a problem (see Section VII), in the opinion of the author the potential of this characteristic has not as yet been sufficiently explored.

In recent years *Neurospora* mutants have not been used for amino acid assays. They have served the excellent purpose of confirming some of the methods that use lactic acid bacteria. The comparative inherent difficulties in measuring growth, lack of sensitivity, and precision of measurement have limited their usefulness.

5. Protozoa

The problem of growing protozoa in a bacteria free culture and the lack of adequate methods for the precise measurement of their growth response has limited their usefulness for amino acid assays. The latter has been somewhat overcome by the development of a colorimetric method employing tetrazolium salts for measuring the response to amino acids.[80] The organism that has been used in bio-assays is *Tetrahymena pyriformis* (*T. geleii*). The organism can be grown in a synthetic medium consisting of a carbohydrate energy source (glucose), inorganic salts, a series of vitamins [including thioctic acid (lipoic acid)], buffer, and a mixture of amino acids.[81–85] The strains differ in their amino acid requirements, but all seem to require arginine, glycine, histidine, isoleucine, leucine, lysine, methionine, phenylalanine, threonine, tyrosine, tryptophan, and valine.[86] They have not been used extensively for the assay of these amino acids and the methods require further development. The chief advantage of protozoal assays is the ability of these organisms to utilize intact proteins.[87–89] This is particularly useful for those amino acids, such as tryptophan, that may suffer large losses during hydrolysis procedures. Rockland and Dunn[87] have used a strain of *Tetrahymena* for the assay of tryptophan in unhydrolyzed casein. More recently there has been some interest in the use of *Tetrahymena* for the determination of protein quality.[88,90] In these instances the unknown protein replaces the entire amino acid mixture. A purified casein can be used as a standard.

[79] A. Z. Hodson and G. M. Krueger, *Arch. Biochem.* **12**, 435 (1947).
[80] M. E. Anderson and H. H. Williams, *J. Nutrition* **44**, 335 (1951).
[81] G. W. Kidder and V. C. Dewey, *Arch. Biochem.* **6**, 425 (1945).
[82] G. W. Kidder and V. C. Dewey, *Arch. Biochem.* **6**, 433 (1945).
[83] G. W. Kidder and V. C. Dewey, *Arch. Biochem.* **20**, 433 (1949).
[84] G. W. Kidder and V. C. Dewey, *Arch. Biochem.* **21**, 58 (1949).
[85] G. W. Kidder and V. C. Dewey, *Arch. Biochem.* **21**, 66 (1949).
[86] G. W. Kidder and V. C. Dewey, *Physiol. Zoöl.* **18**, 136 (1945).
[87] L. B. Rockland and M. S. Dunn, *Arch. Biochem.* **11**, 541 (1946).
[88] H. L. Pilcher and H. H. Williams, *J. Nutrition* **53**, 589 (1954).
[89] M. S. Dunn and L. B. Rockland, *Proc. Soc. Exptl. Biol. Med.* **64**, 377 (1947).
[90] T. Viswanatha and I. E. Liener, *Arch. Biochem. Biophys.* **56**, 222 (1955).

IV. The Lactic Acid Bacteria (Lactobacteriaceae)[91,92]

A. General Requirements for Cultivation

Lactic acid bacteria occur in a wide variety of natural environments and serve a large number of useful functions. They are found in the flora of the mouth and intestinal tract, in milk and other dairy products, and are responsible for a number of useful fermentations such as for fermented milk drinks (buttermilk, kefir, etc.), fermented sausages, cheeses, breads, sauerkraut, pickles, and for the industrial production of lactic acid. There are some industrial problems due to undesirable fermentations of lactic acid organisms including those of beer, wine, frozen concentrated orange juice, meats, etc. None of the *Lactobacillus* or *Leuconostoc* species are known to be pathogenic or to cause food poisoning in man. A few *Streptococcus* species, particularly *S. pyogenes*, are human pathogens.

All of the lactic acid bacteria have relatively complex nutritional requirements in comparison to organisms such as *Escherichia coli* or *Bacillus subtilis* which can be grown in a medium containing an energy source such as acetate, or glucose, and a simple nitrogen source such as ammonium salts and inorganic salts. All of the necessary substances that go to make up the cellular material such as amino acids (proteins), purine, and pyrimidine bases (nucleic acids), carbohydrates, and vitamins (coenzymes) can be synthesized from these simple nutrients by organisms such as *Escherichia coli* and *Bacillus subtilis*. While these organisms have simple nutritional requirements, their synthetic abilities are far more complex than those of the lactic acid bacteria since the latter lack the mechanisms for the synthesis of at least some of the precursors of these cellular substances.

In general, the energy source of lactic acid bacteria is limited to a small number of sugars. The great majority will utilize glucose; some, however, will only ferment pentoses. Many will not grow on even slightly more complex sugars such as the disaccharides sucrose and lactose or on simpler carbon and energy sources such as acetate. Most of the lactic acid bacteria require one or more of the vitamins, and in some cases, a specific form of a particular vitamin. For example, of the vitamin B_6 group, pyridoxine will support the growth of many lactic acid bacteria. Pyridoxamine or pyridoxal will be required by some, and one, *Lactobacillus delbrueckii* (9649), requires a phosphorylated form of the vitamin (pyridoxamine phosphate or pyridoxal phosphate).[93] All require

[91] E. E. Snell, *Wallerstein Lab. Communs.* 11, 81 (1948).
[92] E. E. Snell, *Bacteriol. Revs.* 16, 235 (1952).
[93] V. J. Peters and E. E. Snell, *J. Bacteriol.* 67, 69 (1954).

at least a few amino acids and for many growth is greatly stimulated by the presence of many amino acids. Peptides or materials derived from proteins are stimulatory for some. There are often additional requirements for purine or pyrimidine bases or other nucleic acid precursors or derivatives, acetate or carbon dioxide. In fact some can be cultivated only poorly if at all on synthetic media.

The general emphasis of the discussions in the following sections will be on those organisms that are useful for the microbiological assays of amino acids.

B. Physical and Chemical Factors for Growth (except Amino Acids and Derivatives)

1. Carbon Source and Energy Metabolism

Lactic acid bacteria obtain their energy for growth from the anaerobic fermentation of carbohydrates. The energy yield from the fermentation is relatively small, necessitating the consumption of carbohydrate and the production of acid by this type of energy metabolism at a phenomenally great rate, particularly during rapid growth when the mass of cell substance is doubling once every 40 minutes to 1.5 hours. These two factors enter into the design of assays using these organisms.

The organisms that have been used for amino acid assays fall into both the homofermentative and heterofermentative groups. The homofermentative group, which includes organisms such as the *Streptococci* and some of the *Lactobacilli* such as *L. casei* and *L. delbrueckii*, ferment hexoses nearly quantitatively to lactic acid. The over-all fermentation balance yields 2 lactic acid molecules from 1 glucose molecule and only a small portion of the energy in the glucose molecule. The heterofermentative organisms, such as the genus *Leuconostoc* and some of the *Lactobacilli*, such as *L. brevis* and *L. fermenti*, ferment hexoses only in part to lactic acid (usually about 50%) and the rest to other products such as CO_2, alcohol, glycerol, etc. The energy provided by a heterofermentative metabolism is somewhat greater than that produced by a homofermentative one. However, when compared to the aerobic dissimilation of hexoses to CO_2 the energy yield is still very low. Either D-, L-, or DL-lactic acid may be produced, depending on the particular organism involved. C^{14} labeled D- and L-lactic acids have been produced by fermentation of Lactobacilli.[94] The great bulk of the glucose or other carbohydrate supplied is recovered as the end products of fermentation. Only a very small fraction is assimilated to be used, for example, as pentoses in nucleic acids, polysaccharides, or amino sugars of the bacterial cell wall. The physiological

[94] M. Brin, R. E. Olson, and F. J. Stare, *Arch. Biochem.* **39**, 214 (1952).

division into homo- and heterofermentative groups[95,96] is strictly dependent on the fermentation of hexoses in slightly acid media. Homofermentative organisms can produce other end products of the fermentation at alkaline pH [97,98] or when an energy source other than a hexose is used.[99–103]

The production of acid is related to pH and the buffering capacity of the medium as discussed in Section IV.B.2.

2. Hydrogen Ion Concentration, Buffers, and Osmotic Environment

The optimum initial pH of the lactic acid bacteria is between 5.5 and 6.5. As a general rule the rod shaped organisms (Lactobacilli) grow more slowly, and produce and tolerate more acid than do the cocci (Streptococci, Leuconostoc, and Pediococci). The growth rate of a lactic acid organism slows down as it approaches its growth limiting pH by the production of acid. The addition of buffers helps to keep the pH of the medium within the range of more rapid growth. If the pH is maintained by the continuous addition of alkali, the immense quantities of lactate that will be produced can become toxic to the organism.[104,105] Some of the lactic acid bacteria such as *Streptococcus faecalis* can grow in media of the relatively high pH of 9.6 to 10.5.[106–108] Attempts to take advantage of this for microbiological assays have not as yet been made. The possibility exists that, like changes in growth temperature or gaseous environment (see Sections IV.B.3–5), a high initial pH may change the nutritive requirements of some organisms.

It has been customary to add buffers of acetate, citrate, or phosphate in order to help maintain the pH of an assay medium. Close examination of some of the earlier buffers used reveals that the buffering capacity of the media is not very great at all, especially in relationship to the amounts of lactic acid produced. A typical example is the medium of

[95] A. J. Kluyver and H. J. L. Donker, *Proc. Koninkl. Akad. Wetenschap. Amsterdam* **28**, 297 (1925).

[96] A. J. Kluyver, *Ergeb. Enzymforsch.* **4**, 230 (1935).

[97] I. C. Gunsalus and C. F. Niven, Jr., *J. Biol. Chem.* **145**, 131 (1942).

[98] I. C. Gunsalus and C. F. Niven, Jr., *J. Bacteriol.* **44**, 260 (1942).

[99] J. J. R. Campbell, W. D. Bellamy, and I. C. Gunsalus, *J. Bacteriol.* **46**, 573 (1943).

[100] J. J. R. Campbell and I. C. Gunsalus, *J. Bacteriol.* **48**, 71 (1944).

[101] I. C. Gunsalus, *J. Bacteriol.* **54**, 239 (1947).

[102] R. H. Steele, A. G. C. White, and W. A. Pierce, Jr., *J. Bacteriol.* **67**, 86 (1954).

[103] I. C. Gunsalus and J. J. R. Campbell, *J. Bacteriol.* **48**, 455 (1944).

[104] L. L. Kempe, H. O. Halvorson, and E. L. Piret, *Ind. Eng. Chem.* **42**, 1852 (1950).

[105] R. K. Finn, H. O. Halvorson, and E. L. Piret, *Ind. Eng. Chem.* **42**, 1857 (1950).

[106] J. M. Sherman and P. Stark, *J. Dairy Sci.* **17**, 525 (1934).

[107] J. M. Sherman, J. C. Mauer, and P. Stark, *J. Bacteriol.* **33**, 275 (1937).

[108] W. R. Chesbro and J. B. Evans, *J. Bacteriol.* **78**, 858 (1959).

Stokes and Gunness[20] which contains 6 mg. of sodium acetate and 0.5 mg. of potassium phosphates per ml. This gives a total buffer of only about 0.08 M. There is, however, a very definite limitation on the salt concentration that can be tolerated by the assay organisms. Some organisms, such as *Lactobacillus delbrueckii* (9649), will not grow in a synthetic medium if the salt or buffer concentration is increased.[13] Others, such as *Streptococcus faecalis*, are extremely resistant to high salt concentrations and successful assays have been advantageously done in a medium containing 0.3 M phosphate buffer plus the usually employed amounts of acetate and phosphate[109-111] (Section X). Assays with *Leuconostoc mesenteroides* have been done in the same highly buffered medium.[110]

Production of and sensitivity to acid lead to the necessity for the critical adjustment of initial pH and buffer concentration of both the assay medium and of samples to be assayed (Sections VII and VIII), particularly when media of low buffering capacity are employed. When the buffer concentration of the medium is high, the effect of salts or acidity in samples to be assayed will be reduced.[112]

3. Temperature

The optimal growth temperature of lactic acid bacteria ranges from 20° to 45°C., depending on the specific strain or species.[113] However, almost invariably temperatures between 30° and 37°C. have been used for amino acid assay. Approaching the optimal temperature for the assay organism may somewhat reduce the time required for an assay. For vitamin assays with lactic acid bacteria, the Official Methods of Analysis of the A.O.A.C.[114,115] merely recommends temperatures between 30° and 37° for most assays. Ordinarily, the actual incubation temperature (provided it is maintained constant) has little effect on the over-all assay results. For instance, assay values for valine obtained with *Lactobacillus casei* at 32° and 37° showed no difference at the two temperatures.[17]

There are, however, a few exceptions. Borek and Waelsch[116] have demonstrated that a difference of 2° can make a difference in the phenyl-

[109] G. Toennies and D. L. Gallant, *J. Biol. Chem.* **174**, 451 (1948).

[110] G. Toennies and G. D. Shockman, *Proc. Intern. Congr. Biochem. 4th Congr. Vienna 1958* **13**, 365 (1959).

[111] G. Toennies and G. D. Shockman, *Arch. Biochem. Biophys.* **45**, 447 (1953).

[112] G. Toennies and H. G. Frank, *Growth* **14**, 341 (1950).

[113] R. S. Breed, E. D. G. Murray, and A. P. Hitchens, "Bergey's Manual of Determinative Bacteriology," 6th ed. Williams & Wilkins, Baltimore, Maryland, 1948.

[114] "Revision of Microbiological Methods for the B Vitamins," *J. Assoc. Offic. Agr. Chemists* **41**, 61 (1958).

[115] "Official Methods of Analysis," 8th ed. Assoc. Offic. Agr. Chemists, Washington D. C., 1955.

[116] E. Borek and H. Waelsch, *J. Biol. Chem.* **190**, 191 (1951).

alanine requirement of *L. arabinosus*. Optimal growth was obtained in the absence of this amino acid at 35° but no growth was obtained at 37°. A similar temperature effect was demonstrated for the tyrosine and aspartic acid requirements of this organism.[116] This was, however, related to the deficiency of carbon dioxide at the higher temperatures since the addition of CO_2 at the higher temperatures restored growth in the absence of either aspartic acid or phenylalanine. Similar observations on the temperature dependence of *L. arabinosus* for tyrosine or phenylalanine were made by James.[117]

Therefore, because of the potential result of a temperature change on growth requirements, the recommended incubation temperature should be followed. All of the methods recommended in Section X are done at 37° to 38°C.

Constant incubation temperature, however, is the more important factor for precise microbiological assays. An essentially vibration free constant temperature water bath capable of maintaining an incubation temperature within ±0.5 degrees is recommended for careful work (see Section IX.B).

4. Oxidation-Reduction Potential

Essentially all of the lactic acid bacteria that have been used for amino acid assays are facultative anaerobes or microaerophilic. For practical work this means that no special precautions to either provide or completely exclude oxygen are needed or recommended. Many of the organisms will not grow in thin layers of liquid media or near the surface of solid media. The presence of reducing agents such as cysteine, thioglycollate, or ascorbic acid have been recommended for some assays, mainly when microtechnics that involve a small volume of medium are used.[118-120] For agar diffusion assays or bioautographs a top layer of plain agar has been found to be helpful.[121] Most amino acid assay media have a favorable oxidation reduction potential for these organisms. This has often been attributed to the autoclaving of the carbohydrate containing media. Also, the reducing action of the growing bacteria will rapidly decrease the oxidation-reduction potential of the medium. This may account for the need for high inoculum levels in some instances.

In special cases the oxidation-reduction potential of the medium may affect the growth requirements of the assay organism. An example, taken from vitamin requirements, is the absolute requirement of *Streptococcus*

[117] A. P. James, Ph.D. thesis, Iowa State College, Ames, Iowa, 1949.
[118] O. H. Lowry and O. A. Bessey, *J. Biol. Chem.* **155,** 71 (1944).
[119] L. M. Henderson, W. L. Brickson, and E. E. Snell, *J. Biol. Chem.* **172,** 31 (1948).
[120] G. Agren, *Arkiv Kemi* **1,** 179 (1949).
[121] E. Usdin, G. D. Shockman, and G. Toennies, *Appl. Microbiol.* **2,** 29 (1954).

faecalis for thioctic acid or acetate plus thiamine when grown aerobically as contrasted to the lack of requirement for any of these substances when the cells are grown in the absence of oxygen.[122]

5. CO₂

Increased CO_2 tensions are not ordinarily required for the lactic acid assay organisms, and ordinarily little attention needs to be paid to it. However, in several instances an enriched CO_2 atmosphere may affect the nutritional requirements of an assay organism. Lyman *et al.*[123] have shown that, in an otherwise complete medium, *Lactobacillus arabinosus* required arginine, phenylalanine, and tyrosine when the cultures were incubated in the usual air atmosphere. However, when the same organism was incubated in an atmosphere of air plus 6% CO_2, good growth could occur when these amino acids were individually omitted from the medium. An enriched CO_2 atmosphere has also been shown to enable *Streptococcus faecalis* R to grow when aspartic acid was omitted from the medium,[124] and *Leuconostoc mesenteroides* P-60 to grow in the absence of serine.[125] It should also be recalled that the availability of CO_2 in a culture medium may be related to the incubation temperature (Section IV.B.3).

6. Inorganic Salts

Detailed knowledge of the requirements of lactic acid bacteria for inorganic ions is lacking. The cations Na^+, K^+, Mg^{++}, Mn^{++}, and Fe^{++} are usually provided, sometimes with the addition of Ca^{++}.[126-131] The phosphate, sulfate, and chloride ions are also either provided or present in sufficient quantities in both the organic and inorganic salts present in an assay medium. For assay purposes there are two important facets:

(1) The use of citrate or phosphate buffers may reduce the available amounts of some of these ions by precipitation or by the formation of unavailable complexes.

[122] G. Shockman, *J. Bacteriol.* **72**, 101 (1956).

[123] C. M. Lyman, O. Moseley, S. Wood, B. Butler, and F. Hale, *J. Biol. Chem.* **167**, 177 (1947).

[124] C. M. Lyman, O. Moseley, S. Wood, B. Butler, and F. Hale, *J. Biol. Chem.* **162**, 173 (1946).

[125] J. Lascelles, M. J. Cross, D. D. Woods, *J. Gen. Microbiol.* **10**, 267 (1954).

[126] R. A. MacLeod and E. E. Snell, *J. Biol. Chem.* **170**, 351 (1947).

[127] W. W. Meinke and B. R. Holland, *J. Biol. Chem.* **184**, 251 (1950).

[128] I. J. McDonald, *Can. J. Microbiol.* **3**, 411 (1957).

[129] R. A. MacLeod, *J. Bacteriol.* **62**, 337 (1951).

[130] R. A. MacLeod and E. E. Snell, *J. Biol. Chem.* **176**, 39 (1948).

[131] R. A. MacLeod and E. E. Snell, *Ann. N. Y. Acad. Sci.* **52**, 1249 (1950).

(2) The sodium-potassium balance of the medium[130-136] may be an important factor that can be upset by the addition of the assay sample itself or by its neutralization. Ammonium hydroxide or acetic acid has been used in some cases as a neutralizing agent in order to avoid such imbalance.

7. Miscellaneous

a. *Nitrogen containing compounds.* Repeated observations have been made on the stimulation of lactic acid bacteria by the ammonium ion.[92,137-139] It is easy to visualize the presence of this ion in samples to be assayed. It has not been included in some of the assay media listed in Table VII. However, where such media are to be used for amino acid assays, the addition of the ammonium ion should be routinely considered. Nucleic acid derivatives are usually furnished in the form of purine and pyrimidine bases. For amino acid assays, complications can arise when the assay sample contains a nucleic acid derivative that allows more rapid growth than those provided in the culture medium. The nucleic acid derivatives required by lactic acid bacteria are often interrelated with their vitamin requirements. These will not be discussed in detail since again there is little need to consider these requirements for routine amino acid determinations unless it is known that the samples to be assayed are particularly rich in nucleic acids and their derivatives. Acid hydrolysis of nucleic acids will yield ammonia and glycine.[140] Therefore great caution should be used in the interpretation of assay values for glycine on materials that may contain nucleic acids.

b. *Other miscellaneous compounds.* Fatty acids such as acetic, oleic, or linoleic acids fall into this group. The acetate requirement of *Streptococcus faecalis* was mentioned above (Section IV.B.2) in relation to the gaseous environment. It has been found to be at least stimulatory for a number of lactic acid bacteria. Acetate is often employed as a buffer and provides more than an adequate amount for nutritional purposes. When

[132] M. N. Camien and M. S. Dunn, *Proc. Soc. Exptl. Biol. Med.* **95**, 697 (1957).

[133] M. N. Camien and M. S. Dunn, *Proc. Soc. Exptl. Biol. Med.* **97**, 419 (1958).

[134] G. Lester, *J. Bacteriol.* **75**, 426 (1958).

[135] R. G. Chitre, R. J. Sirny, and C. A. Elvehjem, *Arch. Biochem. Biophys.* **31**, 398 (1951).

[136] R. J. Sirny, O. R. Braekkan, M. Klungsøyr, and C. A. Elvehjem, *J. Bacteriol.* **68**, 103 (1954).

[137] L. R. Hac, E. E. Snell, and R. J. Williams, *J. Biol. Chem.* **159**, 273 (1945).

[138] M. C. Glick, F. Zilliken, and P. György, *J. Bacteriol.* **77**, 230 (1959).

[139] P. György, R. Kuhn, C. S. Rose, and F. Zilliken, *Arch. Biochem. Biophys.* **48**, 202 (1954).

[140] C. F. Crampton, W. H. Stein, and S. Moore, *J. Biol. Chem.* **225**, 363 (1957).

other buffers are used acetate should still be provided. In fact, neutralization of alkaline hydrolysates or extracts with acetic acid will often change the acetate concentration of the culture medium relatively less than the neutralization of acid hydrolysates with sodium and/or potassium hydroxides will change the concentration of the Na or K ions. This can be of great practical importance and is sometimes the difference between a successful and an unsuccessful assay. The requirement for unsaturated fatty acids can again be interrelated with vitamin requirements and sometimes both are provided.[91,92] However, for some organisms, such as *Lactobacillus casei*, certain unsaturated fatty acids may be toxic. The toxicity of linoleic acid can be reversed by an equimolar amount of vitamin D_2.[141-143] Since the inhibitory action may be on concentrative mechanisms of the cell the presence of either or both unsaturated fatty acids and vitamin D_2 (either in the medium or assay sample) may affect the results of an amino acid assay. Serum albumin or a synthetic surface active agent such as Tween 80 or 40 are often used in assay media to detoxify unsaturated fatty acids.[26,144,145]

8. Vitamin and Growth Factor Requirements

One is forced to consider the vitamin requirements of each assay organism for each amino acid assay.[146] This is due to the role that many of the vitamins play (as coenzymes) in the biosynthesis of individual amino acids. For assays, it is of particular advantage to use an organism that requires the amino acid being assayed in the presence of all known vitamins and cofactors. However, some of these interrelationships are quantitative. For instance, the biotin requirement of *L. arabinosus* is about 10 times as high in the absence of aspartic acid as it is in a medium containing ample amounts of this amino acid.[147] An assay set up for aspartic acid using a medium that contains sufficient amounts of biotin by ordinary standards could easily yield an invalid assay. Such an assay could be further complicated by the presence of avidin, a protein that occurs in egg white and that stoichiometrically combines with biotin preventing its uptake by cells. Other examples of vitamin-amino acid in-

[141] E. Kodicek, *Symposia Soc. Exptl. Biol.* **3**, 217 (1949).
[142] E. Kodicek, *J. Gen. Microbiol.* **18**, XIII (1958).
[143] K. McQuillen, *Proc. Intern. Congr. Biochem. 4th Congr. Vienna 1958* **13**, 406 (1959).
[144] J. C. Alexander and D. C. Hill, *J. Nutrition* **48**, 149 (1952).
[145] S. S. Schiaffino and H. W. Loy, *J. Assoc. Offic. Agr. Chemists* **41**, 739 (1958).
[146] S. Shankman, M. N. Camien, H. Block, R. B. Merrifield, and M. S. Dunn, *J. Biol. Chem.* **168**, 23 (1947).
[147] H. P. Broquist and E. E. Snell, *J. Biol. Chem.* **188**, 431 (1951).

terrelationships are serine with the folic acid group, and vitamin B_6 group with a number of amino acids such as cystine and alanine.[148-150] It is therefore of importance to be certain that all of the vitamins are present in *excess* for the particular amino acid assay.

There are a few selected instances where the amino acid is assayed in the absence of a vitamin or a particular form of a vitamin. The outstanding example is the assay of D-alanine in the absence of vitamin B_6. This has been done with *L. casei*[151] and more recently with *L. delbrueckii*.[31] The latter organism has the great advantage in that it requires the phosphorylated form of the vitamin (pyridoxamine phosphate or pyridoxal phosphate) and can only use the nonphosphorylated forms of the vitamin when they are present in very high concentrations.[93] The phosphorylated form is less stable and therefore more easily destroyed or removed from assay samples by heat, acid, or adsorption on activated charcoal. *L. delbrueckii* has been successfully used to assay both D- and L-alanine in hydrolysates of bacterial materials.[13,31]

C. Nutrition and Metabolism of Amino Acids and Derivatives

1. Amino Acid Requirements

As mentioned in Section IV.B.8, the amino acid requirement of individual lactic acid bacteria is dependent upon the growth factors that are supplied in the culture medium. With a very few exceptions, assays are carried out, at least theoretically, with all growth factors in excess. Therefore, only the amino acid requirements in the presence of an excess of all known factors will be considered. Even under these conditions the number and combination of amino acids required for growth are characteristic for the particular organism. In addition, some of the amino acids that are not absolutely required are stimulatory. As an example, *Streptococcus faecalis* 9790 has been used for the assay of 10 amino acids (threonine, leucine, isoleucine, valine, methionine, arginine, histidine, lysine, glutamic acid, and tryptophan).[21,152,153] When we examine the effect of the individual withdrawal of the eleven other amino acids that are normally present in an assay medium (Table III) on the rate of growth and on the maximum growth level, we find that none are completely dispensable and that at least one of these, tyrosine, could be

[148] B. R. Holland and W. W. Meinke, *J. Biol. Chem.* **178**, 7 (1949).
[149] E. E. Snell and B. M. Guirard, *Proc. Natl. Acad. Sci. U. S.* **29**, 66 (1943).
[150] M. L. Speck and D. A. Pitt, *Science* **106**, 420 (1947).
[151] J. T. Holden and E. E. Snell, *J. Biol. Chem.* **178**, 779 (1949).
[152] G. Toennies and D. L. Gallant, *Growth* **13**, 21 (1949).
[153] C. M. Lyman, K. A. Kuiken, L. Blotter, and F. Hale, *J. Biol. Chem.* **157**, 395 (1945).

TABLE III
EFFECT OF OMISSION OF INDIVIDUAL AMINO ACIDS ON GROWTH[a]

Amino acid absent	Initial growth rate		Maximum growth attained	
	Divisions per hour	Per cent of standard	Per cent of standard	Time (hours)[b]
None	1.18	100	100	6.5
Glycine	0.77	65	55	24
Alanine	1.11	94	72	6.5
Serine	0.20	17	15	24
Cystine	0.88	75	62	8.0
Phenylalanine	0.20	17	56	24
Tyrosine	0.07	6	3	24
Proline	1.06	89	100	7.0
Hydroxyproline	1.14	97	96	6.5
Tryptophan	0.17	14	3	24
Aspartic acid	1.21	103	92	6.5
Glutamic acid	0.0	0	0	—

[a] See reference 110.
[b] Size of inoculum AOD 20, i.e., approximately 2% of maximum bacterial density.

classified as completely essential.[110] In addition, most of the data for other organisms are older and to some extent conflicting.[5,17,154,155] Lack of agreement is due to a number of things including various basal media, different strains of the test organisms, varying purity of the medium ingredients (vitamins as well as amino acids), and to differences in test systems. What we are really concerned with is which amino acids can be assayed with a particular test organism and not those that are nutritionally required. Table IV lists some amino acid assays that have been proposed and used. It can rapidly be seen that there is a choice of at least two organisms for each amino acid assay. The recommendations to be given here can be broken down into three classes. (1) Those that have been performed by the author with success and are therefore recommended (Sections X.A–F). (2) Those that have not been performed by the author but are recommended from results reported by others (Section X.G). These are often, but not always, those that have been recommended most often in various compilations, reviews and handbooks.[4-8] (3) Those that are listed in the table without recommendation. These are often useful when, as a check, an assay with a second organism is desired (Section

[154] M. S. Dunn, S. Shankman, M. N. Camien, and H. Block, *J. Biol. Chem.* **168**, 1 (1947).
[155] S. Shankman, *J. Biol. Chem.* **150**, 305 (1943).

TABLE IV

SOME OF THE AMINO ACID ASSAYS THAT HAVE BEEN PROPOSED AND USED[a,b]

	S. faecalis (9790)[c]	Leuc. mesenteroides P. 60 (8042)[d]	L. delbrueckii[e] 780 (9649)[f]	L. arabinosus 17-5 (8014)[f]	Leuc. citrovorum (8081)[g]	L. fermenti (9338)[h]	L. delbrueckii LD 5 (9595)[i]	L. casei (7469)[j]	L. leichmannii 313 (7830)[k]	L. leichmannii 327 (4797)[k]	L. brevis (8257)[l]	L. delbrueckii 3 (4913)[m]
Alanine		+		★[n]	+[n]				+[n]			
Arginine	★	+			+		+	+	+	+		+
Aspartic acid		★					+		+	+		
Cystine		★		+	+				+	+		
Glutamic acid	★	+		+	+		+		+	+		
Glycine		*			+							
Histidine	★	+			+	+			+	+		+
Isoleucine	★	+		+	+		+	+	+	+		
Leucine	★	+		+			+	+	+	+		
Lysine	★	+										
Methionine	★[o]	+		+	+	+[o]			+	+		
Phenylalanine		*		+	+		+	+	+	+		
Proline		★			+						+	
Serine		*					+					+
Threonine	★	+		+	+	+						
Tryptophan	*	+		+			+		+	+		
Tyrosine		*		+	+		+		+	+		
Valine	★	+		+	+		+	+	+	+		

[a] Numbers in parentheses refer to those of the American Type Culture Collection (ATCC), 2112 M Street, N.W., Washington 7, D.C. There has been much confusion in the names of the organisms used for assays. In most cases the most common name is used in the table with the other names given in footnotes.

[b] ★—Recommended assays: performed with good results by the author. *—Preferred method: No first hand experience, but preferred from reference to reports in the literature and discussions with other investigators. Preference is sometimes given where one organism can be used for the determination of more than one amino acid.

[c] Probably identical with *Streptococcus lactic R*. See references 10, 13, 21, 152, 161–170.

[d] Listed by ATCC as *Streptococcus* spp. 8042. See references 10–13, 19, 26–28, 161–163, 165–169, 173–175.

[e] See references 13 and 31.

[f] Listed by ATCC as *Lactobacillus plantarum* 8014. See references 10, 17, 26, 30, 161, 163, 165–167, 170–175.

[g] Listed by ATCC as *Pediococcus cerevisiae* 8081. See references 12, 22, 26, 27, 30, 162.

[h] See references 23–25, 173, 174.

[i] Originally called *Lactobacillus delbrueckii* LD 5. Listed by ATCC as *L. casei* (9595). Both names have been used in the literature. Look for ATCC number. See references 10, 20, 21, 161, 169, 176.

[j] See references 10, 17, 27, 161, 170, 174, 175, 177.

[k] See references 29, 30, 178, 179.

[l] See references 26, 28, 30.

[m] *L. delbrueckii* 3 = ATCC *L. acidophilus* 4913. See references 26, 27, 168.

[n] *L. delbrueckii* (9649) is recommended for the determinations of D- and L-alanine separately. *Leuconostoc citrovorum* has been used for the determination of DL-alanine. See Section XI. *Lactobacillus casei* can also be used for D-alanine. However greater difficulty is encountered with vitamin B6 interference.

[o] *Streptococcus faecalis* used for L-methionine determinations. *Lactobacillus fermenti* has been recommended for the determination of DL-methionine. Refer to Section IV.C.2.

XI). Some of these are not proven assays. That is, they have been proposed and recommended but have not been extensively used. Examples of this group are some of the amino acid assays with *Lactobacillus leichmannii*.

To some extent the choice of organism rests on convenience. It is often better to handle fewer stock cultures. Some organisms are hardier and less subject to mutation or variation than others (Section IV.C.6). These grow more easily, are more easily stored when assays are not regularly being done, and are usually more reliable for the general run of analytical work. With the possible exception of *Leuconostoc mesenteroides* P60, organisms that have more complex requirements are generally more difficult in this respect. For instance, *Lactobacillus delbrueckii* 780 (9649) is often a difficult organism to handle. It has some peptide requirements (for which high concentrations of some free amino acids can substitute) and requires the phosphorylated form of vitamin B_6. It is the latter requirement that makes it the organism of choice for the assay of L- and D-alanine (see Section X.C.4).

A further remark concerns itself with the identity of the various cultures. The identity of cultures used in literature references is not always clear. In Table IV we have attempted to identify the cultures with ATCC (American Type Culture Collection, 2112 M Street, N.W., Washington 7, D. C.) numbers as far as possible. This is by far the preferred method and the preferred source of cultures. Quite often a culture obtained from another laboratory turns out to have properties different from those expected, or the identification is incorrect. The ATCC number is an additional aid since many of the organisms used for amino acid assays were incorrectly identified and more than one name may be encountered in the literature. The name of the organism that is most commonly used in the literature of these assays is used in the heading of the table. Synonyms and the ATCC identification are included in the footnotes. These ATCC cultures are the ones referred to throughout this chapter without repetition of the ATCC number.

In the past, with some organisms, it was necessary to add a crude extract of one sort or another to provide a growth factor. Today, completely synthetic media can be and generally are used for amino acid assays. There are only three exceptions where a protein hydrolysate may be used. In the assay for D-alanine with *L. delbrueckii*[156] acid hydrolyzed and partly hydrolyzed casein can be used if it is free of vitamin B_6. Oxidized peptone (hydrogen peroxide treated) can be used for cystine, methionine and tyrosine determinations[157-160] and acid hydrolyzed casein

[156] E. E. Snell and A. N. Rannefeld, *J. Biol. Chem.* **157,** 475 (1945).

[157] G. D. Shockman, J. J. Kolb, and G. Toennies, *Anal. Chem.* **26,** 1657 (1954).

[158] G. Toennies, *J. Biol. Chem.* **145,** 667 (1942).

[159] C. M. Lyman, O. Moseley, S. Wood, and F. Hale, *Arch. Biochem.* **10,** 427 (1946).

[160] W. H. Riesen, B. S. Schweigert, and C. A. Elvehjem, *J. Biol. Chem.* **165,** 347 (1946).

[161] H. H. Williams, L. V. Curtin, J. Abraham, J. K. Loosli, and J. K. Maynard, *J. Biol. Chem.* **208,** 277 (1954).

[162] H. E. Sauberlich and C. A. Baumann, *Cancer Research* **11,** 67 (1951).

[163] E. V. Cardinal and L. R. Hedrick, *J. Biol. Chem.* **172,** 609 (1948).

[164] I. T. Greenhut, B. S. Schweigert, and C. A. Elvehjem, *J. Biol. Chem.* **162,** 69 (1945).

can be used for tryptophan assays.[34] These are convenient when only one or more of these particular amino acids are being determined. However, for reasons given in Section X.F, the use of oxidized peptone is recommended for cystine assays.

2. D-Amino Acids and Lactic Acid Bacteria

At least three D-amino acids (alanine, glutamic, and aspartic acids) have been found in lactic acid bacteria.[180-183] These have been identified as components of the cell walls of these organisms.[13,31,184,185] A variety of D-amino acids (leucine, serine, valine, etc.) have been found in products of microbial origin (i.e., antibiotics).[186-190] D-Alanine has been found in the blood of the milkweed bug (*Oncopeltus fasiatus*).[191] Octopine, an amino acid that occurs free in scallops, also contains a D-alanine resi-

[165] E. C. Barton-Wright, W. B. Emery, and F. A. Robinson, *Nature* **157**, 628 (1946).

[166] L. M. Henderson and E. E. Snell, *J. Biol. Chem.* **172**, 15 (1948).

[167] E. C. Barton-Wright, *Analyst* **71**, 267 (1946).

[168] R. J. Sirny, I. T. Greenhut, and C. A. Elvehjem, *J. Nutrition* **41**, 383 (1950).

[169] C. E. Graham, H. K. Waitkoff, S. W. Hier, *J. Biol. Chem.* **163**, 159 (1949).

[170] W. Baumgarten, L. Stone, and C. S. Boruff, *Cereal Chem.* **22**, 311 (1945).

[171] B. S. Schweigert, I. E. Tatum, and C. A. Elvehjem, *Arch. Biochem.* **6**, 177 (1945).

[172] B. S. Schweigert, J. M. McIntire, C. A. Elvehjem, and F. M. Strong, *J. Biol. Chem.* **155**, 183 (1944).

[173] S. F. Velick and E. Ronzoni, *J. Biol. Chem.* **173**, 627 (1948).

[174] M. S. Dunn, M. N. Camien, R. B. Malin, E. A. Murphy, and P. J. Reiner, *Univ. Calif. (Berkeley) Publs. Physiol.* **8**, 293 (1949).

[175] S. W. Hier, C. E. Graham, R. Freides, and D. Klein, *J. Biol. Chem.* **161**, 705 (1945).

[176] M. Gunness, I. M. Dwyer, and J. L. Stokes, *J. Biol. Chem.* **163**, 159 (1946).

[177] M. S. Dunn, S. Shankman, and M. N. Camien, *J. Biol. Chem.* **161**, 643 (1945).

[178] H. R. Skeggs, C. A. Driscoll, H. N. Taylor, and L. D. Wright, *J. Bacteriol.* **65**, 733 (1953).

[179] M. N. Yrague, O. B. Weeks, and A. C. Wiese, *J. Bacteriol.* **69**, 20 (1955).

[180] J. T. Holden, C. Furman, and E. E. Snell, *J. Biol. Chem.* **178**, 789 (1949).

[181] M. N. Camien, *J. Biol. Chem.* **197**, 687 (1952).

[182] M. S. Dunn, M. N. Camien, S. Shankman, and H. Block, *J. Biol. Chem.* **168**, 43 (1947).

[183] C. M. Stevens, R. P. Gigger, and S. W. Bowne, Jr., *J. Biol. Chem.* **212**, 461 (1955).

[184] M. Ikawa and E. E. Snell, *Biochim. et Biophys. Acta* **19**, 576 (1956).

[185] M. Ikawa and E. E. Snell, *J. Biol. Chem.* **235**, 1376 (1960).

[186] H. T. Clark, J. R. Johnson, and R. Robinson, "The Chemistry of Penicillin." Princeton Univ. Press, Princeton, New Jersey, 1949.

[187] E. Chain, *Ann. Rev. Biochem.* **17**, 657 (1948).

[188] P. H. Long, ed., *Ann. N. Y. Acad. Sci.* **51**, 853 (1949).

[189] H. Brockmann, N. Grubhofer, W. Kass, and H. Kalbe, *Chem. Ber.* **84**, 260 (1951).

[190] H. Brockmann, G. Bohnsack, and H. Gröne, *Naturwissenschaften* **40**, 233 (1953).

[191] J. L. Auclair and R. L. Patton, *Rev. can. biol.* **9**, 3 (1950).

due.[192,193] An isomer of β-methyllanthionine containing a D-α-amino-n-butyric acid residue has been isolated from yeast.[194] In fact, it is not as yet possible to completely exclude the possibility of the occurrence of small amounts of D-amino acids in animal tissues.[195–197] The natural occurrence of D-amino acids has increased the importance of knowing the configurative specificity of microbiological assays. Where this information cannot be reliably obtained in the literature, the present day commercial availability of high quality D-amino acids makes such a test mandatory. In addition, there are examples of the slow utilization of the D-isomer in the presence of the L-isomer making it imperative to use the L-form as a standard wherever possible.[40] Also, some of the synthetic racemic mixtures can be contaminated. For example, DL-leucine may be contaminated with isoleucine or allo isoleucine.[167]

Organisms that respond equally well to D- or DL-amino acids have been reported. With some exceptions, alanine, methionine, glutamic and aspartic acids are the amino acids involved.[12,22,25,181,182,198] A careful look at the data of most of these reports usually reveals that the response to each of the isomers is not quite equivalent.[166,199,200] This can sometimes be compensated for by a long incubation period for the assay.

Two possibilities exist for the partial or complete utilization of a D-amino acid. (1) The D-isomer may be converted by means of a racemase or via a combination of transaminase and alanine racemase to the L-amino acid. (2) Where the amino acid occurs in the organism (alanine, glutamic, and aspartic acids) in the D-form, it may be assimilated and used as such.

An example of the first possibility would be the utilization of D-methionine by *Lactobacillus arabinosus* 17-5. The organism utilizes D-methionine as well as L- when the medium contains a sufficient level of pyridoxamine or pyridoxal.[198] With pyridoxine or without an added source of vitamin B_6 the utilization of D-*methionine* was negligible.

An example of the second possibility is the utilization of the alanine isomers by *Streptococcus faecalis*. In this organism, like others, D-alanine occurs not in cytoplasmic proteins but in the cell wall.[13,31,184,185] L-Alanine

[192] F. Knoop and C. Martius, *Z. physiol. Chem.* **258**, 238 (1939).
[193] R. M. Herbst and E. A. Swart, *J. Org. Chem.* **11**, 368 (1946).
[194] P. F. Downey and S. Black, *J. Biol. Chem.* **228**, 171 (1957).
[195] F. Kögl and H. Erxleben, *Z. physiol. Chem.* **258**, 57 (1939).
[196] P. Boulanger and R. Osteux, *Compt. rend. acad. sci.* **256**, 2177 (1953).
[197] P. Boulanger and R. Osteux, *Bull. Cancer* **45**, 350 (1958).
[198] M. N. Camien and M. S. Dunn, *J. Biol. Chem.* **182**, 119 (1950).
[199] C. M. Lyman, K. A. Kuiken, L. Blotter, and F. Hale, *J. Biol. Chem.* **157**, 395 (1945).
[200] D. M. Hegsted, *J. Biol. Chem.* **157**, 741 (1945).

is a constituent of both cellular proteins and the cell wall.[13] In the absence of vitamin B_6 the organism requires both isomers.[149,201] Under such conditions the utilization of D-alanine is competitively inhibited by an antibiotic, oxamycin (D-cycloserine), over a wide range of L-alanine concentrations.[202] Oxamycin is a derivative of D-serine (D-4-amino-3-isoxolidone). The two isomers are apparently acting as two different compounds.

In light of present knowledge one would expect a corresponding D-amino acid to be as different from the L-isomer as an L-homolog of an amino acid is from the same amino acid. Even if the difference is only in the intracellular conversion of the D- to the L-form for utilization this would involve at least one additional metabolic step and possibly change either the sensitivity or the character of the response obtained. An example of the response of *Lactobacillus delbrueckii* to D- and L-alanine is given in Fig. 1. It is quite obvious that not only is the shape of the re-

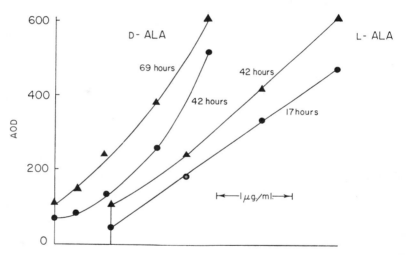

FIG. 1. The response of *Lactobacillus delbrueckii* (9649) to D- and L-alanine. The ordinate is in AOD, an expression of relative bacterial concentration (see Section V.C) for turbidimetric measurements. The abscissa is given by the 1 μg./ml. on the figure. The zero level for the L-alanine assay has been moved to the right in order to separate the curves. The assay levels, as indicated by the circles and triangles, are: 0, 0.33, 0.67, 1.33, and 2.0 μg./ml. for D-alanine and 0, 1.0, 2.0, and 3.0 for L-alanine.

sponse curves different but the sensitivity of the assay to the two isomers differs and the rate of growth (as indicated by the time at which an adequate growth level is reached for the assay) also differs. Results such as these would indicate that, except for the cases where there is only

[201] E. E. Snell, *J. Biol. Chem.* **158**, 497 (1945).

[202] G. D. Shockman, *Proc. Soc. Exptl. Biol. Med.* **101**, 693 (1959).

one isomer in the sample (usually L-) and where the same isomer is used as a standard, results of DL assays can be in considerable quantitative and isomer error. It would therefore be necessary to carefully re-examine the responses to D- and L- isomers for each case.

D-Amino acids can have any of three effects on amino acid assays:[203] (1) they may be inert; (2) they may be utilized by the test organism; or (3) they may be inhibitory.[204]

There are a number of instances of inhibitions caused by the presence of D-amino acids. Two examples of these are: (1) the greater repression of growth of *L. arabinosus* 17-5 and *L. leichmannii* by relatively high concentrations of D-leucine, D-valine, D-methionine, and D-serine as compared with that shown by the corresponding L-amino acids;[205,206] and (2) the inhibition of the utilization of low concentrations of L-tryptophan by D-tryptophan for L. *arabinosus* 17–5 and *Leuconostoc mesenteroides* P-60.[204]

To test for the response of an organism to a D-amino acid it is not always sufficient to merely test the D-isomer alone or to compare the response of the racemic mixture (DL) with that of the L-isomer. A combination of these two, plus tests of various mixtures of the L- and D-isomers, is necessary to be certain that (1) the sample of the D-amino acid was not slightly contaminated with the opposite isomer, (2) the organism will not partially respond to relatively high concentrations of the D-isomer during growth in the presence of the L-isomer, and (3) the presence of relatively high concentrations of the D-isomer does not inhibit the response to the L-isomer. Such thorough tests are rare. An example of such a test is given in Fig. 2 for the response of *Streptococcus faecalis* to methionine.[40] In this case it is clear that relatively large amounts of D-methionine can be used for growth but only in the presence of the L-isomer.

The configuratively specific assays are of particular value for use in conjunction with column chromatography. Only some organisms have been sufficiently checked so that one can be fairly certain that the response is only to the L-isomer or only in a very limited fashion to the D-isomer. This will be discussed in Section X for the particular amino acid assay. Alternate methods involve the selective destruction of one isomer enzymatically followed by a microbiological assay or chemical test for the remaining amino acid or other isomer.

For critical analyses, and where the isomeric form of the amino acid

[203] H. N. Rydon, *Biochem. Soc. Symposia* (*Cambridge, Engl.*) 1, 40 (1948).
[204] J. M. Prescott, B. S. Schweigert, C. M. Lyman, and K. A. Kuiken, *J. Biol. Chem.* 178, 727 (1949).
[205] A. E. Teeri and D. Josselyn, *J. Bacteriol.* 66, 72 (1953).
[206] A. E. Teeri, *J. Bacteriol.* 67, 686 (1954).

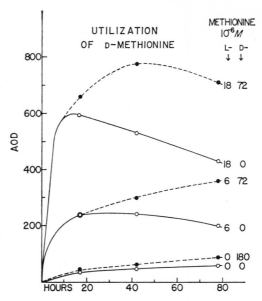

Fɪɢ. 2. Turbidimetric measurement of growth responses of *Streptococcus faecalis* (9790) to ᴅ-methionine in the presence and absence of ʟ-methionine. ᴅ-Methionine equimolar to ʟ- (ᴅʟ-methionine) produces responses approximately half way between those shown for 0 and $72 \times 10^{-6} M$.

is of interest, the use of highly purified ʟ-amino acids plus only those ᴅ-amino acids such as alanine, glutamic, and aspartic acids that have been found in lactic acid bacteria, in place of the customary ᴅʟ-mixtures in the culture medium as well as standards is definitely recommended.

3. Amino Acid Antagonisms

This subject, the previous one, and the following two are all inter-related. Some mention has already been made of the antagonistic effects of ᴅ-amino acids. There are in the literature a number of reports of the inhibition of growth or antagonistic effects by what were then considered to be ʟ-amino acids. However, in many of these instances the ᴅʟ-mixture was used in the experiments and results were calculated for the ʟ-isomer. For assay purposes, antagonistic effects between amino acids are only of importance when a low concentration of an amino acid is being measured in the presence of a relatively high concentration of an antagonist (i.e., 50-fold). Even in some of these cases, a successful assay can be run. Inhibition of the response to ᴅ-alanine was noted in the assay for this amino acid with *Lactobacillus delbrueckii* (9649) by the amino acids glycine and ʟ-alanine.[13] Reducing the levels of these two amino acids in the basal medium made it possible to run assays for

D-alanine in the presence of up to eightfold quantities of glycine and L-alanine. Possibly the use of glycine and L-alanine peptides in the culture medium could have even further reduced the interference of glycine and L-alanine (see Section IV.C.4).[93,207-208] Concentrations as low as 0.6% D-alanine have been measured with a fair degree of precision in the presence of 3.9% L-alanine and 4.5% glycine.

For amino acid assays, some of the amino acid antagonists are useful. An outstanding example seems to be the inhibition of the utilization of D-aspartic acid by cysteic acid [181] for the assay of L-aspartic acid with *Leuconostoc mesenteroides* P-60 (see Section X.E). A similar effect has been noted for DL-ethionine. This antagonist represses the response of *Lactobacillus arabinosus* 17–5 to D-methionine.[209]

4. Peptides

Recently Kihara and Snell [207,210-212] have given us a great deal of new information on the role of both peptides and amino acid antagonisms in the nutrition of lactic acid bacteria. They have demonstrated that the strepogenin requirement of *Lactobacillus casei* is apparently due to the fact that a single peptide may supply several rather than a single amino acid in a form that is available to the organism.[211,212] The "strepogenin" requirement could be fulfilled by appropriate mixtures of simpler peptides such as dipeptides. This is in line with the earlier observations that in some cases a low concentration of a histidine peptide could duplicate the growth-promoting action of either a high concentration of histidine or a partial hydrolysate of casein for *L. delbrueckii* 780.[208] In both of these cases the peptides serve as more efficient sources than do free amino acids themselves, where a particular organism may have an impediment in the utilization of the particular free amino acids. The present-day commercial availability of synthetic peptides of high purity may lead to the use in some assay media of certain peptides to enhance growth while avoiding antagonisms that may be caused by the presence of high concentrations of free amino acids that would otherwise be employed. The better utilization of certain peptides than component amino acids themselves can lead to invalid assays when the assay samples are not completely hydrolyzed. In addition, so that there is little if any effect on the composition of the medium from the addition of the assay sample, assay media must contain moderately high quantities of amino acids and be as sensitive as possible.

[207] H. Kihara and E. E. Snell, *J. Biol. Chem.* **197,** 791 (1952).
[208] V. J. Peters, J. M. Prescott, and E. E. Snell, *J. Biol. Chem.* **202,** 521 (1953).
[209] M. N. Camien and M. S. Dunn, *J. Biol. Chem.* **184,** 283 (1950).
[210] H. Kihara and E. E. Snell, *J. Biol. Chem.* **212,** 83 (1955).
[211] H. Kihara and E. E. Snell, *J. Biol. Chem.* **235,** 1409 (1960).
[212] H. Kihara and E. E. Snell, *J. Biol. Chem.* **235,** 1415 (1960).

The investigations of Shankman and co-workers[213-214] on the utilization of several synthetic peptides of valine and leucine by seven lactic acid bacteria have shown that there is a marked difference in the utilization and inhibition produced by these analogs of leucine and valine. These investigations took into account the optical configuration of the amino acids in the peptides and thus it was found that many of the differences were stereospecific. For example, *Leuconostoc citrovorum* (8081) uses L-L-L-valine as efficiently as it uses L-valine. However, the D-L-L-peptide was not utilized and the L-L-D-tripeptide of valine was inhibitory. Since such peptides are not usually encountered in amino acid assays of hydrolysates of natural products the problem introduced is essentially never of great importance. However, one should be aware of such possibilities particularly when assaying samples that may contain D-amino acids such as synthetic materials or substances of microbial origin.

5. Other Closely Related Amino Acids and Derivatives

Closely related amino acids can affect the results of an assay. Here the outstanding example is the better and quicker utilization of glutamine than of L-glutamic acid by *Lactobacillus arabinosus* 17-5 and *Streptococcus faecalis* (9790).[137,153] Successful assays require the presence of a base level of glutamine in the medium.[153] A lower initial pH has also been recommended.[137] Another example is the need for high proline concentrations in arginine assays and high arginine concentrations for proline assays with *Leuconostoc mesenteroides* P-60.[215] In this assay the test organism responds better to proline containing peptides than it does to free proline reminiscent of some of the results of Kihara and Snell.[207,210-212]

An important example of the response to closely related compounds is that of the response of many lactic acid bacteria to L-tryptophan. Response to this amino acid is complicated by many factors. First, in three different test media D-tryptophan inhibits the response of both *Lactobacillus arabinosus* 17-5 and *Leuconostoc mesenteroides* P-60 to low concentration of the L-compound. However, D-tryptophan does not noticeably affect the growth of *Streptococcus faecalis* on low concentrations of L-tryptophan.[204,216] Secondly, with some test organisms such as *Lactobacillus arabinosus* 17-5 [34,204,217,218] but not with others such as

[213] S. Shankman and Y. Schvo, *J. Am. Chem. Soc.* **80**, 1164 (1958).

[214] S. Shankman, S. Higa, H. A. Florsheim, Y. Schvo, and V. Gold, *Arch. Biochem. Biophys.* **86**, 204 (1960).

[215] R. J. Sirny, L. T. Cheng, and C. A. Elvehjem, *J. Biol. Chem.* **190**, 547 (1951).

[216] S. A. Koser and J. L. Thomas, *J. Infectious Diseases* **101**, 168 (1957).

[217] E. E. Snell, *Arch. Biochem.* **2**, 389 (1943).

[218] R. D. Green and A. Black, *Proc. Soc. Exptl. Biol. Med.* **54**, 322 (1943).

Streptococcus faecalis (9790),[21,217,219,219a] indole and anthranillic acid can partially replace tryptophan and therefore must be removed by extraction with ether and toluene. The third problem with tryptophan is destruction by acid hydrolysis and racemization on alkaline hydrolysis unless proper precautions are taken. The racemization can cause difficulties (such as drift and false values) with assays using organisms that are inhibited by the D-isomer.

Fortunately, these are the exceptions. The assay organisms can usually differentiate between two very similar amino acids. Leucine and isoleucine, L-alanine and L-alanine are the prime examples of this (see Sections X.A and X.C).

6. Stability of the Test Organisms in Relationship to Their Amino Acid Requirements

Compared to many bacteria, such as *Escherichia coli*, lactic acid bacteria are extremely stable and apparently have a very low mutation rate. Using the penicillin technic, serious efforts to produce auxotrophic mutants of *Streptococcus faecalis* (9790) that had additional growth requirements have met with complete failure (Shockman, unpublished observations). There are only a few examples in the literature of mutations or variations of lactic acid bacteria and few of these are in regard to amino acid nutrition. Cultures have been known to change in their requirements over a number of years of cultivation in one laboratory. Just how many of these changes are due to the organism (mutation, selection, etc.) and how many are due to the investigators (wrong culture, contamination, etc.) is often a difficult question to answer. It is a particularly good idea to start with a culture from the ATCC if possible and when experimental results disagree with the literature to check first with another ATCC culture.

Not all lactic acid bacteria are of equivalent stability. *Lactobacillus casei* and *L. delbrueckii* seem to be among the less stable ones. A mutant of *L. casei* requiring D-lactic acid was obtained by Camien and Dunn[219b] while looking for a mutant requiring D-alanine, by means of ultraviolet radiation. *Lactobacillus arabinosus* 17-5 seems to be able to mutate, although infrequently, in regard to its aromatic amino acid requirements.[220] A tyrosine independent mutant of *L. arabinosus* has been obtained. However, apparently, growth of the mutant is inhibited in the presence of the parent strain. In addition the mutant grows more slowly than the

[219] W. Baumgarten, A. N. Mather, and L. Stone, *Cereal Chem.* **23**, 135 (1946).
[219a] K. A. Kuiken, C. M. Lyman, and F. Hale, *J. Biol. Chem.* **171**, 551 (1947).
[219b] M. N. Camien and M. S. Dunn, *J. Biol. Chem.* **201**, 621 (1953).
[220] D. E. Atkinson and S. W. Fox, *Arch. Biochem.* **31**, 212 (1951).

parent strain. The mutant has a generation time of 173 to 176 minutes as compared to 79 minutes for the parent strain.[221]

The ability to utilize small amounts of histidine rather than a requirement for either large amounts of histidine or smaller amounts of histidine peptides can develop in cultures of *L. delbrueckii* 780 when they are grown in media containing only low levels of this amino acid.[208] This change in the nutritional requirement of this assay organism appears to be a true mutation and apparently does not affect the other nutritional requirements of the organism.

It is indeed fortunate for both amino acid and vitamin assayist who use these organisms that mutations are rare.

V. Measurement of Response to Amino Acids

A. General Considerations

A number of different methods can be used to measure the response of an organism to a substance without which it either cannot grow or grows very slowly. Many of these methods have been reviewed in earlier chapters. For the determination of amino acids with lactic acid bacteria, however, primarily only acidimetric and turbidimetric methods have been used.

B. Acidimetric Methods

Acidimetric methods depend on the amount of acidity produced as an end product of the organism's energy metabolism. Rate of growth and acid production do not run exactly parallel courses. The maximum growth will have been reached considerably before maximum acid production has occurred. In fact, the total amount of acid produced would be a function of the cellular mass and the time that this cellular mass has had to produce the acid. If the incubation is for a sufficiently long period of time, the amount of acid produced is closely related to the cellular mass. For this reason, then, these methods often require 72 or more hours incubation time versus 16–48 for turbidimetric method.

For measuring acidity, a number of instruments can be employed. At one extreme is an automatic titrating and recording apparatus that automatically changes the sample,[222] an automatic titrating machine such as the Cannon Automatic Dispenser Titrator (International Instrument Company, P.O. Box 7781, Los Angeles 15, California).[223] In addition,

[221] A. P. James, *J. Bacteriol.* **60,** 719 (1950).

[222] C. H. Eades, Jr., B. P. McKay, W. E. Romans, and G. P Ruffin, *Anal. Chem.* **27,** 123 (1955).

[223] M. D. Cannon, *Science* **106,** 597 (1947).

there are the Coleman, Metrohn, Beckman, and similar titrators, or an ordinary buret operated manually. At the other extreme is simply the measurement of the final pH and, from a titration curve, calculating the amount of acidity produced.

C. Turbidimetric Methods

Turbidimetric methods depend on the absorption of light by the cellular mass as determined with a suitable instrument such as a photoelectric colorimeter or spectrophotometer. Nephelometers, which measure the light-scattering (Tyndall effect) of the number of cells present, have also been used. They have not been used as extensively as spectrophotometers and colorimeters mainly because of difficulties in standardization and their extreme sensitivity (see Section IX.C).

For measuring turbidity, the minimum equipment required is somewhat more expensive than a buret. The photometers used have ranged from home made to expensive recording spectrophotometers. Any sufficiently sensitive photoelectric densitometer is suitable.[224] The precision obtainable with the numerous commercially available instruments is a variable, as is their sturdiness, dependability, sensitivity, and convenience for day-to-day usage. For rapid, precise, and critical work the instrument should hold at least two cuvettes and have a sufficiently long (over 6 inches minimum) direct reading or expandable scale. Null-point type instruments are too slow and cumbersome for assay work. A system for using a particular wavelength of light is extremely desirable, particularly to eliminate color effects of culture media or assay sample. Ordinary light filters are usually satisfactory for this purpose. A narrower spectrum is often desirable and either interference filters or a spectrophotometer (grating or prism) can be used. A properly selected narrow light band reduces color interference. A particularly annoying type of interference occurs when the color of the assay medium darkens during the incubation period. Such color development can be different in the tubes containing growth and those that are used as uninoculated blanks. In some cases inoculated tubes which have been centrifuged show less color than uninoculated ones.[109] Various wavelengths have been used ranging from 400 to 750 mμ, depending in some cases on instruments and filters available. Culture media and interfering colors are usually yellowish and have a minimum absorption between 600 and 750 mμ. Readings within this range will require the minimum of color correction, therefore introducing the minimum error. Turbidity readings will increase as the wavelength of light used is decreased. Therefore, the use of lower wavelengths will increase the sensitivity of the assay.

[224] L. G. Longsworth, *J. Bacteriol.* **32**, 307 (1936).

Because of light scattering of the particles in suspension, turbidity of bacterial suspensions will not follow Beer's law, and while it is not imperative it is certainly convenient and advantageous to calibrate the instrument. Various methods have been used. Dry weight and nitrogen per unit are the most common parameters that have been employed. Per unit may refer to either a per cent transmission or optical density scale. Nitrogen standardization has the disadvantage that the nitrogen content of an assay organism may not be constant during the growth cycle.[225] Older cells may have less nitrogen per unit mass than young cells. Dry weight measurements are laborious and of limited precision. Either a microbalance or a large sample of cells is required. Dry weight calibration is recommended in those instances where a comparison between two or more turbidimetric instruments is required (i.e., two or more different laboratories).

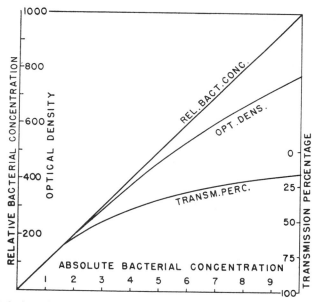

Fig. 3. Relations between relative bacterial concentration, optical density, and transmission percentage. The relation between concentration and density is based on the method described in Section V.C (see also reference 226). The ordinates have been arranged so that the plotted values of the readings of the first tenth of the abscissa coincide. Optical density = observed optical density times 1000.

When absolute magnitudes are not required, a relative bacterial concentration scale can be used. Figure 3 illustrates how relative bacterial concentration is related to the optical density and per cent transmission

[225] G. D. Shockman, J. J. Kolb, and G. Toennies, *J. Biol. Chem.* **230**, 961 (1958).

scales of a photoelectric densitometer.[226] It is clear that with increasing turbidity there is a greater deviation of the optical density scale from proportionality with relative bacterial concentration. For convenience optical density is expressed as $1000 \times \log I_0/I$. The arithmetic scale of transmission gives an even greater degree of distortion. The relative bacterial concentration scale can be empirically constructed quite easily for any turbidimetric instrument. A thick suspension of cells can be used to make a series of closely spaced dilutions. These dilutions are read in the turbidimetric instrument and the corresponding values on either a per cent transmission or optical density scale will thus be obtained. Since the correction will be smaller from an optical density scale, its use, if possible, is to be preferred. The correction can also be made using mathematical formulae.[226] Once the correction is made, turbidimetric readings can be translated to relative bacterial concentration by any of several means. A relative bacterial concentration scale can be used in the turbidimeter, the correction can be made graphically or a table can be constructed such as that shown in skeleton form in Table V, which will convert optical density readings into relative bacterial concentration. The term Adjusted Optical Density (AOD) has been used for the corrected values[226] and this term will be used in this chapter.

TABLE V

OPTICAL DENSITY ADJUSTMENT TABLE FOR MEASUREMENTS IN THE COLEMAN SPECTROPHOTOMETER OF SUSPENSIONS OF LACTIC ACID BACTERIA[a]

NOD[b]	0	10	20	30	40	50	60	70	80	90
					AOD					
0	0	10	20	30	40	50	60	71	81	91
100	101	112	122	132	142	153	163	173	184	194
200	204	215	225	236	246	257	267	278	289	300
300	311	322	334	345	357	368	380	392	404	416
400	428	440	452	465	477	490	503	516	529	543
500	556	569	583	597	612	626	640	655	669	684
600	699	714	729	745	760	777	793	810	828	845
700	862	880	898	917	936	956	976	996	1017	1038

[a] NOD = net optical density. The optical density read minus the blank. AOD = adjusted optical density. Net optical density adjusted to be a measure of relative bacterial concentration.

[b] An NOD value of 550 corresponds to an AOD value of 626.

In our experience the correction needs to be made with only one organism. Apparently it will hold for a variety of organisms, both rods

[226] G. Toennies and D. L. Gallant, *Growth* **13,** 7 (1949).

and cocci, of approximately the same size. This includes all of the lactic acid bacteria that are used for amino acid assays. Theoretical considerations indicate that the correction should be made for the instrument to be used since the deviation from Beer's law is dependent on the relationship of the light scattering of the suspension to total transmission.[227-230] The extent to which light scattering will contribute to the observed measurements will depend on factors determined by the construction of the instrument. For instance, the distance between the suspension of bacteria in a cuvette and the photoelectric cell is one such important factor. However, our experience indicates that a correction table of the type shown in Table V, which was made for a Coleman model 14 spectrophotometer at 675 mμ, can be used to correct optical density readings obtained on a wide variety of commercial spectrophotometers such as the Coleman models 11, 14, and 6, Beckman B and DU, Bausch & Lomb Spectronic 20, Unicam SP 600 and SP 300, and the Hilger Spekker Absorptiometer. However, since these instruments are of varying sensitivities, the actual turbidities obtained on the same suspension of bacteria may differ.

There are several advantages of the adjustment of observed turbidity readings (either transmission percentage or optical density) to compensate for deviations from Beer's law. The practical aspects of a linear dose response in microbiological assays are obvious and have been discussed in previous chapters. By such adjustments linear relationships that are not apparent from unadjusted readings would be disclosed. In addition, since the relative bacterial concentration (AOD) is directly proportional to the length of the path of light through the cell suspension (Lambert's law), with the aid of an optically determined ratio of tube diameters, interconversion of turbidities obtained in tubes of different diameter is greatly simplified. A third benefit of adjustment of turbidity is the simplification of the study of growth rates.

It should be remembered that in all cases where per cent transmission, optical density, adjusted optical density, or any turbidity measurement that is not calibrated on an absolute scale (i.e., dry weight) are used, they should be specified in terms of the length of the light path, wavelength at which measured, and the instrument upon which the measurements were made in order to be most widely useful.

An additional factor to be aware of when using turbidimetric methods is the flow bifringence shown by rod shaped organisms when first poured into a cuvette or when shaken or swirled in a cuvette. This will show itself as a slow drift in the reading of the photometer and will necessitate

[227] H. Mestre, J. Bacteriol. **30**, 335 (1935).
[228] A. Dognon, Rev. opt. **22**, 9 (1943).
[229] A. Dognon, Sci. et inds. phot. **16**, 193 (1945).
[230] L. G. Longsworth, J. Bacteriol. **32**, 307 (1936).

some standardization of reading procedure. Usually readings are taken when movement has stopped but other standardized procedures work equally well.

Turbidities can be read by shaking and pouring the contents of the growth tubes into a calibrated cuvette. Devices for filling, flushing, and emptying samples in a photometer are also available. It is often far more convenient to grow the cultures in calibrated tubes that can be inserted directly into the instrument. In this way, growth curves or multiple readings can be done on a single tube (or pair of tubes). Calibrated culture tubes, however, are expensive and very often not as accurately calibrated as desired. A relatively simple and accurate method of calibrating, matching, and standardization is given in Section IX.C.2.

D. Other Methods

Because of the nature of the nutritional, chemical, and physical requirements of the lactic acid bacteria (see Section IV), there has been only very limited use of other methods for the quantitative determination of amino acids. For instance, because the organisms are microaerophilic, special conditions must be employed for the use of an agar diffusion method. However, a disk or cup plate assay is feasible if the oxygen tension is controlled. Such a method has been proposed [121] for amino acid assays with *Streptococcus faecalis*. In this instance oxygen tension was controlled by the use of a top layer of plain agar. Triphenyl-tetrazolium chloride was included in the medium to aid in the detection of growth zones and to increase the sensitivity of the method.

Agar plate methods for amino acids are qualitatively useful in conjunction with paper chromatography for the identification of amino acids and growth factors.[231-234] The method has not been widely used for amino acids particularly since there are more convenient methods available.[9]

E. Choice of Method

Titrimetric methods have been employed more widely than turbidimetric methods for amino acid assays. To a great extent the methods are interchangeable. However, some organisms produce smaller amounts of acid and in such cases turbidimetric methods would definitely be preferred. Some organisms tend to clump during growth in some media, making uniform suspension difficult. In these cases titration of acid would be the preferred method. The theoretical aspects of these two methods

[231] W. A. Winsten and E. Eigen, *Proc. Soc. Exptl. Biol. Med.* **67**, 513 (1948).
[232] W. R. Lockhart, *Appl. Microbiol.* **3**, 153 (1955).
[233] A. L. Bacharach, *Nature* **160**, 640 (1947).
[234] J. C. Foster, J. A. Lally, and H. B. Woodruff, *Science* **110**, 507 (1949).

have been discussed in earlier chapters. In the opinion of the author, turbidimetric methods are preferred except for organisms that are very difficult to resuspend or when samples to be assayed cause nonbacterial turbidity to form during incubation or contain colored materials that cannot be compensated for by a judicious selection of the wavelength of the light source.

The primary factor for the preference of turbidity methods is the added controls for specificity and interfering materials that can be introduced. If the turbidity of the assays are routinely read at more than one time interval (which can be done with a titration method only by introducing a complete set of additional tubes for each time period), not only can the "drift" in assay values of different sample levels be observed, but also the "drift" in time can be observed. This will be discussed further in Sections VI.A.5 and 6.

VI. Validity, Specificity, Reliability, and Calculation of Results of Amino Acid Assays

A. Validity

Over the last 17 years or so the validity of microbiological assays has been proven innumerable times. The criteria most often used are (1) recovery of added amino acids, (2) agreement with other methods, (3) agreement between assays with different test organisms, (4) consistent values on repeated assays, and (5) agreement of values calculated from various assay levels. To these we can add a sixth criterion, agreement in time. Well-executed assays can meet all of these requirements. However, it must be kept in mind that all are negative criteria. That is, when the assay does not fulfill any one of them, it is invalid. If it fulfills all, it is still not necessarily a valid determination.

1. Recovery of Added Amino Acids

It is quite obvious that when a known quantity of the particular amino acid that is being determined is added to a sample it must be quantitatively recovered in order to have a valid determination. In fact, recoveries of 98 to 102% of amino acids added to samples before hydrolysis are far from rare.[4,5] However, since the amino acid added is not actually part of the protein that is being hydrolyzed, complete recovery of an added amino acid is not a true control. Low recoveries could indicate (1) instability of the amino acid to hydrolysis procedure (Section VIII), (2) incomplete or irregular hydrolysis of the same amino acid in the protein, or even more important, (3) that the response of the test organism is being inhibited by one or more substances in the sample. High recoveries of an added amino acid could only indicate the stimulation of the growth response of the organism by a substance in the test sample. Examples of stimulatory substances would include the presence of (1) a peptide that is utilized better than the free amino acid by the assay organisms (Section IV.C.4), (2) a high concentration of a vitamin or particular form of a vitamin that can substitute for the required amino acid, or (3) a derivative or hydrolytic product of an amino acid (such as antranillic acid and indol for replacing tryptophan) that can replace the required amino acid. Both high or low recoveries could therefore indicate an invalid assay.

2. Agreement with Other Methods

The advent of ion exchange chromatographic methods has greatly extended our ability to determine the validity of a microbiological assay for an amino acid.[10,13] In the past it had been possible to determine only a few amino acids by chemical methods. Where this has been possible, such as for tryptophan[34,35] or methionine,[10] agreement of chemical and microbiological determinations done on the same test sample has been good. Table II shows a comparison of results obtained by Stein and Moore ion exchange chromatography and by microbiological assay by the turbidimetric methods recommended in this chapter.[13] It can readily be seen that in almost very case, excepting those instances where D-amino acids were found to be present, agreement between the two widely different methods is excellent.

The additional example of the agreement between L-lysine determined by microbiological assay and by L-lysine decarboxylase in a wide variety of protein containing material should be mentioned.[10]

3. Agreement between Microbiological Assays with Different Test Organisms

This is a less rigid criterion of validity than the previous one. There are numerous examples of agreement of assay values of identical or similar test samples that have been obtained with various lactic acid bacteria as test organisms.[4-6,10] However, since these test organisms are often quite similar both nutritionally and physiologically, the assays could all suffer from the same error. A more positive criterion of validity would be the agreement between two assays with two widely different test organisms such as between an assay with a lactic acid bacterium and a *Neurospora* or Protozoan. Since the more strict criterion of agreement with other methods can be fulfilled, the importance of this criterion is thus reduced. In fact lack of agreement between two microbiological assay methods where one method may be in good agreement with another type of method may merely be taken to mean that only the second method is, for one reason or another, invalid.

4. Consistency of Values on Repeated Assays

This again is a requirement of any analytical procedure. It does act as a control on the influence of the day to day variable conditions that might not otherwise be apparent. This test can be made to be more rigorous by the addition of controlled variations to the assay technic. McMahan and Snell [17] have shown (Table VI) that the valine content of

TABLE VI

CONSTANCY OF RESULTS ON REPEATED ASSAY:
EFFECT OF ALTERED CONDITIONS[a]

Assay no.	Valine present in casein (%)		
1	6.72		
2	6.66	Turbidimetric:	6.69
3	6.90	Acidimetric:	6.63
4	6.46	Incubated 32°C.:	6.58
5	6.80	Incubated 37°C.:	6.60
6	6.62		
	Av: 6.69 ± 0.23		

[a] Assay organism: *Lactobacillus casei*.[17]

casein can be consistently assayed with *Lactobacillus casei*. In fact, purposefully changing the incubation temperature or the measurement of response either turbidimetrically or acidimetrically gave exactly the same result. This criterion should be taken more as an indication of the reliability of the procedures used than as a criterion of validity of the assay values obtained.

5. Agreement of Values Calculated from Various Assay Levels

The percentage of an amino acid in the sample being assayed should be constant at more than one concentration level of the sample, providing the concentration level is well within the range of the assay standard curve. There should be no consistent trend, either up or down, of the assay values as the amount of the sample added to the assay tube is increased. The tendency of assay values to change in one direction or another with an increasing amount of sample has been termed "drift." Absence of "drift" would indicate that the assay organism is responding in exactly the same way to both the pure amino acid standard and the sample. In some instances the extent of the "drift" that is encountered must be evaluated (Section VI.D). If other simultaneous assays on similar materials fail to show "drift," if the "drift" is somewhat irregular, or if it is well below the standard deviation of the assay, it may be considered not to be significant.

"Drifts" of this type as well as those to be described below (Section VI.A.6) are usually due to substances in the test sample that either stimulate or inhibit the growth of the test organism. This particular form of "drift" is due to the variability of the stimulatory or inhibitory effect with the level of sample present. Actually, lack of "drift" with increasing levels of the sample is a test for specificity of the assay. When "drift" occurs, standard and sample are not yielding exactly the same end result in growth of the test organism and therefore the substance in the sample is, in whole or in part, different from the standard. For amino acids this can often mean incomplete hydrolysis, the presence of stimulatory or inhibitory peptides in the sample, or an amino acid or growth factor in the sample that is not present in the medium in adequate concentration. An additional possible cause of drift can be the amount of salt added with the sample or a difference in pH of the tubes containing the sample and those containing the standard due to acidity or alkalinity of the test sample. In one way or another these causes of drift (salt, pH, incomplete hydrolysis) can usually be corrected.

6. Agreement in Time

Not only should the percentage of an amino acid in a sample be constant at more than one concentration level of the sample, but for each concentration level the percentage of the amino acid present should be, within limits, independent of the time at which the assays are read. In other words, the character of the response produced by a sample should in every way be identical to that produced by the pure amino acid standard. This means that not only should the acid produced or the total turbidity attained at the end of a relatively long incubation period be identical, but that during the period of growth and acid production the rates and changes in rates should be identical. However, the measurement of rates and rate changes is impractical from an assay standpoint. The limits imposed are those of the very rapid growth and rate of acid production that will occur during the logarithmic growth phase of the test organism. Logarithmic increases are for relatively short periods of time, and the changes that occur subsequently are much slower and smaller. However, these subsequent changes (or lack of change as the case may be) can be followed and used to assess the validity and specificity of an amino acid assay. This is an advantage of turbidity measurements that are made directly on the tubes in which the assays are incubated (Section

V). The best way to indicate what is involved is by means of an illustrative example. For this purpose we will use an assay for L-valine with *Streptococcus faecalis* as described in Section X.A.

Figure 4 illustrates a series of growth curves of *S. faecalis* on growth

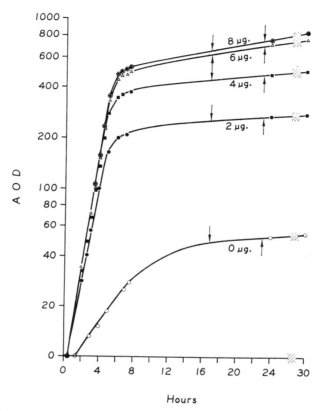

FIG. 4. Turbidimetric growth curves of *Streptococcus faecalis* (9790) on growth limiting quantities of L-valine. The arrows indicate the time period at which the assay readings are usually made.

limiting concentrations of L-valine. These curves are obtained by starting with exponentially (logarithmically) growing inocula, as described in Section IX.G.2.*b*, so that there is essentially no lag before growth commences. Logarithmic growth ceases when the amount of L-valine present is exhausted from the medium.[111] In this experiment the L-valine content of the assay medium was higher than that usually obtained and as can be seen gives significant growth in the zero level tubes. However, after logarithmic growth has ceased, "growth" (increased turbidity) continues in a pattern that is characteristic for the growth limitation of L-valine in this particular medium. When turbidity readings are taken at 23 hours and compared to those taken at 16 hours after inoculation, it is clear that turbidity has risen by a small amount (about 5%) that is characteristic for valine growth curves. If the growth limitation were L-threonine (Fig. 5) or L-isoleucine (Fig. 8), for instance, the increase in observed

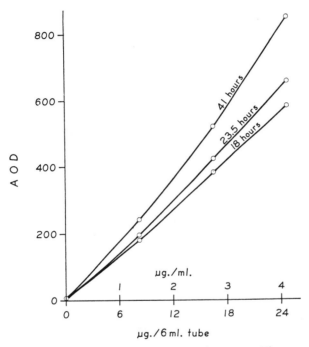

FIG. 5. Turbidimetric assay curve for an L-threonine assay. The concentrations of L-threonine are given in both μg./ml. and μg./6-ml. tube.

turbidity from 17 to 23 hours of incubation would be far greater and would continue for a longer incubation time.

When the amount of L-valine in a test sample is measured and compared to a standard, the amounts calculated at 23 hours should be the same as the amounts calculated at 17 hours of incubation despite an increase (or decrease) of both tubes containing sample and standard. Lower or higher values of the sample would indicate an inhibitory or stimulatory effect on the rate of growth during either the logarithmic or postlogarithmic growth phases. It is obvious that all readings must be taken at a time that is well past the end of the more active growth phases in order to eliminate the effect of small differences in the size of the inoculum. Since overnight incubation is most convenient, it has been our practice to read assay tubes at 16 to 17 hours, 22 to 24 hours, and occasionally at 40 to 43 hours after inoculation. An example of this method is given in Section VI.D.

B. Specificity of Microbiological Amino Acid Assays

Several of the criteria for validity, particularly numbers 2, 3, 5, and 6, are also indicative of the specificity of microbiological amino acid assays. In addition, reference should be made to Sections IV.C.2–5 for information concerning the extreme specificity of microbiological methods that is sometimes not possible to obtain by any other method. The stereospecificity of some of the assays is such an example.

C. Reliability

The reliability that can sometimes be obtained can be judged by criteria 2 and 4 under validity (Section VI.A). The importance of good quantitative and microbiological techniques should also be mentioned here.

D. Results and Their Calculation

A standard curve is drawn by plotting the response of the test organism (AOD) against increasing concentrations of the amino acid standard. Some examples of such a curve are shown in Figs. 1, 5, 8, and 9. The amount of amino acid in any given sample is then obtained by interpolation of the response obtained in the tubes containing the sample onto the standard curve. The dilution factors for the sample tubes must then be accounted for to obtain the value for the undiluted sample. Several sample levels that will fall within the range of the standard curve should always be employed in order to determine the specificity of the assay. This method of calculation can be used for both turbidimetric and titrimetric assays.

The data, results, and calculations of a typical turbidimetric assay for L-threonine are given in Table VII. The relative bacterial concentration (AOD) (Section V.B) is given for each assay tube. From these data, a standard curve such as that shown in Fig. 5 is drawn and the amount of L-threonine per milliliter in each assay tube is obtained by finding the corresponding value of the standard for that turbidity. To determine the amount of L-threonine in the original sample hydrolyzed, the calculations that are indicated at the bottom of the table are made. If these data for both standard and sample are plotted as the logarithm of the response against the logarithm of the dose,[235] straight and parallel lines would be obtained, indicating that by this criterion the assay is valid and reliable. However, arithmetic plotting, interpolation, and calculations can be done more precisely and simply.

[235] E. C. Wood, *Analyst* **72**, 84 (1947).

When the values presented in Table VII, part B, are inspected closely, it can be seen that there is a slight downward "drift" in assay values both with increasing amounts of sample and with time. The μg./ml./ml. of hydrolysate for 3 ml. of sample read at 41 hours is lower than for 1 ml. of sample read at 18 hours. The amount of "drift" is quite small, would not be apparent from a log-log plot, and should be considered to be below the level of significance. This particular assay was selected to illustrate what "drift" would look like.

Titrimetric assays may be graphed and calculated in the same manner. The vertical axis would then be in milliliters of the alkali used for titration.

VII. Preparation of Sample for Assay

The determination of amino acid content is done only on very limited types of materials. With the exceptions of free amino acids in tissues,[236] artificial mixtures of amino acids, and the one example of an ester bound amino acid in the cell wall of certain microorganisms,[237,238] amino acids occur naturally in structures that have peptide bonds such as proteins and peptides. Microbiological assay procedures, with the exception of those utilizing protozoa and *Neurospora,* require complete, or at least nearly complete, hydrolysis of the peptide bonds and therefore are subject to the same problems and losses as are chemical and chromatographic methods. Three types of hydrolytic procedures using acids, bases, and enzymes have been used.

A. Acid Hydrolysis

Because of destruction during acid hydrolysis, this type of hydrolysis is not suitable for use in determining the amino acids tryptophan and tyrosine.[167] Other amino acids may also suffer losses during acid hydrolysis, and in some cases corrections can be made. Hydrochloric, sulfuric, hydriodic, and hydrochloric:formic acids have been used for protein hydrolysis. Sulfuric acid has the advantage that excess acid can be removed with $Ba(OH)_2$ as an insoluble precipitate. However, almost invariably hydrochloric acid or hydrochloric:formic acid hydrolyses have been used for microbiological assays. While chromatographic methods require absolutely complete hydrolysis regardless of destruction so that

[236] A. Meister, "Biochemistry of the Amino Acids," p. 49. Academic Press, New York, 1957.

[237] J. J. Armstrong, J. Baddiley, J. C. Buchanan, and B. Carss, *Nature* **181,** 1692 (1958).

[238] J. J. Armstrong, J. Baddiley, J. G. Buchanan, B. Carss, and G. R. Greenberg, *J. Chem. Soc.* **882,** 4344 (1958).

TABLE VII

L-Threonine Assay of Lyophilized Bacteria

A. *Assay results*

ml.	μg./ml.	Blank	18 hours OD[a]	18 hours NOD[b]	18 hours AOD[c]	23.5 hours OD	23.5 hours NOD	23.5 hours AOD	41 hours OD	41 hours NOD	41 hours AOD
1. Standard											
0	0	7	12	5	5	11	4	4	15	8	8
0	0	5	10	5	5	11	6	6	15	10	10
1.0	1.4	3	175	172	175	192	189	193	232	229	235
1.0	1.4	3	180	177	180	197	194	198	236	233	239
2.0	2.8	3	360	357	377	393	390	416	472	469	515
2.0	2.8	6	371	365	386	404	398	425	488	482	532
3.0	4.1	2	525	523	588	575	573	659	700	698	859
3.0	4.1	3	520	517	579	574	571	656	695	692	848
2. Sample											
1.0		8	158	150	153	172	164	167	202	194	198
1.0		8	160	152	155	172	164	167	201	193	197
2.0		10	318	308	320	342	332	348	411	401	429
2.0		9	323	314	327	343	334	350	415	406	435
3.0		11	453	442	480	486	475	523	595	584	675
3.0		12	460	448	488	498	486	537	598	586	679

[a] OD = observed optical density (× 1000).

[b] NOD = observed optical density minus the blank.

[c] AOD = NOD adjusted to agree with Beer's law (see Section V.C).

TABLE VII (Continued)

B. Calculations

ml. of hydrolysate	18 hours			23.5 hours			41 hours		
	AOD	μg./ml.	μg./ml. per ml. hyd.	AOD	μg./ml.	μg./ml. per ml. hyd.	AOD	μg./ml.	μg./ml. per ml. hyd.
1.0	153	1.18	1.18	167	1.17	1.17	198	1.16	1.16
1.0	155	1.19	1.19	167	1.17	1.17	197	1.16	1.16
2.0	320	2.34	1.17	348	2.32	1.16	429	2.32	1.16
2.0	327	2.37	1.19	350	2.34	1.17	435	2.36	1.16
3.0	480	3.44	1.14	523	3.38	1.13	675	3.43	1.14
3.0	488	3.48	1.16	537	3.46	1.15	679	3.45	1.15
Av.			1.17 ± 0.02 μg. (2%)			1.16 ± 0.01 μg. (1%)		1.16 ± 0.02 μg. (2%)	1.16 ± 0.01 μg. (1%)

(a) 6 ml. per assay tube
(b) 48.1 mg. hydrolysate made to 25.0 ml.
(c) 7.0 ml. of (b) diluted to 50 ml.

$$\frac{(a)\ (b)\ (c)\ (100)}{(1.16)\ (2.23)}$$

$$\frac{(6)\ (25/48,100)\ (50/7)}{ }$$

= 2.23

= 2.59 ± 0.05% L-threonine in sample

a good base line and only separate peaks identifiable as amino acids are obtained, for microbiological assays only a degree of hydrolysis sufficient to liberate in a free form all of the amino acids being assayed is required. For instance, as shown in Table VIII, 8-hour hydrolysis with 3 N HCl in

TABLE VIII

LENGTH OF HYDROLYSIS AND AMINO ACID CONTENT
OF DRIED *Streptococcus faecalis*[a]

Length of hydrolysis (hours)	Amino acid content found			
	Valine (%)		Leucine (%)	Isoleucine (%)
8	4.43	4.39	4.76	4.36
16	4.35	4.38	4.72	4.39
24	4.39			

[a] Hydrolysis with 3 N HCl in 40% formic acid in an evacuated sealed tube (Section VII.A.2).

40% formic acid at 120°C. in an evacuated sealed tube is sufficient to completely liberate all of the valine, leucine, and isoleucine in dry bacterial substance.[111] A much longer hydrolysis period with stronger acid (6 N HCl, 120°C. 24 hours, sealed tube) is required for satisfactory chromatographic determinations of these amino acids. During acid hydrolysis, serine, threonine, cystine, methionine, and tyrosine may suffer significant losses.[239–242] Where precise figures are desired for these amino acids, it is sometimes necessary to hydrolyze for several different time periods including extended ones (i.e., 72 hours) so that there will be extensive destruction of the amino acid, and then to extrapolate the assay values back to zero time of hydrolysis.[241–243] Such procedures leave much to be desired, but are the best that can be done where the losses that occur during hydrolysis cannot be avoided. Empirical corrections such as that of Rees for a 5% loss of threonine and a 10% loss of serine during a 20-hour hydrolysis in 6 N HCl [239] have been widely employed. Since two protein preparations cannot be expected to behave identically under different conditions of temperature, purity of acid, ratio of acid to protein, etc., this sort of empirical correction probably is not justified.

[239] M. W. Rees, *Biochem. J.* **40,** 632 (1946).

[240] C. F. Crampton, W. H. Stein, and S. Moore, *J. Biol. Chem.* **225,** 363 (1957).

[241] C. H. W. Hirs, W. H. Stein, and S. Moore, *J. Biol. Chem.* **211,** 941 (1954).

[242] R. J. Block and K. W. Weiss, "Amino Acid Handbook." C. C Thomas, Springfield, Illinois, 1956.

[243] E. J. Harfenist, *J. Am. Chem. Soc.* **75,** 5528 (1953).

For acid hydrolysis a choice of two procedures with variations is given here.

1. It has been customary to reflux proteins with 2.5 to 5000 times their weight of 6 N HCl for 18 to 24 hours. A large excess of acid is claimed to minimize amino acid losses.[244] For use with assays too large an excess of acid can, if the acid is not first removed, introduce too high a concentration of acid or (if neutralized) salt into the assay tubes. HCl hydrolysates can be evaporated to dryness *in vacuo* over NaOH to reduce the amount of acid and salt present. This procedure is recommended for routine use. The dry sample can be redissolved in water, humins removed by filtration, and either neutralized with NaOH, KOH, or NH₄OH or made to volume and used without neutralization for at least some assays.

2. A more convenient method for the small samples of protein used for microbiological assays is to hydrolyze in a sealed tube. A minimum volume of 0.5 ml. 6 N HCl, 1 ml. 3 N HCl or 0.3 ml. 3 N HCl in 40% formic acid should be used with small (2–20 mg.) samples in a 16 × 120 to 18 × 150 mm. heavy walled Pyrex tube. With larger samples or larger tubes, more acid should be used. The minimum acid volume may be more important than ratios of acid to protein. The tubes are then frozen in dry ice-ethanol, drawn out in a flame so that near the center of the tube the diameter is small enough to be easily sealed in a flame (i.e., 1- to 3-mm. bore), evacuated with an oil pump and sealed. Hydrolysis is then carried out in either a 110° oven or in the autoclave at 121° for the required period of time. Hydrolysis in 3 N and 6 N HCl are usually carried out for 16 to 24 hours and in 3 N HCl in 40% formic acid for 8 to 24 hours. We have found it convenient to use a small domestic type pressure cooker for hydrolysis. The sealed tubes containing samples and acid can be placed in the pressure cooker along with a shallow layer of water. The pressure cooker can then either be brought to pressure over a flame and placed in a 110° or 120° oven or placed in the oven cold allowing sufficient additional time for the samples to reach the proper temperature. After hydrolysis the tubes are cooled in dry ice-ethanol, and opened with caution. The tubes can be scratched with a file and cracked open by touching with a heated glass rod. This should be done behind a protective screen, mask, or goggles since pressure may have built up in the tubes during hydrolysis. If more than one time period is used, it is often convenient to start the longest time period earlier so that the hydrolysates will be ready to be opened at the same time. Because of the possibility of amino acid losses after hydrolysis is completed, it is advisable to remove excess acid as soon as possible. This can be done by placing the opened

[244] J. P. Dustin, C. Czaikowska, S. Moore, and E. S. Bigwood, *Anal. Chim. Acta* **9**, 256 (1953).

tubes in a desiccator over NaOH and carefully evacuating to avoid bumping of the hydrolysates. In most cases it is equally satisfactory to neutralize with alkali. Humin should be removed by filtration or centrifugation as soon as possible to avoid difficulties with the assays.[245] Hydrolysates that have been dried over NaOH can be dissolved in water, the humin filtered off, brought to pH 6.8 to 7.0, made to volume, and added to the assay tubes. Care must be taken to avoid microbial contamination of hydrolysate solutions. Storage at subzero temperatures is the preferred method of storage although volatile preservative may also be used.

Constant boiling glass redistilled HCl is often recommended and should be used for hydrolysis.[241,246]

B. Alkaline Hydrolysis

Alkaline hydrolysis has been used for the liberation of tryptophan and tyrosine because of the destruction of these two amino acids during acid hydrolysis. During alkaline hydrolysis amino acids are racemized so that care must be taken to be sure that the response of the test organism is specific for the L-isomer and not interfered with by the D-isomer (see Section IV.C.2). The figures given by the assay will also have to be doubled to give the correct value for the L-amino acid in the sample.[21] Tyrosine is relatively stable to alkaline hydrolysis, but small losses after exposure to 5 N NaOH for 14 to 30 hours at 100° to 120° may occur.[242,247] Alkali may cause not only the partial and erratic destruction of tryptophan, but racemization is often not complete.[34,248] Since the test organism only responds to the L-isomer, an indefinite degree of racemization can lead to false values. In addition, two of the degradation products of tryptophan, indole, and anthranillic acid are active for some test organisms.[34,204,217,218] The conventional alkaline hydrolysis is given below with a modification that is said to avoid the destruction of tryptophan during hydrolysis.

1. Conventional Alkaline Hydrolysis[7,10,21]

Twenty to 50 ml. of 5 N NaOH are used per gram of dry sample. These are then hydrolyzed in an autoclave at 121°C., either in a tube with a loose cover or in sealed tubes (as in acid hydrolysis) for 8 to 16 hours. The pH is adjusted to 6.8 to 7.0 with HCl (acetic acid can also be used) and the sample is centrifuged to remove the precipitate. The precipitate

[245] M. J. Horn, A. E. Blum, C. E. F. Gersdorff, and H. W. Warren, J. Biol. Chem. 203, 907 (1953).

[246] E. Smith and A. Stockell, J. Biol. Chem. 207, 501 (1954).

[247] D. Bolling and R. J. Block, Arch. Biochem. 2, 93 (1943).

[248] I. T. Greenhut, B. S. Schweigert, and C. A. Elvehjem, J. Biol. Chem. 165, 235 (1946).

is usually washed twice, the washings added to the supernatant and made up to volume with distilled water. The samples can be stored under toluene or frozen until assayed.

2. Modified Alkaline Hydrolysis

Kuiken et al.[219a] recommend the following procedure to avoid the destruction to tryptophan during alkaline hydrolysis. These authors claim that oxidation is the main cause of tryptophan destruction. One hundred milligrams of L-cysteine HCl and 16 ml. of 4 N NaOH are autoclaved at 121°C. for 1 hour. While still hot, another 100 mg. of L-cysteine HCl and the tryptophan or sample are added and the hydrolysis is continued for 16 hours at 121°C. in the autoclave. Quantitative recovery of tryptophan and good results with proteins were obtained with this method.

C. Enzymatic Hydrolysis

Because of the difficulties with the alkaline hydrolysis procedures for tryptophan, Wooley and Sebrell [34] developed an enzymatic method for liberating tryptophan from proteins. Microbiological assays with both *Lactobacillus arabinosus* and *Eberthella typhosa* on materials digested enzymatically showed close agreement with the chemical method of Horn and Jones.[35]

To 1 gm. of protein in a 100-ml. Erlenmeyer flask, 50 ml. of 0.1 N H_2SO_4 and 10 mg. of pepsin (1:10,000 potency) are added. The flask is then incubated overnight at 37°C. Three grams of $K_2HPO_4 \cdot 12H_2O$ are then added and the pH is adjusted to 8.4. Ten milligrams of trypsin (1:300 potency) are added and the material is incubated at 40°C. for 12 to 24 hours. The pH is then adjusted to 7.8 with dilute acid and 100 mg. of erepsin are added, and incubation at 40°C. is continued for 2 more days. After this final digestion period the pH is adjusted to 6.8 and the volume is brought to 100 ml. with water. The digestions should be carried out under toluene and the flasks should be shaken occasionally during each incubation period. The recommended test organism, *Streptococcus faecalis*, does not respond to indole and anthranilic acid (see Section X.A) so that extraction with ether or toluene is not required. The digest may be filtered before use if necessary.

Since there will be some liberation of tryptophan, as well as stimulatory peptides,[249] as a result of the autodigestion of the enzymes, it is necessary to determine the degree of stimulation and the amount of tryptophan entering the assay via this route. This can be accomplished by setting up a parallel digestion using ten times as much enzyme and

[249] D. E. Kizer, L. Hawkin, M. L. Speck, and L. W. Aurand, *J. Dairy Sci.* **38,** 303 (1955).

omitting the protein sample. This material is assayed along with the test material and the values obtained from the enzymes are subtracted from the final results. However, there is no reason to believe that the extent and type of autodigestion in the absence of protein is the same as in its presence.

Evidence of complete digestion is indicated by lack of "drift" of the assay values (see Sections VI.A.5 and 6). An upward drift in the assay will indicate either incomplete digestion of the sample or the presence of stimulatory substances. A downward drift will indicate the presence of inhibitory substances.

D. General Comments

The limitation of the procedures used to liberate amino acids from their bound forms are among the limitations of microbiological assay methods as they are for all other methods that require the presence of the amino acids in the free form. Another limitation that again is not unique to the microbiological methods, but that should be mentioned since it is one that is often overlooked, is the initial weight of the dry sample. Often proteins and protein containing materials are hygroscopic and special precautions are necessary to be certain that weights are comparable.[250] If necessary, moisture determinations can be made on separate but identical samples.

The methods of hydrolysis given in the preceding sections are far from ideal, particularly for the liberation of some of the amino acids from some types of materials. Among the "problem" amino acids are methionine, cystine, and tryptophan. For these amino acids it is particularly necessary to follow the time course of hydrolysis to be certain that complete liberation is obtained, and if there is a significant loss the proper correction can be made. It is also recommended, particularly for these three amino acids, that more than one method of hydrolysis be tried. In this way the maximum amounts liberated by one method can be compared with the maximum for another. For instance, for tryptophan liberation, alkaline hydrolysis (methods 1 or 2) can be compared with the enzymatic digestion procedure. In this connection Greenhut et al.[248] claim to obtain higher tryptophan values on proteins by digesting with pancreatin and hog mucosa with shaking than those that have been obtained with either alkaline hydrolysis or the enzymatic digestion method recommended in Section VII.C.

While generalizations concerning liberation and corrections can and have been made, overmechanization of procedures can lead to false or inconsistent values, particularly for the "problem" amino acids.

[250] A. C. Chibnall, M. W. Rees, and E. F. Williams, *Biochem. J.* **37**, 354 (1943).

VIII. Assay Media

In general, completely synthetic media have been used for the assay of amino acids. In earlier days, natural sources, such as tomato juice extract, etc., were used as a source of growth factors.[2,3] Today, the vitamin requirements of the commonly used lactic acid bacteria are known[91,251] so that the individual pure vitamins can be and are added to assay media. In a few special cases hydrolysates of proteins can be used to replace amino acid mixtures. An example would be the assays for cystine, methionine, and tyrosine with *Streptococcus faecalis* on a medium containing oxidized peptone as a source of most of the amino acids.[157,159] Of the recommended methods (Section X), only the assay for cystine (Section X.F) is done on a nonsynthetic medium.

Media of many compositions have been used and many have been listed in reviews and other compilations.[4–10] The underlying principle is to have a medium that is adequate in all respects except that of the substance being assayed. Thus, the considerations here are no different than for any other assay. However, one precaution must be observed. That is that the concentrations of nutrients are not so far in excess as to be toxic to the assay organism or, in the case of the amino acid nutrients, to lead to competitive antagonisms (Section IV.C).

The compositions of some assay media that have been used in recent years for amino acid assays are compared in Table IX. The concentrations are expressed as per milliliter final concentration in the assay tubes. There is considerable confusion in the literature as to the concentration of medium ingredients, and often it is necessary to go through various calculations in order to determine the actual concentrations that have been used. The amounts can be found to be expressed in terms of final concentration, in terms of double strength basal medium, or as dry weight of a mixture, per tube or per volume of medium that happens to be one batch in that particular laboratory. However, (1) the volume in an assay tube can be anywhere from 0.2 to 10 ml., (2) it may be desirable to make up other quantities of medium, and (3) basal medium certainly is not of necessity double strength. In addition to this, a degree of uncertainty exists in the actual amount of a substance that has been used when the commercially available products are most commonly hydrates, hydrochlorides, hydrochloride hydrates, etc. The footnotes to Table IX indicate such information where it has been given. In most of these cases the differences will be small, and since generous excesses of these nutrients are usually employed, difficulties will probably not be encountered. There

[251] R. P. Tittsler, C. S. Petterson, E. E. Snell, D. Hendlin, and C. F. Niven, Jr., *Bacteriol. Revs.* **16,** 227 (1952).

TABLE IX
Composition of Assay Media[a,b]

μg./ml.	A	B	C	D	E	F	G	H
Alanine L-	—	100	25	—	—	—	—	—
DL-	200	200[c]	—	—	200	200	1000	200
D-	—	—	25	—	—	—	—	—
Arginine L-	200	200[d]	200	—	200	200	200[d]	100
Asparagine L-	5	100	100	200	400	—	—	400
Aspartic L-	100	100	500	—	100	—	—	100[e]
DL-	—	—	—	—	—	200	—	—
Cysteine L-	—	50	—	—	50	—	1000	—
Cystine L-	200	50	100	—	—	200	—	200
Glutamic acid L-	300	500[f]	1000	—	300	—	100	400
DL-	—	—	—	—	—	200	1000	—
D-	—	—	30	—	—	—	—	—
Glutamine L-	5	—	5	—	—	—	—	—
Glycine	200	200	12.5	—	100	200	100	100
Histidine L-	200	200[d]	1250[d]	—	62[d]	200	100	100
Hydroxyproline L-	200	100	—	—	—	200	—	—
Isoleucine L-	100	100	100	—	—	—	100[g]	100
DL-	—	—	—	—	250	200	—	—
Leucine L-	100	100	100	—	—	—	100	100
DL-	—	—	—	—	250	200	—	—
Lysine L-	100	100[d]	200	—	200	100	200[h]	—
Methionine L-	100	100	100	—	—	—	100[g]	200
DL-	—	—	—	400	100	200	—	100
Norleucine L-	—	100	—	—	—	—	—	—
DL-	—	—	—	—	—	200	—	—
Phenylalanine L-	100	100	100	—	—	—	100[g]	50
DL-	—	—	—	—	100	200	—	—
Proline L-	200	100	100	200	100	200	100	50
Serine L-	100	100	100	—	—	—	100[g]	50
DL-	—	—	—	—	50	200	—	—
Threonine L-	100	100	100	100	—	—	100[g]	100
DL-	—	—	—	—	200	200	—	—
Tryptophan L-	200	100	100[g]	—	—	—	100[g]	100
DL-	—	—	—	400	40	400	—	—
Tyrosine L-	200	100	100	400	100	200	100	100
Valine L-	100	100	100	—	—	—	100[g]	100
DL-	—	—	—	—	250	200	—	—
Peptone—oxidized	—	—	—	5000	—	—	—	—
Ascorbic acid	—	—	300	250	—	—	—	—
Biotin	0.005	0.0025	0.01	0.001	0.001	0.0001	0.01	0.01
Ca pantothenate	0.40	0.50	0.80	0.50	0.50	0.20	1.0	0.50
Folic acid (Pteroyl glutamic acid)	0.05	0.005	0.1	0.01	0.01	0.001	0.01	0.01
Folinic acid (Leucovorin)	—	—	—	—	0.01[i]	—	—	—
Nicotinamide	1.0	—	2.0	—	—	—	—	—
Nicotinic acid	—	1.0	—	1.0	1.0	0.2	1.0	1.0
p-Aminobenzoic acid	0.04	0.1	0.08	0.1	0.1	0.04	0.2	0.125
Pyridoxal	—	1.0	—	—	0.3[d]	—	0.2	0.5
Pyridoxamine	0.40	1.0	—	0.2[d]	0.3[d]	0.4	—	—
Pyridoxine	—	—	—	—	1.0[d]	—	—	0.5
Riboflavin	0.20	0.5	0.4	0.5	0.5	0.2	1.0	0.5
Thiamine	0.20[d]	0.5	0.4[d]	0.5	0.5[d]	0.2[d]	1.0	0.5
Adenine	30	10	20	10[j]	10[j]	10	10[j]	10
Guanine	30	10	20	10[j]	10[j]	10	10[j]	10
Uracil	30	10	20	10	10	10	10	10
Xanthine	—	—	—	—	10	—	10	10
KH₂PO₄	442	500	500	500	600	500	—	500
K₂HPO₄	305	500	500	500	600	500	5000	500
CaCl₂	—	—	5	—	—	—	—	—
FeSO₄·7H₂O	10	10	20	10	10	10	20	20
MgSO₄·7H₂O	200	200	800	200	200	200	400	400
MnSO₄·4H₂O	10	10	20	10	10	10	80	80
NaCl	10	10	20	10	10	10	—	20
Sodium oleate	—	—	20	—	—	—	—	—
Tween 40	—	—	2000	—	—	—	—	—
Thymidine	—	—	4	—	—	—	—	—
mg./ml.								
Na acetate	6[k]	10–20[l]	10	20	20	6	1	20
Na citrate	—[l]	—	—	—	—	—	20	[n]
Na₂HPO₄	25.65[n]	—	—	—	—	—	—	—
NaH₂PO₄·H₂O	16.45[n]	—	—	—	—	—	—	—
NH₄Cl	—	—	—	—	3	—	3	—
(NH₄)₂SO₄	0.6	—	—	—	—	—	—	—
Glucose	20	10–20[l]	10	20	25	10	20	20
Final pH	6.5	6.8	6.5	6.5	6.8	6.8	6.8	—

can be exceptions to this generalization primarily with substances that are present in high concentrations and/or serve as buffers. For instance, where such information is not specified, one must assume that sodium citrate concentrations are that of the salt itself and not that of the commonly available Na citrate·2H$_2$O or 2Na citrate·11H$_2$O.

In general, amino acid assay media have all been of very similar composition. All contain an assortment of amino acids and vitamins, purine and pyrimidine bases, a buffer which may be one or a combination of sodium acetate, sodium or potassium citrate, or sodium or potassium phosphate, small amounts of inorganic salts and glucose as an energy source. The only exception to this general rule can be found in medium D in Table IX which contains oxidized peptone instead of some of the amino acids usually used.

The assortment of amino acids and their concentrations are again

---

^{a} Medium A.[13,110,225] This has been used for the determination of L-arginine, L-glutamic acid, L-histidine, L-lysine, L-isoleucine, L-leucine, L-methionine, L-threonine, L-tryptophan, and L-valine with *Streptococcus faecalis* (9790) (Section X.A) It may also be used with *Leuconostoc mesenteroides* P-60 for the determination of L-aspartic acid and L-proline and could be used for the assay of the other amino acids required by *L. mesenteroides*. Better results have been obtained with this organism when the potassium phosphate buffer was used (Sections X.D and E).

Medium B. This has been used as a uniform medium for the following assays[10]: with *Streptococcus faecalis* (9790), arginine, histidine, isoleucine, leucine, lysine, methionine, threonine, tryptophan, and valine; with *Lactobacillus arabinosus* 17-5, isoleucine, leucine, methionine, and phenylalanine; with *Leuconostoc mesenteroides* P-60, histidine, lysine, methionine, and phenylalanine; with *Lactobacillus casei* (7469), arginine, isoleucine, leucine, phenylalanine, and valine; with *L. delbrueckii* LD5, phenylalanine.

Medium C. This has been used for the determination of D- and L-alanine with *L. delbrueckii* (9649).[13] It is the modified medium of Snell *et al.*[31] which contained both acid hydrolyzed and partially hydrolyzed casein (Section X.C)

Medium D.[157] This is a modification of the oxidized peptone medium of Riesen *et al.*[160] It has been used for the determination of cystine and could also be used for methionine and tyrosine assays with *Leuconostoc mesenteroides* P-60 (Section X.F).

Medium E. This is the medium of Steele *et al.*[12] and has been used for the determination of 18 amino acids with *L. mesenteroides* P-60 and *L. citrovorum* (8081) including DL-alanine with the latter organism.

Medium F. This is the medium of Stokes *et al.*[21] for the determination of the following amino acids: with *Streptococcus faecalis*, arginine, histidine, isoleucine, leucine, lysine, methionine, threonine, tryptophan, and valine; with *Lactobacillus delbrueckii* LD5, phenylalanine.

Medium G. This is the uniform medium of Henderson and Snell[166] which has been used for the following assays: with *L. arabinosus* 17-5, glutamic acid, leucine, phenylalanine, valine, and threonine; with *Streptococcus faecalis*, arginine, histidine, methionine, and threonine; with *Leuconostoc mesenteroides* P-60, aspartic acid, glycine, histidine, isoleucine, lysine, proline, and tyrosine. *Lactobacillus casei*, *L. delbrueckii* 3, *L. delbrueckii* 5, and *L. fermenti* also grow well and give good acid production on this medium. The complete medium or the amino acid mixture of the medium has been widely used for the amino acid assays.[253-255] The following assays can be done with the organism listed: *L. arabinosus*: glutamic acid, isoleucine, leucine, phenylalanine, tyrosine, and valine; *Streptococcus faecalis*: arginine, histidine, tryptophan, and threonine; *Leuconostoc mesenteroides* P-60: aspartic acid, cystine, glutamic acid, glycine, histidine, lysine, methionine, proline, serine, and tyrosine; *L. citrovorum* (8081): alanine, glycine, and tyrosine; *Lactobacillus leichmannii* 327: alanine; *L. brevis*: proline; *L. delbrueckii* LD5: phenylalanine. For methionine determinations oxidized peptone (7.5 mg./ml.), cystine, tryptophan, and tyrosine at the amounts indicated replace the amino acid mixture.

^{b} The concentrations given are for the free substances, except for thiamine HCl and Ca-pantothenate. Where hydrochlorides or hydrates, etc., are used, calculations for the free substances should be made and the amounts so compensated.

^{c} Recommended substitution for L-alanine.

^{d} HCl salt used.

^{e} Recommended addition. (*Leuconostoc mesenteroides* requires aspartic acid and according to Schweigert *et al.*[30] asparagine substitutes only poorly for aspartic acid.)

^{f} DL-glutamic acid was used for methionine assays.

^{g} Two hundred micrograms if DL-form is used.

^{h} HCl·H$_2$O.

^{i} Folinic acid substituted for reticulogen used by Steele *et al.*[12]

^{j} Adenine sulfate·H$_2$O; guanine·HCl·2H$_2$O.

^{k} Na acetate is 0.073 M. When potassium acetate is used it should be equivalent (i.e., 7.18 mg.).

^{l} Ten milligrams of glucose and Na acetate are used for *Streptococcus faecalis* and 20 mg. of each are used with *Leuconostoc mesenteroides* or *Lactobacillus arabinosus*.

^{m} For assays with *Streptococcus faecalis* R sodium acetate and potassium phosphate salts were replaced by 25 mg./ml. Na citrate and 5 mg./ml. K$_2$HPO$_4$.

^{n} Potassium phosphates can be substituted for the sodium salts (Section X.A).

quite similar. In some media the amino acid amide, asparagine, is not included. In one, medium H, asparagine is present in high concentration, but aspartic acid is omitted. Glutamine is included in only two of the media listed (media A and C) and in one of these (medium C) was not included when partially hydrolyzed and acid hydrolyzed casein was used as a source of amino acids.[31] In view of the better utilization of glutamine than of glutamic acid, particularly at low concentrations,[137] and the need to provide glutamine to prevent the lag in the standard curve for glutamic acid assays[153] (see also Section X.A), the more widespread use of the amide in recently employed amino acid assay media would be expected. Its addition, as a sterile filtered solution, at least in low concentrations (5 μg./ml.), is definitely recommended.

The glucose concentrations listed are either 10 or 20 mg./ml. The amount used is apparently dependent on the assay organisms used, and to a lesser extent on the method of growth measurement employed. Ammonium salts are included in only a few media. Ammonium salts may be present in acid hydrolysates and various reports indicate the stimulation of growth of lactic acid bacteria by ammonium salts.[5,91,137,138] Therefore, the presence of at least low concentrations of ammonium salts in the assay medium should be considered to be highly desirable.

In some media, particularly for assays with *S. faecalis*, acetate has been omitted and citrate has been substituted. This may lead to difficulties since acetate or lipoic acid may be required by a variety of lactic acid bacteria and in the absence of either one or the other of these substances this test organism will be sensitive to the gaseous environment[122] (see Section IV.B). For this reason the presence of some acetate in all assay media using a lactic acid organism is desirable.

The amounts of vitamins used in the media listed in Table IX should be adequate. The concentrations of pantothenate, biotin, and folic acid were increased in medium A over that previously employed (medium F) when it was found that the supply of these vitamins was inadequate to support very high growth levels.[225] The same is true for the usually employed adenine, guanine, and uracil levels. The presence of xanthine is apparently superfluous and appears to be a carryover from earlier media used for vitamin assays. Folinic acid is used only in media for *Leuconostoc citrovorum*. The form of a vitamin supplied can also be important with other assay organisms. This is particularly true of the vitamin B_6 group. At least one of either pyridoxal or pyridoxamine should be supplied. The exception here is the recommended assay for D- or L-alanine where the vitamin B_6 group should be entirely omitted (Section X.C). In some of the media both pyridoxal and pyridoxamine are supplied, or pyridoxine and one or both of the others are supplied. Comments on the buffers, inorganic salts, and the sodium-potassium balance of assay media have already been made (Section IV.B.6). These are important factors and should not be overlooked.

A variety of media not listed in Table IX or variations of the media listed in Table IX have been used. CaCl₂ has been included in some media to counteract the inhibition of high levels of threonine for *Lactobacillus casei* and *L. delbrueckii* 3.[27] Tween 80 (polyoxyethylene sorbitan mono-oleate) has been added to assay media to eliminate drift, decrease the blank, or eliminate the lag in the standard curve in some assays.[27,144] There are several reports favoring the substitution of potassium salts (citrate, phosphate, acetate) for sodium salts.[132,135,136] At least in some cases these modifications seem to have been beneficial particularly when difficulties have been encountered in the assay of a particular type of material.

As a general rule the composition of the assay medium will not affect the reliability of an assay. There are numerous examples in the literature of excellent agreement of the assay of the same material using different media and in some cases different assay organisms.[4,5] The composition, however, will affect maximum growth or acid production with a particular assay organism, "drift" and of course whether or not the assay organism will grow and have a requirement for the amino acid.

The use of media containing all L-amino acids plus possibly D-alanine, D-glutamic acid, and D-aspartic acid must currently be viewed as conservative and is recommended (see Section IV.C). For amino acid assay standards it would now seem to be quite imperative to use the same isomer as the one that is being determined. The principal reasons for this have been given in Section IV.C. An additional word of caution is in order concerning adequate knowledge of the stereospecificity of the assay being used, where the material being assayed may contain more than one isomer (i.e., substances of microbial origin).

In this connection, the assay for the content of the D-isomers of alanine, glutamic, and aspartic acids has, in recent years, increased in importance. The assays for both L- and D-alanine with *L. delbrueckii* have been quite successful.[13,31] Assays for D-glutamic and D-aspartic have been less promising and even dual microbiological methods, that is, assays with one organism for the total amount of the amino acid minus a separate assay for the L- form,[252] have not been widely used with any great

[252] M. N. Camien, *Proc. Soc. Exptl. Biol. Med.* **77**, 578 (1951).

[253] B. S. Schweigert, B. T. Guthneck, H. R. Kraybill, and D. A. Greenwood, *J. Biol. Chem.* **180**, 1077 (1949).

[254] W. H. Riesen, H. H. Spengler, A. R. Robblee, L. V. Hankes, and C. A. Elvehjem, *J. Biol. Chem.* **171**, 731 (1947).

[255] B. S. Schweigert, B. A. Bennett, and B. T. Guthneck, *Food Research* **19**, 219 (1954).

success. Somewhat greater success has been obtained by ion exchange chromatography for the total amounts of glutamic and aspartic acids in conjunction with stereospecific assays for the L-isomers of these amino acids[13,185] (see also Sections X.A and E).

IX. Apparatus, Instruments, General Methods, and Techniques

A. Glassware, Closures

The quantitative microbiological determination of amino acids with lactic acid bacteria requires lots of *clean* glassware. Large numbers of volumetric and graduated pipets and culture tubes are required. If turbidity is to be measured directly in the culture tubes, calibrated tubes are required (see Section IX.C.2). The usual supplement of volumetric flasks, graduated cylinders, funnels, beakers, and Erlenmeyer flasks are required. Glass-stoppered graduated cylinders are particularly useful for preparing, dissolving, and bringing to volume the assay medium. Six-, 7-, 8-, and 9-ml. pipets (available from Bellco Glass, Inc., Vineland, New Jersey) are also extremely useful. The size of the culture tubes will depend on the volume of the assay and the method of measurement. For the turbidimetric methods given, 18 × 150 mm. lipless culture tubes are used. Borosilicate glassware is usually recommended. Many people have found that culture tubes and pipets made of noncorrosive borosilicate glass (Kimble N. C. glass) are less subject to etching by detergents and sterilization than are Pyrex, Kimax, or ordinary flint glass. Pipets that are made from noncorrosive borosilicate glass are now available from Bellco Glass, Inc., Vineland, New Jersey.

In former years the usual recommendation was for washing all glassware in chromic-sulfuric acid, autoclaving in distilled water to remove traces of acids and toxic ions,[109] and then drying. The availability of detergents (such as Haemo-sol, Meinecke and Company Inc., 225 Varick Street, New York 14, New York) that are highly soluble and that rinse off glassware completely makes this the recommended procedure for all but the rare exceptions. Automatic glass washing equipment has been found to be suitable providing the rinsing time and volume are sufficient. The glassware washing problem for amino acid assays is not as severe as it is for vitamin assays since the amounts of substance to be assayed are larger. Culture tubes should be cleaned so that there is little or no liquid adhering to the side of the tube after mixing. Thorough rinsing of all glassware in high quality distilled water is imperative. Since deionized water may contain resin, its use as a final rinse should be avoided.

For aseptic techniques, pipets and other glassware may be sterilized by conventional dry heat methods. Pipet cans and long glassine paper bags are useful. Autoclaving will merely increase the rate of etching of the glass. A supply of sterile glassware, particularly various sizes of pipets, flasks, and culture tubes, will be found to be convenient.

While not always made of glass, some sort of cover is required for sterile cultures and assays. Cotton plugs, polyurethane foam plugs, glass, polypropylene, aluminum and stainless steel caps, and screw capped tubes have all been used. With the exception of the polyurethane foam plugs, which have been found to liberate an inhibitor of the growth of some of the assay organisms (unpublished observations), any of the above are recommended. Screw capped tubes and flasks are particularly useful for stock cultures because of the lessened drying and spilling. The stainless steel caps (Bellco Glass, Inc., Vineland, New Jersey), while more expensive than some of the other types of caps or plugs, are found useful for assay tubes. They are not as prone to fall off as are aluminum or glass caps and cotton hairs are not a problem. When cotton plugs are used the cotton should be of high grade long fiber type (Johnson and Johnson Red Cross cotton has been found to be suitable). Some cottons will not only yield fibers into flasks and tubes but also tend to leach oils of various sorts into the medium.

B. Constant Temperature Water Bath

An air incubator of the most modern, expensive, and finest type will not maintain temperatures throughout the incubator with sufficient uniformity for truly precise, accurate, and uniform assays. The use of a constant temperature water bath that can maintain ±0.1°C. at 30° to 39°C. is therefore recommended.

Commercial water baths are available that can maintain ±0.05°C. One can also be easily assembled using an ordinary glass aquarium (stainless steel frame), a variable temperature mercury actuated thermoregulator with electrical relay (Philadelphia Thermometer Company, Philadelphia, Pennsylvania), or a mercury plunger relay (H-B Instrument Company, Philadelphia, Pennsylvania), a continuous duty stirrer (Lightnin model L, Mixing Equipment Company, Rochester 11, New York), and a 300- to 500-watt stainless steel clad immersion type heater. To avoid vibration the stirrer should not be mounted directly on the water bath. Stainless steel test tube racks are essential in order to avoid corrosion problems during incubation in the water bath (Norwich Wire Works, Norwich, New York or Harford Metal Products Inc., Aberdeen, Maryland).

An aquarium 30 inches long, 16 inches deep, and 13 inches wide will hold five 10 × 4 stainless steel racks with 1-inch openings. A larger tank that is 36 inches long holds six such racks. Approximately the outer 1 inch of water in the bath serves as an insulator and is not used for the incubation of assay tubes because of the larger temperature variation. This means that 160 assay tubes can be incubated in the smaller bath and 192 in the larger. A 300-watt heater is large enough to maintain constant temperature. Auxiliary heaters (500–1000 watts) can be used to bring the bath from room to incubation temperature more quickly.

A constant temperature unit such as the Haake Unitherm (available in the United States from Brinkmann Instruments, Inc., Great Neck, New York), can be placed in an aquarium or other suitable vessel to serve as a constant temperature water bath.

C. Turbidimeter and Calibration of Cuvettes

1. Choice of Turbidimetric Instrument

There is no commercially available instrument that can be unqualifiedly recommended. The choice is totally dependent on use. We have used a Coleman model 14 spectrophotometer for a number of years and all of our results presented here are based on that instrument. The instrument has been available without improvement for a long time and by some standards is somewhat antiquated. However, when carefully operated, precise results can be obtained.

Other suitable spectrophotometers include the Beckman model B which is clumsy, somewhat skittish, and slow to use for assay purposes. The light path is at a 90° angle to the operator and cuvettes are difficult to align in the light path. The Bausch & Lomb Spectronic 20 is useful for routine assays because of its simplicity and despite its somewhat short direct reading scale.

By the use of interference filters, which in some cases have a narrower band pass than do some spectrophotometers, a colorimeter or photometer can also be used. A narrower spectrum of light waves will reduce color blanks and variations. The Lumitron model 402, Evelyn, and others including homemade ones are suitable. Some colorimeters and spectrophotometers, such as the Coleman Jr spectrophotometer (model 6) are not particularly useful because of their very limited sensitivity.

The range of any of these instruments can be extended for more accurate readings of high turbidity, at least to some extent, by the use of neutral density filters (American Optical Company). If the instrument is zeroed against a neutral density filter of 50% transmission or 0.300 optical density, then 0.300 is added to the optical density read to obtain the optical density of the suspension. The limitation of this is the sensitivity of the photocell and the intensity of light available in the particular instrument being used.

Care should be taken to know the minimum volume, in tubes of known diameter, that can be used in the particular turbidimetric instrument. We have done our assays in 6-ml. volumes in 18-mm.-diameter tube because about 5.5 ml. is the smallest volume that can be used in the Coleman 14 with this diameter tube. The scaling down of turbidimetric assays is therefore somewhat difficult. Smaller diameter tubes with adapters can be used. However, decreasing the volume in the assay tubes also decreases the light path through the cuvette, thereby decreasing the turbidity read by the instrument.

The use of a nephelometer for routine assays is not recommended. Such instruments measure the light reflected by particles in suspension ("Tyndall effect"). While nephelometers are usually more sensitive than are spectrophotometers and colorimeters, their value is limited by lack of good day-to-day standardization. The added sensitivity is not often required and can introduce many problems of cleanliness. Cotton hairs, bits of dust, etc., are far greater problems with nephelometry than they are with densitometry. Nephelometric measurements will be proportional to the number of organisms present[256] while densitometric measurements will be proportional to the mass of the culture.[226]

Whenever possible, it is recommended that the optical density scale rather than a transmittance scale be used. Optical density measurements will deviate less from proportionality to the mass of the culture and with a proper correction can be made proportional to the mass of the culture (Section V). This will yield more linear standard curves.

The accuracy of densitometric readings could be considerably improved by providing a linear rather than a logarithmic scale. Null point measurements have this advantage which is, however, outweighed by the increased time required to determine the turbidity of assay tubes.

2. Calibration of Cuvettes for Turbidimetry

The following method for calibrating tubes to be used for turbidimetric measurements in a Coleman 14 spectrophotometer has been in use for a number of years and has been found to relatively conveniently yield precisely calibrated tubes.

A solution of copper sulfate in dilute sulfuric acid can be used for the calibration of tubes to be used for assay. It is easily prepared, inexpensive, and gives a stable colored solution. This solution has a maximum adsorption peak at 750 mμ, gives an optical density of approximately 0.400 with the instrument and conditions described below, and has a temperature coefficient of 0.4% per degree centigrade. An 0.04 M copper sulfate solution in 0.04 N sulfuric acid is made by dissolving 119.5 gm. $CuSO_4 \cdot 5H_2O$ in water, adding 13.3 ml. concentrated sulfuric acid, and diluting to 12 liters. This solution is then filtered through a medium porosity sintered glass filter and is ready for use.

[256] J. Stárka and J. Koza, Biochim. et Biophys. Acta 32, 261 (1959).

For all amino acid assays 18 × 150 mm. lipless culture tubes can be used. A mean for the calibration of these tubes can be established by taking a random sample of 100 tubes from a number of cases. From these a group of tubes that fall within ±1% can be selected. Later, when a sufficient number of tubes has been examined, 20 tubes that agree with each other within ±0.001 optical density units are retained as the standard reference tubes.

On occasion, even in original packages of 4 dozen, the tint of the glass varies from yellow to green. Consequently, the tubes must be selected on a color basis before calibration.

Experience with calibration has been limited to the use of a Coleman model 14 spectrophotometer equipped with an electronic power supply. The description that follows is therefore limited to the use of this instrument. However, adaptation of the procedure to another instrument should include precautions similar to those described.

Both the spectrophotometer and the power supply must warm up for 30 minutes before use. Then the setting of the instrument at zero transmission is verified, the wavelength calibration of the instrument is checked against the didynium calibration filter, and the value obtained is recorded. The adsorption maximum of the copper sulfate solution is verified, and the wavelength indicator is set at 750 mμ. The positioning of the two-tube carrier of this instrument is then checked to be certain that the light is going through the center of the cuvettes, and, if necessary, the positioning screws are suitably adjusted.

The standard tubes and the tubes to be calibrated are numbered with a glass marking pencil which will withstand washing and filled to a depth of 1.5 to 2.0 inches with the copper sulfate solution. Tubes and solution are equilibrated at the temperature of the room where the calibrations are to be done. Care must be taken that subsequently little change in temperature occurs. All tubes are dipped in water and wiped dry with a lintless towel to remove any fingerprints or dirt that might be on the outside of the tubes. The standard tubes, containing the copper sulfate solution, are then read against a tube containing water (water blank), the average of the readings is then taken as the reference value for that particular calibration period. The zero reading should be checked after every 5 or 10 tubes, depending on the stability of the line voltage, power supply, and instrument at the time. Then the tubes to be calibrated are read. After reading the optical density of 40 tubes, every fifth tube of this group is rechecked to determine the reproducibility and consistency of the readings.

For round cuvettes the direction and positioning of the tubes can be an important factor. A point on the tube is selected. For example, the top of the diamond on Kimble tubes, the small P below the Corning trademark, or the top of the bell on Bellco tubes. This point is positioned in the same manner in the center of the cuvette wall for all tubes. Acceptable tubes are those that agree with the standard tubes within ±1% (0.004 o.d. in 0.400 o.d.).

The selected tubes are then inscribed with a reference line to indicate their position in the spectrophotometer and numbered serially along the long axis of the tube. The last two digits should be 2 inches below the mouth of the tube so that they will not be hidden by cotton plugs or aluminum or stainless steel caps. If the tubes are to be used in a Beckman B the reference line should be placed at 90° to that used on the Coleman 14 so that it will be visible to the operator. Other suitable arrangements can be made for other turbidimeters. A small hand grinder and a suitable jig will be found to facilitate such markings.

Tubes that have been selected at 750 mμ using $CuSO_4$ will also agree within ±1% of the mean when standardized against $CoCl_2$ at 520 mμ.

Tubes calibrated in this manner are available from Mr. J. J. Kolb, 111 W. Wyoming Avenue, Philadelphia, Pennsylvania.

D. Acidimetric Instruments

Some of these have been mentioned in Section V. The Cannon automatic titrator seems to be the one most commonly used. Good burets of a suitable size or an accurate and sensitive pH meter are just as suitable.

Even for turbidimetric methods there is need to have available a good pH meter in order to accurately check or adjust the pH of assay media and other solutions.

E. Other Instruments or Equipment That Are Required, Convenient, or Useful

Here we can list:

(1) An assortment of balances such as a good analytical balance for weighing out standards; a 5-gm. and a 250- or 500-mg. capacity balance of the torsion or other type would be extremely convenient for the rapid weighing of medium ingredients; a triple beam balance of about 300-gm. capacity is convenient for weighing the larger medium constituents such as buffers and sugars. A single pan, direct reading analytical balance such as the Mettler model H15 or B5 is so fast and simple to use that it can with the exception of the triple beam balance substitute for all of the above balances.

(2) A refrigerator or cold room for storage of reagents, cultures, and media used for assay purposes only. Avoid the storage of solvents or other volatile materials in the same unit.

(3) A centrifuge suitable for spinning down cells that are being harvested and washed.

(4) An automatic pipetting machine or a mechanical pipetting device

that can be sterilized and used aseptically. This will be found to save a great deal of time if a large number of assays are to be run.

F. Reagents

1. General Considerations

All reagents should be the finest grade obtainable. Over the past number of years the cost of the necessary medium ingredients has decreased while their quality (from at least a number of sources) has increased. It is false economy to save a few cents (or even dollars) to find out later that blanks are overly high or that the assay is invalid. Even with top quality commercial chemicals a constant watch is required to be sure that they are of adequate quality for the particular assay. (Methods of testing for contamination of amino acids are given in Section IX.F.2.) Also, in the early days of amino acid assays it was generally recommended that the synthetic DL-amino acids be used wherever possible. It was thought that because they were synthetic they were more likely to be pure than those isolated from natural materials. The D-isomer was thought to be inactive or, in a few cases, as active as the "natural" L-isomer. Some difficulties have been encountered in the purity of synthetic DL- forms such as the contamination of DL-leucine with isoleucine or allo isoleucine. In addition, factors of amino acid antagonism must be considered (Section IV.C.3). Media composed of all L-amino acids with the exception of DL-alanine have been used successfully for about 4 years with extremely low blank values for the determination of a number of amino acids. Standard curves differed only very slightly from those previously obtained in media containing racemic, aspartic, and glutamic acids, phenylalanine, serine, isoleucine, leucine, methionine, threonine, tryptophan, and valine. Results of determinations on protein hydrolysates were identical.

Particular care should be taken in the selection of amino acid standards. In the United States there are several sources of amino acids that can be recommended. Two of these are the California Corporation for Biochemical Research (3625 Medford Street, Los Angeles 63, California) and Mann Research Laboratories Inc. (136 Liberty Street, New York 6, New York). In addition, the United States Pharmacopoeia offers for sale 11 L-amino acids in their list of U.S.P. Reference Standards (U.S.P. Reference Standards, 46 Park Avenue, New York 16, New York).[256a] These are claimed to be of at least 99% purity as determined by solubility

[256a] The following U.S.P. references standards are obtainable from U.S.P. Reference Standards, 46 Park Avenue, New York 16, N. Y.: L-arginine monohydrochloride, L-cystine, L-histidine, L-isoleucine, L-leucine, L-lysine monohydrochloride, L-methionine, L-phenylalanine, L-threonine, L-tryptophan, L-tyrosine, and L-valine.

analysis and other physical and purity tests. No data are available on the microbiological assay purity of these amino acids. No single source or individual amino acid product should be completely relied upon as a medium constituent or particularly as a standard.

2. Purity of Amino Acids

It is not the purpose of this section to detail methods for the purification of commercially available amino acids that are not up to the required high standards. If such products are not available or if a particular sample is found to be unsatisfactory by the criteria described, and such purification procedures are deemed necessary, reference should be made to various pertinent publications.[9,257,258]

The usual criteria of purity, such as elemental analysis, melting point, absence of inorganic salts, optical rotation, amino nitrogen and carboxyl content, are quite often not sufficient to detect the low level of impurities that may be responsible for an elevated zero level or other interference in an amino acid assay. In these instances we are almost always dealing with vitamin or amino acid impurities in one or more of the ingredients of the culture medium. For instance, medium A (Table IX) contains 300 μg./ml. of L-glutamic acid. If this were contaminated with 0.1% L-histidine (0.3 μg.), a significant turbidity (or acid production) in a zero tube would be observed.

Only three methods—enzymatic, paper chromatographic, and microbiological—are available for detecting traces of contaminating amino acids. The enzymatic methods are of particular usefulness in determining optical purity.[9] Various paper chromatographic methods have been devised,[9,259,260] some of which have been claimed to be able to detect 0.2% of an amino acid impurity.[259,260] Thorough and properly designed chromatographic procedures have the advantage over microbiological assay methods in that more than one impurity in a particular sample may be detected with one procedure. The statement of "chromatographically pure" however, is not meaningful, and could easily be viewed with scepticism in regards to trace (less than 0.5%) impurities unless the limits of detection of the method are known. For instance, by a chromatographic method devised in Dunn's laboratory 500 μg. of amino acid are spotted on the paper to detect impurities at the 0.2% level.[259,260] Such information is not usually forthcoming from commercial sources and

[257] M. S. Dunn and L. B. Rockland, *Advances in Protein Chem.* **3**, 296 (1947).

[258] "Criteria for Biochemical Compounds," Natl. Acad. Sci., Natl. Research Council, Washington, D. C., 1961.

[259] M. S. Dunn and E. A. Murphy, *Anal. Chem.* **32**, 461 (1960).

[260] M. S. Dunn and E. A. Murphy, *Anal. Chem.* **33**, 797 (1961).

doubts as to the sensitivity of the procedures employed are usually warranted.

It has been stated by Dunn and Rockland in 1947 that "It is noteworthy that the degree of purity of most amino acids required as standards in the microbiological assay of amino acids can be determined with sufficient accuracy for this purpose only by microbiological assay." [257] The advent of paper chromatographic methods has somewhat modified but not eliminated this as an ultimate criterion of purity. No other method can detect amino acid impurities of the order of 0.02%.

It is by no means to be recommended that the purity of all amino acids to be used either in a medium or as a standard need be extensively examined for impurities to this level. When an assay works well and the blanks are low such extensive testing would be a waste of time. When difficulties are encountered the following procedure can be employed.

The substance or solution that is responsible for a high blank can be identified by the simple procedure of adding additional amounts of each of the individual medium ingredients (or solutions) to a series of zero level tubes. One-tenth to 0.4 ml. of the solutions used to make up the medium can be conveniently added in most cases without severely upsetting the balance of the assay medium. A further increase in the amount of growth in one or more of such zero level tubes will identify the guilty solutions. The substances included in the solution can then be tested individually in the same manner to identify the contaminating chemical. This simple method cannot be used when extremely high levels of contamination are encountered (i.e., extremely high zero level tubes). When such is the case the chromatographic methods of Dunn and Murphy[259,260] can be used. Once the contaminating substance is identified (usually an amino acid), two paths are then open. First, a different lot from the same manufacturer or a sample from a different source can be tried in the same system; alternatively, the contaminated substance can be purified by an appropriate method. In our experience with the finest grades of amino acids that are available in the United States, the first method usually proves to be sufficient and more convenient.

There are occasions when an estimation of the level of contamination as well as the identity of the contaminant is desirable. For such cases the following procedure is recommended.

The assay methods given in Section X.A are used. In this case the samples to be assayed are the "pure" amino acids. Known quantities can be added to zero level tubes. The amounts added will be dependent on the limits of detection desired and the sensitivity of the assay system to excessive amounts of the product being tested. Using *Streptococcus faecalis* in medium A-1 (Section X.A), large amounts of at least some amino acids have been successfully assayed for amino acid contaminants.

Examples would be 2.5 mg./ml. of L-glutamic acid, and 1.7 mg./ml. of L-isoleucine or L-histidine. When these very large amounts of amino acids are employed, contamination with 0.01% of most of the other amino acids is detectable. More frequently lesser amounts have been assayed. Table X shows some results of such contamination assays that were done

TABLE X

PER CENT CONTAMINATIONS OF VARIOUS AMINO ACIDS[a]

Possible contaminant	In L-lysine		In L-glutamic acid		In L-phenyl-alanine		In L-histidine		In L-leucine		In L-methio-nine	
	A	B	A	B	A	B	A	B	A	B	A	B
L-Arginine-HCl	0.14	0.1	0.0	0.04	0.05	0.03	0.0	0.0	0.01	0.0	0.03	0.0
L-Lysine-HCl	—	—	0.07	0.0	0.25	0.0	0.0	0.0	0.0	0.0	0.4	0.01
L-Threonine	0.0	0.0	0.0	0.0	0.02	0.0	0.0	0 02	0 0	0.02	0.0	0.0
L-Histidine-HCl	0.0	0.04	0.0	0.02	0.06	0.02	—	—	0.0	0.0	0.0	0.0
L-Isoleucine	0.0	0.0	0.0	0.02	0.0	0.0	0.0	0.0	0.0	0.0	0.0	0.0
L-Methionine	0.0	0.02	0.0	0.35	0.05	0.0	0.0	0.0	0.0	0 02	—	—
L-Leucine	0 0	0.0	0.0	0.04	0.14	0.0	0.0	0.0	0.0	0.0	0.02	0.01
L-Valine	0.0	0.0	C.0	0.18	0.0	0.0	0.0	0.0	0.0	0.0	0.01	0.0

	In L-isoleucine		In glycine		In L-serine		In L-tyrosine		In L-hydroxy-proline		In L-trypto-phan	
	A	B	A	B	A	B	A	B	A	B	A	B
L-Arginine-HCl	0.03	0.0	0.17	0.0	0.02	0.0	0.0	0.0	0.0	0.0	0.01	0.01
L-Lysine-HCl	0.01	0.78	0.6	0.0	0.0	0.25	0.0	1.0(?)	0.01	0.0	0.0	0.1
L-Threonine	0.0	0.01	0.01	0.0	0.0	0.0	0.0	0.0	0.0	0.0	0.0	0.01
L-Histidine-HCl	0.0	0.0	0.01	0.04	0.05	0.05	0.04	0.09	0.03	0.05	0.02	0.02
L-Isoleucine	—	—	0.0	0.01	0.0	0.0	0.0	0.0	0.01	0.01	0.01	0.02
L-Methionine	0.01	0.0	0.0	0.0	0.02	0.07	0.09	0.13	0 01	0.02	0.02	0.01
L-Leucine	0.0	0.0	0.01	0.01	0.0	0.0	0.0	0.1	0.0	0.03	0.0	0.0
L-Valine	0.04	0.0	0.0	0.0	0.0	0.0	0.0	0.0	0.0	0.0	0.02	0.0

	In L-arginine	
	A	B
L-Arginine-HCl	—	—
L-Lysine-HCl	0.01	0.5
L-Threonine	0.0	0.01
L-Histidine-HCl	0.0	0.0
L-Isoleucine	0.0	0.0
L-Methionine	0.0	0.0
L-Leucine	0.0	0.01
L-Valine	0.01	0.0

[a] "A" and "B" refer to samples of amino acids from two different commercial sources. 0.0 indicates less than 0.01% contamination.

at the Institute for Cancer Research on commercially available products in 1953. For these tests 50–300 μg./ml. of each amino acid was tested. The amount used was governed by the amount usually employed in medium A-1. Since the organism is less sensitive to L-lysine than it is to the other amino acids, tests for this amino acid as a contaminant were the least sensitive.

It is clear from the tests described in Table X that small amounts of amino acid contaminants can be detected by microbiological assay. In addition, it is also clear that most of the products tested were quite pure as judged by this criterion.

In this connection it is certainly worth mentioning that over the past 10 or so years we have experienced continuously decreasing blanks in most of our assays. For example, in 1951 the zero tube of an L-valine assay had an AOD of about 20 to 25. Valine assays in 1958 had zero level AOD's of 3 to 6. This of course would be indicative of the general increase in purity of the various reagents used over this time period.

G. Stock Cultures and Inocula

1. Stock Cultures and Subcultures

It has been found to be most suitable to keep all assay organisms as lyophilized cultures. When obtained from the ATCC the cultures are grown and immediately a portion is lyophilized for future reference. Depending on the organism, the working subcultures are kept in different ways. *Streptococcus faecalis* has been found to be stable when carried as liquid cultures in the complete assay medium for long periods of time. It can be routinely carried in this manner. Working cultures of other organisms such as *Leuconostoc mesenteroides* P-60 and *Lactobacillus delbrueckii* (9649) can be carried as agar stab cultures in any one of a variety of rich organic media that contain glucose, meat and/or yeast extract, and a suitably hydrolyzed protein. Both Bacto Micro Assay Culture agar (Difco) or L-Agar (Baltimore Biological Laboratory, Inc., 2201 Aisquith Street, Baltimore 18, Maryland) have been found to be suitable. Subcultures of these can be made in rich organic liquid medium such as Bacto Micro Inoculum Broth (Difco) or APT Broth (BBL).

Lyophilized stock cultures can be stored for long periods of time; i.e., years. Working subcultures (stabs or broth) should be stored in the refrigerator after overnight growth or when the cultures are sufficiently turbid. They should be transferred and regrown at least every 2 to 4 weeks and with some organisms more often if possible.

2. Preparation of Inocula from Stock and Subcultures

a. *General considerations.* Inocula can be taken either directly from lyophilized cultures to the inoculum medium or indirectly through a preliminary growth in a rich organic medium. In many cases, for example, where *Streptococcus faecalis* or *Leuconostoc mesenteroides* is the test organism, inocula can be successfully grown in the synthetic assay medium. Particularly in combination with the methods described in Sections IX.G.3.a and b, such procedures offer definite advantages over growth of the inocula in richer media. For example: (1) since the organism is completely adapted to the assay medium the growth response will be more rapid and reproducible; (2) at least in some cases it is possible to grow the inoculum in growth limiting quantities of the amino acid to be assayed and in this manner eliminate the need to wash the inoculum before use. The highest assay level of the amino acid can be used in the inoculum tube and 1 drop (about 0.05 ml.) of such an inoculum per 6-ml. tube will not significantly contribute to the blank. In fact, the blank will be the same or less than that obtained from inocula grown on a complete medium and washed once. This is particularly convenient when the content of only one amino acid is being assayed.

Some assay organisms, such as *Lactobacillus delbrueckii*, apparently cannot be reliably successively subcultured in a synthetic assay medium. Inocula for assays with such organisms must therefore be grown in a rich organic medium.

b. *The preparation of Streptococcus faecalis inocula.* Routinely, *S. faecalis* can be very satisfactorily maintained on the complete assay medium or as a stab culture in the complete assay medium plus 1.5% agar. When carried in this manner cultures should be stored in the refrigerator and transferred at least monthly. No difficulties have been encountered by this method of routine maintenance in over 10 years. Standard curves currently obtained are virtually superimposable on those obtained 10 years ago by different individuals. The inoculum for the assay can be grown in the assay medium.

For an assay two inoculum tubes are grown: one for inoculating the assay and one for future cultures. When more than one amino acid is to be assayed the inoculum should be grown on the complete medium and washed twice in the centrifuge with sterile distilled water and suspended in 6 ml. of sterile distilled water immediately before inoculation. Log phase inoculum, as described in Section IX.G.3, can be used.

The standard conditions for the standard curves and results of the assays in Sections X.A and B is a starting AOD of 1. This is approximately 1,000,000 cells/ml. (0.4 μg./ml. dry weight). Inocula are grown to AOD 120 and 1 drop (about 0.05 ml.) per 6.0-ml. tube is used. All of the assays in Section X are expressed in micrograms per milliliter of amino acids and these responses were obtained with an 18-mm. light path through the tube (18 $\times$ 150 mm. dimensions) in a Coleman model 14 spectrophotometer set at 675 mμ.

c. *Preparation of inocula for other test organisms.* (1) *Leuconostoc mesenteroides* P-60. For assays with this organism in medium A-2 (Sections X.D and E) and in medium B (Section X.F) inocula can be grown in the same manner as that indicated above for *Streptococcus faecalis*. However, the organism should be maintained as a stab culture in one of the media indicated in Section IX.G.1. Excellent assay responses will be obtained when the inoculum from such an organic medium is grown 2 or more times in the assay medium. When an assay medium other than those listed above is to be used, it should be checked first in order to be certain that the organism can be successfully subcultured in that particular medium. If this is not possible then the alternative method described below for other test organisms should be used.

(2) *Lactobacillus delbrueckii* and other assay organisms. Since not all assay organisms can be continuously subcultured in a synthetic medium that is suitable for an amino acid assay, the inocula must be grown on a different medium. A rich organic medium such as those recommended for the maintenance of working subcultures can be used. There is no apparent advantage to supplementing a synthetic medium with one or more organic constituents of unknown composition excepting where this can be done in vitamin quantities. Even in these instances nearly all the advantages of log phase inocula (Section IX.G.3) can be utilized.

3. Standardization of Inocula

a. Methods available. A number of different methods have been used for growing and standardizing inocula. Lyophilized inocula[261] and overnight cultures[5,78,17] have been used most often. Both of these methods have obvious disadvantages in both standardization and reliability. A precise and defined quantity of cells of known physiological state can be used for the inoculation of assay tubes if exponentially growing cells are employed. Some of the advantages are:

(1) The cells are growing at their maximum rate when inoculated into the assay tubes and will either continue to do so if the inoculum has been grown in the assay medium or will more rapidly return to their maximum growth rate if it is found necessary to use another medium for inoculum growth.

(2) During exponential growth the maximum number of cells in the population are viable and growing. This in essence increases the effective inoculum size.

(3) The standardized amount of inoculum can be so controlled that it will vary very little from assay to assay.

(4) The size of the inoculum can be quite precisely controlled and varied. A precise fraction or multiple of the standardized quantity of inoculum can be used depending on need. In this manner, timing can be controlled so that one assay to be read at 20 hours can be made to be directly comparable to another that had been read at 16 hours. Some of these advantages will become more apparent with the description of

---

[261] F. E. Volz and W. A. Gortner, *Arch. Biochem.* **17,** 141 (1948).

the method of growing and using log phase inocula as it has been in practice for over 10 years with a number of organisms and for both amino acid and vitamin assays.[110,262]

The greatest benefit will be derived from these methods if the turbidimetric instrument is calibrated either in terms of dry weight of cell substance or, as in the example to be given, in terms of adjusted optical density units (AOD) (Section V).

b. *Standardized procedure for growing and using log phase inocula of Streptococcus faecalis.* When cells are taken from 24-hour organic broth culture, put into a synthetic medium, and the turbidity of the cultures is followed, a typical growth curve will ensue. There will be a lag phase until the cells become "adjusted" to the medium, the growth rate will increase until it is constant and maximum for that culture, and then the rate of growth will decrease until it becomes zero. For the purposes of assay these are the phases as obtained from turbidity (mass of the population) but not necessarily from cell numbers. If a subculture is taken from this tube, its initial rate of growth will be dependent on when the subculture was taken. If taken from an older (i.e., 24-hour) culture a lag and the complete growth curve similar to the first one will occur. However, if the cells are transferred from a younger culture, the lag will be shorter or completely eliminated. With successive subcultures, cells growing in the log phase will continue to initiate growth in log phase and the rate of log phase growth (division time) will increase to a maximum and remain constant during log growth for the particular conditions employed. An example of such successive subcultures is given in Fig. 6. This is for *S. faecalis* grown in medium A-1 (Table IX). This medium has a high buffering capacity and log phase will continue until the medium is exhausted of a nutrient that is essential for growth of the organism.[110] In nonlimited medium it will continue to very high densities (AOD 3000, 1.2 mg./ml. dry weight).[225] Between transfers log phase cultures should be kept in an ice bath. When transfers of cultures that are growing at their maximum rate are made on successive days the rate of growth will be nearly the same as the previous day. If there are several days between transfers the rate of growth of the first subculture will be somewhat less than the previous one and there may be a short (1–3 hours) lag (curve 1 or 2, Fig. 6), but the second or third subculture (curve 2 or 3, Fig. 6) will return to the maximum growth rate. With other test organisms it may sometimes require more than three transfers to regain maximum rate after several days of rest in the cold.

Using this method, exact predictions of densities reached at definite times can be made. Conventionally, inocula can be grown to AOD 120

[262] G. Toennies, H. G. Frank, and D. L. Gallant, *Growth* **16,** 287 (1952).

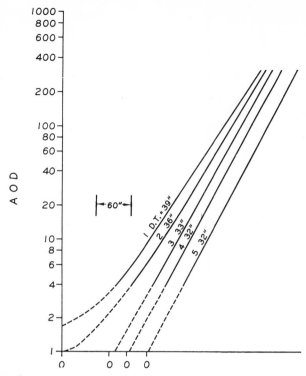

Fig. 6. Growth curves for successive subcultures of *Streptococcus faecalis* in a highly buffered synthetic medium. The starting time of each subculture is indicated by the zero time on the abscissa. The time scale is indicated by the 60 minute mark on the figure. The dotted portions of the curve have been estimated by interpolation between the starting time and the first readings of any degree of precision. The division time is indicated for each curve (D.T.).

and when 0.05 ml. is used to inoculate a 6-ml. assay tube the initial optical density in the inoculated tube is 1 (0.05 ml. × 120 AOD/6 ml.). Assays can be conveniently read at 16 to 18 hours for the first reading (i.e., assay tubes inoculated at 4 P.M., turbidity read at 9 A.M. the following day).

For purposes of the following discussion a division time of exactly 30 minutes will be used. (The division time of *S. faecalis* as described above is usually 31–35 minutes.) If it were more convenient to obtain the equivalent of a 16 hour reading at, say, 22 hours, the inoculum will have to be set back by 6 hours or the equivalent of 12 divisions. This would be equivalent to an initial optical density of 0.00025 per milliliter of the tube. The inoculum would therefore have to be diluted 1:4000 or the equivalent so that the 22-hour readings would be the same as the 16-hour readings. Inocula levels can be increased and incubation time decreased in the same manner. In such cases if the inoculum was not thoroughly washed the level of the assayed substance introduced might become a factor. Lower or higher levels of inocula than the example given can be used.

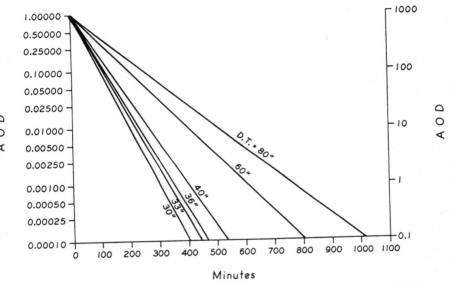

FIG. 7. The type of graph that can be used to adjust the timing of assays by the adjustment of the level of log phase inocula depending on the division time of the assay organism. The construction and use of such a graph is described in Section IX.G.3b.

An accurate graph, such as that illustrated in Fig. 7, or a table such as that shown in Table XI, can be used to calculate inocula levels. For the example given, one would simply look at the graph or table for the 30-minute division time and see that to set the inoculum back by 6 hours (360 minutes) an inoculum of AOD 0.00025 would be required. Graphically, intermediate times that are not equivalent to 1 division can be easily and conveniently estimated. For example, to set the inoculum back by 5 hours an AOD of 0.001 would be used. Interpolation of the tabulated data would be required. The same graph can be used to plot higher inoculum levels by using the scale on the right hand side of the graph. Accurately made dilutions have consistently resulted in the predicted turbidities within 1 division time.

The same method can be used to predict when an inoculum tube will reach a desired turbidity. When shorter times are available, a larger inoculum will give the desired turbidity in a shorter predictable time. The same is true when longer times are desired. For instance, slow grow-

TABLE XI

TIME REQUIRED FOR LOG PHASE CULTURES TO REACH
PREDICTED TURBIDITIES FROM DILUTED INOCULA[a]

Num- ber of divi- sions	Initial concen- tration (AOD/ml.)	Division time of inoculum (minutes)						Initial concen- tration (AOD/ml.)
		30	33	35	40	60	80	
	1							1000
1	0.5	30	33	35	40	60	80	500
2	0.25	60	66	70	80	120	160	250
3	0.13	90	99	105	120	180	240	125
4	0.063	120	132	140	160	240	320	62.5
5	0.032	150	165	175	200	300	400	31.8
6	0.016	180	198	210	240	360	480	15.9
7	0.008	210	231	245	280	420	560	7.9
8	0.004	240	264	280	320	480	640	4
9	0.002	270	297	315	360	540	720	2
10	0.001	300	330	350	400	600	800	1
11	0.0005	330	363	385	440	660	880	0.5
12	0.00025	360	396	420	480	720	960	0.25
13	0.00013	390	429	455	520	780	1040	0.13
14	0.00006	420	462	490	560	840	1120	0.063
15	0.00003	450	495	525	600	900	1200	0.032
16	0.000016	480	528	560	640	960	1280	0.016
17	0.000008	510	561	595	680	1020	1360	0.008
18	0.000004	540	594	630	720	1080	1440	0.004
19	0.000002	570	627	665	760	1140	1520	0.002
20	0.000001	600	660	700	800	1200	1600	0.001

[a] The numbers in the body of the table are in minutes. The column of initial concentrations on the left can be used to determine the dilution required to reach AOD 1 in the period of time indicated in the body of the table for the actual division time of the culture. The column on the right can be used in the same manner to obtain the required dilution to reach AOD 1000. For intermediate levels the time required for the necessary number of divisions can be subtracted.

ing inocula can be grown from dilute inocula so that the predicted turbidity can be obtained overnight.

A somewhat arbitrary limit is that below about 1000 cells per inoculation (AOD 0.001 or in a 6-ml. tube AOD 0.00017/ml.) the inoculum level becomes erratic because of the nature of bacterial suspensions.

Inoculum grown in the assay medium containing a growth limiting quantity of the amino acid to be assayed eliminates the need to wash the inoculum before use in the assay. The level of amino acid introduced with the inoculum is reduced to an insignificant level. The highest quantity to be used in the assay is usually employed.

The method has been found to be applicable as described with other lactic acid bacteria such as *Leuconostoc mesenteroides* P-60. It has also been successfully employed for growing inocula of *Escherichia coli* and other organisms for the assay of nucleic acid derivatives[263] and for growing *Streptococcus faecalis* and *Lactobacillus casei* for folic acid assays.[264]

However, some organisms such as *L. delbrueckii* will not grow adequately upon successive transfers in the assay medium. In such instances inoculum can be grown into the log phase in another medium (i.e., an organic medium). Washing is then required and while exponential growth will not immediately occur in the assay medium, the length of the lag will be significantly reduced. When log phase inoculum is used and washing is required (i.e., from organic media or from media containing large amounts of the amino acid to be assayed) all washings should be done rapidly and in the cold.

X. Recommended Turbidimetric Assays

A. L-Arginine, L-Glutamic Acid, L-Histidine, L-Isoleucine, L-Leucine, L-Methionine, L-Threonine, L-Tryptophan, and L-Valine with *Streptococcus faecalis* (9790)

1. The Assay Medium

Medium A-1 or A-2 can be used for the assay of all of these amino acids. Although they differ only in the phosphate buffer used medium A-1 is preferred on the basis of greater experience. The medium can be prepared entirely from stock solutions, entirely from dry ingredients, or from dry ingredients plus a few stock solutions.

a. *Media A-1 and A-2 prepared from stock solutions.* These media can be prepared from the stock solutions listed in Table XII. The amounts listed are for 100 ml. of basal medium which is double strength as compared to the finished medium in the assay tubes. The medium, when made as indicated in Table XII, is sterile and may be kept for weeks in a refrigerator. When setting up assay tubes, 3 ml. of basal medium is added aseptically to 3 ml. sterile solution to give a total volume of 6 ml. of finished medium per tube.

The pH of the basal medium is 6.5 and the pH of the finished medium after sterilization is 6.7. There is no need to adjust the pH.

Stock solutions should be stored at 4° to 5°C. Amber glass bottles are recommended, particularly for riboflavin and folic acid solutions.

[263] H. K. Miller, *Methods of Biochem. Anal.* **6**, 31 (1958).
[264] G. Toennies and H. G. Frank, *Growth* **14**, 341 (1950).

TABLE XII
PREPARATION OF MEDIA A-1 AND A-2 FROM STOCK SOLUTIONS

	Per 100 ml. basal medium		Per milliliter finished medium (μg.)
	(ml.)	(mg.)	
1. KH_2PO_4—4.42 gm. K_2HPO_4—3.05 gm. } in 50 ml. H_2O	1	88.4 61	442 305
2. $(NH_4)_2SO_4$—6.0 gm. in 50 ml. H_2O	1	120	600
3. L-Amino acid mixture	4		

L-aspartic acid—500 mg. in 1.6 ml. of
     2.5 N NaOH; add 25 ml. H_2O; heat
     until dissolved; then add:
		20	100

L-phenylalanine —500 mg		20	100
L-serine —500. mg.		20	100
L-proline —1.0 gm.		40	200
L-hydroxyproline—1.0 gm.		40	200
glycine —1.0 gm. diluted to 100 ml.		40	200
4. L-Leucine—500 mg. in 100 ml. H_2O; heat until dissolved	4	20	100
5. L-Glutamic acid—1.50 gm. in 4.8 ml. of 2.5 N NaOH and 10 ml. H_2O; heat until dissolved; diluted to 50 ml.	2	60	300
6. DL-Alanine—1.0 gm. in 50 ml. H_2O	2	40	200
7. L-Tyrosine—1.0 gm. in 2 ml. of 5 N NaOH and 10 ml. H_2O; dissolve; then add H_2O to 100 ml.	4	40	200
8. L-Isoleucine—500 mg. in 50 ml. H_2O (heat)	2	20	100
9. L-Methionine—500 mg. in 50 ml. H_2O (heat)	2	20	100
10. L-Threonine—1.0 gm. in 50 ml. H_2O	1	20	100
11. L-Arginine—1.2 gm. of L-arginine·HCl in 50 ml. H_2O (1.0 gm. of free base should be used)	2	40 (A.A.)	200
12. L-Cystine—1.0 gm. in 20 ml. of 2 N HCl; dissolve; dilute to 100 ml.	4	40	200
13. L-Histidine—1.35 gm. of L-histidine·HCl in 50 ml. H_2O	2	40 (A.A.)	200
14. L-Tryptophan—1.0 gm. in 50 ml. of 0.2 N HCl	2	40	200
15. L-Valine—1.0 gm. in 50 ml. H_2O	1	20	100
16. L-Lysine—1.25 gm. of L-lysine·HCl in 50 ml. H_2O	1	20 (A.A.)	100
17. L-Asparagine—25 mg. in 25 ml. H_2O	1	1.0	5
18. Riboflavin—8 mg. in 100 ml. of 0.02 N acetic acid	0.5	0.04	0.2

TABLE XII (Continued)

| | Per 100 ml. basal medium | | Per milliliter finished medium |
	(ml.)	(mg.)	(μg.)
19. Vitamin mixture	0.5		
p-aminobenzoic acid— 4 mg.		0.008	0.04
thiamin·HCl — 20 mg.		0.04	0.2
Ca pantothenate — 40 mg.		0.08	0.4
nicotinamide —100 mg. in 250 ml. H_2O		0.2	1.0
20. Biotin—5 mg. in 1 ml. 95% ethanol; dilute to 50 ml. with 0.01 N HCl; dilute 5 ml. of this to 50 ml. with 0.01 N HCl	0.1	0.001	0.005
21. Folic acid—20 mg. of pteroylglutamic acid in 100 ml. of 0.01 N NaOH; dilute 10 ml. of this to 100 ml. with H_2O	0.5	0.01	0.05
22. Pyridoxamine·2HCl—23 mg. in 100 ml. H_2O (= 16 mg. pyridoxamine)	0.5	0.08	0.4
23. A.G.U. solution	6		
adenine—450 mg. adenine·H_2SO_4·$2H_2O$		6.0	30
guanine—340 mg. guanine·HCl·H_2O		6.0	30
uracil —250 mg. dissolve with heat in 25 ml. of 2 N HCl; dilute to 250 ml.		6.0	30
24. Na_2HPO_4—5.13 gm. (0.181 M) (A-1)	—	—	25,700
(or K_2HPO_4—5.59 gm. (0.161 M) (A-2)	—	—	(28,000)
25. NaH_2PO_4·H_2O—3.29 gm. (0.119 M) (A-1)	—	—	16,500
(or KH_2PO_4—3.8 gm. (0.139 M) (A-2)	—	—	(19,000)
26. Sodium acetate (anhyd.)—1.2 gm.	—	—	6,000
27. Salts B	1		
$MgSO_4$·$7H_2O$—4.0 gm.		40	200
NaCl —200 mg.		2	10
$FeSO_4$·$7H_2O$ —200 mg.		2	10
$MnSO_4$·$4H_2O$—200 mg.; dissolve in 2.4 ml. 1.0 N HCl; dilute to 100 ml. (the salts B solution is added last to prevent undue precipitation)		2	10

At this stage the medium is diluted to 65 to 70 ml. and is left to stand in the refrigerator overnight. It is redissolved with warming, made to 75 ml., and filtered through a medium or fine sintered glass filter. It is then autoclaved at 15 p.s.i. for 3 minutes.

	(ml.)	(mg.)	(μg.)
28. Glucose—4 gm. in 24 ml. H_2O autoclave separately at 15 p.s.i. for 3 minutes; cool, and combine with medium	—	—	20,000
29. Glutamine—25 mg. in 25 ml. H_2O. Filter through a sterile ultrafine sintered glass filter. Add aseptically as a sterile solution to the autoclaved basal medium	1	1	5

Folic acid has been shown by Dick et al.[265] to be much more stable in alkaline than in acid solutions. Biotin has been shown by Gallant and Toennies[266] to be more stable in acid solutions.

Large amounts of stock solutions should not be made, nor should they be stored for long periods of time. With care in handling stock solutions we have not found it necessary to add a preservative. If desired, a volatile preservative[267] can be used. Care in handling stock solutions involves *not* pipetting directly from stock solution bottles but pouring some out into clean tubes and measuring from that. Stock solutions should not be out of the refrigerator longer than absolutely necessary. Solutions are visually inspected before use and visibly contaminated solutions are discarded. Solutions 3, 13, and 17 are the most prone to contamination. Hot, boiled water is an aid in dissolving substances and considerably reduces the contamination problem.

b. Media prepared from dry ingredients. This type of medium can also be used;[10,174,268] however, this necessitates the preparation of 10 separate dry amino acid mixtures which is not usually found to be convenient. In addition, the solubility of some of the dry constituents, particularly cystine, is sometimes a problem. The use of stock solutions of the essential amino acids, glutamine, and cystine is more feasible and is given here.

Media A-1, A-2, and A-3 (see Section X.B for L-lysine) can be made by combining the following dry ingredients:

(A) Salts—sufficient for 50 liters finished medium

		Medium (gm.)		
1.	NaCl	0.50	0.50	0.50
2.	$MnSO_4 \cdot 4H_2O$	0.50	0.50	0.50
3.	$FeSO_4 \cdot 7H_2O$	0.50	0.50	0.50
4.	$MgSO_4 \cdot 7H_2O$	10.0	10.0	10.0
5.	K_2HPO_4	15.25	15.25	15.25
6.	KH_2PO_4	22.1	22.1	22.1
7.	Na acetate (anhyd.)	300	300	300
8.	$NaH_2PO_4 \cdot H_2O$	823	—	17.27
8a.	KH_2PO_4	—	952	—
9.	Na_2HPO_4	1283	—	80.0
9a.	K_2HPO_4	—	1400	—

[265] M. I. B. Dick, I. T. Harrison, and K. T. H. Farrer, *Australian J. Exptl. Biol. Med. Sci.* **26**, 231 (1948).

[266] D. L. Gallant and G. Toennies, *Anal. Chem.* **21**, 1427 (1949).

[267] S. H. Hutner, A. Cury, and H. Baker, *Anal. Chem.* **30**, 849 (1958).

[268] J. J. Mayernik and S. M. Ewald, *J. Am. Pharm. Assoc. Sci. Ed.* **40**, 462 (1951).

(B) Vitamins, etc.—sufficient for 1000 liters finished medium (A-1, A-2, or A-3)

	Medium	
1. Biotin	5.0	mg.
2. p-Aminobenzoic acid	40	mg.
3. Pteroylglutamic acid	50	mg.
4. Ca pantothenate	400	mg.
5. Riboflavin	200	mg.
6. Thiamin HCl	200	mg.
7. Nicotinamide	1.00	gm.
8. Adenine sulfate·H_2O	54.0	gm.
9. Guanine·HCl·H_2O	40.8	gm.
10. Uracil	30.0	gm.
11. L-Asparagine	5.0	gm.
12. Pyridoxamine·2HCl	575	mg.

(C) Amino acids—sufficient for 50 liters finished medium (A-1, A-2, or A-3)

	Medium (gm.)
1. L-Phenylalanine	5.0
2. L-Serine	5.0
3. L-Tyrosine	10.0
4. L-Aspartic acid	5.0
5. Glycine	10.0
6. L-Hydroxyproline	10.0
7. L-Proline	10.0
8. DL-Alanine	10.0
9. $(NH_4)_2SO_4$	30.0

The mixtures must be ground and well mixed with the aid of a ball mill or other similar device. To make basal medium A-1 from the dry constituents the following procedure is employed: for each 100 ml. basal medium 9.82 gm. A-1, 26.4 mg. B, and 380 mg. C are dissolved with the aid of heat in about 40 ml. of water. To this is added 4 ml. of the cystine stock solution (10 mg./ml.) and 9 of the 10 essential amino acids (all except the one to be assayed). Cystine is omitted from the dry medium because of difficulties in dissolving it completely. Individual amino acid mixtures can be either added as stock solution or as dry mixtures for each individual assay.

This mixture is filtered through a medium sintered glass funnel and made up to 75 ml. and autoclaved for three minutes at 15 lb./sq. inch in a preheated autoclave. Glucose (4 gm. in 24 ml. water) is made up and autoclaved for the same time period in a separate flask. The contents of the two flasks are combined (glucose and the other medium ingredients) and to this is added 1 ml. of the sterile glutamine solution (1 mg./ml.). Glutamine would be partly destroyed even by this short autoclaving time.

For medium A-2, which is potassium phosphate buffered medium, 10.8 gm. of mixture A-2 is substituted for 9.82 gm. of mixture A-1.

For medium A-3, which is a low buffer, low pH medium (pH 5.9) used in lysine determinations (Section X.B), 1.78 gm. of mixture A-3 is substituted for mixture A-1.

c. Comments on the preparation of basal medium. Basal medium is prepared (from stock solutions or dry ingredients plus some stock solutions) with all of the ingredients *except the one amino acid to be assayed.* The prepared sterile basal medium can be stored in the refrigerator for several weeks until ready to use. It is added aseptically (3 ml.) to previously prepared and autoclaved tubes containing standard or sample (3 ml.).

Media A-1 and A-2 have the advantage of high buffering capacity, which helps to maintain a nearly constant pH and eliminates the need to check and adjust the pH each time the medium is prepared. It also yields higher growth levels and greater constancy of growth than does the same medium without the high amount of phosphate buffer. Exponential growth will continue until AOD 3000 or higher (see Section IX.G.3). This is equivalent to 1.2 mg./ml. dry weight of cells.[225]

Separate sterilization of glucose greatly reduces the color of the medium and the color formed during the incubation period. The finished medium is water white. While the aseptic addition of medium to each tube is somewhat more laborious (particularly to the unskilled) than the non-sterile addition the benefits derived are well worth the effort. The same batch of sterile medium can be used in different assays weeks apart. The absence of glucose in the medium mixture during sterilization prevents the reactions of the sugar with amino acids, etc., preserving the synthetic nature of the medium. This is particularly important for the assay of cystine (Section X.F.1).

The short autoclaving time is probably made possible by the high salt concentration of the medium. Contamination of aseptically handled medium is not a problem. In addition, this short autoclaving time is apparently not at all deleterious to the medium since no significant differences have been noted in either growth rate or total growth in media that have been aseptically filtered rather than autoclaved as described.

---

[269] L. R. Hac and E. E. Snell, *J. Biol. Chem.* **159**, 291 (1945).

[270] J. J. McGuire, S. S. Schiaffino, and H. W. Loy, *J. Assoc. Offic. Agr. Chemists* **43**, 34 (1960).

TABLE XIII

STANDARD SOLUTIONS FOR AMINO ACID ASSAYS

Recommended methods	Assay range (μg./ml.)		mg./200 ml.	Turbidimetric			Titrimetric		
	Turbidimetric	Titrimetric		ml.	Final volume (ml.)	Conc. of standard (μg./ml.)	ml.	Final volume (ml.)	Conc. of standard (μg./ml.)
Streptococcus faecalis (9790)									
L-Arginine[b]	4.5	10[c]	121[d]	9	500	9	10	250	20
L-Histidine[b]	2	5	135[e]	4	500	4	10	500	10
L-Isoleucine[b]	7	10[c]	100	7	250	14	10	250	20
L-Leucine[b]	8	10[c]	100	8	250	16	20	250	20
L-Lysine[b]	15	20[c]	125[d]	15	250	30	10	250	40
L-Methionine[b]	3	5[c]	100	6	500	6	10	500	10
L-Threonine[b]	5	10[c]	100	5	250	10	5	250	20
L-Tryptophan		2[c]	80				10	500	4
L-Valine[b]	7	10[c]	100	7	250	14	10	250	20
L-Glutamic acid[b]	12		120	10	250	24			
Leuconostoc mesenteroides P-60 (8042)									
L-Aspartic acid[b]	8	15[f]	100	8	250	16	15	250	30
L-Cystine[b]	0.225	3.6[g]	72	5	4000[h]	0.45	5	250	7.2
L-Proline[b]	2.5	10[i]	100	5	500	4	10	250	20
Glycine		10[i]	100				10	250	20
L-Phenylalanine	2.5[j]	10[i]	100	5	500	5	10	250	20
L-Serine		10[i]	100				10	250	20
L-Tyrosine	2[j]	10[i]	100	4	500	4	10	250	20
Lactobacillus delbrueckii (9649)									
L-Alanine[b]	3		120	5	500	6			
D-Alanine[b]	2		80	5	500	4			

[a] High level given; all assays are from 0 to this level.
[b] Turbidimetric assays which have been used in author's laboratory.
[c] Reference 10.
[d] HCl salt used.
[e] HCl·H2O salt used.
[f] Reference 269.
[g] Reference 254.
[h] Cystine is diluted 5 to 200; then this is diluted 10 to 200.
[i] Reference 12.
[j] Reference 270.

2. Standard Solutions

The standard solutions of amino acids are made in volumetric flasks according to Table XIII. The highest grade amino acids available are used (Section IX.F). L-Amino acids are used for all standards. The hydrochlorides (arginine and lysine) and hydrochloride hydrate (histidine) are taken into account.

The concentration ranges for the turbidimetric assays are related to the conditions employed (i.e., turbidimetric instrument, diameter of tubes, etc.) and are given for the conditions that we have employed (see Section IX.G.2). Higher or lower concentrations of standards may be necessary depending on these factors.

3. Inoculum

Stock cultures, subcultures, and inocula are handled as recommended in Section IX.G. Log phase inocula, either individually grown at the highest concentration of amino acid used for each amino acid assay can be used (without washing) or an inoculum grown in complete assay medium (with washing) can be used. This would of course depend on the number of amino acids to be assayed.

Conventional inocula, grown overnight, can also be employed. Responses may then be a little slower, less reproducible, and predictable, and possibly somewhat less regular.

4. Preparation of Sample

a. Liberation of amino acids. With a very few exceptions, such as free amino acids in cells and tissues, it is necessary to release amino acids from the peptide bound polymers. This can be done by acid, alkaline, or enzymatic hydrolysis. Acid hydrolysis can be used to liberate all of the amino acids in this group except for tryptophan.

Method 2 in Section VII.A is the preferred method of acid hydrolysis for the microbiological assay of this group of amino acids. Method 1 (Section VII.A) can be used when microbiological assays are used in conjunction with other procedures such as Stein and Moore chromatography. The length of the time period of hydrolysis is not only dependent on the stability of the amino acid being assayed but also on the particular materials that are to be hydrolyzed. It is therefore recommended that more than one time period of hydrolysis be used until the proper conditions can be established. Except for threonine, all of the amino acids in this group are relatively stable to acid hydrolysis so that it is unlikely that it will be necessary to apply any sort of correction once the time required to completely liberate these amino acids is established. It would be anticipated that this would be in the range of 8 to 24 hours of hydrolysis by method 2 (see Table VIII). Since threonine is more labile to acid hydrolysis, it is necessary to either routinely employ more than one hydrolysis period with back extrapolation to zero time, or make an empirical correction for threonine destruction (see Section VII.A).

For tryptophan determinations, the enzymatic hydrolysis method given in Section VII.C is recommended. Since the assay organism, *Streptococcus faecalis*, does not respond to either indole or anthranillic acid,[21] the extraction of the enzymatic hydrolysate with ether or toluene can be eliminated. However, it is imperative to run the necessary controls and blanks.

Occasionally it might be desirable or more convenient to use alkaline hydrolysis for tryptophan (i.e., when assays for tyrosine are also being done). The method given in Section VII.B.2 is then preferred. In such cases controls for destruction and racemization of tryptophan should be utilized. Results obtained by alkaline hydrolysis should be checked against those obtained enzymatically with similar samples.

b. The use of the hydrolysates and other samples for assay. With this assay organism and this assay medium there is no necessity to neutralize acid hydrolysates that are dried over NaOH after hydrolysis. The buffering capacity of the medium is great enough to take care of even larger amounts of acid than would be introduced in this way.

The enzymatic hydrolysates for tryptophan do not require any further treatment before assay but when alkaline hydrolysis is used neutralization with hydrochloric or acetic acid is required.

After the digest is suitably dissolved or neutralized, suitable aliquants of the digest and standard are added to the culture tubes in duplicate or triplicate. Where the approximate level of amino acid in the sample is known and three different levels will be within the assay range, three levels of hydrolysate and standard in duplicate have been found to be adequate for precise determinations. A greater number of levels will be required depending solely on the expected range of amino acid content of the sample. The tubes are set up as indicated in Table XIV. Sample and distilled water are added to make the volume in the tubes to 3.0 ml. The tubes are capped and autoclaved for 3 minutes at 15 lb./sq. inch in a preheated autoclave. Three milliliters of the sterile double strength basal medium (medium A-1) is then added aseptically to all of the tubes which are then shaken and blanked in the turbidimeter against a tube containing water. Tubes that are to remain as uninoculated color blanks are temporarily removed from the rack and the remainder of the tubes are inoculated with previously grown inoculum (see Section IX.G.2). When media of low color and low color development, such as media A-1 and A-2 used here and A-3 used for L-lysine assays (Section X.B), are used, tubes Nos. 11, 12, 21, and 22 (uninoculated blanks) (Table XIV) are not re-

TABLE XIV
Assay Protocol

Standard (valine, 12 μg./ml.)

Tube no.	ml. std.	μg./ml.	ml. H_2O	Basal medium	Blank readings	Time (17 hours, etc.)		
						OD	Blank	AOD
1	0	0	3.0	3.0				
2	0	0	3.0	3.0				
3	0.5	1.0	2.5	3.0				
4	0.5	1.0	2.5	3.0				
5	1.0	2.0	2.0	3.0				
6	1.0	2.0	2.0	3.0				
7	2.0	4.0	1.0	3.0				
8	2.0	4.0	1.0	3.0				
9	3.0	6.0	0	3.0				
10	3.0	6.0	0	3.0				
11[a]	0	0	3.0	3.0				
12[a]	0	0	3.0	3.0				

Sample (most concentrated) dilution (1)[b]

	ml. sample	ml. H_2O	Basal medium
13	0.5	2.5	3.0
14	0.5	2.5	3.0
15	1.0	2.0	3.0
16	1.0	2.0	3.0
17	2.0	1.0	3.0
18	2.0	1.0	3.0
19	3.0	0	3.0
20	3.0	0	3.0
21[a]	3.0	0	3.0
22[a]	3.0	0	3.0

Sample diluted 1–10 (if necessary)

23	0.5	2.5	3.0
24	0.5	2.5	3.0
25	1.0	2.0	3.0
26	1.0	2.0	3.0
27	2.0	1.0	3.0
28	2.0	1.0	3.0
29	3.0	0	3.0
30	3.0	0	3.0

[a] Uninoculated tubes for color blanks.

[b] This gives a 60-fold range. Quantities can usually be estimated closer than that. If not, a preliminary assay of a few widely varying dilutions can be run using only one tube per level.

quired. All of the tubes are then placed in a 37°–38°C. (±0.1°) constant temperature water bath in stainless steel racks for the required incubation period.

5. Responses to the Amino Acids

The assays can be read in the turbidimeter after 16 to 17 hours, 22 to 24 hours, and 40 to 43 hours. All of the assay tubes are removed from the water bath, dried, and well shaken. A mixer such as the Vortex Jr. (Scientific Industries, Inc., Springfield 3, Massachusetts, and available from a number of laboratory supply houses) will be found to be convenient to agitate the cultures thoroughly without spilling. The turbidity in the tubes is then read with a turbidimeter. Care must be taken to ensure that the tubes are clean and dry, that there are not any air bubbles entrapped in the medium, and that the cells in the tubes are uniformly suspended. The length of time that it takes to read all of the tubes (providing it is of reasonable length), that is, the time that the tubes are out of the water bath, will not significantly affect the accuracy of the readings. Growth of the cultures is so slow at any time after 8 or 10 hours of incubation under these conditions of incubation temperature and inoculum condition and size that the variation of one hour in reading time would be insignificant.

Standard curves for leucine, isoleucine, valine, methionine, arginine, histidine, and glutamic acid are shown in Fig. 8. It should be remembered that it is essential that the 5 μg./ml. level of glutamine be included in the medium for the assay of glutamic acid and that it is responsible for the somewhat high blank observed.

With the exception of methionine, *Streptococcus faecalis* responds only to the L-isomers of these amino acids and the D-isomers are completely inactive even in the presence of the L-form.[21,111] *Streptococcus faecalis* does not respond to D-methionine alone but will give greater turbidities when grown in mixtures of L- plus D-methionine (Fig. 2).[40]

6. Calculations and Results

The method for calculating results is detailed in Section VI.D and can be used for all of these assays. Remember, if alkaline hydrolysis has been used, multiply by 2.

B. The Assay of L-Lysine with *Streptococcus faecalis*

The lysine assay is carried out in the exact manner of the other assays with *S. faecalis*. Growth on limited amounts of lysine in the medium containing the 0.3 *M* phosphate buffer leads to lysis.[271,272] This can be over-

[271] G. Toennies and D. L. Gallant, *J. Biol. Chem.* **177**, 831 (1949).
[272] G. D. Shockman, M. J. Conover, J. J. Kolb, P. M. Phillips, L. Riley, and G. Toennies, *J. Bacteriol.* **81**, 36 (1961).

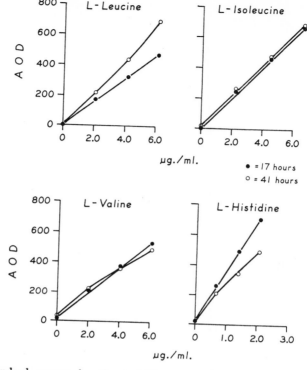

Fɪɢ. 8. Standard curves for the turbidimetric assay of ʟ-leucine, ʟ-isoleucine, ʟ-valine, ʟ-histidine, ʟ-arginine, ʟ-glutamic acid, and ʟ-methionine with *Streptococcus faecalis* (9790). Only two time intervals, 17 and 41 hours, are shown. See page 655 for continuation of Fig. 8.

come by the combined use of a reduced initial pH and a medium containing smaller amounts of phosphate buffer. Medium A-1 has been so modified by replacing the $0.3\ M$ sodium phosphate buffer with $0.014\ M$ sodium phosphate buffer. When 69 mg. of $NaH_2PO_4 \cdot H_2O$ ($0.005\ M$) and 320 mg. of Na_2HPO_4 ($0.023\ M$) are used, the final pH of the basal medium will be 5.9 (medium A-3) (Section X.A.1.*b*). Under these conditions lysis does not occur and standard curves such as are shown in Fig. 9 will be obtained with good assay results on hydrolyzed samples. The inoculum can be grown in either medium A-1 or A-3 with equally good results.

C. The Assay of D- and L-Alanine with *Lactobacillus delbrueckii* 780 (9649)

1. *Medium*

Medium C. Table IX (Section VIII), is used for these assays. ᴅ-Alanine is added to the basal medium for an assay of ʟ-alanine and ʟ-alanine is

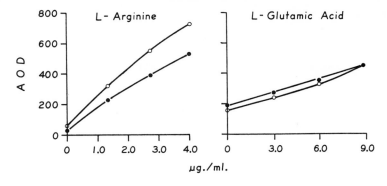

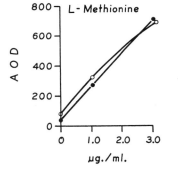

FIG. 8 *continued*

for the assay of D-alanine. The organism is extremely sensitive to the sodium ion. Therefore, KOH is used for stock solutions and for neutralizing the basal medium.

a. Stock solutions. The medium is made from the following stock solutions:

(1) Glycine, 16 mg.; L-histidine, 1.56 gm.; L-lysine, 250 mg.; L-arginine, 250 mg.; and 125 mg. each of L-isoleucine, L-leucine, L-methionine, L-proline, L-asparagine, L-threonine, L-valine, and L-serine are dissolved in 50 ml. of distilled water—use 8 ml./100 ml. basal medium.

(2) L-Aspartic acid, 625 mg.; L-glutamic acid, 1.25 gm.; L-phenylalanine, 125 mg.; and D-glutamic acid, 38 mg. are dissolved in 7.5 ml. of 2.5 N KOH and made to 50 ml.—use 8 ml./100 ml. basal medium.

(3) L-Tyrosine, 125 mg. is dissolved in 0.75 ml. of 2.5 N KOH and diluted to 25 ml.—use 4 ml./100 ml. basal medium.

(4) Thymidine, 16 mg. in 10 ml.—use 2 ml./100 ml. basal medium.

(5) K_2HPO_4 and KH_2PO_4, 5 gm. each in 50 ml.—use 1 ml./100 ml. basal medium.

Solutions 12, 14, 18, 19, 20, 21, and 23 used for medium A can be used

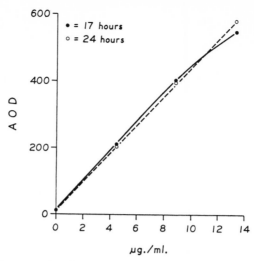

FIG. 9. Standard curves for the turbidimetric assay of L-lysine with *Streptococcus faecalis* (9790).

for this medium in the following amounts per 100 ml. basal medium: 12—L-cystine 2 ml.; 14—L-tryptophan, 1 ml.; 18—riboflavin, 1.0 ml.; 19—vitamin mixture, 1 ml.; 20—biotin, 1.0 ml.; 21—pteroylglutamic acid, 1.0 ml.; 23—A.G.U., 2 ml.

(6) Salts, $MgSO_4 \cdot 7H_2O$, 8 gm.; $MnSO_4 \cdot 4H_2O$, 1.2 gm.; $FeSO_4 \cdot 7H_2O$ and NaCl, 200 mg. each and $CaCl_2$, 50 mg. are dissolved in 2.4 ml. 1 N HCl and made to 100 ml.—use 2 ml./100 ml. basal medium.

b. Preparation of the basal medium. To make 100 ml. of basal medium the following are added to the indicated quantities of the above solutions: 2 ml. of a solution consisting of 2 mg. sodium oleate and 200 mg. Tween 40 per milliliter; Na acetate, 2 gm.; ascorbic acid, 60 mg.; glucose, 2 gm.; and 5 mg. L- or D-alanine. The pH is adjusted to 6.7 with KOH and the medium is made to volume and filtered through a medium sintered glass funnel. Tubes of standard solution (0–2 μg./ml. D- and 0–3 μg. L-alanine) and samples are prepared in duplicate and made to 3 ml. with distilled water. Three milliliters of the basal medium is added and the tubes are autoclaved for 3 minutes in a preheated autoclave at 15 lb./sq. inch.

2. Inoculum

The organism will not grow for more than two subcultures in the assay medium. Therefore inoculum is grown in an organic medium such as Difco Micro-inoculum broth. Log phase inoculum is used (Section IX.G.2). The inoculum is washed twice with sterile water and suspended in 2 ml. sterile glutamine solution (1 mg./ml.) plus 4 ml. sterile water. The inoculum level used for this assay is higher than that for the ones previously described. One-tenth milliliter of inoculum is required to have the stated level of glutamine. The inoculum is grown to 2 or 3 times the turbidity level of that used in previous assays (AOD 240–360) so that an initial optical density of 4 to 6 (1.6–2.4 μg./ml. dry weight) is obtained.

3. Preparation of Sample

Samples are prepared in the same manner as for the other assays (Section VII). Acid hydrolysis (method 2, Section VII.A) is used for alanine. However, since alanine, particularly D-alanine, occurs in other than protein or other peptide-linked substances, samples prepared in other ways (extracts, etc.) can be used. In these instances the preparation of the sample would be entirely dependent on the material. Since the medium is not highly buffered, acid hydrolysates after drying over NaOH should be neutralized with KOH or NH₄OH so that the pH of the medium will not be affected. When other methods of sample preparation are used it should be remembered that the organism is sensitive to the sodium ion and to changes in the salt concentration of the medium. Often potassium or ammonium salts rather than sodium salts or acetic acid rather than HCl can be used and thereby retain a good K:Na balance. Neutralization with acetic acid will affect the acetate content of the medium relatively less than HCl neutralization would affect the inorganic constituents.

4. Response to the Amino Acid

As pointed out previously (Section IV.C.2), the response of the organism to the D- and L- forms is different and characteristic. Both assays are sensitive; the assay range for D-alanine being 0–2 μg./ml. and for L-alanine 0–3 μg./ml. It should be noted that both the glycine content and the L- (or D-) alanine content of the medium have been reduced. This is to increase the sensitivity of the assay since both of these amino acids have an inhibitory effect on the response of the organism to the other form of alanine. A similar effect was noted by Camien and Dunn for *Leuconostoc mannitopoeus*.[273] Under these conditions, it is possible to assay for L- or D-alanine in the presence of as much as eightfold quantities of glycine or the other alanine isomer.

The alanine assays with this organism have an advantage over such assays with *Lactobacillus casei*, which has previously been employed for D- and L-alanine determinations.[151] *Lactobacillus delbrueckii* 780 requires either phosphorylated forms of vitamin B₆ or both alanine isomers, while

[273] M. N. Camien and M. S. Dunn, *J. Biol. Chem.* **185**, 553 (1950).

L. casei can use unphosphorylated pyridoxal or pyridoxamine. Since vitamin B_6 phosphates are far less stable than the unphosphorylated forms of the vitamin they should be completely destroyed by acid hydrolysis. Only extremely high levels of pyridoxamine or pyridoxal (100–300 mg./ml.) can substitute for phosphorylated B_6 or both alanine isomers. Care should be taken to read the assays at different time periods and different assay levels so that any deviation from the pattern of the standard (drift) would be observed (Sections VI.E and F). The response to B_6 phosphates differs from that of either D- or L-alanine and the presence of the vitamin in a sample would cause drift in the assays.

Samples that might contain vitamin B_6 phosphates (i.e., unhydrolyzed samples) can be treated with activated charcoal to remove vitamin B_6[201,202] as follows. A 2:1 mixture by weight of activated charcoal (Carboraffin or Darco G60) and sample is stirred for 30 minutes with warming and then filtered. The charcoal is activated by heating and must be used immediately.

Since L-alanine might be contaminated with vitamin B_6 active compounds,[202] it should also be treated with charcoal before it is used either as a medium ingredient or a standard.

Response of D- and L-alanine standards is illustrated in Fig. 1. Growth of the organism in response to these levels of D-alanine is relatively slow and it should be noted that, like the previously described assays (Section X.A.6), the response to L-alanine is read after 17, 23, and 42 hours, but the response to D-alanine is read after 24, 42, and 48 hours. For reasons that are not yet known, occasionally the response to D-alanine is even slower and assays are not read until the 42-hour interval.

Recently we have found that, by extending the assay range of the D-alanine standard to 4.0 μg./ml., a more consistent response, particularly at the 24 hour time interval is obtained. This also enables a reduction in the level of inoculum used by one-half (e.g., inoculum is suspended in 4.0 ml. of sterile glutamine solution plus 8.0 ml. sterile water) which in turn reduces the turbidity obtained in the zero level tubes. Since the D-alanine standard curve is pronouncedly concave, a sufficient number of standard points are required so that the curve may be adequately defined. For this reason, most recently we have used 0, 1.0, 2.0 and 3.0 ml. of a solution containing 4.0 μg./ml. D-alanine and 2.0 and 3.0 ml. of a solution containing 8.0 μg./ml. of D-alanine per 6.0-ml. assay tube (in duplicate) as our standard series.

5. Results and Calculations

The method used is identical to that used for the other assays (Section VI.D). Typical results of D- and L-alanine assays are shown in Table XV.

TABLE XV

RESULTS OF TYPICAL ASSAYS FOR D- AND L-ALANINE WITH *Lactobacillus delbrueckii* 780[a]

ml.	L-Alanine			D-Alanine		
	17	23	43	42	48	69
		(hours)			(hours)	
1	4.04	3.90	4.44	3.60	3.57	3.45
1	3.84	3.72	4.68			
2	4.20	3.72	4.14	3.57	3.56	3.45
2	4.11	3.90	4.38			
3	4.20	4.08	4.12	3.64	3.68	3.72
3	4.13	4.24	4.28	3.56	3.58	3.73
Av.	4.09	3.93	4.34	3.59	3.60	3.59
	±0.10	±0.16	±0.16	±0.03	±0.04	±0.14

$$= 3.16\% \pm 0.1 \qquad\qquad = 3.07\% \pm 0.06$$

$$\text{D} + \text{L ALA} = 6.23\% \qquad \text{chromatography DL} = 6.2\%$$

[a] Values are $\dfrac{\mu g.}{ml.} \times \dfrac{6\ ml.}{ml.\ hyd./tube}$.

Agreement of the sum of D- and L-alanine determined by these assays with total alanine by chromatography is illustrated in Table II.

D. The Assay of L-Proline with *Leuconostoc mesenteroides* P-60

1. *Medium*

The medium used here is medium A-2 without proline and with 0.3 M potassium phosphate buffer substituted for the sodium phosphate buffer of medium A-1. Apparently *L. mesenteroides* is more sensitive to high sodium-to-potassium ratios than is *Streptococcus faecalis*.[135,136] This is shown by the fact that *Leuconostoc mesenteroides* will not grow on successive subculture in medium A-1 but will in medium A-2. The organism grows somewhat more slowly than does *Streptococcus faecalis* (division time: about 65 minutes for *Leuconostoc mesenteroides*, about 33 minutes for *Streptococcus faecalis*). However, the reading time of the turbidity of the assays is about the same. A medium of reduced pH (pH 6)[26,166] has been recommended to eliminate the lag in the standard curve at low proline levels. Only a small lag at 17-hour readings has been observed with this medium. The lag is pronounced, however, when a sodium phosphate buffered medium is used.

2. Inoculum

The inocula for an assay can be prepared by first inoculating a tube of complete medium A-2 and growing the organism overnight. It should then be transferred to a second tube of either complete medium A-2 or medium A-2 with only 2 μg./ml. proline and grown until it is well into logarithmic growth. This can then be used to inoculate the assay. If grown in the 2 μg./ml. proline medium, the inoculum will not require washing. If grown in the complete medium, it will be necessary to wash the inoculum twice with distilled water. The standardization of log phase inoculum is also recommended (Section IX.G.3). Incubation is at 37°–38°C. with turbidity readings after 17, 24, and 42 hours.

The remaining procedures are identical with those for the assays using *S. faecalis* (Section X.A).

E. The Assay of L-Aspartic Acid with *Leuconostoc mesenteroides* P-60

Medium A-2 without aspartic acid and asparagine is used for L-aspartic acid assays. The organism does respond to D-aspartic acid but in a manner different from its response to L-aspartic acid. However, the response to the D-isomer can be essentially eliminated by the inclusion of 0.1% cysteic acid and a base level of 18 μg./ml. of D-aspartic acid in medium A-2.[13,181] Commercially obtained cysteic acid should be recrystallized in 50% ethanol.[181]

Using these conditions (0.1% cysteic acid and 18 μg./ml. D-aspartic acid in the medium), the D-isomer is less than 10% as active as the L-isomer. Standard curves are shown in Fig. 10. Incubation is at 37°–38°C. with readings at 17, 22, and 42 hours. The remaining procedures are identical with those given for the *Streptococcus faecalis* assays in Section X.A.

F. The Assay of L-Cystine with *Leuconostoc mesenteroides* P-60

For the assay of this amino acid, medium A-2 could probably be used. However, the method was developed with medium D which contains oxidized peptone as a source of most of the amino acids. The same medium can also be used for the determination of methionine and tyrosine. Since the data of Riesen *et al.*[254] and Schiaffino *et al.*[274] indicate that the relative sensitivity of the assay would probably be decreased with a synthetic medium under these conditions, the method using the oxidized peptone in the medium will be given here. There has been much written on the destruction or the decreased utilization of cystine in autoclaved

---

[274] S. S. Schiaffino, J. J. McGuire, and H. W. Loy, *J. Assoc. Offic. Agr. Chemists* **41,** 679 (1958).

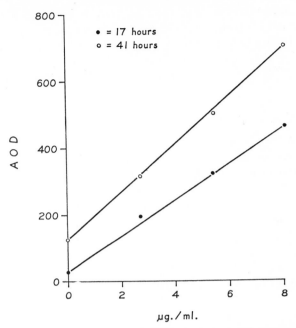

FIG. 10. Standard curves for the turbidimetric assay of L-aspartic acid with *Leuconostoc mesenteroides* P-60. The medium contains potassium phosphate buffers (medium A-2), 0.1% recrystallized cysteic acid, and 18 μg./ml. D-aspartic acid. The presence of 9 μg./ml. of additional D-aspartic acid in the highest level tubes will increase the turbidity readings approximately 50 AOD units at both time reading intervals. The readings of the zero tubes will also be correspondingly higher in the presence of the same additional quantity of D-aspartic acid.

media.[275] Apparently the disadvantages are overcome with this recommended method by autoclaving the glucose for 3 minutes in a preheated autoclave separately from the rest of the medium and the assay sample. This may account for the increased sensitivity of this method compared to others.

1. Medium

The composition of the medium is given in Table IX (medium D). Many of the stock solutions given in Section X.A.1 can be used and others need to be adjusted to fit the formula of this medium. The pH of this medium must be adjusted (pH 6.8). Separate and short autoclaving procedures for sample, glucose, and the rest of the medium ingredients is of particular importance for this assay. This is true not only because of the low color blanks but because of the instability of cystine. It should be

[275] L. E. Lankford, J. M. Ravel, and H. H. Ramsey, *Appl. Microbiol.* **5**, 65 (1957).

noted also that ascorbic acid (250 μg./ml.) is included in this medium. This highly reducing environment greatly increases the speed and amount of growth response to cystine.[157] The sterile basal medium will darken with age and therefore will not keep as long as the A group of media. Incubation is at 37°–38°C. and turbidity is read after about 17, 23, and 42 hours.

2. Preparation of the Oxidized Peptone

This is prepared according to the method of Lyman et al.[159] Fifty grams of Bacto Peptone (Difco) is dissolved in 460 ml. of water. Forty-two milliliters of concentrated HCl and 5.1 ml. of 9.82 M H_2O_2 is added and the mixture is allowed to stand overnight (about 16 to 18 hours). It is then heated either in a steam sterilizer or in a boiling steam bath for 30 minutes, cooled, neutralized with 40 ml. of 12 N NaOH and steamed again for 1 hour. The pH of the neutralization should be checked with indicator paper. It is then diluted to 1 liter with distilled water and toluene or a volatile preservative[267,276] should be added. The oxidized peptone will keep for months in a stoppered bottle in a refrigerator.

3. Inoculum

Log phase inoculum is grown in the assay medium in the same manner as for the previous methods (Section IX.G.2.b). If the inoculum is grown several times in the assay medium and arrested during the log growth phase it can be used without washing, at the rate of 1 drop per 6 ml. finished medium. Because of a decreased growth rate of the assay organism on low concentrations of cystine, twice the highest concentration of cystine used in the assay (0.225 μg./ml.) is used in the inoculum tubes (0.45 μg./ml.).

4. Preparation of Sample

For cystine the HCl-formic acid evacuated sealed tube hydrolysis method is recommended (method 2, Section VII.A). Cystine is one of the amino acids that is subject to destruction by acid hydrolysis. This is apparently the best method to keep such destruction at a minimum, at least for some materials.[277] The use of several time periods of hydrolysis and if necessary the method of extrapolation back to a hydrolysis time of zero is recommended.

5. Response to the Amino Acid

Table XVI shows the turbidity readings of a typical set of standards

[276] S. M. Hutner and C. A. B. Jerknes, *Proc. Soc. Exptl. Biol. Med.* **67**, 393 (1948).
[277] J. J. Kolb and G. Toennies, *Anal. Chem.* **24**, 1164 (1952).

TABLE XVI
TYPICAL READINGS OF CYSTINE STANDARD

	Cystine (μg./ml.), finished medium			
Readings[a]	0	0.075	0.150	0.225
At 17 hours	84	319	491	658
	103	320	498	652
At 22 hours	95	340	531	678
	112	338	539	678

[a] Adjusted optical density.

after 17 and 22 hours of incubation at 38° in a water bath. A precision of 2 to 3% on hydrolyzed materials has been obtained.[157] The activity of D-cystine in substituting for the L-isomer has not been thoroughly tested. Results are calculated in the same manner as for the arginine, etc., assays (Section X.A.6).

G. Glycine, L-Serine, L-Phenylalanine, and L-Tyrosine with *Leuconostoc mesenteroides* P-60

These are the only ones of a group of recommended assays with which first-hand experience has not been had. The oxidized peptone medium has been recommended for tyrosine assays. There is no reason why these assays cannot be done successfully using either media A-2 (Section X.A.1b), B, E, or H (Table IX) since all of these are very similar. This is not, however, a proven turbidimetric procedure used for these assays. The record of only two of these (phenylalanine and tyrosine) having been done by turbidimetric procedures has been found.[274,278] However, there is no reason why the other two cannot be done similarly.

Methods of inoculum, standard solutions, incubation, and calculation of results are identical with the other assays with this organism (Sections X.A.6, X.F.1 and 3).

1. *Medium*

Medium B or H can be used for all of these assays by eliminating from the medium the amino acid (glycine, serine, phenylalanine, or tyrosine) being assayed. The oxidized peptone medium (D) without tyrosine can be used for the assay of tyrosine. It is recommended that L-aspartic acid be included in medium H. *Leuconostoc mesenteroides* requires aspartic

[278] S. S. Schiaffino, J. J. McGuire, and H. W. Loy, *J. Assoc. Offic. Agr. Chemists* **41,** 420 (1958).

acid [12] and according to Schweigert *et al.*[30] asparagine only very ineffi-
ciently substitutes for aspartic acid.

2. Preparation of Sample

Acid hydrolysis in the evacuated sealed tube is recommended for these
amino acids except tyrosine (method 2, Section VII.A). Tyrosine is partially destroyed on acid hydrolysis so the alkaline hydrolysis method
(method 1 or 2, Section VII.B) should be used. Since tyrosine will be
racemized by alkaline hydrolysis, consideration of this should be made
in the calculations of the results. It should also be kept in mind that
serine is one of the amino acids that is subject to minor losses during acid
hydrolysis (Section VII.A).

3. Response to the Amino Acids

a. Glycine. Some difficulties have been encountered with high blanks in
the assay for glycine. If high blanks are found, other amino acids should
be carefully checked for contamination with glycine. Treatment of the
amino acids that go into the basal medium with activated charcoal at pH
3 can be of assistance. Such contamination may be a vitamin B_6 derivative rather than glycine.[202] The assay range given for the titrimetric
method is 0–5 μg./ml. It can be expected that by a turbidimetric method
the sensitivity of the assay would be at least doubled.

b. Phenylalanine and tyrosine. The assay of both of these amino acids,
in a medium very similar to the one recommended here, by both turbidimetric and titrimetric methods has been recently compared by McGuire
et al.[270] Turbidimetry was found to be four times as sensitive for phenylalanine and 5 times as sensitive for tyrosine as the titrimetric methods.

c. Serine. The assay range given for titrimetric methods is 0–5 μg./ml.
L-serine. Turbidimetric methods would be expected to be more sensitive.
Some difficulties have been encountered in that vitamin B_6 derivatives
contribute to the blank values[12,30] giving high blanks and glycine will
stimulate growth in the presence of low serine concentrations and an enriched CO_2 atmosphere or folinic acid.[125] In addition, there is some evidence that large amounts of threonine inhibit the response to serine.[26,279]
The assays can be done either in the presence or absence of pyridoxine
and pyridoxal.[12,30] The higher forms of vitamin B_6 (pyridoxamine,
pyridoxal, and their phosphates) may be encountered in test samples but
are usually destroyed by acid hydrolysis. Folinic acid is also destroyed by
acid hydrolysis. The amounts of threonine required to inhibit the response
to serine are quite large. The availability of pure L-threonine and the
possibility that the inhibition was due to the D-form (Section IV.C.2) are
of interest.

[279] W. W. Meinke and B. R. Holland, *J. Biol. Chem.* **173**, 535 (1948).

XI. Recommended Titrimetric Assays

A. L-Arginine, L-Glutamic Acid, L-Histidine, L-Isoleucine, L-Leucine, L-Lysine, L-Methionine, L-Threonine, L-Tryptophan, and L-Valine with *Streptococcus faecalis* (9790)

1. General

Titrimetric methods are presented for those few instances where a turbidimetric method may not be satisfactory. Examples would be when large amounts of color or extraneous turbidity is present in the samples to be assayed. In principle, the assays are carried out in exactly the same manner as are the turbidimetric assays for these amino acids, except that the incubation period is longer (72 hours), the amounts of standard amino acid in the tubes are larger, and the assay tubes are titrated with alkali.

2. The Assay Medium

Medium B (Table IX) is recommended for the titrimetric determination of these amino acids. It should be proposed here that medium A-1 could also be used. It would be anticipated that the use of this medium, containing a high concentration of phosphate buffer, would yield greater acid production by *Streptococcus faecalis* at a shorter incubation period [110] as well as more precise assay results and more sensitive assays. However, this medium is not recommended for routine titrimetric assays without the experimental results of actual assays.

Medium B can be prepared from either stock solutions or dry ingredients in a manner essentially identical to that shown in Section X.A for the turbidimetric assays. Suitable adjustments should be made for the differences in concentration of some of the medium constituents. For use with this test organism the medium should contain the lower quantities of glucose and sodium acetate (10 mg./ml.). Additional recommendations are: the inclusion of D-alanine in the medium and increasing the concentrations of adenine, guanine, uracil, folic acid, and biotin to those of medium A-1.

3. Standard Solutions

The standard amino acid solutions should be made up as indicated in Section X.A.2. For titrimetric assays the concentration of the standard is dependent on (1) the assay range desired, (2) the volume used in the assay tubes, and (3) the concentration of the alkali used for titration.

4. Inoculum

Stock cultures, subcultures, and inocula are handled as indicated in

Section IX.G. The use of log phase inocula is recommended for titrimetric assays. Since the incubation period is longer, the active and rapid growth of such inoculum is not of any great benefit. However, the increased degree of standardization is of some importance. The use of log phase inocula requires the use of a turbidimetric instrument which may not be available when titration is the assay method being used. In such instances inocula grown overnight in the assay medium or, even better, for 4 to 6 hours after the first visual appearance of turbidity, can be used with extremely satisfactory results.

5. Preparation of Sample

The acid and alkaline hydrolysis procedures recommended are exactly the same as those used for turbidimetric assays (Sections X.A.4 and VII). It should be remembered that assays by this method are somewhat less sensitive than the turbidimetric assays so that somewhat larger amounts of materials will have to be hydrolyzed.

Neutralization without first drying the acid hydrolysates may prove to be satisfactory in most instances. However, since the medium is not highly buffered, it is recommended that acid hydrolysates be dried over NaOH to decrease the amount of acid present before neutralization. Alkaline hydrolysates should also be neutralized before use in the assay.

6. The Assays

After digestion and neutralization, suitable aliquants of the hydrolysate and standard are added to clean culture tubes in duplicate or triplicate. Since the standard curves will be curvilinear throughout their lengths when plotted arithmetically, a sufficient number of points will be necessary to adequately define the curve. Therefore, 5–8 levels of amino acid standard plus a zero-level tube will be required (i.e., 0, 0.5, 1.0, 2.0, 2.5, 3.0, 4.0, and 5.0 ml. of the particular standard for assays done in 10-ml. final volume). For the same reason, one should aim for 4 levels of sample to be within the range of the standard. After the addition of samples and standard to the tubes, the tubes are made up to half their final volume with distilled water, and then an equal volume of double strength basal medium is added to each tube. Assays are most conventionally performed in 10-ml. final volume (in 16×150 to 19×150 mm. tubes) so that in this case the samples and standards would be made to 5.0 ml. and then 5.0 ml. of basal medium would be added. Tubes are then capped or plugged, and autoclaved at 15 lb./sq. inch for 8 to 10 minutes. After the tubes have cooled they are inoculated with 1 drop (about 0.05 ml.) of the inoculum and incubated at 37° for 72 hours. At the end of this period the tubes are autoclaved for 5 minutes at 15 lb./sq. inch and, since growth has been stopped by this procedure, they can be stored in the cold until titrated. Acid production can be titrated with 0.05 to 0.075 N NaOH either directly in the tubes using 0.1% bromthymol blue as an indicator or by means of an automatic titration setup.

A constant temperature water bath is recommended for the incubation of these assays for the reasons given in Section IX.B. Since the color of the medium is of no importance for acid production, it is neither necessary nor of benefit to separately sterilize the medium and glucose.

The method as given is for a 10-ml. volume in 16 to 19 × 150 mm. culture tubes. It should be remembered that if the assays are to be scaled down to smaller volumes, tubes of smaller diameter should be used.

7. Responses to the Amino Acids and Calculation of Results

Titrimetric assays can be read at only one time interval for each set of assay tubes. The differences between the two methods can be summed up as (1) the inherent difference in the method of measurement, (2) the difference in incubation period, and (3) the relative sensitivities of the assays. The procedures for calculating results from a standard curve are identical to those given in Section VI.D. Recent results indicate that the two methods (turbidimetric and titrimetric) are interchangeable with results from both methods being in good agreement.[270]

B. The Assay of L-Cystine with Leuconostoc mesenteroides P-60

The titrimetric assay for L-cystine should be carried out as indicated in Section X.F for the turbidimetric method. Medium D, containing oxidized peptone, should be used. Separate autoclaving of glucose, the assay medium, and the standards or sample for short time intervals (15 lb./sq. inch for 3 minutes) is proposed. In this way the sensitivity of the titrimetric method should be increased at least fourfold (i.e., from a range of 0 to 3.6 μg./ml. to 0 to about 0.9 μg./ml.). In view of the low cystine content of some proteins (i.e., casein, elastin, zein, hemoglobin), increased sensitivity would seem to be of value. Results with the titrimetric method using these modifications have not as yet been obtained. However, it would be expected that such modifications of a titrimetric method would be comparable to the same modifications of a turbidimetric assay.

In the absence of the separate and short autoclaving, the factors involved in the destruction and decreased utilization of cystine in autoclaved media[254,275] should be kept in mind.

The remaining procedures are identical to those described in Section XI.A except that 0.1 N NaOH should be used for titration.

The instability of cystine to acid hydrolysis should be kept in mind (Sections VII.A and X.F.4).

C. The Assay of L-Proline, L-Aspartic Acid, L-Serine, L-Phenylalanine, L-Tyrosine, and Glycine with *Leuconostoc mesenteroides* P-60

Although the titrimetric assays for these amino acids could undoubtedly be successfully carried out using medium A-2, its use is not yet proven. Therefore, medium B is recommended for the titrimetric assay of these amino acids. The assays should be carried out as specified in Section X with the modifications imposed by the titrimetric method of growth measurement that are indicated in Section XI.A.

TABLE XVII
ASSAYS FOR AMINO ACIDS AND THEIR SENSITIVITY*

Alternative methods

	Assay range[a] (μg./ml.)			Assay range[a] (μg./ml.) Titrimetric
	Turbidimetric	Titrimetric		
Leuconostoc mesenteroides P-60			*Lactobacillus delbrueckii* LD5 (9595)	
1. L-Arginine		20[i]	14. L-Aspartic acid	80[l]
2. L-Glutamic acid	12	30[i]	15. L-Phenylalanine	8[l]
3. L-Histidine	1.5[i]	5[i]	16. L-Serine	25[l]
4. L-Lysine	7.5[i]	25[i]		
5. L-Methionine	1.0[i]	5[i]	*Leuconostoc citrovorum*	
6. L-Threonine	2.5[i]	10[i]	17. DL-Alanine	25[i]
			18. Glycine	10[i]
Lactobacillus arabinosus 17-5 (ATCC 8014)			19. L-Cystine	2.5[i]
			20. L-Tyrosine	10[i]
7. L-Glutamic acid	8[k]	8[k]		
8. L-Isoleucine	2.5[i]	10[i]	*Lactobacillus brevis* (8257)	
9. L-Leucine	2.5[i]	10[i]	21. L-Proline	13[m]
10. L-Methionine		2[c]		
11. L-Phenylalanine		10[c]	*Lactobacillus fermenti* 36 (9338)	
12. L-Tryptophan	4[i]	10[i]	22. DL-Methionine	10[n]
13. L-Valine	2.5[i]	10[i]		
			Lactobacillus casei	
			23. L-Arginine	10[c]
			24. L-Cystine	3.5[g]

* Footnotes *a, c–j* are the same as for Table XIII.
[k] Hac *et al.*[137]
[l] Stokes *et al.*[20,21]
[m] Dunn *et al.*[28]
[n] Dunn *et al.*[25]

A lag in the standard curve at low proline concentrations may be encountered and this may be overcome by the reduction of the initial pH of the assay medium to pH 6[26,166] or the use of potassium phosphate buffered medium (Section X.D). Cysteic acid 0.1% and a base level of D-aspartic acid should be included in media for the assay of L-aspartic acid if stereospecificity is desired (Section X.E). The amount of D-aspartic acid to be included in the medium would have to be determined. Comments on the response of *L. mesenteroides* P-60 to serine, phenylalanine, tyrosine, and glycine that appear in Section X.G are equally applicable to the titrimetric method.

Titration, after 72 hours of incubation at 37° to 38°C., should be with 0.1 N NaOH.

XII. Alternative Methods for Amino Acid Assays

Some alternate assays and the assay organisms are listed in Table XVII. Most of these are proven titrimetric assays and many have been extensively used. They are often useful for confirming the results of the recommended methods with another test organism. For many, the stereochemical specificity is either not adequately known or unsatisfactory for materials that might contain more than one isomer. Comment has already been made on the specificity of some of these test organisms in Section IV.C. Two examples here are (1) the rather widely used assay for glutamic acid with *Lactobacillus arabinosus* 17-5 which also partially responds to D-glutamic acid [137,280,281] and (2) the almost equal response of *L. delbrueckii* LD5 [20] to D- and L-aspartic acid which would confuse assays where both isomers are present in the sample. Also, *L. delbrueckii* LD5 is apparently less sensitive for aspartic acid than is *Leuconostoc mesenteroides*.

Many of the organisms and assays listed have not as yet been used for turbidimetric amino acid determinations.

Table XVIII lists some of the lactic acid bacteria that have been used in the last 12 years (1949–1961) for the determination of amino acids. *Leuconostoc mesenteroides* P-60 is apparently the most widely used test organism for at least most of the amino acids that it requires. Except for alanine, all of the amino acids listed can be very satisfactorily determined with either *Streptococcus faecalis* or *Leuconostoc mesenteroides*. Where a choice between the two organisms is possible, factors such as stereospecificity and relative sensitivity must be considered, and where these are equal or not of particular importance the choice is often sub-

[280] J. C. Lewis and H. S. Olcott, *J. Biol. Chem.* **157**, 265 (1945).

[281] M. S. Dunn, M. N. Camien, L. B. Rockland, S. Shankman, and S. C. Goldberg, *J. Biol. Chem.* **155**, 591 (1944).

TABLE XVIII

SOME LACTIC ACID BACTERIA THAT HAVE BEEN USED IN THE LAST 12 YEARS TO DETERMINE AMINO ACIDS[a]

	S. f.	L. m.	L. c.	L. a.	L. d.	LD5	L. c.	L. b.	L. f.	L. l.	L. p.
L-Alanine	—	—	—	—	h	—	—	—	—	—	—
D-Alanine	—	—	—	—	d, h	—	—	—	—	—	—
DL-Alanine	—	—	c–e, g, i, k, l	—	—	—	—	—	—	c	—
Aspartic acid	—	c, e, g–i, k, l	d	—	—	—	—	—	—	—	—
Arginine	b, c, e, f, g, k, l	e, j, k	k	—	—	e	b, g, i	—	—	—	—
Cystine	—	c, e, h, i, k	e, k	g	—	—	—	—	—	—	—
Glutamic acid	d, h, k	c, k	e, k	c, e, g, i, l	—	e	—	—	—	—	d
Glycine	—	c, e, i, k, l	c, e, k	—	—	—	—	—	—	—	—
Histidine	b, c, e–h, j, k	b, c, e, g, k, l	k	—	—	e	—	—	i	—	—
Isoleucine	b, f, h	g, k, l	k	b, c, i, j	—	—	b	—	—	—	—
Leucine	b, f, h	k, l	—	b, c, g, i, j	—	—	b	—	—	—	—
Lysine	b, d, f, h, k	b–d, g, i–l	—	—	—	—	—	—	—	—	—
Methionine	b, f, k	b, c, i–l	k	b, g	—	—	—	—	i	—	—
Phenylalanine	—	b, c, g, i–l	k	b, c	—	b, c, f	b, g	—	—	—	—
Proline	—	c, d, g, h, k, l	c	—	—	—	—	c, e, k	—	—	—
Serine	—	c, e, k, l	—	—	—	—	e, g, i	—	—	—	—
Threonine	b, c, f, h, j–l	k	k	—	—	—	—	—	g, i	—	—
Tryptophan	b, c, f, j, k	k, l	—	b	—	—	—	—	—	—	—
Tyrosine	—	c, e, g, k, l	c, e, k	c	—	e	—	—	—	—	—
Valine	b, f–h	k	k	b, c, i, j, l	—	—	b	—	—	—	—

ATCC no.

[a] S. f. —*Streptococcus faecalis*   (9790)
    L. m.—*Leuconostoc mesenteroides* P-60   (8042)
    L. c. —*Leuc. citrovorum*   (8081)
    L. a. —*Lactobacillus arabinosus* 17-5   (8014)
    L. d. —*L. delbrueckii*   (9649)
    LD5 —*L. delbrueckii* LD5   (9595)
    L. c. —*L. casei*   (7469)
    L. b. —*L. brevis*   (8257)
    L. f. —*L. fermenti* 36   (9338)
    I. l. —*L. leichmanii* 327   (4797)
    L. p. —*L. pentosus* 124-2   (8041)
[b] See reference 10.
[c] See references 30, 255, and 282.
[d] See references 31 and 184.
[e] See references 26, 27, and 168.
[f] See reference 268.
[g] See reference 120.
[h] See references 13, 111, 157, 225, and 283.
[i] See references 28, 252, and 284–292.
[j] See reference 293.
[k] See references 12, 22, and 162.
[l] See reference 294.

[282] D. A. Greenwood, H. R. Kraybill, and B. S. Schweigert, *J. Biol. Chem.* **193,** 23 (1951).

[283] G. D. Shockman, *J. Biol. Chem.* **234,** 2340 (1959).

jective. Next in choice as a test organism comes *Lactobacillus arabinosus* 17-5, one of the first organisms used for amino acid assays. These are certainly proven assays and can be well utilized within their limitation, although the preference for the assays with *Streptococcus faecalis* or *Leuconostoc mesenteroides* is usually fairly clear cut. The remaining organisms seem to be most useful for special situations. D- or L-alanine with *Lactobacillus delbrueckii* or DL-alanine with *Leuconostoc citrovorum*, DL-methionine with *Lactobacillus fermenti* 36 are certainly examples of these. An additional useful situation would be as a check on the primary assay method being used.

XIII. Inhibition Assays

Recently an agar diffusion assay for phenylalanine has been described by Guthrie and co-workers.[295-298] The method is based on the inhibition of growth of *Bacillus subtilis* by β-2-thienylalanine and its competitive reversal by phenylalanine. This assay has been proposed as a screening test for phenylketonuria in young infants. The assay differs in principle from those mentioned above (Section V) in that a growth requirement is created in a relatively nonfastidious organism by the incorporation of a specific antimetabolite in the growth medium. At first glance, this approach would seem to have some advantages over the use of a more fastidious organism such as a member of the lactic acid group in that a less complex culture medium and an easier to handle organism can be used (e.g., a stable suspension or dried preparation of spores of *B. subtilis*). However, decreased specificity of the assay similar to that encoun-

---

[284] S. H. Lovett and M. S. Dunn, *Proc. Soc. Exptl. Biol. Med.* **97**, 240 (1958).

[285] M. N. Camien, A. Yuwiler, and M. S. Dunn, *Proc. Soc. Exptl. Biol. Med.* **94**, 137 (1957).

[286] M. N. Camien and M. S. Dunn, *Science* **124**, 3233 (1956).

[287] Y. Chang and M. S. Dunn, *Food Research* **22**, 182 (1957).

[288] P. F. Salisbury, M. S. Dunn, and E. A. Murphy, *J. Clin. Invest.* **36**, 1227 (1957).

[289] M. N. Camien and M. S. Dunn, *J. Biol. Chem.* **183**, 561 (1950).

[290] M. S. Dunn, E. R. Feaver, and E. A. Murphy, *Cancer Research* **9**, 306 (1949).

[291] E. A. Murphy and M. S. Dunn, *Cancer Research* **17**, 567 (1957).

[292] M. S. Dunn, M. N. Camien, R. B. Malin, P. J. Reiner, and J. Tarbet, *Univ. Calif. (Berkeley) Publs. Physiol.* **8**, 327 (1949).

[293] C. M. Lyman and K. A. Kuiken, *Texas Agr. Expt. Sta. Bull.* **708**, 1949.

[294] H. A. Harper and M. D. Morris, *Arch. Biochem. Biophys.* **42**, 61 (1953).

[295] R. Guthrie and H. Tieckelmann, *Proc. London Conf. on Sci. Study of Mental Deficiency* p. 672 (1960).

[296] R. Guthrie, Letters *J. Am. Med. Assoc.* **178**, 863 (1961).

[297] R. Guthrie, *Proposal to the Children's Bureau* Jan. (1962).

[298] Press release, *U. S. Dept. Health, Education and Welfare, Children's Bur.* (Feb. 5, 1962).

tered in the use of auxotrophic mutants may then be introduced. For example, depending on the organism and the inhibitor (or mutant) one or more precursors of the required amino acid may relieve the inhibition (or metabolic block). Of course, as in the previously mentioned instances, a mole for mole relationship among the various compounds that will satisfy a growth requirement is seldom, if ever, found to exist. Thus, the assay of mixtures is difficult, if not impossible (see also the chapter on the assay for members of the folic acid group). In the phenylalanine inhibition assay, proline, phenylpyruvic acid, and phenyllactic acid will reverse thienylalanine inhibition.[295] In cases where great specificity or accurate quantitation is not of great importance, this type of assay might prove to be extremely useful. For example, a suitable application may well be in a preliminary screening procedure for the *detection* of a metabolite.

The method of Guthrie and co-workers for phenylalanine is at present being tested as a simple screening procedure for phenylketonuria. The method as described is far from a proven procedure. In fact, its usefulness is somewhat controversial.[299] Therefore it will not be described in detail here. However, it should be mentioned that since the discovery of the relationship of the excretion of phenylalanine in the urine and mental retardation there has been increasing interest in the relationship of other amino acids and/or their metabolites excreted in the urine, and metabolic diseases and their manifestations. For instance, the excessive excretion of histidine and the related metabolite, imidazolepyruvic acid, may be related to speech retardation;[300] abnormal excretion of isoleucine, leucine, cystine, histidine, tryptophan, aspartic acid, glutamic acid, and glycine has been discovered in ten children with petit mal epilepsy.[301] Microbiological assays for certain amino acids in urine or blood may well prove to be useful as inexpensive screening procedures for the early discovery or suspicion of some of these metabolic diseases in infants and children.[301,302] Such procedures, particularly the agar diffusion methods using paper disks, lend themselves to routine application. In addition, the amino acids in general are quite stable. Therefore, samples can be accumulated as dry spots on filter paper later to be assayed in larger numbers.

The value of inhibition assays for any specific use remains to be adequately evaluated. Based on the experiences of the past in the development of such methods, most probably this process will take at least several years. However, the more straightforward methods utilizing one or

[299] C. Scheel and H. K. Berry, *J. Pediatrics* **61**, 610 (1962).

[300] V. H. Auerbach, A. M. DiGeorge, R. C. Baldridge, C. D. Tourtellotte, and M. P. Brigham, *J. Pediatrics* **60**, 487 (1962).

[301] J. G. Millichap and J. A. Ulrich, *Proc. Staff Mayo Clinic* **37**, 307 (1962).

[302] H. K. Berry, C. Scheel, and J. Marks, *Clin. Chem.* **8**, 242 (1962).

more of the lactic acid bacteria are available and certainly utilizable. Amino acid agar diffusion assays with the use of proper conditions for this microaerophilic group of organisms along with the use of triphenyl-tetrazoleum chloride to increase the sensitivity and ease of detection of small growth zones should offer no real problem. The specificity of such assays is much more of a known quantity. The agar diffusion methods,[121] as described in Chapter 7.5, The Folic Acid Group, using a synthetic medium and test organism suitable for the amino acid to be determined (Section X) would, in the opinion of this author, be a good approach. It would seem that the slight additional cost of the more complex medium required should not be of great consequence. Concommitently, several amino acids could thus be semiquantatively determined. Acceptable quantities of each amino acid in the urine or blood of "normal" individuals of the age group to be tested would, of course, have to be determined. The addition of specific competitive antagonists might then be used to increase the minimal easily detectable levels of the amino acid being determined to that considered to be idiopathic. Obviously, even the more well-established microbiological assay methods for amino acids would require some investigative effort before they could be established as routine clinical procedures.

CHAPTER 9

Glucose

FREDERICK KAVANAGH

Eli Lilly and Company, Indianapolis, Indiana

Smith *et al.*[1] devised a method for assaying for glucose which used the ability of *Lactobacillus casei* to ferment glucose selectively in a mixture of sugars found in starch conversion liquors. The response was determined by titrating the acid with standard alkali and the dose-response curve was linear up to 4 mg. of glucose. A very heavy inoculum was used to reduce the incubation period and the necessity for strict aseptic precautions.

I. Organism

Lactobacillus casei (ATCC 7469) is maintained by frequent transfers into glucose-yeast agar medium. Incubate the stabs at 37°C. for 18 to 24 hours. Use a freshly incubated stab to prepare the inoculum for the assay.

II. Inoculum

To each of four 50-ml. centrifuge tubes add 20 ml. of basal medium, 4 ml. of glucose solution (18 w/v%), 4 μg. of riboflavin, and distilled water to a final volume of 40 ml. Cap or plug the tubes, sterilize at 121°C. for 20 minutes, cool, inoculate, and incubate at 37°C. for 24 hours. Centrifuge the cell suspension at 3000 r.p.m. for 15 minutes and discard the supernatant. Suspend the cells from the four tubes in 25 ml. of sterile 0.9% sodium chloride solution, centrifuge, and discard the supernatant. Suspend

[1] M. D. Smith, M. W. Radomski, and J. J. Kagan, *Anal. Chem.* **32**, 678 (1960).

TABLE I

Buffer solution

NaCl	8.0 gm.
Na acetate (anhy.)	1.0 gm.
Salt solution A	0.5 ml.
Water, distilled	100 ml.
Acetic acid to pH 6.8	

Add toluene and store in the refrigerator

Basal medium

Peptone	10 gm.
Yeast extract	2 gm.
L-Cystine solution	200 ml.
L-Asparagine	0.25 gm.
L-Glutamic acid	0.25 gm.
Salt solution A	20 ml.
Salt solution B	20 ml.
Water, distilled, to make	1000 ml.

Fill 100 ml. into 125-ml. Erlenmeyer flasks, plug, and sterilize at 121°C. for 20 minutes

Salt solution A

KH_2PO_4	25 gm.
K_2HPO_4	25 gm.
HCl, conc.	5 drops
Water, distilled, to make	500 ml.

Salt solution B

$MgSO_4 \cdot 7\ H_2O$	10 gm.
NaCl	0.5 gm.
$FeSO_4 \cdot 7\ H_2O$	0.5 gm.
$MnSO_4 \cdot 4\ H_2O$	0.5 gm.
HCl, conc.	5 drops
Water, distilled, to make	500 ml.

L-Cystine solution

Dissolve 1 gm. L-cystine in 20 ml. of HCl (1:3) and dilute to 1000 ml., add toluene to preserve

the cells in 2 ml. of sterile saline solution. Add 2 drops (total about 0.08 ml.) of the heavy suspension to each assay tube.

III. Glucose Assay

Pipet a sample of less than 4 mg. of glucose into a 15 × 150 mm. test tube, add 1 ml. of the buffer solution and enough distilled water to make the total volume 10 ml. With each run, assay 0, 3, and 4 mg. of glucose in duplicate to obtain the standard dose-response curve. Heat the tubes at 121°C. for 5 minutes, cool, inoculate with 2 drops of the heavy inoculum prepared above, and incubate in a water bath at 37°C. for 2 hours, and cool. Pour the contents of each tube into a beaker, rinse with two 10-ml. volumes of distilled water. Titrate electrometrically with 0.005 N sodium hydroxide to pH 6.75 ±0.05 using a 10-ml. buret graduated in 0.01-ml. divisions.

The relation between glucose and standard alkali is expressed by $G = 0.54T$ where G is glucose in milligrams and T is milliliters of alkali corrected for the blank. The blank titrations ranged from 0 to 0.4 ml. The results obtained from the microbiological assays agreed (no significant difference at the 5% probability level) with chemical assays. Forty unknowns can be assayed in one working day. Because of its sensitivity, the method was used to correlate hydrolytic glucose yields with molecular structure in the starch family of polysaccharides.

The compositions of the various basal media for conducting this assay are shown in Table I.

Author Index

Numbers in parentheses are footnote numbers. They are inserted to indicate the reference when an author's work is cited but his name is not mentioned on the page.

A

Abraham, E. P., 2, 265, 314, 315 (1)
Abraham, J., 590 (161), 591
Adamson, D. C. M., 90
Agren, G., 584, 670 (120)
Albert, A., 64
Albrecht, A. M., 442
Alexander, A. E., 16, 38
Alexander, J. C., 574, 587, 590 (27), 625 (27, 144), 670 (27)
Alicino, J. F., 340
Allfrey, V., 484
Alper, T., 144, 146, 183
Andrew, M. L., 255
Anderson, M. E., 579
Anderson, R. C., 442
Angier, R. B., 433, 441, 442 (14), 484
Armstrong, J. J., 613
Arret, B., 252, 328, 329, 332
Arth, G. E., 442
Asheshov, I. N., 266, 295, 296
Ashworth, J. N., 452
Asselineau, J., 577
Atkin, L., 490, 498, 504
Atkins, L., 498, 503 (2)
Atkinson, D. E., 599
Auclair, J. L., 592
Aurand, L. W., 619
Avery, C. M., 452, 478 (88)
Avi-Dor, Y., 149
Axelrod, A. E., 422

B

Bach, M. K., 536, 552
Bacharach, A. L., 80, 489, 605
Bacher, F. A., 550, 556
Baddiley, J., 613
Bakay, B., 571, 572 (13), 583 (13), 588 (13), 590 (13), 592 (13), 593 (13), 594 (13), 596 (13), 607 (13), 623 (13), 625 (13), 626 (13), 660 (13), 670 (13)

Baker, B. W., 490
Baker, H., 180, 442, 453, 461, 462, 465, 466, 467 (98), 485, 486, 488 (43), 503, 506, 532, 533, 536 (11), 646, 662 (267)
Baldridge, R. C., 672
Baldwin, E., 56
Barber, M., 67
Bardos, T. J., 443
Barker, H. A., 532
Barton, A. D., 576
Barton-Wright, E. C., 413, 519, 569, 589 (7), 590 (165, 167), 592, 593 (167), 613 (167), 618 (7), 621 (7)
Baumann, C. A., 443, 569, 574, 590 (12, 22, 162), 591, 593 (12, 22), 623 (12), 649 (12), 664 (12), 670 (12, 22, 162)
Baumgarten, W., 590 (170), 592, 599
Beadle, G. W., 578
Beavan, G. H., 554, 555 (16)
Beckner, C. W., 574, 590 (27), 625 (27), 670 (27)
Bellamy, W. D., 582
Benedict, R. G., 170, 182 (38)
Bennett, B. A., 574, 590 (30), 623 (30, 255), 625, 664 (30), 670 (30, 255)
Berkman, S., 62, 64
Berridge, N. J., 124
Berry, H. K., 672
Berry, L. J., 432
Bessell, C. J., 552
Bessey, O. A., 438, 519, 520, 584
Bicking, J. B., 522
Bigwood, E. S., 617
Billimoria, H. S., 432
Binkley, B. S., 441
Binkley, S. B., 433, 448
Bird, O. D., 433, 441, 446, 448, 450, 499, 500, 502, 506
Bird, T. J., 576
Black, A., 598, 618 (218)
Black, S., 593

Bliss, C. I., 203, 399
Bliss, E. A., 64
Block, H., 574, 587, 589, 592, 593 (25, 182), 668 (25)
Block, R. J., 616, 618
Bloom, E. S., 433, 441, 446, 448
Blotter, L., 568, 573 (3), 588, 593, 598 (153), 621 (3), 624 (153)
Blum, A. E., 618
Bohnsack, G., 592
Bohonos, N., 441, 445
Boley, A. E., 550, 556
Bolling, D., 618
Bond, C. R., 35, 37, 169
Bond, J. M., 266
Bond, T. J., 443
Boniece, W. S., 370
Booth, J. H., 484
Boothe, J. W., 433, 441, 442 (14)
Borek, E., 583, 584 (116)
Boruff, C. S., 590 (170), 592
Borzani, W., 57
Boulanger, P., 593
Bowne, S. W., Jr., 592
Boxer, G. E., 340, 341
Boyd, M. J., 575
Braekkan, O. R., 586, 625 (136), 659 (136)
Brand, E., 578
Bratton, A. C., 484, 486 (109)
Breed, R. S., 583
Brian, P. W., 388, 389
Brickson, W. L., 584
Brigham, M. P., 672
Brimblecombe, R. W., 266
Brimley, R. C., 22, 24
Brin, M., 581
Brockman, J. A., Jr., 444, 445
Brockmann, H., 592
Brody, S., 42, 47
Broome, M., 415
Broquist, H. P., 423, 442, 443, 444, 445, 587
Brown, A. M., 169
Brown, G. M., 499, 506, 509
Brown, R., 388, 399
Brown, R. A., 433, 441, 446
Brownlee, K. A., 73, 76, 77, 90
Brunzell, A., 272
Buchanan, J. C., 613
Burch, H. B., 520
Burget, G. E., 498

Burkholder, P. R., 541
Burlet, E., 500
Burn, J. H., 73
Buskirk, H. H., 498
Butenandt, A., 442
Butler, B., 585
Butler, E. T., 578

C

Calkins, D. G., 441, 446, 448
Camien, M. N., 569, 574, 586, 587, 589, 590 (11, 19, 23, 24, 25, 174, 177), 592, 593, 597, 599, 625, 646 (174), 657, 660 (181), 668 (25), 669, 670 (252, 285, 286, 289, 292), 671
Campbell, C. J., 441, 446
Campbell, J. A., 513, 552, 556 (10)
Campbell, J. J. R., 582
Campbell, L. L., 465
Cannon, M. D., 600
Cardinal, E. V., 590 (163), 591
Carlile, M. J., 47
Carpenter, F. H., 490
Carpenter, O. S., 413
Carss, B., 613
Caswell, M. C., 574, 575 (21), 588 (21), 590 (21), 599 (21), 618 (21), 623 (21), 653 (21), 668 (21)
Cerecedo, L. R., 418
Chaiet, L., 442, 445 (33)
Chain, E., 2, 314, 315 (1), 592
Challenger, F., 576
Chandler, C. A., 64
Chang, Y., 670 (287), 671
Charney, J., 555
Cheldelin, V. H., 501, 521
Cheng, L. T., 598
Chesbro, W. R., 582
Chibnall, A. C., 620
Chitre, R. G., 586, 625 (135), 659 (135)
Choucron, N., 577
Clark, A. J., 57
Clark, H. T., 592
Clark, P., 145
Clark, W. M., 166
Clutterbuck, P. W., 432
Codner, R. C., 266
Coghill, R. D., 75
Cohen, B., 166
Cohn, E. J., 57, 452

Subject Index

DATE DUE